Introduction to Logistics Systems Management

Wiley Series in
OPERATIONS RESEARCH AND MANAGEMENT SCIENCE

Advisory Editors • *Optimization Models*
Lawrence V. Snyder
Lehigh University
Ya-xiang Yuan
Chinese Academy of Sciences

Founding Series Editor
James J. Cochran
Louisiana Tech University

Operations Research and Management Science (ORMS) is a broad, interdisciplinary branch of applied mathematics concerned with improving the quality of decisions and processes and is a major component of the global modern movement towards the use of advanced analytics in industry and scientific research. *The Wiley Series in Operations Research and Management Science* features a broad collection of books that meet the varied needs of researchers, practitioners, policy makers, and students who use or need to improve their use of analytics. Reflecting the wide range of current research within the ORMS community, the Series encompasses application, methodology, and theory and provides coverage of both classical and cutting edge ORMS concepts and developments. Written by recognized international experts in the field, this collection is appropriate for students as well as professionals from private and public sectors including industry, government, and nonprofit organization who are interested in ORMS at a technical level. The Series is comprised of three sections: Decision and Risk Analysis; Optimization Models; and Stochastic Models.

Introduction to Logistics Systems Management

With Microsoft® Excel® and Python® Examples

Gianpaolo Ghiani
University of Salento, Italy

Gilbert Laporte
HEC, Montréal, Canada
University of Bath, United Kingdom

Roberto Musmanno
University of Calabria, Italy

Third Edition

To Laura, Allegra and Vittoria
To Ann, Cathy and Xavier
To Maria Carmela, Francesco and Andrea

Contents

Foreword

Logistics is concerned with the organization, movement, and storage of material and people. The term logistics was first used by the military to describe the activities associated with maintaining a fighting force in the field and, in its narrowest sense, describes the housing of troops. Over the years the meaning of the term has been gradually generalized to cover business and service activities. The domain of logistics activities is providing the customers of the system with the right product, in the right place, at the right time. This ranges from providing the necessary subcomponents for manufacturing, having inventory on the shelf of a retailer, to having the right amount and type of blood available for hospital surgeries. A fundamental characteristic of logistics is its holistic, integrated view of all the activities that it encompasses. So, while procurement, warehouse management and distribution are all important components, logistics is concerned with the integration of these and other activities to provide the time and space value to the system or corporation.

Excess global capacity in most types of industry has generated intense competition. At the same time, the availability of alternative products has created a very demanding type of customer, who insists on the instantaneous availability of a continuous stream of new models. So the providers of logistics activities are asked to do more transactions, in smaller quantities, in less time, for less cost, and with greater accuracy. New trends such as mass customization will only intensify these demands. The accelerated pace and greater scope of logistics operations has made planning-as-usual impossible.

Even with the increased number and speed of activities, the annual expenses associated with logistics activities in the United States have held constant for the last several years, around ten per cent of the Gross Domestic Product. Given the significant amounts of money involved and the increased operational requirements, the management of logistics systems has gained widespread attention from practitioners and academic researchers alike. To maximize the value in a logistics system, a large variety of planning decisions has to be made, ranging from the simple warehouse-floor choice of which item to pick next to fulfill a customer order to the corporate-level decision to build a new manufacturing plant. Logistics management supports the full range of those decisions related to the design and operation of logistics systems.

There exists a vast amount of literature, software packages, decision support tools and design algorithms that focus on isolated components of the logistics system or isolated planning in the logistics systems. In the last two decades, several companies have developed enterprise resource planning (ERP) systems in response to the need of global corporations to plan their entire supply chain. In their initial implementations, the ERP systems were primarily used for the recording of transactions rather than for the planning of resources on an enterprise-wide scale. Their main advantage was to provide consistent, up-to-date and accessible data to the enterprise. In recent years, the original ERP systems have been extended with advanced planning systems (APS). The main function of APS is for the first time the planning of enterprise-wide resources and actions. This implies a coordination of the plans among several organizations and geographically dispersed locations.

So, while logistics management requires an integrated, holistic approach, their treatment in courses and books tends to be either integrated and qualitative or mathematical and very specific. This book bridges the gap between those two approaches. It provides a comprehensive and modelling-based treatment of the logistics processes. The major components of logistics systems—storage and distribution—are each examined in detail. For each topic the problem is defined, models and solution algorithms are presented that support computer-assisted decision-making, and numerous application examples are provided. Each chapter is concluded with case studies that illustrate the application of the models and algorithms in practice. Because of its rigorous mathematical treatment of real-world management problems in logistics, the book will provide a valuable resource to graduate and senior undergraduate students and practitioners who are trying to improve logistics operations and satisfy their customers.

Marc Goetschalckx
Georgia Institute of Technology
Atlanta, United States

Preface

In the last few decades, and in particular during the pandemic and post-pandemic eras, logistics has become pivotal in the global economy. The exponential growth of e-commerce has forced companies to manage multi-channel strategies to supply their markets. The "new normal" requires that consumer requirements at multiple locations be met in a timely fashion with a variety of transportation modes. This level of service can be accomplished by setting up and managing complex and flexible logistics systems that can adapt easily and economically to unexpected circumstances. Such complexity and flexibility is made possible by advanced information technology solutions, capable of ensuring the sustainability of logistics processes as well as their efficiency. In this context, the extensive use of *data*, *descriptive analytics*, *predictive models* and *optimization techniques* prove to be invaluable to assist the decisions and actions of logistics and supply chain managers.

This book grew out of a number of undergraduate and graduate courses on logistics that we have taught to engineering, computer science and management science students. The goal of these courses is to give students a solid understanding of the analytical tools necessary to reduce costs and improve service level in logistics systems. The lack of a suitable book had forced us in the past to make use of a number of monographs and scientific papers which tend to be beyond the level of most students. We therefore committed ourselves to developing a quantitative book written at a level accessible to most students.

In 2004 we published with Wiley a book entitled "Introduction to Logistics Systems Planning and Control", which was widely used in several universities around the world. The 2004 edition of the book received the "Roger-Charbonneau" award from HEC Montréal, as the best pedagogical book of the year. In 2013 we published a revised edition entitled "Introduction to Logistics Systems Management", in which more emphasis was put on the organizational context in which logistics systems operate. After the success of the first two editions, both in terms of number of copies sold and the very positive feedback received from readers, we accepted Wiley's invitation to prepare the current third edition entitled "Introduction to Logistics Systems Management, with `Excel` and `Python`".

This edition deeply revises the content of the two previous editions in several respects. First, it covers new organizational concepts and techniques that have recently emerged in the field of logistics and were not covered in the previous editions. In addition, new numerical examples and problems have been added to each chapter.

A further novelty is that their solutions are illustrated in great detail by using a spreadsheet in `Microsoft Excel` and a `Python` code, giving the reader the opportunity to replicate step by step the modelling and solution approach adopted in the book. In this way, readers can verify their understanding of each concept before moving on to the next one.

The book targets both academic and practitioner audiences. On the academic side, it should be appropriate for advanced undergraduate and graduate courses in logistics and supply chain management. It should also serve as a methodological reference for consultants and industry practitioners. We make the assumption that the reader is familiar with the basics of Operations Research and Statistics, and we provide a balanced treatment of forecasting, logistics system design, procurement, warehouse management, and freight transportation management.

Gianpaolo Ghiani (gianpaolo.ghiani@unisalento.it)
Gilbert Laporte (gilbert.laporte@cirrelt.net)
Roberto Musmanno (roberto.musmanno@unical.it)

Teaching Material and Website

The book's website,

    http://www.wiley.com/go/logistics_systems_management3e,

completely renewed, provides readers, among others, with:

- the `Microsoft Excel` spreadsheets and `Python` source files used in the solution of the problems presented in the boxes;
- the data of the numerical problems proposed at the end of each chapter.

For the latter problems, instructors can obtain, upon request from the authors, the `Microsoft Excel` spreadsheets and the `Python` source files used in their solution. They can also get the LaTeX source files containing the formulae, optimization models, tables and algorithms described in the book, as well as the pdf files of the figures. In this way, they can easily compose their own LaTeX course slides.

Acknowledgements

The publication of this book was made possible in part thanks to the financial contribution of the Laboratory of Technologies for Simulation and Optimization (TESEO), Department of Mechanical, Energy and Management Engineering, ▆ University of Calabria.

The authors wish to acknowledge the reviewers and all the individuals who have helped to produce this book, in particular Annarita De Maio for her scientific and technical support.

About the Authors

Gianpaolo Ghiani is Professor of Operations Research at University of Salento (Italy), where he teaches Automated Planning, Decision Support Systems, Business Analytics, and Logistics courses at the Department of Engineering for Innovation.

Gilbert Laporte is professor emeritus at HEC Montréal (Canada), professor at the University of Bath (United Kingdom), and adjunct professor at Molde University College (Norway). His main research interests lie in the field of Distribution Management.

Roberto Musmanno is Professor of Operations Research at University of Calabria (Italy), where he teaches, among others, a Logistics course at the Department of Mechanical, Energy and Management Engineering.

List of Abbreviations

1-BP	one-bin packing
1PL	first-party logistics
2-BP	two-bin packing
2PL	second-party logistics
3-BP	three-bin packing
3PL	third-party logistics
4PL	fourth-party logistics
5PL	fifth-party logistics
AGV	automated guided vehicle
AH	air hub
AHP	analytical hierarchy process
ANN	artificial neural network
AP	assignment problem
APM	A.P. Moller-Maersk
ARP	arc routing problem
AS/RS	automated storage and retrieval system
ASIN	Amazon standard identification number
ATO	assembly to order
ATSP	asymmetric travelling salesman problem
B2B	business to business
B2C	business to consumer
BF	best fit
BFD	best-fit decreasing
BFGS	Broyden–Fletcher–Goldfarb–Shanno
C2C	cash-to-cash
CDC	central distribution centre
CL	carload
CL-NRP	node routing problem with capacity and length constraints
C-NRP	node routing problem with capacity constraints
COFC	container on flatcar
COT	cut-off time
CPFR	collaborative forecasting and replenishment program
CPL	capacitated plant location

CPP	Chinese postman problem
CPR	Canadian Pacific Railway
CRM	customer relationship management
CRP	continuous replenishment program
C-VRP	vehicle routing problem with capacity constraints
CWC	Central Warehousing Corporation
DC	distribution centre
DDAP	dynamic driver assignment problem
DWT	deadweight
EAN	European article number
EDA	exploratory data analysis
EDI	electronic data interchange
EFC	e-fulfillment centre
ELC	European logistics centre
EOQ	economic order quantity
EPAL	European Pallet Association
EPP	European Pallet Pool
ERP	enterprise resource planning
ETO	engineering to order
EVPI	expected value of perfect information
FAA	Federal Aviation Administration
FBF	finite best fit
FCNDP	fixed charge network design problem
FDA	Food and Drug Administration
FEU	forty-foot equivalent unit
FF	first fit
FFD	first fit decreasing
FFF	finite first fit
FIFO	first-in, first-out
FR	fill rate
GIS	geographic information system
GMA	Grocery Manufacturers Association
GMROI	gross margin return on investment
GPS	global positioning system
GSE	ground servicing equipment
GTIN	global trade item number
GVW	gross vehicle weight
H&M	Hennes & Mauritz
IATA	International Air Transport Association
ICP	inbound crossdocking point
ICT	information and communication technology
IDOS	inventory days of supply
IoT	Internet of Things
IP	integer programming
IRP	inventory routing problem
ISBN	International Standard Book Number
ISO	International Organization for Standardization

IT	inventory turnover
KPI	key performance indicator
LB	lower bound
LFCNDP	linear fixed-charge network design problem
LIFO	last-in, first-out
LMCFP	linear single-commodity minimum-cost flow problem
LMMCFP	linear multi-commodity minimum-cost flow problem
LNG	liquefied natural gas
LP	linear programming
LPG	liquefied petroleum gas
LSP	logistics service provider
LTL	less-than-truckload
MAE	mean absolute error
MAPE	mean absolute percentage error
MCTE	multi-commodity two-echelon location problem
MDP	material decoupling point
MFC	material flow control
MIP	mixed integer programming
MIS	management information system
MMCFP	multi-commodity minimum-cost flow problem
MMR	mass market retailer
MRO	maintenance, repair, and overhaul
MRP	material requirements planning
MSrTP	minimum spanning r-tree problem
MSE	mean squared error
MTO	make to order
MTS	make to stock
NOOS	never out of stock
NPV	net present value
NRP	node routing problem
OCT	order-cycle time
PLC	programmable logic controller
POR	perfect order rate
PRC	pneumatic refuse collection
QR	quick response
RDC	regional distribution centre
RFID	radio-frequency identification
RMG	rail-mounted gantry
RNG	pseudo-random number generator
Ro-Ro	roll-on/roll-off
RPP	rural postman problem
RTG	rubber-tired gantry
RTSP	road travelling salesman problem
S/R	storage and retrieval
SC	set covering
SC	shipment centre
SC/AS	shipment centre/air stop

SCM	supply chain management
SC-NRP	set covering, node routing problem
SCOE	single-commodity one-echelon
SCTE	single-commodity two-echelon
SHAS	special handling at source
SKU	stock keeping unit
SLA	service level agreement
SMA	selling and market area
SNDP	service network design problem
SPL	simple plant location
SQI	supplier quality index
SRM	supplier relationship management
SSCC	serial shipping container code
SSE	sum of squared errors
STSP	symmetric travelling salesman problem
SVM	support vectors machine
TAP	traffic assignment problem
TEU	twenty-foot equivalent unit
TL	truckload
TMS	transportation management system
TOFC	trailer on flatcar
TS	tabu search
TSP	travelling salesman problem
TW-NRSP	node routing and scheduling problem with time windows
UB	upper bound
ULCC	ultra-large crude carrier
ULCV	ultra-large container vessel
ULD	unit load device
UPC	universal product code
VAP	vehicle allocation problem
VIMS	visual interactive modelling system
VLCC	very large crude carrier
VMI	vendor-managed inventory
VRP	vehicle routing problem
VRDP	vehicle routing and dispatching problem
VRPMT	vehicle routing problem with multiple trips
VRPPD	vehicle routing problem with pickups and deliveries
VRSP	vehicle routing and scheduling problem
WCS	warehouse control system
WMS	warehouse management system
WORM	write-once-read-many
ZIO	zero inventory ordering

1

Introducing Logistics

1.1 Definition of Logistics

According to a widespread definition, *logistics* is the discipline that studies, in an *organization* (such as a private company, a public administration, a non-profit association, a military corps), the management and implementation of the operations concerning the flow of tangible goods (materials, food and medical supplies, refuse, equipment, weapons, etc.) from their sources (suppliers, mines, crop fields, etc.) to their points of utilization or consumption or disposal (retailers, landfills, army units, etc.) to meet the objectives of the organization. To this end, logistics requires the collection, integration, and processing of data from several sources in order to plan, organize, and control activities such as material handling, production, packaging, warehousing, and distribution.

The words "logic" and "logistics" both come from the Greek term *lógos*, which means, among other things, "order". However, while "logic" derives directly from Greek, "logistics" first passed into Middle French as "logis", meaning "lodging", and then into English.

Defence Logistics. The origins of logistics are of a strictly military nature. In fact, the discipline arose as the study of methodologies to guarantee the correct supply of troops with victuals, ammunitions, fuel, etc. Indeed it was the Babylonians, in the distant twentieth century BCE, who first created a military corps specialized in the supply, storage and delivery of soldiers' equipment. The relevance of logistics became apparent during the American Revolutionary War (1775–1783) when the lack of adequate supplies for the 12 000 British soldiers overseas during the first six years devastated the troop morale and contributed to their final defeat. In modern times, logistics played a key role in World War II where it helped the Allied powers to succeed. In modern times, the key concept in defence logistics is that of *supply chain*, defined as the set of processes, infrastructure, equipment and personnel ensuring that a specific vehicle or weapon is fully functioning in the theatre of operations.

Industrial Logistics. Only in the twentieth century, were logistics principles and techniques extended to manufacturing companies. In industrial logistics, a supply chain resembles a military one and is defined as the network of organizations (suppliers, carriers, logistics providers, wholesalers, and retailers, etc.), resources, activities, and

Introduction to Logistics Systems Management: With Microsoft® Excel® and Python® Examples, Third Edition. Gianpaolo Ghiani, Gilbert Laporte, and Roberto Musmanno.
© 2022 John Wiley & Sons Ltd. Published 2022 by John Wiley & Sons Ltd.

information built around a company to produce and distribute a specific product to a specific market. Here, the goal of logistics is to manage the flow of materials and information from the extraction, harvesting or purchase of raw materials and components up to the delivery of the finished products to customers. In this sector, logistics activities are traditionally classified depending on their location with respect to the production and distribution processes. In particular, *procurement logistics* comes before the manufacturing process and consists of supplying raw materials and components to support the company's production plan. *Internal logistics* is about material handling and storage in production plants in order to feed production lines and the subsequent product packaging and shipment. Finally, *distribution logistics* falls after the production plants and before the market, and aims to supply sales points or customers. In this framework, procurement logistics and distribution logistics are collectively called *external logistics*.

Service Logistics. Logistics issues are also increasingly present in the service sector, for example in postal services, in urban solid waste collection, in the post-sales activities of manufacturing companies as well as in humanitarian organizations. *Logistics service providers* (LSPs), performing transportation or warehousing activities for other organizations, including manufacturing companies, also fall into this category.

Integrated Logistics and Logistics Alliances. Logistics activities may be carried out entirely by a specific function of the organization (see Section 1.5 for details). Otherwise, they may be jointly performed by multiple departments of the organization such as production, marketing, etc. (*integrated logistics*) or even in collaboration with different partner organizations (*logistics alliances*). Logistics alliances can be implemented in two different forms. The *efficiency-oriented approach* relies on contracts of a strictly operative nature that do not modify the organization's own strategy but simply tend to create synergies or economies of scale with the primary objective of minimizing costs. On the other hand, in the *differentiation-oriented approach* the company tries to forge exclusive alliances with some partners, not replicable by competitors, to generate an added value with respect to the competition.

An efficiency-oriented logistics alliance was implemented by SkyTeam, the second global airline alliance in the world, that in 2021 counted 19 members (Aeroflot, Aerolíneas Argentinas, Aeroméxico, Air Europa, Air France, Ita Airways, China Airlines, China Eastern, Czech Airlines, Delta Airlines, Garuda Indonesia, Kenya Airways, KLM Royal Dutch Airlines, Korean Air, Middle East Airlines, Saudi Arabian Airlines, TAROM Romanian Air Transport, Vietnam Airlines, and Xiamen Airlines). The alliance allows the collaboration among airlines in different forms: creating synergies in timetable design and ticket pricing, sharing information about customers, operating ground services, managing frequent flyer programmes, and airplane maintenance. In addition, customers of the SkyTeam airlines can benefit from a larger number of flights, with more destinations and connections as well as a larger number of lounges located within the network's airports. In 2021 SkyTeam transported about 675 million passengers over

15 500 daily flights reaching about 1000 destinations in 170 countries. The cargo branches of 11 of the 19 air companies cited above have also signed a strategic alliance, called SkyTeam Cargo, for freight transportation. The members of SkyTeam Cargo share airplanes and cargo buildings (see Section 6.3.2) located in 76 air cargo terminals worldwide (e.g., the cargo building located in the Vienna Airport is shared among China Airlines Cargo, Korean Air Cargo and Aerflot Cargo).

An example of a differentiation-oriented logistics alliance has been set up in 2019 between Unilever, a global Dutch–British consumer goods company owning the Algida ice cream brand, and Ferrero, a world-renowned Italian manufacturer of branded chocolate and confectionery products, including the Kinder brand of chocolate products and Nutella. The agreement concerned the launch of a new Kinder Ice Cream (whose recipe was created by Ferrero), produced and distributed by Unilever in various European markets (Germany, France, Italy, Austria, etc.). The partnership has clear mutual benefits. Ferrero may take advantage of Unilever's experience in the ice cream sector to take its Kinder brand to new attractive markets, without the cost of investing in a frozen food supply chain. On the other hand, Unilever may take advantage from the Kinder brand power to increase its sales and enlarge its product portfolio.

1.2 Logistics Systems

A *logistics system* is a set of interacting infrastructures, equipment, and human resources whose objective is, as a whole, the execution of all the functional activities determining the flow of materials among a number of *facilities*. Facilities may be plants, warehouses, landfills, sorting centres, air, and ground hubs where either production or assembly, disposal, consolidation, storage, packaging, distribution, etc. is carried out.

It is concerned with the flow of materials among facilities. For example, in a waste collection system, materials flow from households to waste recycling plants and landfills through a number of facilities such as transfer points and mixed waste sorting plants. In a postal system, letters and packages flow from the pickup points to the delivery addresses through mail sorting sites, air and ground hubs, regional distribution centres (DCs), etc. In a manufacturing system, materials flow from suppliers to production plants and then reach the distributions system and the retailers (or directly the final customers). The distribution phase may rely, depending on the cost structure and customer expectations, on a single layer of DCs or on a central distribution centre (CDC) and a number of regional distribution centres (RDCs). In any case, at each facility the flow of materials is temporarily interrupted, generally in order to change their physical or chemical composition (production, assembly), appearance (packaging), availability (storage), or ownership. Such logistics activities, along with transportation and material handling, constantly add value to the product, as it draws nearer the final customer.

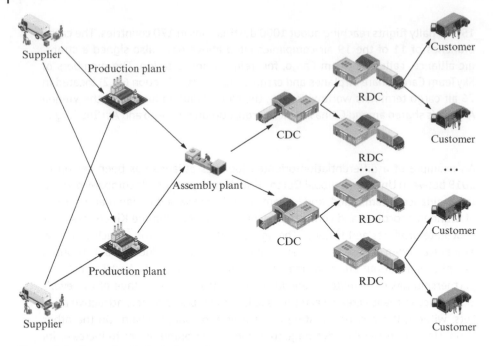

Figure 1.1 Example of a logistics system.

An exception to the downstream flow from suppliers to markets is when a defective product, or a product at the end of its life cycle, is returned to be repaired or disposed of (see Section 1.7.1).

Figure 1.1 shows a schematic representation of a logistics system in which the manufacturing process of the finished goods is divided into a transformation phase and an assembly phase, performed in different facilities. At the start are the suppliers of materials and components which feed the final manufacturing process. The end part represents a two-level distribution system with a tree structure. The CDCs are directly supplied by the production plants, while each RDC is connected to a single CDC which has the task of serving the retailers or directly the customers.

A logistics system can be represented as a directed graph $G = (V, A)$, where V is the set of facilities, and A is a set of arcs representing material flows. In a multi-graph there can be several arcs between a pair of facilities, representing alternative transportation services or different routes or products.

In addition to materials, there is a flow of information between facilities. This can happen in different ways, including emails or more sophisticated data exchange platforms. In general, information can flow in both directions between two facilities even if material flows are unidirectional. A logistics system including information flows is also called a *logistics network* and can be itself represented as a directed graph, in which some arcs represent information flows (e.g., sales figures, orders, inventory updates, etc.) between a pair of nodes. Figure 1.2 provides a logistics network representation of the logistics system illustrated in Figure 1.1.

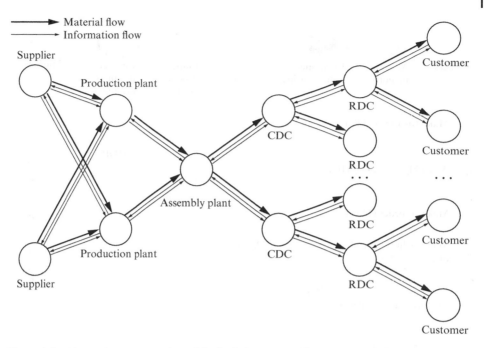

Figure 1.2 Network representation of the logistics system illustrated in Figure 1.1.

1.3 Supply Chains

Supply Chain Management (SCM) is concerned with the coordination and management of the supply chains of an organization. Note the plural in "chains": organizations may operate multiple supply chains that work together to serve different segments. For example, in defence logistics, the *maintenance, repair, and overhaul* (MRO) of the military systems in a theatre of operations can be managed through distinct supply chains. Similarly, in industrial logistics, there may be a specific supply chain for different combinations of markets, consumers, products or even seasons.

1.3.1 Logistics Versus Supply Chain Management

The difference between logistics and SCM is subtle (and controversial among scholars and practitioners). Based on the definition adopted in the book, logistics has a wider scope than SCM, since it also encompasses logistics systems that are not supply chains (e.g., solid waste collection systems or postal delivery systems).

1.3.2 A Taxonomy of Supply Chains

As shown schematically in Figure 1.3, in industrial logistics, supply chains may be configured in a variety of shapes. When the demand of single products can be predicted accurately, all the activities of procurement, manufacturing, assembly, and distribution can be planned in advance, based on forecasts of finished product demand (*make-to-stock* (MTS) supply chain). On the other hand, when finished products come in a very

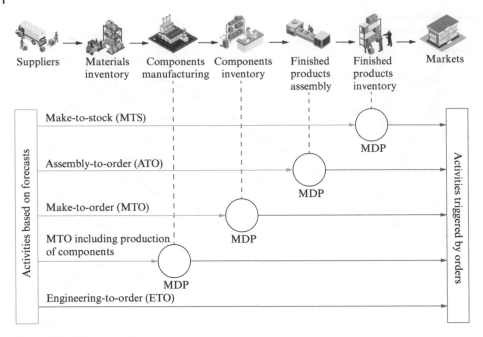

Figure 1.3 Supply chain taxonomy.

large number of variants and only the aggregate demand can be predicted accurately, it is reasonable to produce common components in advance (based on aggregate forecasts) and then assemble the products when orders arrive (*assembly-to-order* (ATO) supply chains). Of course, a hybrid MTS–ATO supply chain can also be implemented by holding a small inventory of the most demanded finished products in either DCs or retailers if customers expect to receive the goods in a timely fashion. If even the aggregate demand of an entire range of products is hard to be predicted, all the activities of manufacturing, assembly and distribution (and sometimes procurement) should be triggered by customers' orders (*make-to-order* (MTO) supply chains). Again, a hybrid MTO–ATO–MTS supply chain may prove a good compromise between the needs to keep inventories low and to provide a reasonable level of service. Finally, some products are so unique that even the design of the product is done on the basis of the customers' specifications (*engineering-to-order* (ETO) supply chains). The separation between the activities based on planning (*push* subsystem) and the activities triggered by orders (*pull* subsystem), which is often a significant stock-holding point, is called *material decoupling point* (MDP). Products are pushed to the MDP and pulled from it.

1.3.3 The Bullwhip Effect

Managing a supply chain is relatively simple for standard products with a stable demand. This is the case, for example, for some household cleaning products that come in a small number of variants with a predictable demand. On the other hand, SCM can be complex when dealing with short life cycle products having a highly uncertain demand. This is the case, for example, of the fashion apparel and consumer electronics industries. Forecasting errors in the fashion industry may have an order of magnitude of 10% at the start of season, 20% 16 weeks ahead and 40% 26 weeks ahead. Similarly, in consumer electronics the equivalent rule of thumb is that the forecasting error is 5% one month ahead, 20% two months ahead and 50% three months ahead. Since manufacturing is not instantaneous, poor planning can result in a stockout or in a very large surplus (followed by a seasonal sale at knock-down prices) or, even more dramatically, in the so-called *bullwhip effect*: many stockouts alternating with very large surpluses in the same selling season.

SCM is even more challenging when facilities are located thousands of kilometres away from each other, in different countries or even in different continents. The mass migration of manufacturing from the developed world to emergent economies is motivated mainly by fewer regulatory controls and significantly lower wages. This has significantly increased the global trade. While sea cargo shipments are relatively cheap, they are also relatively slow. For instance, the shipment of a container from a supplier in China to an assembly plant in the USA by a combination of truck, train, and sea shipments may take weeks. These longer lead times (see Table 1.8 and Section 5.12) may amplify the bullwhip effect.

The beer distribution game, also known as the *beer game*, is an educational game invented in 1960 by J. W. Forrester at the MIT Sloan School of Management. It is commonly used to teach supply chain principles and, in particular, to demonstrate the bullwhip effect. In the game, more teams composed of four players each, acting as the factory manager, the distributor, the wholesaler and the retailer, respectively, are involved in a role-play simulation of a beer supply chain (see Figure 1.4).

The common goal of the participants is to produce and deliver beer to the final consumers at minimum cost (total cost being the sum of reorder, shortage, and holding costs along the entire planning horizon, see Section 5.12.1 for more details). The team that achieves the minimum total cost in the simulation of the supply chain wins. The game is played by each team in a certain number of rounds and in each round:

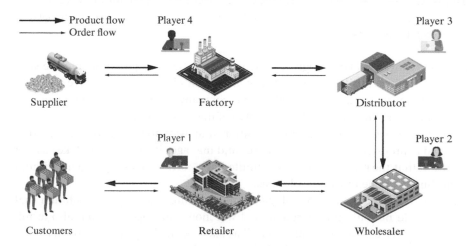

Figure 1.4 Supply chain of the beer distribution game.

- the number of beer packs demanded by the consumers in the last period becomes known to the retailer;
- the retailer updates its inventory and may decide to issue an order to the wholesaler;
- the wholesaler is instantaneously informed about the new (possible) order, updates its inventory and may decide to issue an order to the distributor.

A distinctive feature of the beer game is that orders are not delivered instantaneously, but only after a given number of time periods (corresponding to the lead time).

It is worth noting that, in the original game, players do not have a full picture of what is happening in the supply chain, each player being only aware of the orders received from the downstream player of the same team and its inventory. When playing the game under such conditions, the simulation shows a considerable inventory oscillation (i.e., a bullwhip effect) even for small variations of market demand (see Problem 5.17 for further insight). On the other hand, supply chain performance is greatly improved if the players are allowed to share information and coordinate their decisions.

1.4   Logistics Service Providers

While some manufacturers operating in geographically concentrated markets may still possess and operate their own logistics resources (facilities, vehicles, crews, etc.), nowadays the complex needs of global supply chains are often contracted out to external service providers. LSPs, introduced in Section 1.1, allow manufacturing companies to outsource some specific logistics operations (warehousing, sea freight forwarding, rail freight transportation, intermodal transportation, last-mile distribution, etc.), or even the entire SCM, to external organizations. Outsourcing may involve only a subset of the logistics activities, leaving some products or operations to the in-house logistics (if this works better or is cheaper than an external provider). There are five main logistics outsourcing paradigms:

- *First-party logistics* (1PL). Logistics needs, such as the transportation and the storage of goods, are fulfilled by using the organization's internal resources (e.g., trucks and warehouses bought or rented by the company). Hence, a 1PL provider is a department or a division (see Section 1.9.2) of the organization itself.
- *Second-party logistics* (2PL). Transportation and warehousing activities are outsourced to an external LSP that possesses and manages its own assets. A key aspect of 2PL is that the provider's service is limited to a portion of a supply chain without including integrated logistics solutions.
- *Third-party logistics* (3PL). A LSP guarantees an integrated logistics solution that may include not only transportation and warehouse management, but also terminal operations, customs clearance as well as package tracking. A 3PL provider is not involved in the design of the supply chains. Examples of 3PL providers include freight forwarders and courier companies.

- *Fourth-party logistics* (4PL). A 4PL provider deals with all the aspects of the client's supply chains, including their design. Hence, the client company simply communicates to the 4PL provider the needs of its supply chain (e.g., amount of goods to be stored and delivered). Then, the 4PL provider manages the entire supply chain process for the client company. As a rule, 4PL providers entrust the execution of operational activities to subcontractors (3PL providers). For this reason, sometimes they often do not own physical assets. Due to its technological and integration capabilities, a 4PL provider is also called a *logistics integrator*.
- *Fifth-party logistics* (5PL). 5PL providers are the same as 4PL providers, except that they have an extensive focus on e-business solutions and provide additional services such as *customer relationship management* (CRM), online payments and so on.

Finally, it is worth noting that, broadly speaking, any company offering some type of logistics service for hire defines itself a 3PL provider.

UPS Supply Chain Solutions is an American 4PL company which offers, in partnership with its customers, a variety of services including global supply chain design, logistics and distribution, customs brokerage, and international trade organization. In particular UPS, through its warehousing and freight distribution services, promotes the adoption of an outsourced solution for e-commerce business that eliminates the need for its partners to hold inventory or pick, pack, and ship freights.

1.5 Logistics in Service Organizations

Logistics principles may be even more important in service organizations than in production firms. The following subsections provide two relevant examples.

1.5.1 Logistics in Solid Waste Management

Solid waste management (SWM) is concerned with the collection, treatment and disposal of solid materials that have served their purpose or are no longer useful. Municipal solid waste includes waste from residential, commercial, and institutional (e.g., schools, government offices) sources. Other materials that are frequently disposed in landfills include construction and demolition waste as well as non-hazardous industrial process waste.

Every citizen generates a fairly large amount of solid waste every day (on average, 2.58 kg in the USA, 2.11 kg in Germany, 0.52 kg in Indonesia, just to mention a few statistics). This creates the need to manage a complex logistics system that can be often divided into two major subsystems: a municipal collection system and a regional management system.

Each municipality is in charge of its own curbside garbage collection, using either its own fleet or a contracted service. The collection is done on the basis of a weekly schedule, depending on the type of refuse (organic waste, paper, plastic, etc.) and district. Depending on the local regulations and policy, the residential waste is collected directly at home (*door-to-door collection system*), or has to be taken to a collection site by

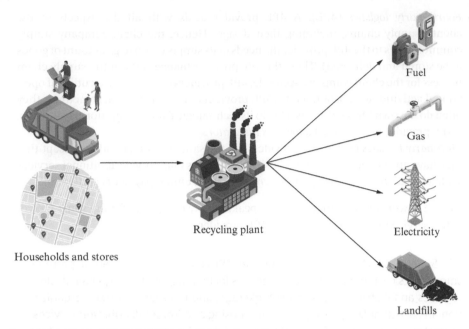

Households and stores

Recycling plant

Fuel

Gas

Electricity

Landfills

Figure 1.5 Logistics in a Solid Waste Management System.

the citizens (where it is later collected by specialized vehicles). Once the waste has been collected, it is possibly shredded and taken directly to a recycling plant or to a interme- diate transfer station. Finally, fuel, gas and electricity (as well as other materials such as compost) are generated from the waste. The residuals of the process are then disposed in a landfill (see Figure 1.5).

Building a new treatment or disposal facility may take one to four years, while its operating life is around 15–30 years. Consequently, designing or redesigning a regional SWM system is a strategic decision having long-lasting effects.

Another issue to be considered is the large percentage of total solid waste man- agement cost related to waste collection (about 75%) due to the equipment and the workforce. Therefore, partitioning the service territories into districts, allocating waste flows among the facilities selected at the strategic level, choosing collection days for each district and for each waste type, determining the composition of the fleet and the crew assignment could result in significant cost savings.

1.5.2 Humanitarian Logistics

Every year, there are approximately 150 000 deaths and 200 million people affected by sudden onset disasters, such as hurricanes, tsunami or earthquakes, and slow onset disasters, such as famine or droughts. Humanitarian logistics aims to provide support in the form of medicines, water, food, and shelter to the affected population in a timely and cost-effective fashion.

As far as sudden onset disasters are concerned, humanitarian logistics is funda- mentally different from commercial SCM. Some warehouses, often located in donor countries or at strategic points in the world, such as the United Nations Humanitarian Response Depot (UNHRD) in Brindisi (Italy), store goods (blankets, spare parts, equipment, tools, etc.) for a long time (months or even years) until they are needed. Only then, are they transported near the heart of affected zones, displaced or distributed to the population.

Transportation plays a fundamental role in making search and rescue teams, doctors, equipment and supplies available in a timely way to the affected regions. Its use must consider the level of urgency, the total cost as well as the geographical characteristics of target areas. Another peculiar feature of humanitarian logistics is the need to coordinate several players, including governments, military forces, aid agencies, donors, non-governmental organizations, suppliers, and private service providers (e.g., airlines for freight transportation).

1.6 Case Studies

This section aims at introducing, through case studies, the fundamentals of logistics systems management. In each case study, a logistics system is sketched based on articles published in either general or specialist media. It is out the scope of this book to be exhaustive (especially because some information is kept confidential by companies and may change rapidly over time). Rather, a flavour of the variety of challenges and problems faced by logistics managers in today's global economy is given. The purpose is to make the methodological treatment of the subsequent sections easier to understand. In no way are the following case studies intended to be accurate and updated business reports.

1.6.1 Apple

Apple is an American multinational company, headquartered in California (USA), that designs, develops, and sells consumer electronics, computer software, and online services. Apple markets high-end high-value electronic devices (including the *iPhone* smartphones, the *iPad* tablet computers, the *Mac* personal computers, the *iPod* portable media players, the *Apple Watch* smart-watch, the *Apple TV* digital media player), as well as software (including the *OS X* and *iOS* operating systems, the *iTunes* media player and the *Safari* Web browser). The company also operates online services such as the iTunes Store, the App Store and the iCloud operating system.

As a competitive strategy, Apple constantly seeks a high rate of innovation in product and service design, as well as excellence in customer service, which allows it to charge customers relatively high selling prices. In 2019, the iPhone generated $ 142.3 billion in revenue, which is approximately 55% of the company's total revenue for that year. A look at the supply chain (see Figure 1.6) shows that the company makes very little of its own hardware products: components and subsystems are manufactured by dozens of suppliers around the globe and sometimes even from direct competitors. In particular, for the iPhone, Texas Instruments (USA) makes the touch screen controller, Micron (USA) the flash memory, Cirus Logic (USA) the audio controller, Dialog Semiconductors (Italy) the power management components, ST Microelectronics (Taiwan) the accelerometers and gyroscope, Infineon (Germany) the phone network components, Murata (Japan) Bluetooth and Wi-Fi components while Samsung makes the memory and application processor. This approach lowers

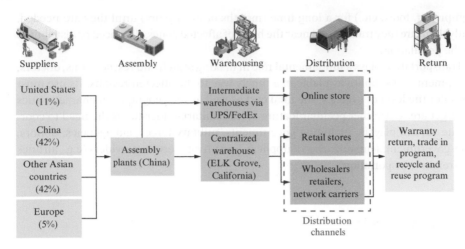

Figure 1.6 Apple iPhone supply chain.

the capital investment for Apple but requires a strict coordination of the contractors along the supply chain. In extreme cases, when a supplier proved unreliable, the company cut it off and quickly procured the same components from another manufacturer. Components are shipped by air from the suppliers to China where the finished products are assembled by the Taiwanese company Foxconn. Air transportation is substantially more expensive than sea shipments: around $ 1.21 per ton-mile versus $ 0.04 per ton-mile (see Chapter 6 for more details). However, this is not a major problem for Apple: profit margins are relatively high and the supply chain priority is being responsive opposed to being efficient. A look at a few figures will clarify this point. The retail price of the *iPhone 11 Pro Max* is nearly $ 1099 while the estimated cost of its components is approximately $ 490.50 per phone (the Samsung battery unit costs around $ 10.50, the triple camera $ 73.50 while other equipment such as the processor, modem, and circuit boards costs approximately $ 159 per phone). While it is difficult to determine the actual profit per unit (R&D, marketing, sales, and administrative costs are not included in the $ 490.50 estimate), it is clear that a pricey but timely air shipment service is worth using.

To mitigate the risk of supply shortage or disruption, Apple pools inventory and other capacity resources. The Apple's chief executive officer, Tim Cook, is convinced that the inventory of high-end high-value electronic devices depreciates very quickly, losing 1–2% of their value each week. "Inventory is fundamentally evil", he stated in a famous interview, "you kind of want to manage it like you're in the dairy business".

Apple operates multiple distribution channels: an online store, wholesalers, retail stores and network carriers (such as Bell, Verizon, Vodafone, Wind 3). Products bought from the Apple's online store are shipped to the customers via couriers (such as FedEx and UPS) on the basis of an outsourcing service level agreement (SLA). These companies hold their own iPhone stocks in order to provide timely shipments. For the other distribution channels, Apple ships products from its centralized warehouse in Elk Grove (California, USA). Finally, at the end of product's life, customers can return their devices to the nearest Apple stores or to dedicated recycling facilities.

1.6.2 Adidas AG

Adidas AG is a German multinational corporation that designs and produces sport shoes, sport clothing (including men's and women's t-shirts, jackets, hoodies, pants, leggings) and accessories (such as eyewear, bags, caps, socks as well as deodorants, perfumes, aftershave, and lotions). It is the largest sportswear manufacturer in Europe and the second largest in the world, after Nike. It is part of the Adidas Group which also includes the Reebok sportswear company and Runtastic, a fitness technology company. Adidas' revenue was around $ 21 billion in 2018.

In 2017 the supply chain of Adidas included 296 suppliers, 79% of which where in Asia while 11% were in the Americas, 9% in Europe, and 1% in Africa. In particular, 97% of the total footwear volume (nearly 403 million pairs) was made in Asia (44% in Vietnam, 25% in Indonesia, and 19% in China). As for apparel production, the total production equalled 404 million units that were manufactured mainly in China (23%), Cambodia (22%), and Vietnam (18%). Finally, the total volume of hardware (such as balls and bags) amounted to 110 million units and were produced mainly in China (40%), Pakistan (18%), and Turkey (15%). Adidas uses ocean freight to transport goods to distributions centres around the world. The company owns more than 2500 retail shops and sells through approximately 13 000 mono-branded franchise stores and 150 000 wholesalers around the globe.

To cope with this complexity, retailers are segmented into homogeneous groups and several supply chain models are used. As far as sportswear is concerned, the retailer segments are the following:

- *"Mi Adidas" e-commerce segment.* The Adidas online shop is devoted to those looking for a high performance gear. Consumers can design their sporting goods (e.g., running shoes) to meet their expectations, choosing from a variety of colours, styles, and features. Products are then manufactured and delivered to consumers in 21 days.
- *Specialist store segment.* Customers shopping at a specialist retailer are serious about their sport and are willing to pay a premium price for the latest products. Unsurprisingly, price is not considered an order winner in this segment.
- *Sports generalist segment.* These retailers are patronized by customers who want to keep fit and look for reasonably priced products of acceptable quality.
- *Lifestyle generalist segment.* The fourth retailer segment sells the latest fashion items at premium prices. It guarantees high margins based on brand perception, innovation, and exclusivity of the product range.

Adidas uses the following supply chain models to cope with the short-life cycles and unpredictable demand of its sportswear garments.

- *Never out of stock (NOOS) model.* Small retailers can place any size order and have the stock delivered from a DC in two to three days. Adidas combines the orders from the

NOOS retailers to reach minimum order quantities with the producers and benefit from volume discounts. This supply chain model proved to be able to increase sales by a factor of five with retailers in the lifestyle generalist segment.

- *Special handling at source (SHAS) model.* For major customers, batches are prelabelled at production sites and shipped in ocean containers directly to DCs and then sent to stores with no rework.
- *Consignment model.* The consignment model allows retailers to return unsold product to Adidas without charge, thus removing retailer risk. This approach allowed Adidas to improve its market position in specialist running stores.
- *FLASH model.* The FLASH model aims to revive consumer interest by constantly refreshing products at retail shops. Packages are shipped to stores every six weeks while unsold stocks are returned to Adidas. This model was first used by the Spanish apparel retailer Zara, who developed new ranges of products every six to eight weeks to create a sense of urgency with consumers and encourage them to buy before the stock disappears.
- *"Mi Adidas" model.* The large number of design variations required by customers in the "Mi Adidas" segment needs buffers of raw materials and components to be maintained in the factories. Semi-finished products are kept in stock and are completed according to the customers' desired colour, size, and features, only when orders are received. Products bought online can be returned to nearby stores in order to get a refund or a different product.

1.6.3 Galbani

Galbani, a company of the French Lactalis group, is a leader in Italy in the dairy products sector and one of the main actors in the pressed pork market. Galbani is currently made up of three independent operational branches. One of them, biG Logistics, has the mission to manage the logistics activities of the whole company. The logistics system is organized in such a way as to guarantee an efficient synchronization of the internal production and distribution processes of the products, both for *mass market retailers* (MMRs) and traditional retail shops. The distribution network of the company is organized on two levels: there are, between the production plants and the destination markets, a central warehouse and 11 distribution platforms. This solution promotes a rapid delivery of the products (within 12 hours in Italy and 24 hours abroad), strictly respecting the requirements of the *cold chain*. The daily products are dispatched directly by the production plants to the central warehouse, located in the area of Ospedaletto Lodigiano, considered a barycentric position with respect to the national market. The central warehouse of $185\,000$ m^3, highly automated and constantly kept at a temperature of 4 °C, serves, in turn, the second level platforms with the orders mixed according to their destination (see Figure 1.7). The platforms receive the incoming flow of goods from the central warehouse and supply both the DCs of the MMRs and the so-called satellites. The satellites are small-size warehouses from which a fleet of distinctive yellow vans (operating as truly travelling stores) delivers to retailers. There are 108 satellites distributed throughout the national territory, each covering roughly a province.

Galbani determines the daily production plan for each plant on the basis of forecasts derived from historical data. These data are gathered at the 11 logistics platforms,

Figure 1.7 Location of the central warehouse and the 11 distribution platforms in the logistics system of Galbani.

MMRs and the satellites. Then this information is shared with the central warehouse and the production plants.

1.6.4 Pfizer

The Pfizer Pharmaceuticals Group is the largest pharmaceutical corporation in the world. Its mission is "to discover, develop, manufacture and market innovative, value-added products that improve the quality of life of people around the world and help them enjoy longer, healthier, and more productive lives". The Pfizer range of products also includes self-care and well-being products for livestock and pets.

Founded in 1849 by Charles Pfizer, the company was first located in a modest red-brick building in the Williamsburg section of Brooklyn, New York (USA), that served as office, laboratory, factory, and warehouse. The firm's first product was Santonin, a palatable antiparasitic which was an immediate success. In 1942 Pfizer responded to an appeal from the US Government to expedite the manufacture of penicillin, the first real defense against bacterial infection, to treat Allied soldiers fighting in World War II. Of the companies pursuing mass production of penicillin, Pfizer alone used the innovative fermentation technology. Pfizer manufactures some of the most effective and innovative active ingredients including atorvastatin, whose medicine is the most prescribed cholesterol-lowering one in the USA, amlodipine, belonging to the calcium channel blocker dihydropyridine class, used as an anti-hypertensive, azithromycin, the most-prescribed brand-name oral antibiotic in the USA, and sildenafil citrate, a breakthrough treatment for erectile dysfunction. With a portfolio that includes five of the world's 20 top-selling medicines (including Advil, Viagra, Xanax, Zoloft and the Comirnaty COVID-19 vaccine), Pfizer sets the standard for the pharmaceutical industry. Ten of its medicines are ranked first in their therapeutic class in the US market, and eight earn a revenue of more than one billion dollars annually. Research, development and innovation represent the lifeblood of Pfizer business that supports the world's largest biomedical research laboratory, with 12 000 scientists worldwide and a financial investment of six billion dollars annually.

The Pfizer logistics system comprises 58 manufacturing sites around the world (see Table 1.1), producing pharmaceutical, veterinary, and cosmetic products for more than 150 countries. Because manufacturing pharmaceutical products requires highly specialized and costly machines, each Pfizer plant produces a large amount of a limited number of pharmaceutical products for the international market (see Table 1.2).

The attention will be now focused on the Pfizer supply chain of a cardiovascular product, fictitiously named Alpha10. The product is packaged in blisters, each containing 20 tables of five or 10 mg. Alpha10 is produced in a unique European plant (EUPF) for an international market including 90 countries (see Figure 1.8). Every year the plant produces over 117 million blisters. The product expires 60 months after its production and must be stored at a temperature varying between 8 °C and 25 °C. The main component of Alpha10 is a particularly active pharmaceutical ingredient, based on a Pfizer property patent, manufactured in a North American plant. Its packages are transferred

Table 1.1 Geographical distribution of the manufacturing sites of Pfizer.

Continent	Number of sites
Africa	7
Asia	13
Australia	2
Europe	16
America	20

Table 1.2 Features of some Pfizer plants in Europe. The productivity rate is measured in millions of items per year.

Country	Number of plants	Number of products	Productivity rate
Belgium	1	29	6.5
France	1	14	2.4
Germany	1	3	11.4
Italy	3	182	87.1
United Kingdom	1	8	5.0

Figure 1.8   Markets for Pfizer product Alpha10.

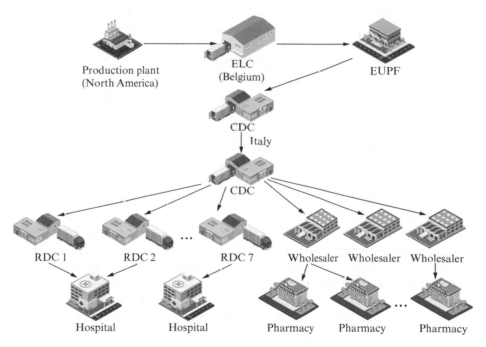

Figure 1.9   Pfizer supply chain for product Alpha10 in Italy.

by air to the European logistics centre (ELC), located in Belgium, which in turn replenishes the EUPF on a monthly basis (see Figure 1.9). Freight transportation between the ELC and the EUPF is performed by overland transportation providers (e.g., Danzas). The EUPF plant manufactures Alpha tablets which are subsequently packaged into 120 blister boxes and sent weekly to a third-party CDC.

In Italy, Alpha10 is distributed, together with other products of Pfizer, both to hospitals and pharmacies, using two different channels (see Figure 1.9). Hospitals (about

2000) are supplied directly by the company, throughout a CDC and seven RDCs. Hospitals may be supplied by more than one warehouse, depending on stock levels available at the RDCs. Transportation is performed by specialized haulers in refrigerated vans. Pharmacies (about 16 000) are supplied through (about 200) wholesalers. Wholesaler orders are collected directly by Pfizer and shipped weekly by the CDC. The CDC is able to deliver the product to any Italian location within at most 60 hours. Wholesalers receive orders from pharmacies very frequently (up to four times a day). Pharmacies expect the wholesalers to deliver medicines within four to 12 hours (it is worth noting that, in Italy, pharmacies have a high contractual power over the wholesalers). Therefore, the average revenue of the wholesalers is low, due to the high logistics costs that are incurred to guarantee a high service level to pharmacies. The Pfizer logistics network and its wholesalers share data (orders, delivery times, etc.) with an information system named Manugistics.

1.6.5 Amazon

Amazon is an American multinational technology company headquartered in Seattle (USA). Its core business is based on cloud computing, digital streaming and, above all, e-commerce. Initially born as a online marketplace for books, Amazon now sells a multitude of product categories in more than 200 countries, including electronics, hardware, software, video games, apparel, furniture, food, toys, accessories, and jewellery. Amazon carries its business on an international scale, operating complex logistics activities 24–7 to guarantee fast deliveries to its customers all over the world (see Section 1.7.2). The logistics system is composed of three main levels: *procurement & fulfilment*, *distribution*, and *last mile* (see Figure 1.10). The three levels are described in the following with reference to North America.

- *Procurement & fulfilment.* The first level is focused on procurement, with the main objective of stocking products to fulfil customer orders promptly even if demand is uncertain, as well as the preparation of packaging and parcels for the subsequent delivery stages. In North America, this level is composed of 10 *inbound crossdocking points* (ICPs) and 187 *e-fulfilment centres* (EFCs), each of which named after the International Air Transport Association (IATA) code of the closest airport, followed by a digit (e.g., MIA5 is the code adopted for the e-fulfilment centre located in Miami, Florida). The ICPs are usually located in proximity of major intermodal terminals and corridors to facilitate trans-shipment. The EFCs, having an average size of 79 500 m^2 each, are usually located in suburban areas with an easy access to highways. In order to face the high variability and frequency of online orders, a great part of the EFCs is completely automated, guided by algorithms (developed by *Amazon Robotics*, a subsidiary company of Amazon) that optimize the routes and schedules of pickers and *automated guided vehicles* (AGVs, see Section 5.6.2). The procurement strategy is based both on low-cost manufacturers making products offered under the Amazon brand, and *associated* vendors that sell their own products on the Amazon e-marketplace. The associated vendors usually have to send a share of their inventory to the EFCs before the actual sale, in order to guarantee a fast-delivery service (e.g., *Amazon Prime* deliveries).

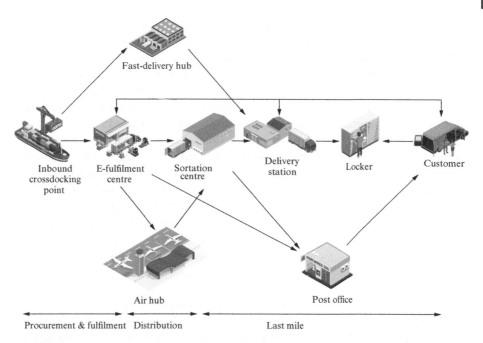

Fast-delivery hub

Inbound crossdocking point

E-fulfilment centre

Sortation centre

Delivery station

Locker

Customer

Air hub

Post office

Procurement & fulfilment   Distribution   Last mile

Figure 1.10 Amazon's supply chain. For simplicity, only one node for each type of facility is depicted.

- *Distribution.* The second level involves the distribution of parcels to downstream facilities, up to a centre close enough to the final destination. The two main facilities at this level are *air hubs* (AHs) and *sortation centres*. The distribution through AHs is performed by *Amazon Air*, a cargo airline, set up in 2015 to transport Amazon parcels, operating a fleet of 42 airplanes. Amazon Air serves six AHs and 23 airports in North America, with a *hub-and-spoke* structure. The airports are chosen for their proximity to metropolitan areas, in order to move freight directed to the sortation centres with the use of trucks. Sortation centres represent the first layer of a city logistics framework (see Section 1.7.3). Nowadays, there are 47 sortation centres in North America, with an average size of 29 900 m^2 each. A sortation centre is located in an area with a minimum level of demand and its role is to split the loads into parcels to be delivered directed to *delivery stations* (see the *last mile* level) or to other destinations, such as post offices and households.
- *Last mile.* The third level involves the transportation of parcels to final destinations. In this case the challenges faced by the company are those of city logistics (see Section 1.7.3) with a variety of operational constraints related to traffic congestion, presence of limited traffic zones, lack of parking spaces, etc. Excluding 3PL services, this level is managed by using two different types of facilities: *fast-delivery hubs* and *delivery stations*. Fast-delivery hubs are facilities which specialize in the fast-delivery market segment, for managing specific company services like: Amazon Prime deliveries within the same day or 48 hours, depending on the destination), Amazon Fresh (grocery and perishable) and Amazon Pantry (household and cleaning). The network is made up of 53 Amazon Prime hubs of an average size of 3700 m^2 each, that represent a compromise between reduced lead time and costs; 24 Amazon Fresh and Pantry

hubs usually co-located with Prime hubs. A delivery station is the most common type of facility. Indeed there are around 250 delivery stations inside North America with an average size of 8500 m². In this case, the main strategy for the last mile is building a series of horizontal collaborations. In particular, Amazon defined a series of pickup points located in affiliated physical bookshops, news vendors, shopping centres and drug stores. In particular, in September 2011 the company launched the Amazon Locker program in major US cities where self-service kiosks (called *lockers*, see Section 1.7.2) were installed in affiliated stores. The service was then extended to Canada, France, Spain, Germany, and Italy. In June 2018, Locker was available in over 2800 locations in 70 cities around the world. Finally, Amazon is experimenting with other innovations for last-mile delivery: Amazon Prime Air and Amazon Flex services. Amazon Prime Air is a drone delivery service launched in 2013 specifically for parcels of at most 5 kg that have to be delivered over small distances in around 30 minutes. The Amazon Flex platform was launched in 2015 in 14 American metropolitan areas, introducing one of the first crowdshipping services (see Section 1.7.2) to encourage ordinary citizens to carry out deliveries on the last mile.

As is common in the e-commerce sector, Amazon stipulates specific contracts with 3PL service providers such as DHL, FedEx, and UPS (see Section 1.4) to operate long-haul transportation from EFCs directly to customers, or to move freight by truck among the different nodes of the supply chain.

1.6.6 FedEx

FedEx is an American multinational delivery and SCM company servicing more than 220 countries and territories with over 500 000 team members, 180 000 cars, vans, tractors, motorcycles, 670 cargo planes, and hundreds of freight terminals around the world. Every day the company delivers more than 16 million packages.

Major competitors include the United States Postal Service (USPS) and UPS, along with national postal services and carriers such as Canada Post (and its subsidiary Purolator), Deutsche Post (and its subsidiary DHL), Royal Mail, and Poste Italiane.

FedEx deals with two distinct (air and ground) networks (see Figure 1.11). Both are based on a hub-and-spoke strategy in which parcels are consolidated through hub terminals connected with multiple decentralized terminals (spokes) along the network. Conversely, a transportation network may connect directly every pair of origin-destination pairs (*point-to-point network*) but this is only rarely economically viable.

Air parcels (often classified as urgent shipments) are driven to a nearby airport facility (*express station*), successively flown to an air hub (either the major hub in Memphis, USA, or to a regional hub), transported to the destination airport, and then delivered by truck. It is worth observing that FedEx air shipments are typically overnight. This allows a customer to ship a package, for example, from Grenoble (France) on Wednesday to Seattle (USA) by 8:30 on Thursday. To achieve this result, FedEx operates one of the largest cargo airlines in the world, with more than 1950 express stations at 13 air hubs, and flies to over 800 destinations worldwide.

Non-urgent packages in North America or in Europe, when the distance is compatible, are picked up, dropped off at a ground centre, sent out to a ground hub, forwarded

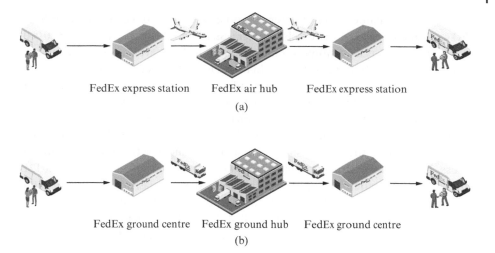

FedEx express station   FedEx air hub   FedEx express station

(a)

FedEx ground centre   FedEx ground hub   FedEx ground centre

(b)

Figure 1.11   FedEx's multiple hub-and-spoke (a) air and (b) ground networks.

to a destination hub by truck, driven to a destination centre and finally shipped by truck to the final destination. FedEx operates more than 600 ground facilities in 39 ground hubs.

FedEx uses independent contractors for many deliveries (depending on the country and service level) while other competitors (such as UPS) use their own employees for all services. The company deals daily with all the facets of international logistics, including customs regulations.

1.6.7 A.P. Moller-Maersk

A.P. Moller-Maersk (APM) is a Danish company operating in the transportation, logistics and energy sectors, with subsidiaries and offices in 130 countries. Its operating units include APM Terminals and Maersk Line. Both operate in the sea freight transportation sector (see Section 6.2.2).

Since the 1970s, the majority of long-haul transportation makes use of *containerization*, a system based on the use of intermodal containers of standardized dimensions (see Section 5.4.5 for more details). This allows productivity gains to be achieved through simplified cargo handling: containers are loaded and unloaded, stacked and transported efficiently over long distances. In particular, containers are moved from one mode of transportation to another (e.g., container ships, rail flatcars, and semi-trailer trucks) without being opened. Container ship capacity is measured in *twenty-foot equivalent units* (TEU), representing the volume of a 20-foot-long (around 6.1 m) container.

With a staff of 83 000 employees worldwide, APM owned 307 ships and was chartering some 417 ships in 2021. It accounted for around 17% of the world's merchant container fleet. In terms of TEUs, APM is the world's largest container-shipping company. It has ships with an overall capacity of around 4 000 000 TEU (2 300 000 TEU of owned capacity and about 1 800 000 TEU of chartered capacity).

Maersk operates a global container network with a hub-and-spoke topology: medium-size freight ships usually under 3000 TEU (called *feeder ships*, see Section

6.2.2) transport containers from different ports (*spokes*) to central container terminals (*hubs*) to be loaded onto larger vessels (*mother vessels*) with a capacity up to over 20 000 TEU. The hub-and-spoke configuration, introduced in 1985, avoids very large mother ships deviating from their round-the-world routes to visit ports whose volumes are not large enough to justify direct calls. In particular, hub ports link the major East–West services with one another, and to the North–South services (see Figure 1.12). Some *transshipment* hubs, such as Algeciras (Spain) and Gioia Tauro (Italy), serve no hinterland, their only attraction being their location close to the shortest intercontinental routes. The update of the Maersk's hub-and-spoke container network is constantly in progress. In 2021 the network was based on 343 ports and terminals operating in 121 countries, and the routes were split over six areas: Africa, Asia Pacific, Europe, Latin America, North America, and West Central Asia. By using the company website, it is easy for potential customers to recover detailed shipping information for any route to and from each area. For example, for the FEW1 route (see Figure 1.13), corresponding

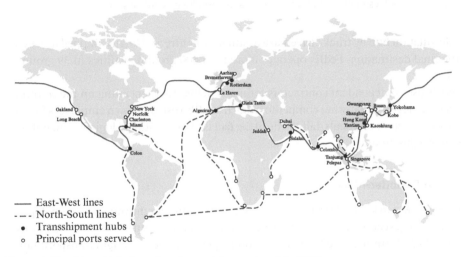

Figure 1.12 Maersk's hub-and-spoke container network in 2003.

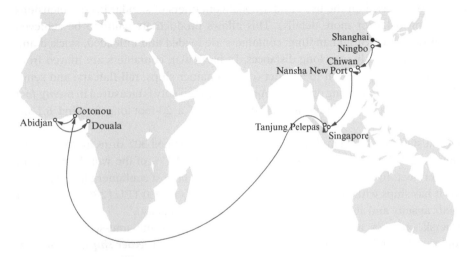

Figure 1.13 Maersk's FEW1 westbound route.

Table 1.3   Timetable of the Maersk's FEW1 westbound route.

Port	Country	Transit time	Arrival day	Departure day
Shanghai	China	–	–	day 1
Ningbo	China	1 day	day 2	day 2
Chiwan	China	3 days	day 4	day 4
Nansha New Port	China	3 days	day 4	day 4
Tanjung Pelepas	Malaysia	10 days	day 11	day 11
Singapore	Singapore	11 days	day 12	day 12
Cotonou	Benin	33 days	day 34	day 34
Abidjan	Ivory Coast	36 days	day 37	day 37
Douala	Cameroon	41 days	day 42	day 42

to the westbound service from Asia Pacific to Africa, and in particular, from Shangai port (China) to Douala port (Cameroon), the timetable in Table 1.3 is shown.

1.6.8   Canadian Pacific Railway

Canadian Pacific Railway (CPR) hauls freight over 20 100 km of track in six provinces of Canada and into the USA. Its transcontinental railroad system stretches from Montreal to Vancouver, from Edmonton to Milwaukee, Detroit, Chicago, and Albany (see Figure 1.14).

Rail is the cheapest freight transportation mode in inner territories which makes it the mode of choice for low-value and less time-sensitive goods. In North America it costs approximately $ 0.025 to move a tonne of freight over one kilometre, which means that a tonne of freight can be moved on a train from New York to Los Angeles for about $ 100. In Europe the cost is nearly double. The competitors to trains are trucks that can haul a tonne of freight for about $ 0.13 per km.

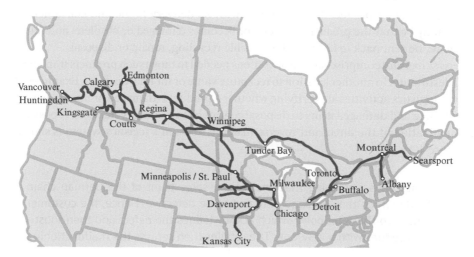

Figure 1.14   Canadian Pacific Railway network.

In North America the rail network is quite extensive and is owned mostly by private companies. The governments subsidize passenger transportation. Seven major rail operators, each with its own territory, exist: Union Pacific Railroad, BNSF Railway, CSX Transportation, Kansas City Southern Railway, Norfolk Southern Railway, Canadian National and CPR. None of these companies covers the entire continent. So for most journeys a single operator cannot get a load shipped from origin to destination. Hence, rail companies need to cooperate: rail cars are trans-shipped from the convoys of an operator to the convoys of others at special terminals.

CPR primary business is transporting intermodal grain (24% of 2016 freight revenue), intermodal freight (22%), and coal (10%) as well as chemicals (12%), automotive parts and automobiles. Coal is shipped in unit trains from coal mines to terminals in British Columbia, from where it is shipped inland. Grain is hauled from the prairies to ports on the Atlantic and Pacific coasts, where it is then shipped overseas. Over half of CPR's freight traffic is in western Canada. The company also provides logistics services via partners. Passenger services were terminated in 1986.

CPR runs very long freight trains (as long as 4000 m). It usually takes a crew of only two people to run such a train. Long routes are divided into segments: for example, the route from Chicago to Seattle is partitioned into 10 segments, each of which assigned to a different crew. Crews get on a train at the origin point of a segment (coinciding with the endpoint of the preceding segment), run the train (for at most 12 hours in the USA) up to the other endpoint where they are taken over by another crew. They then drop off and rest in a hotel, and drive another train along the same segment in the opposite direction.

1.7 Trends in Logistics

This section reviews the trends that are expected to transform the global logistics industry in the near future.

1.7.1 Reverse and Sustainable Logistics

The life cycle of a product does not finish with its delivery to the end consumer. In fact, it may be that the product is unsold, or becomes damaged or obsolete, and must, therefore, be sent back to its origin for possible recycling, repair, or disposal.

Reverse logistics comprises all the operations needed to move such products from their final destination to another location to recapture value or for final disposal. Examples of reverse logistics activities are verifying whether a product is damaged or not; transporting obsolete and damaged items to disposal centres or to secondary markets. A possible schematization of the direct and reverse material flows in a logistics system is shown in Figure 1.15.

A customer buys a washing machine from a sales point of the German chain MediaMarkt which they subsequently find to be defective. Hence, the customer takes it back to the sales point which acknowledges the defect and then substitutes the product with a new one. The retailer then returns the malfunctioning

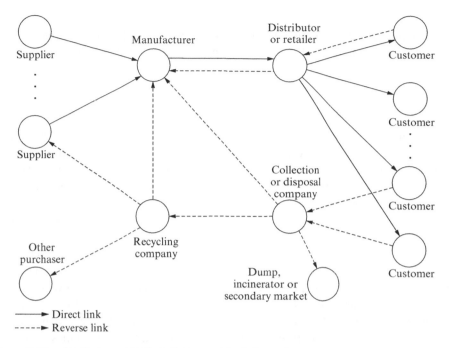

Figure 1.15 Graph representation of a reverse logistics system.

machine to an appropriate collection centre. The product is finally sent to the manufacturer which repairs it and sends it to a secondary market. In this way the manufacturer obtains an added value from the defective product.

Reverse logistics is intrinsically aligned with environmental sustainability and can therefore be seen as an important part of *sustainable logistics* which aims to lower the ecological footprint of logistics operations, such as CO_2 emissions, noise pollution, consumption of raw materials and energy, etc., balancing financial performance measures and environment care.

Hennes & Mauritz AB (more commonly referred to as H&M) is a Swedish multinational retail company known for its fast-fashion clothing. In 2012, H&M innovated the concept of reverse logistics by collecting old and unwanted clothes of any brand in any condition. The initiative started in Sweden and was later extended worldwide through the "Garment Collecting Program", a global initiative for a sustainable future of fashion. In 2019, the company collected 29 005 tons of fabrics, equivalent to approximately 145 million T-shirts. The garments collected are divided into three different categories: items that can be reworn; items reused for the production of new clothes; items recycled in order to make fabrics for non-fashion markets. The initiative rewards participating customers with a discount on their next purchase with the aim to make customers respectful of the environment as well as reducing the cost of raw materials. The ultimate goal of H&M is to manufacture all its products with 100% recycled materials by 2030.

1.7.2 E-commerce Logistics

Retail e-commerce sales have reached $ 4 trillion in 2020 and are expected to constantly increase during the next few years. The massive use of e-commerce implies a lot of advantages: reaching customers living in remote areas, the creation of new markets for niche products, the possibility to sell products 24 hours per day, etc. On the flip side, managing logistics assets becomes much more complex. For this reason, e-commerce companies are required to have a distribution system that is as organized, fast, efficient, rational, and as optimized as possible.

E-commerce logistics is obviously very different from traditional logistics because the dynamics of physical stores differ from those of e-shops in many aspects. Indeed, a large number of activities like procurement, shipping, storage, deliveries, reverse logistics, and customer care cannot be managed in the same way inside the two channels. For example, traditional LSPs try to ship many parcels within the same delivery in order to exploit reductions in transportation costs deriving from economies of scale. On the other hand, in e-commerce distribution, logistics providers need to manage lots of shipments individually which may require a completely different approach. Broadly speaking, e-commerce changed the dynamics both in *business-to-consumer* (B2C) and *business-to-business* (B2B) sectors.

In e-commerce, the speed and reliability of the shipments are crucial: customers who buy online expect a delivery within a short amount of time, without additional costs, and with a high level of personalization (e.g., the possibility to choose the delivery place and time). Other aspects that increase the complexity of e-commerce logistics are the extreme variability and uncertainty of the demand as well as the complexity of the whole logistics process (which includes online ordering, billing, shipment tracking, and providing updates to customers).

Impact of E-commerce on Logistics Performance Measures
The adoption of the e-commerce paradigm has a significant impact on all the logistics performance measures (Section 1.8), including costs and service level provided to customers. In general, e-commerce largely improves product availability and variety since companies can concentrate their inventories in a few centralized warehouses, without the need to distribute them across a large number of stores. This implies lower overall inventory levels as well as larger facilities, usually automated, with reduced unit warehousing costs. The benefits derived from this aggregation are lower for products with high demand and low variability, while they are higher for low demand products with a high level of variability. E-commerce also allows quick adaptation to changing market conditions (e.g., demand lower than expected) by means of real-time changes in pricing and promotions. Moreover, direct deliveries allow costs to be cut by removing intermediaries from the value chain and, in the case of new products, to reduce the time-to-market, that is the length of time required from a product being conceived until its commercialization.

The other side of the coin is that delivery time may become longer than the travel time to a shop. In addition, the impossibility of trying the products can be, in some sectors, a minus. Except for digital goods, the aggregation of stocks generates longer trips to reach the customers with higher delivery costs. Moreover, an e-company needs to build and maintain a suitable *information and communication technology* (ICT) infrastructure (an online store, a payment system, a tracking or tracing system, etc.). Finally, in the e-commerce sector, the return of unsatisfactory products is much higher than in traditional supply chains (even 25–40% in the fashion and hi-tech sectors versus 8–9% in traditional stores). Hence, reverse logistics (see Section 1.7.1) has to handle a very large number of small parcels. Since the return process is often free of charge for the final customer (companies that decide to charge their customers for returning goods have a significant reduction in revenue) and returned products can rarely be resold at the original price, reverse logistics can generate large profit losses.

Issues in E-commerce Logistics

E-commerce companies face a number of peculiar issues.

- *Outsourcing of logistics activities.* Several companies outsource their e-commerce logistics to 3PL, 4PL, or even 5PL providers (see Section 1.4). These providers have the resources and expertise to face the challenges posed by e-commerce supply chains. Of course, this approach requires a high level of integration between the information systems of the providers and the company's order-entry system and *warehouse management system* (WMS).
- *Big data analytics.* E-commerce allows companies to collect a very large amount of data about their customers' buying behaviour. These data can be used to enhance customer service, define personalized pricing strategies and promotions, facilitate collaboration with logistics partners, produce accurate forecasts, etc. (see Section 1.10 for more details).
- *Horizontal, vertical, and lateral integration.* These supply chain collaboration strategies allow costs and risks to be shared with different partners. *Horizontal integration* consists in sharing resources (such as warehouse space or production capacity) at the same level in the supply chain. *Vertical integration* occurs at different levels of the supply chain (such as the integration between a producer and a retailer chain) in order to optimize service level or cost. *Lateral collaboration* combines the benefits of both vertical and horizontal integration (such as intermodal transportation).
- *Last-mile logistics.* The last mile is the least efficient part of most supply chains and generates costs which account for up to 30% of the total transportation cost. Many factors generate inefficiencies, such as traffic in urban areas and the absence of customers at the destination to receive parcels. For this reason, e-commerce companies have developed alternative approaches to the classical home delivery: pickup points and *crowdshipping*. In the first case, the company delivers the goods to specific pickup points where the customer can go and collect the order at a later time. This solution saves transportation costs (as the company delivers parcels to a limited number of points) and allows customers to collect the goods during their leisure time. A variant of this approach makes use of lockers as pickup points. Lockers are automated distribution machines that allow the collection and delivery of goods 24 hours a day. When a parcel has been placed in a locker by a delivery crew, the company sends a

code to the customer which can be used to open the locker. In the second case, companies rely on a distribution strategy based on the principles of *sharing economy*. Crowdshipping can be defined as a system in which delivery is outsourced to occasional drivers who travel inside an area as a part of their daily routine. Crowdshipping uses an internet-based platform to match the needs of customers with drivers having some spare capacity in their schedule or in their vehicles. Drivers are rewarded by the company for their service.

> WalMart, an American retail company operating worldwide, launched in 2013 a plan to improve its transportation operations for delivering online orders. The company offered to in-store customers the possibility to deliver packages to other customers located along their route home. The system implemented by WalMart can be considered one of the first examples of crowdshipping. As an incentive, WalMart offered a discount on the customers' shopping bill covering the cost of their gas in return for the delivery of packages.

1.7.3 City Logistics

City logistics is concerned with freight distribution in urban areas with the aim of mitigating externalities such as congestion, emissions, and noise pollution, in addition to pursuing the traditional objectives of logistics (see Section 1.8). City logistics, which can be seen as a part of sustainable logistics (see Section 1.7.1), has gained increasing relevance in the past few decades due to the progressive population migration from small towns to large cities and the rise of e-commerce (see Section 1.6.5). Nowadays, the distance covered by delivery vehicles in urban areas represents 15–20% of all road freight transportation. The first city logistics initiatives have been taken in Europe and Japan, followed more recently, by North America and Australia.

A key feature of city logistics is a new organizational model, not considering each player (logistics providers, customers, municipal police, etc.) individually. Rather, all the stakeholders and activities are seen as part of a unique and integrated logistics system within the city. As a consequence, city logistics focuses on a strong coordination between players at all levels and, in particular, consolidation of the loads of different customers and shippers inside the same (low-emission) vehicles. In the following, the distinctive features of city logistics are reviewed.

City DCs. City DCs are intermodal platforms whose aim is to provide coordination in freight movements inside the urban area. City DCs work as crossdocking points or warehouses, receiving the inbound freight that has to be delivered within the city from long-haul multimodal transportation. In the city DC the orders are consolidated and loaded onto smaller vehicles to their final destination in the urban area. The city DCs also deal with the outbound freight flows (e.g., those generated by returned items) from the city to destinations outside the urban area. Systems where freight flows from a single city DC directly to the city centre are defined as *single-tier*. In large cities, a central city DC located in the outskirts of the urban area may serve a set of smaller city DCs (defined as *satellites*) in different city neighbourhoods (*two-tier* systems).

Ad hoc and clean vehicles. City logistics makes use of different types of vehicles. In a two-tier system, urban trucks move freight from the city outskirts to the satellites and vice versa, along selected corridors around the urban area, traversing the city centre only when necessary (e.g., when vehicles are close to their destinations). Then, smaller vehicles (called *city freighters*) are used to deliver goods from satellites to the final destinations. City freighters may be zero-emission vehicles, cargo bikes, electric vehicles, drones, etc.

Massive use of ICT. The complexity of city logistics may take advantage of ICT to coordinate the various players and control operations. Such systems create an ecosystem of connected devices and computing mechanisms to exchange data and information. Collected data include:

- vehicle positions through GPS-based devices;
- the status of deliveries through barcode (see Section 5.7.3) or *radio-frequency identification* (RFID) systems (see Section 5.7.6);
- traffic updates from sensor networks or mobile phone networks;
- planned shipments.

These data are then used to make forecasts, extract demand patterns, control operations, and update customers. Moreover, a Web-based platform is often used to plan vehicle routes and loading operations, reoptimizing routes in real time when traffic conditions vary or accidents occur, tracking last-mile deliveries, booking loading and unloading bays. In the near future, the adoption of the *Internet of things* (IoT) paradigm will make this approach more and more pervasive.

> The municipality of Donostia-San Sebastián (Spain) has recently carried out a city logistics project to make city freight distribution more environment-friendly. A cycle logistics micro-hub has been located in the city centre (with an area of $500 \ m^2$) as a city DC from which a fleet of pedal-assisted electric cargo bikes, with a cargo capacity of 180 kg for each, serve the urban area. In addition, a Web-based platform has been set up to allow logistics providers to reserve loading and unloading bays in the city centre, to plan shipments at night and to coordinate distributors, local shops and municipal police. In a single year, the cargo bikes performed about 21 000 deliveries, saving around 27 000 km in van and truck routes and highly reducing vehicle emissions.

Alternative delivery schemes. In a city logistics context, other paradigms based on the concept of sharing and integration have gained in popularity. These approaches aim to reduce costs, to increase the vehicle utilization rate, and to optimize routes. One of the most common approaches is the implementation of a *van-sharing* system, operated by specialized companies or consortia of logistics providers, through Web-based applications. The service allows the creation of a collective rental of minivans. Another common approach is the integration of public and freight transportation. In this case, the spare capacity of transit systems (e.g., metro lines) is used for freight transportation. Other delivery schemes are peculiar to e-commerce (see Section 1.7.2).

ILOS (Intelligent Freight Logistics in Urban Areas) is a city logistics project developed in Vienna (Austria) to reduce traffic congestion and air pollution caused by freight flows inside the city. ILOS has developed a freight routing optimization tool that suggests the best routes to LSPs and provides information about the associated time and cost savings. The system implements a massive collection of data from GPS-based devices on board the vehicles. The data input a simulation model which, based on real time traffic conditions, estimates travel and delivery times. The project has been able to achieve a 60% reduction in travel time, a 15% reduction in distance travelled, a 20% reduction in fuel and a 30% reduction in cost.

1.8 Logistics Objectives and KPIs

The ultimate goal of logistics management is to move goods effectively and efficiently. The concept is often summarized by stating that "logistics aims to get the right goods, in the right quantity, in the right condition, at the right place, at the right time, to the right destination, at the right cost". Nowadays, logistics is seen by organizations as a fundamental tool to gain a *competitive advantage* and its objectives are more and more closely tied to the global organizational strategy (see Section 1.9.2). This means that logistics is no longer regarded as a mere "cost minimizer" (as it used to be in the past), but rather aims to achieve a suitable trade-off among the following aspects, in accordance with the organization's mission and plans:

- minimize investment (capital) or maximize return on investment;
- minimize logistics cost;
- maximize customer service level.

In practice, logisticians use a number of *key performance indicators* (KPIs) and *performance measures*. The two terms are sometimes used interchangeably but they are not synonyms. Indeed, the word "key" in KPI is crucial. KPIs express the truly vital and strategic objectives of the organization (see Section 1.9.2). They must be few and well chosen to get focus from top managers. Performance measures, on the other hand, are chosen by departments and may be numerous in order to track specific aspects of operational progress. The main KPIs related to logistics are described in the remainder of this section, while logistics performance measures are illustrated in Section 1.9.3.

1.8.1 Capital-related KPIs

This family of KPIs is related to the amount of capital invested in the logistics system, both as infrastructures, equipment, inventory, or cash. In particular, in commercial supply chains, it includes measures of how well cash flow (money coming in and out) is managed. Its primary objective is to avoid the company running low on (or even out of) cash. The main KPIs of this family are listed in the following.

- *Cash-to-cash time*, C2C. This measures the lag between the time when a company pays its suppliers and when it receives cash from its customers. Several studies have proved empirically the direct correlation between shorter C2C times and greater profitability.
- *Gross margin return on investment*, GMROI. This is defined as the ratio between the gross profit generated by a product and its average inventory investment. GMROI shows which products are poor performers and which are more valuable.
- *Inventory days of supply*, IDOS. This is defined as the number of days it would take a company (or a DC) to run out of stock if it was not replenished. A large value of IDOS indicates that the company has excess inventory with respect to sales.
- *Inventory turnover*, IT. More formally defined and described in Section 5.8, IT measures the number of times a company sells and replaces its stock of goods during a specified time period (e.g., a year). Similarly to IDOS, the IT is an indicator of the logistics management efficiency. In general, the higher the IT, the better. On the other hand, a low inventory turnover shows difficulties in turning the stock into revenue.

1.8.2 Cost-related KPIs

The logistics cost can be broken down into three broad cost components: inventory-carrying costs, transportation costs, and logistics administration costs, each containing several fixed and variable cost subcomponents. Figure 1.16 and Table 1.4 report a summary of the logistics costs in the USA in 2012. The main cost-related KPI is the *logistics cost as a percentage of sales*. It shows the overall relevance of logistics in a given business, or sector, and highlights the potential of improvements in logistics efficiency. If more granularity is needed, this KPI can be split into *warehousing costs as a percentage of sales* and *transportation costs as a percentage of sales*.

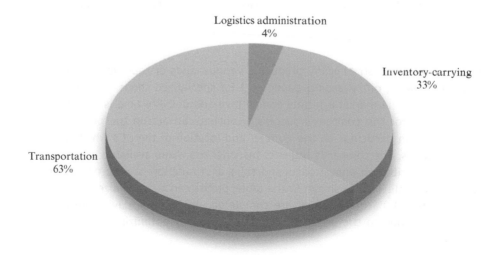

Figure 1.16   Magnitude of logistics costs in the USA.

Table 1.4 Incidence of different cost components in inventory-carrying and transportation costs in USA.

Cost category	Components	[%]
Inventory-carrying	Taxes, obsolescence, insurance	66
	Warehousing	26
	Interest	8
Transportation	Truck intercity	50
	Truck local	27
	Railroads	6
	Logistics administration	8
	Water	4
	Air	4
	Oil pipelines	1

1.8.3 Service Level-related KPIs

Broadly speaking, the service level encompasses the overall degree of customer satisfaction and depends on numerous factors (indicated collectively as the 4Ps of *marketing mix*): product features, price, promotional offers, and place (distribution). The latter defines the logistics component of the service level: as a rule, an effective and efficient organization of logistics yields an increase in market share and profits, at least to some extent. However, if the logistics service level further increases, it produces a fast rise in costs while sales and profits grow more slowly, making the profit fall. Hence, logisticians determine the service level for maximum profit (difference between revenues and costs over a given period). An alternative approach is to minimize the logistics cost (e.g., on a yearly basis) while keeping the service level unchanged.

Ecopaper is a Turkish company producing various kinds of paper (glossy paper, newsprint paper, gift wrapping paper, etc.) for specialized shops, mass market retailers, etc. The market is highly competitive and delivery time is seen to be a key factor for the company's success. Ecopaper can act on the distribution system and, in particular, on the number and location of the DCs which affect delivery times. The company's logistics manager has made available some estimates of distribution annual costs and sales as a function of different service levels (see Table 1.5). As shown in the table, profits can be maximized by guaranteeing that 80% of deliveries are completed within three days from order date. This is due to the possible savings obtained by optimizing the number of DCs.

Table 1.5 Annual estimate of sales, costs and profits (in M€) for Ecopaper.

	Orders dispatched within three days [%]				
	60	70	80	90	95
Sales	4.00	5.00	7.00	9.00	10.50
Costs	1.80	3.00	3.50	6.00	7.10
Profits	2.20	2.00	3.50	3.00	3.40

The logistics component of the service level is often expressed by means of the following indicators.

- *Fill rate*, FR. Also indicated as *demand satisfaction rate*, FR is the percentage of customer demand that is met through stock availability, without backorders or lost sales. Empirical studies have showed that the FR is strongly correlated to sales volume.
- *Perfect order rate*, POR. This is defined as the percentage of orders delivered incident-free, that is, with no inaccuracies, damages, or delays. This KPI has a direct impact on customer retention and loyalty. The overall POR can be computed by multiplying the POR of each stage of the logistics system. For instance, if four stages are performing at 99%, the entire logistics process will achieve only a $0.99^4 = 96\%$ POR.
- *Order-cycle time*, OCT. This measures the average time required to fulfil a customer order, from a customer placing an order to its delivery. The OCT contains several components depending on the type of business. It may include order processing, producing and assembling the goods (or, more simply, order picking in a warehouse), order packaging, and shipping time. In general, the duration of each of these operations is not known accurately, given that several internal and external random factors affects the company's logistics system. For this reason, each component of the OCT can be seen as a *random variable* of unknown probability distribution, whose *statistics* can be estimated from historical data. Statistics provide information about the distributional tendencies that one encounters when observing samples of random variables. The two most significant statistics are the *expected value* (or *mean*) and the *standard deviation* whose estimates are the *sample mean* and the *sample standard deviation*, respectively. As a sum of random variables, the OCT is a random variable itself and its statistics can be estimated from those of its individual components.

[`MobilTrust.xlsx`, `MobilTrust.py`] The OCT of the British company MobilTrust is made up of two components: assembly time and transportation time. Five hundred observations of assembly time and 252 samples of transportation time are available. The minimum observed assembly time is 2.3 days while the maximum is 15.9 days. The minimum transportation time is 6.9 days while the maximum is 13.2 days. For simplicity, both parameters are discretized and expressed as an integer number of days (see Table 1.6).

Table 1.6 Historical assembly and transportation times (in days) in the MobilTrust problem.

Assembly		Transportation	
Time (x_i)	Number of observations (h_i)	Time (y_j)	Number of observations (k_j)
2	1	7	19
3	4	8	27
4	4	9	54
5	18	10	65
6	38	11	48
7	56	12	25
8	69	13	14
9	96		
10	72		
11	68		
12	41		
13	18		
14	12		
15	2		
16	1		

Let X and Y be the independent random variables associated with the assembly time and transportation time, respectively. The set of (discrete) realizations of the assembly time is indicated as Ω_X ($\Omega_X = \{2, \dots, 16\}$). Moreover, h_i denotes the number of observations recorded for realization $x_i \in \Omega_X$ (e.g., $h_3 = 4$). Similarly, Ω_Y indicates the observed transportation times ($\Omega_Y = \{7, \dots, 13\}$) and k_j the number of observations recorded for every realization $y_j \in \Omega_Y$. The sample mean $\overline{X}$ and the sample standard deviation S_X of X are

$$\overline{X} = \frac{\sum\limits_{i \in \Omega_X} h_i x_i}{\sum\limits_{i \in \Omega_X} h_i} = 9.13 \text{ days,}$$

$$S_X = \sqrt{\frac{\sum\limits_{i \in \Omega_X} h_i (x_i - \overline{X})^2}{\sum\limits_{i \in \Omega_X} h_i - 1}} = 2.3 \text{ days.}$$

Similarly, the sample mean and the sample standard deviation of transportation time are computed as

$$\overline{Y} = 9.9 \text{ days,}$$

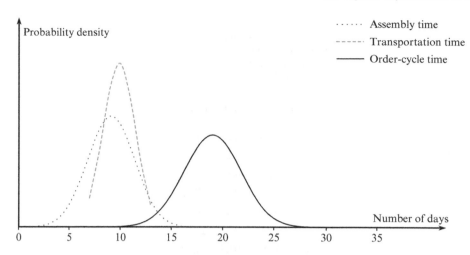

Figure 1.17 Plot of the probability density functions of assembly time, transportation time, and OCT in the MobilTrust problem.

$$S_Y = 1.55 \text{ days.}$$

Hence, the OCT $Z = X + Y$ has mean and standard deviation estimated as

$$\bar{Z} = \bar{X} + \bar{Y} = 19.03 \text{ days,}$$

$$S_Z = \sqrt{S_X^2 + S_Y^2} = 2.77 \text{ days.}$$

Figure 1.17 shows the plot of the probability density functions of assembly time, transportation time, and OCT (in number of days), obtained in Excel by using the NORM.DIST function, in which the mean and the standard deviation of each random variable are replaced with the corresponding sample mean and sample standard deviation, respectively.

The coefficient of variation of the OCT, which is defined as the ratio of its standard deviation and expected value, can be used as a measure of the *variability* involved in customer service. The lower the value of this coefficient, the lower the variability.

[Lugan.xlsx, Lugan.py] Lugan is a Chilean chain of supermarkets aiming to select the carrier with the least variability for its distribution system. For the best two LSPs A and B, Lugan examined 20 observations each, corresponding to travel times (in hours) between a sample sales point and its DC (see Table 1.7). The sample means and the sample standard deviations of the transportation times T_A and T_B of the two providers, computed by using the Excel functions named AVERAGE and STDEV.S, came up to be

Table 1.7 Observed transportation times (in hours) for providers A and B in the Lugan problem.

Provider A		Provider B	
7.74	6.63	5.22	7.11
6.20	10.11	3.02	7.98
5.55	7.78	4.85	9.21
7.16	7.56	9.87	7.92
8.61	6.65	7.86	4.14
7.61	7.74	8.06	10.86
7.08	7.28	9.05	7.17
8.30	9.40	5.92	7.66
7.40	7.55	12.17	4.51
8.32	8.80	7.37	6.28

$$\overline{T}_A = 7.67 \text{ hours,} \tag{1.1}$$

$$S_{T_A} = 1.07 \text{ hours,} \tag{1.2}$$

$$\overline{T}_B = 7.31 \text{ hours,} \tag{1.3}$$

$$S_{T_B} = 2.31 \text{ hours.} \tag{1.4}$$

On the basis of the coefficients of variation, estimated as $V_{T_A} = S_{T_A}/\overline{T}_A = 0.14$ and $V_{T_B} = S_{T_B}/\overline{T}_B = 0.32$, respectively, the company decided to select carrier A.

1.9 Logistics Management

The principles of business management comprise four basic steps: planning, organizing, leading, and controlling, commonly known as the *P-O-L-C framework*. *Planning* sets the objectives of the organization and determines a course of action to achieve them. Organizational strategies are then converted into more specific objectives for the various functional areas (including logistics). As far as logistics is concerned, Section 1.8 describes the most common objectives and the associated KPIs. *Organizing* amounts to developing an organizational structure and allocating human resources to guarantee the accomplishment of the objectives set in the planning phase. *Leading* stimulates employees to participate in achieving the organization's goals. *Controlling* is the ongoing process of evaluating the execution of the plan (and making adjustments if necessary) to ensure that the organizational objectives are met. These four functions of management constitute a process where each function builds on the previous one. In the remainder of this section, planning, organizing, and controlling will be described in the context of logistics management, whereas the in-depth study of leadership is left to the reader, since it is beyond the scope of this book.

1.9.1 Logistics Planning

Logistics planning can be organized at three different decision-making levels: strategic, tactical, and operational.

Strategic decisions have a long-term effect (multiple years) on the logistics system and typically involve major financial investments. They are generally based on forecasts relative to aggregated data (e.g., on the predicted demand of groups of similar products at a regional level). A fundamental strategic decision is whether logistics (or parts of it) has to be outsourced to an external service provider (see Section 1.4) on the basis of a suitable SLA. Alternatively, logistics is managed by the organization itself. In this case, a key strategic decision is the design of the logistics system, including the choice of the number, size and location of facilities, and a rough assignment of demand to stocking points, the design of layout and material handling for each facility, the selection of the mode of transportation for each link of the logistics system as well as fleet sizing.

Tactical decisions concern the detailed implementation of the long-term strategy, usually with a medium term impact (a quarter, six months, a year) on the logistics system. They may include aggregate production planning, seasonal freight flow assignment, the decision to rent space in public warehouses to cover seasonal demand peaks, space utilization and inventory policies in warehouses, the definition of safety stock levels, etc.

Operational decisions concern the definition of short-term (e.g., weekly or daily) work plans for human resources and equipment. They are mainly based on actual data, e.g., orders issued by customers, product inventory levels, real-time location of delivery vehicles, etc. Such decisions include machine scheduling in production plants, order consolidation, vehicle dispatching, repositioning of idle vehicles and containers, etc.

Logistics planning affects five areas: order processing, procurement (see Chapter 4), warehousing and inventory (see Chapter 5), and transportation (see Chapter 6).

1.9.2 Logistics Organizational Structures

The organizational structure is usually represented by an *organizational chart*, also called *organigram*, which provides a graphical representation of the chain of command. It determines the duties and responsibilities of each individual in the organization, as well as the manner in which these duties should be carried out. Organizing at the corporate level amounts to deciding how to departmentalize the organization, i.e., divide the organization into departments to facilitate the achievement of the objectives set in the planning phase. There are many ways to departmentalize an organization, such as organizing by function, product, or geographical area.

The organizational structure determines the distribution of responsibilities and tasks within the company itself. It is influenced by factors such as sector, culture (company's shared values, goals, attitudes, and practices), technology (the greater the use of advanced technology, the slimmer the organizational structure) and size (small companies typically have a sole decision maker, whereas in medium to large ones several responsibilities must be delegated). The organizational structure can be *functional, divisional* or *matricial*.

The Functional Model

The functional model is based on principles of labour division and specialization. In such a framework, activities requiring similar skills and resources are grouped within the same *function*. A function is associated with a *department*, headed by a director. In a functional structure, top management deals with strategic decisions, whereas tactical and operational decisions are delegated to departments. The functional structure is particularly suitable for small- and medium-size organizations, especially those with a low degree of diversification that use a consolidated technology in a stable economic environment. The functional model allows an efficient use of the organizational resources (in particular, staff, equipment, and budget). Another strength is the simplicity which facilitates control from top management. In contrast, the interdependence among departments may result in inefficiencies due to contrasts in departments' objectives. Moreover, a functional structure (which by nature is not very flexible) is often incapable of dealing with product diversification and economic turbulences. Finally, an increase in the organization's size may slow down the decision-making process because each department becomes overwhelmed by responsibilities. Figure 1.18 depicts a functional-type organizational chart in which the logistics function is carried out by a corresponding department which depends directly on top management. Each department is directed by a functional manager who reports to the general manager.

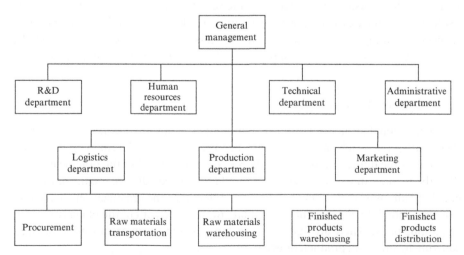

Figure 1.18 Functional-type organizational chart where the logistics department depends directly on general management.

> The Swiss company Akira Foods specializes in the production and marketing of food products. The organizational chart resembles that illustrated in Figure 1.18. Because of its great importance for the organization, logistics is autonomous from other functions and coordinates the following activities: raw materials supply, raw materials warehousing, order processing, inventory and production scheduling, packaging, and transportation. This kind of organizational structure allows the company to guarantee a high level of service to its customers in the Swiss market.

The functional structure of a company may not explicitly contain a logistics function, in which case logistics activities are carried out by one or more departments. This solution is adopted by several food, chemical, and clothing companies and, more generally, by those companies for which the distribution of finished products is critical, as it often happens in mature markets.

In another common variant of the functional model, logistics activities are performed by the materials department (see Figure 1.19). This is the preferred choice for companies manufacturing complex products based on orders. After products have been manufactured, distribution simply amounts to shipping them to customers, with no or little warehousing. In contrast, there are plenty of raw materials and components whose supplies have to be synchronized with production and assembly.

The Divisional Model

Divisions are completely autonomous operational structures that behave like independent businesses. They plan, create, and market products or services autonomously. Each division is organized entirely according to a functional-type logic. The subdivision of the organizational chart into divisions can be operated on the basis of products or services, geographical areas, or markets. The general management is responsible for strategic decisions, especially for product portfolio selections. Each single division makes strategic, tactical, and operational decisions for its products or services. This

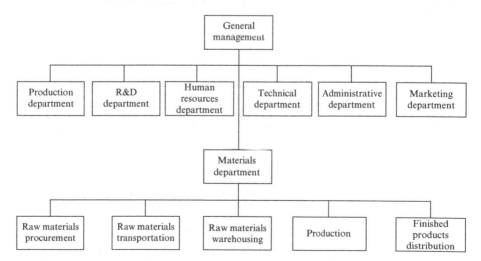

Figure 1.19 Functional-type organizational chart where the logistics activities depend on the materials department.

structure is suitable for large companies, even if the market is turbulent since divisions are able to react quickly to shifts in demand, production costs, availability of raw materials, etc.

On the flip side, this organizational model creates information silos which make it difficult to share knowledge among divisions. Moreover, the duplication of job positions not only generates an increase in fixed costs, but can also stimulate a harmful competition among divisional managers, leading to an excessive focus on short-term (rather than long-term) goals.

As far as the logistics function is concerned, each division adopts its own organizational structure.

Varsth is a multinational company, registered in Denmark, made up of three divisions, one for each the following geographic areas: Europe, North America, and South America. In Europe, the company produces and markets three lines of products: washing machines, television sets, and small electrical appliances (respectively referred to as P_1, P_2, and P_3). Each of the three lines of products is produced independently, whereas the marketing channels and the logistics system are shared (see Figure 1.20).

The Matrix Model

The matrix organizational structure is adopted by large-size companies manufacturing a variety of complex (often high-tech) products with a short- or medium-life cycle. The design, launch, production, and marketing of each product is managed like an individual *project*. The matrix structure combines features of the functional and divisional organizational structures. Technicians and specialists from different functions are in fact assigned to one or more project groups, coordinated by a *project manager*. Each employee involved in a project team reports simultaneously to a functional manager and to a project manager. There are therefore two simultaneous lines of authority (function and project) even if one dimension usually proves dominant.

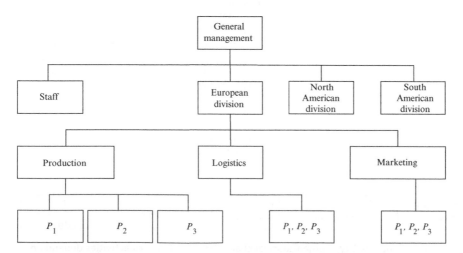

Figure 1.20 Divisional-type organizational chart of Varsth.

The matrix model is well suited for turbulent economic markets and favours elements like motivation and staff development. The dual hierarchical dependence can, however, generate confusion and conflicts because of the presence of different lines of command. Therefore, companies need to establish efficient and effective coordination mechanisms to favour cooperation among managers, which may result in increased operating costs.

In a matrix organization, logistics resources are shared among projects. The functional manager is responsible for the overall logistics system but shares authority with the various project managers.

Elifly is a French–German consortium designing, manufacturing, and delivering industry-leading commercial and military helicopters. The company adopts a matrix organizational model in which a specific project is coordinated by a manager for each product. At present three products (called EXcopter, VLcopter and GRcopter) are marketed. Figure 1.21 depicts Elifly's organizational structure.

1.9.3 Controlling

Controlling deals with monitoring that KPIs and performance measures do not deviate from predefined standards. It consists of three steps, which include: (1) selecting appropriate KPIs and performance measures and establishing their performance standards; (2) comparing actual performance against standards; (3) taking corrective action when necessary. Performance standards are often stated in monetary terms such as revenue, costs, or profits but may also be stated in other terms, such as units produced, number of defective products, or level of service. Effective controlling requires the existence of plans, since planning provides the necessary performance standards. Controlling also requires a clear understanding of where responsibility for deviations from standards lies. For the control of the efficiency and effectiveness of

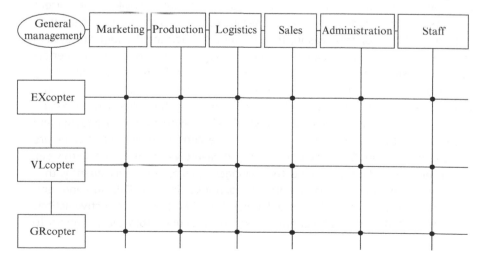

Figure 1.21 Elifly matrix organizational chart.

the logistics system, managers use tailored performance measures described in the following.

Logistics Performance Measures and the Control Panel

As noted in Section 1.8, unlike KPIs, logistics performance measures are used to track specific results related to logistics operations. Each such performance measure generally belongs to two different categories:

- measures of *effectiveness*, that express the capability of the logistics system to achieve the organizational objectives;
- measures of *efficiency*, that express how well logistics resources are utilized to achieve the organizational objectives.

Furthermore, the performance measures are usually classified into *families*, defined in accordance to the company needs, or more generally, corresponding to the decision making areas identified in the planning phase (see Section 1.9.1), such as procurement, storage and distribution (see Table 1.8 for a non-exhaustive list). Some of them correspond to warehouse efficiency measures that will be illustrated in detail in Section 5.8.

The frequency at which performance measures are monitored depends on the logistics decision to be taken: in some cases, the frequency can be monthly or even shorter (e.g., total sales in the last quarter). In other cases, it can be daily or even more frequently: e.g., this may be the case of the sales of a limited-edition series of products on an e-commerce website.

It is good practice to periodically revise the number and type of performance measures used by the logistician by inserting new indicators that capture uncovered relevant aspects or by eliminating obsolete ones. Finally, it is worth noting that an excessive number of performance measures can generate confusion for the decision maker and lead to results that are difficult to interpret (*performance measure overload*).

[Cardena.xlsx, Cardena.py] Cardena is a Romanian company producing and commercializing perforated bricks. During the eighth week of the current year, its logistics manager noticed numerous complaints from customers due to delivery delays. For this reason, they decided to monitor the logistics system starting from the first week of the year, through the use of a specific performance measure referred to as *punctuality*, defined as the percentage of weekly orders delivered on time. A sufficient level of punctuality is set by the manager to the value of 95%. The required data for computing the performance measure have been extrapolated from the Cardena database with a weekly frequency and theresults obtained from the first week to the current (twenty-fifth) week are illustrated in Figure 1.22. The performance measure values from the sixth week to the eighth, in fact, confirmed the manager's perception. Consequently, they decided to implement a series of corrective actions to improve fleet size and vehicle routing (see Chapter 6 for more details). The effects of the corrective actions were evaluated in the subsequent weeks. In particular, from the twentieth to the current week, the performance measures were systematically above the fixed threshold, which motivated the logistics manager to confirm the adoption of the corrective actions in the coming weeks.

Table 1.8 List of performance measures related to procurement, storage, and distribution.

Family	Performance measure	Formula	Category
Procurement	Average volume	Total supply volume/ number of orders	Efficiency
	Unit cost per invoice	Total invoice cost/number of invoices	Efficiency
	Rejected supplies	Rejected orders/total orders	Effectiveness
	Supplier compliance	Orders received after due date/ total orders	Effectiveness
	Lead time	Order delivery date − order issuing date	Effectiveness
Storage	Warehouse utilization rate	Occupied storage locations/ total number of storage locations	Efficiency
	Surface utilization rate	Surface used/ total warehouse surface	Efficiency
	Inventory turnover index	Outgoing freight flow/ average inventory level	Efficiency
	Productivity	Orders fulfilled/working days	Efficiency
	Unit inventory cost	Total inventory cost/total number of items at stock	Efficiency
	Obsolescence	Value of obsolete inventory/ total inventory value	Efficiency
	Out-of-stock rate	Unfulfilled orders/total orders	Effectiveness
	Inventory accuracy	Incorrect product codes/total codes	Effectiveness
	Punctuality	Orders delivered on time/ orders delivered	Effectiveness
	Lead time deviation	Actual lead time/expected lead time	Effectiveness
Distribution	Cost per km	Total transportation cost/ km travelled	Efficiency
	Average delivery	Delivered amount/ number of deliveries	Efficiency
	Trip saturation	Amount dispatched per vehicle/ vehicle capacity	Efficiency
	Punctuality	On time deliveries/total deliveries	Effectiveness
	Delivery accuracy	Incorrect deliveries/total deliveries	Effectiveness

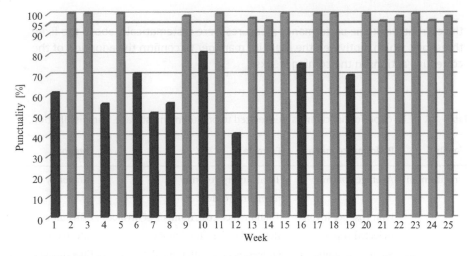

Figure 1.22 Weekly observations measuring delivery punctuality during the last 25 weeks in the Cardena problem.

If a company needs to check multiple performance measures (computed with the same frequency), it is worth using a *control panel*, an information tool which provides a synthetic overview of the entire set of performance measures. Since indicators are measured on different scales, a *min–max normalization* is required to make all the measures homogeneous and easy to compare. With this technique, indicators are shifted and rescaled so that they end up ranging between *min* and *max* (*min* < *max*, see Section 2.5.7 for more details). This operation transforms all the performance measures to a notionally common normalized scale.

[Borg.xlsx, Borg.py] Borg is a Canadian company producing wooden utensils. Following a recent organizational restructuring, the new logistics manager is in charge of monitoring the most critical supply chain activities every month. When designing a control panel, it was found that the most significant problems were connected to the large number of complaints received about errors in dispatched orders, frequent delivery delays, an incorrect policy of inventory management, overstaffing in the warehouse and inefficiencies in the transportation system.

The logistics manager therefore identified 19 performance measures, subdivided into five families calculated with a monthly frequency: two of them (storage and delivery) are representative of typical logistics activities; the others (order dispatch, etc.) are defined considering the Borg specific needs. These measures are described in Table 1.9, in which the calculation method used for each of them is also indicated.

Table 1.9 Performance measures for the Borg logistics system.

Family	Performance measure	Computing method
Dispatch of orders	Number of orders	Orders received in a month/working days in the month
	Complaints	Number of complaints in a month
	Extent of completeness	Order lines dispatched in a month/order lines received in the month
	Errors in order lines	Order lines with errors in a month/order lines inserted in the month
	Errors in orders	Number of orders with errors in a month/orders fulfilled in the month
Staff	Warehouse employees	Monthly average of number of employees
	Effective employees	Monthly average number of effective employees/monthly average of employees
	Productivity of warehouse employees	Monthly average number of daily handling operations/monthly average number of warehouse employees
Storage	Material handling	Monthly average number of daily handling operations
	Pickup operations	Monthly average number of pickups/number of order lines monthly dispatched
	Savings	Monthly cumulated value of economic-financial savings deriving from cost-cutting initiative
	Inventory value	Overall average monthly value of inventory
Delivery	Deliveries per vehicle trip	Monthly average number of customers served by a vehicle trip
	Trip saturation	Monthly average amount of goods dispatched per vehicle/vehicle capacity
	Trip forecast	Trips planned in the month/effective trips in the month
	Deliveries dispatched within delivery time window	Number of lines dispatched in the month within delivery time window/total number of lines dispatched in the month
	Value of deliveries dispatched within delivery time window	Value of deliveries dispatched in the month within delivery time window/total value of deliveries dispatched in the month
	Reliable deliveries	Monthly order lines delivered correctly on time/total monthly order lines
Costs	Budget	Average monthly total cost/monthly budgeted cost

The performance measures were computed in March and were normalized in the interval $[1, 10]$, using the *min–max* normalization procedure on the time series composed of the last 12 data entries available for each measure (from April of the previous year to March of the present year). The values obtained are shown in Table 1.10.

The logic of the normalization procedure is that for each measure, independently of its meaning, a normalized value equal to 1 corresponds to the worst possible outcome (e.g., the minimum number of orders or the maximum number of complaints per month), whereas 10 is the best one. Every performance measure within its own family was weighed in a suitable way (the sum of the weights equals 1), so as to find a single efficiency value for every family. Table 1.11 shows these values, together with the weights chosen for each measure in the month considered.

The logistics manager has determined that the minimum value to be achieved for each performance measure should be 6, and the objective should be 10. The control panel was constructed using a *radar chart* (see Figure 1.23) which has a great visual impact. The control panel allowed the logistics manager to identify the areas needing priority corrective action (those with a value lower than 6).

Table 1.10   Normalized performance measures for the Borg logistics system in the considered month.

Family	Performance measure	Value
Dispatch of orders	Number of orders	4.72
	Complaints	6.21
	Extent of completeness	5.77
	Errors in order lines	7.83
	Errors in orders	6.34
Staff	Warehouse employees	5.40
	Effective employees	7.23
	Productivity of warehouse employees	6.42
Storage	Material handling	6.15
	Pickup operations	5.05
	Savings	4.15
	Inventory value	6.24
Delivery	Deliveries per vehicle trip	7.28
	Trip saturation	6.34
	Trip forecast	7.22
	Deliveries dispatched within delivery time window	6.33
	Value of deliveries dispatched within delivery time window	4.21
	Reliable deliveries	4.87
Costs	Budget	4.88

Table 1.11 Performance measures for the Borg logistics system in the considered month.

Family	Performance measure	Value	Weight	Family performance
Dispatch of orders	Number of orders	4.72	0.20	
	Complaints	6.21	0.25	
	Extent of completeness	5.77	0.18	6.13
	Errors on order lines	7.83	0.17	
	Errors on orders	6.34	0.20	
Staff	Warehouse employees	5.40	0.35	
	Effective employees	7.23	0.30	6.31
	Productivity of warehouse employees	6.42	0.35	
Storage	Material handling	6.15	0.20	5.24
	Pickup operations	5.05	0.30	
	Savings	4.15	0.30	
	Warehouse value	6.24	0.20	
Delivery	Deliveries per vehicle trip	7.28	0.30	
	Trip saturation	6.34	0.10	
	Trip forecast	7.22	0.10	
	Deliveries dispatched within delivery time window	6.33	0.15	
	Value of deliveries dispatched within delivery time window	4.21	0.25	6.03
	Reliable deliveries	4.87	0.10	
Costs	Budget	4.88	1.00	4.88

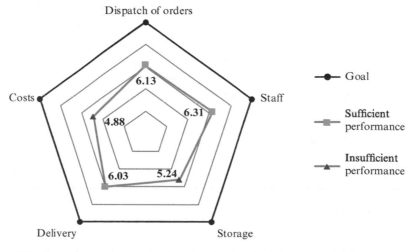

Figure 1.23 Control panel by a radar chart for the Borg logistics system in the considered month.

1.10   Data Analytics in Logistics

In the past, logistics managers made decisions based on pure judgement and experience, whereas today they rely more and more often on factual data. Indeed, the data flow generated every day by logistics operations is very large. *Logistics analytics* is the discipline that uses such data, along with information retrieval, statistical analysis, mathematical optimization, and simulation models, to help managers analyse and coordinate logistics systems in order to ensure the smooth running of operations in a timely and cost-effective manner. Analytics are customarily divided into *descriptive analytics*, *predictive analytics* and *analytics* (see Figure 1.24).

1.10.1   Descriptive Analytics

Descriptive analytics is the interpretation of historical data to better understand changes that have occurred in the logistics system. It takes raw data and makes use of information retrieval and descriptive statistics to allow logistics managers to make sense of them. The output is formatted as a set of charts representing the evolution of the logistics *outcomes* (KPIs or performance measures, see Section 1.8) over time. Nowadays, descriptive analytics are often associated with the term *big data*, an expression that indicates the technologies and algorithms that cope with data that are too large or complex to be dealt with by traditional approaches. In particular, big data applications require a parallel and distributed framework (e.g., MapReduce) that processes data across multiple servers. The peculiarities of big data applications are often referred to as the "three Vs": *volume* (the quantity of data), *variety* (the type and nature of the data; for example, structured data coming from relational databases and unstructured data in a natural language coming from social media like *Facebook* and *Twitter*), *velocity* (the speed at which the data are generated).

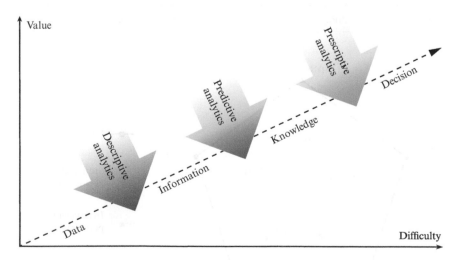

Figure 1.24   Taxonomy of analytics.

A remarkable example of big data application in logistics is Amazon (see Section 1.6.5), that handles millions of customer orders every day. This generates a multitude of shipment tracking updates coming from the sensors attached to AGVs in DCs, air terminals, airplanes and delivery vans. It also generates an enormous number of inventory updates and transactions with third-party sellers. Amazon relies on three Linux databases (with a total capacity of the order of several terabytes) running on a cluster of 28 servers with an Oracle database software.

1.10.2 Predictive Analytics

Predictive analytics aims to make forecasts about unknown future outcomes (e.g., sales trends and customer behaviour patterns). In logistics it is mainly used to predict the demand for finished products over time (as well as the associated material requirements).

Predictive analytics may be based on quantitative techniques such as *extrapolation methods* from *time series analysis*. Alternatively, it may be based on *explanatory methods*, such as *regression* or *supervised learning* methods that automatically capture the relationships between some explanatory variables and the outcome variable, and exploit them to make forecasts. See Chapter 2 for an in-depth discussion of these methods.

Amazon Forecast is a time series data forecasting service, operated by Amazon. It allows forecasting in supply chain planning, in particular, for product demands and travel demands, and, in addition, in financial planning. Amazon Forecasts exploits the potential of *machine learning* (ML) technology (and deep learning, in particular).

1.10.3 Prescriptive Analytics

Finally, prescriptive analytics suggests feasible actions (i.e., actions satisfying budget, logical, temporal and technological constraints) that optimize a given outcome z, based on available data. In the simplest case, there is a limited number n of feasible alternatives to choose from. If the outcome z_i of each alternative $i = 1, \dots, n$, is *deterministic*, that is, it depends on data known at decision time, then the optimal alternative i^* is trivially

$$i^* = \arg \min_{i = 1,\dots,n} \{z_i\},$$

assuming conventionally that the decision maker is interested in minimizing the outcome. The *weighted scoring method*, applied in Section 3.5 for locating facilities and in Section 4.4 for selecting suppliers, is based on this approach.

Timor is a grocery wholesaler that purchases, at the beginning of every week, a number of packs, typically between 1600 and 2000, of organically grown salad, which is a highly perishable product, at € 1.00 each. During the week, the company distributes the salad packs to some retailers, at € 1.50 each, on the basis of received orders. At the end of the week, the unsold packs are discarded since they have deteriorated. The problem faced by Timor is to determine how many salad packs to purchase in order to maximize its profit. There are five alternatives shown in Table 1.12. The profit (in €) for each choice is reported in the third column of the same table under the hypothesis that next week the retailers' orders will amount to 1780 packs. The best alternative is the third one with a purchase cost equal to € 1800, a revenue equal to 1780 × € 1.50 = € 2670 and hence a profit of € (2670 − 1800) = € 870.

Table 1.12 Profit (in €) corresponding to the five alternatives in the Timor problem.

Alternative (i)	Number of salad packs to purchase	Profit (z_i)
1	1600	800
2	1700	850
3	1800	870
4	1900	770
5	2000	670

The Bayes Criterion

On the other hand, if the outcome of alternative $i = 1, \ldots, n$, depends on data that will only be disclosed in the future, but are not known at the moment of the decision, the outcome can be seen as a function f_i of one or more random variables, corresponding to unknown data (for the sake of simplicity, a single random variable Θ is assumed). Hence, the outcome is a random variable itself, denoted as Z_i:

$$Z_i = f_i(\Theta), i = 1, \ldots, n.$$

If the probability distribution of each Z_i variable is known for $i = 1, \ldots, n$, the decision maker may be interested in selecting the alternative i^* of least expected outcome (*Bayes criterion*), that is,

$$i^* = \arg \min_{i = 1, \ldots, n} \{E_\Theta[Z_i]\}.$$

If Θ is discrete with realizations $\theta_1, \ldots, \theta_k$ and the mass probability function $Pr(\Theta = \theta_j) = p_j$ (where $\sum_{j=1}^{k} p_j = 1$), the expected outcome of alternative $i = 1, \ldots, n$, is

$$E_\Theta[Z_i] = \sum_{j=1}^{k} p_j z_{ij},$$

where z_{ij} is the outcome of alternative i when $\Theta = \theta_j$.

If Θ is continuous with probability density function $p_\Theta(\theta)$, the expected outcome of alternative $i = 1, \dots, n$, is

$$E_\Theta[Z_i] = \int_{-\infty}^{\infty} p_\Theta(\theta) z_i(\theta) \, d\theta,$$

where $z_i(\theta)$ is the outcome of alternative i if Θ is equal to realization θ.

In the Timor problem, it is now assumed that the number of salad packs ordered by the retailers at the beginning of the week can be modelled as a discrete random variable Θ, whose realizations are 1600, 1700, 1800, 1900 and 2000, with estimated probabilities $Pr(\Theta = 1600) = 0.1$, $Pr(\Theta = 1700) = 0.2$, $Pr(\Theta = 1800) = 0.3$, $Pr(\Theta = 1900) = 0.2$ and $Pr(\Theta = 2000) = 0.2$. The associated outcome matrix is reported in Table 1.13. As shown in the rightmost column, the optimal solution is to purchase 1800 salad packs, with an expected profit equal to € 840 given by

$$0.1 \times € \, 600 + 0.2 \times € \, 750 + 0.3 \times € \, 900 + 0.2 \times € \, 900 + 0.2 \times € \, 900.$$

Table 1.13 Values of profit (in €) $z_{ij}, i = 1, \dots, 5, j = 1, \dots, 5$, by considering the alternatives i of salad packs purchase and the different realizations j of salad packs ordered by retailers in the Timor problem.

Alternative (i)	$j = 1$	$j = 2$	$j = 3$	$j = 4$	$j = 5$	$E_\Theta[Z_i]$
1	800	800	800	800	800	800
2	700	850	850	850	850	835
3	600	750	900	900	900	840
4	500	650	800	950	950	800
5	400	550	700	850	1000	730

The Expected Value of Perfect Information

It is now worth introducing the concept of *expected value of perfect information* (*EVPI*), which corresponds to the improvement, in terms of expected outcome, that might be achieved if the realization of the random variable Θ was known in advance, at decision time.

For the sake of simplicity, the discussion is limited to the discrete case of Θ. If the realization known in advance were θ_j, the optimal outcome would clearly be $\min_{i = 1, \dots, n} \{z_{ij}\}$. Unfortunately, under perfect information, the decision maker knows in advance the realization of the uncertain parameters, but cannot choose their values. Hence, each realization θ_j still occurs with a given probability p_j. Consequently, in the long run, the decision maker should expect an outcome (*expected outcome under perfect information*) equal to

$$\sum_{j=1}^{k} p_j \min_{i=1,\dots,n} \{z_{ij}\}.$$

Therefore, the expected value of perfect information is

$$EVPI = \min_{i=1,\dots,n} \sum_{j=1}^{k} p_j z_{ij} - \sum_{j=1}^{k} p_j \min_{i=1,\dots,n} \{z_{ij}\},$$

where the right-hand side is always non-negative. *EVPI* represents the improvement in the expected outcome if the realization of the random variable Θ were known in advance, as compared to choosing the alternative before the realization of Θ becomes known. Since any forecasting method (see Chapter 2 for more details) to predict the value taken by the random variable Θ is based on imperfect information, then *EVPI* represents an upper bound on the outcome improvement which can be achieved when using any forecasting method.

In the Timor problem, the *EVPI* can be computed as follows. The expected profit under perfect information corresponds to € 910, given by

$$0.1 \times € \, 800 + 0.2 \times € \, 850 + 0.3 \times € \, 900 + 0.2 \times € \, 950 + 0.2 \times € \, 1000.$$

The maximum expected profit, previously computed, is € 840 (recall that in the Timor problem the outcome is a profit to be maximized) and *EVPI* is equal to

$$EVPI = € \, 910 - € \, 840 = € \, 70.$$

Hence, the additional expected profit that can be obtained by using any forecasting technique to predict the weekly demand of salad packs is no greater than € 70. Therefore, if Timor was offered a demand forecasting tool at a cost greater than *EVPI* (i.e., € 80 per week), it should refuse.

Indifference Zone Selection

If the mass probability function (or probability density function) of Θ is not known, the choice of the action having the best expected outcome is more complex since the expected outcomes $E_\Theta[Z_i]$ of alternatives $i = 1,\dots,n$, are not known and can only be estimated from a sample $z_{i1},\dots,z_{im_i}$ of finite size m_i. A point estimate of $E_\Theta[Z_i]$ is provided by the *sample mean*

$$\overline{Z}_i = \frac{1}{m_i} \sum_{j=1}^{m_i} z_{ij}.$$

In most cases, the decision maker also computes a *confidence interval* for the expected outcome, i.e., the interval in which $E_\Theta[Z_i]$ falls, with a prescribed *confidence level* $(1-\alpha)$, where α is the probability that $E_\Theta[Z_i]$ lies outside the confidence interval (e.g., $(1-\alpha) = 0.95$, hence $\alpha = 0.05$). This is because a point estimate does not necessarily coincide with the real expected value, while a confidence interval is more reliable.

The confidence interval at $(1 - \alpha)$ level of $E_\Theta[Z_i], i = 1, \dots, n$, is defined as

$$Pr\left(\overline{Z}_i - t_{\alpha/2, m_i-1}\frac{S_i}{\sqrt{m_i}} \leq E_\Theta[Z_i] \leq \overline{Z}_i + t_{\alpha/2, m_i-1}\frac{S_i}{\sqrt{m_i}}\right) = 1 - \alpha,$$

where $t_{\alpha/2, m_i-1}$ is the quantile of order $(1-\alpha/2)$ of the Student's t-distribution with $m_i - 1$ degrees of freedom and S_i is the *sample standard deviation*

$$S_i = \sqrt{\frac{\sum_{j=1}^{m_i}(z_{ij} - \overline{Z}_i)^2}{m_i - 1}}.$$

Therefore, the confidence interval $(1 - \alpha)$ level of $E_\Theta[Z_i], i = 1, \dots, n$, is

$$\left[\overline{Z}_i - t_{\alpha/2, m_i-1}\frac{S_i}{\sqrt{m_i}}, \overline{Z}_i + t_{\alpha/2, m_i-1}\frac{S_i}{\sqrt{m_i}}\right].$$

[Lugan.xlsx, Lugan.py] In the Lugan problem (see Section 1.8), recall that the number of observations of transportation times T_A and T_B is equal to $m_0 = 20$. Assuming $(1 - \alpha) = 0.95$, hence, $\alpha/2 = 0.025$, $(1 - \alpha/2) = 0.975$ and, consequently, $t_{0.025,19} = $ T.INV(0.975;19) $ = 2.0930$, where T.INV(0.975;19) is the Excel function which returns the left-tailed inverse of the Student's t-distribution with probability equal to 0.975 (corresponding to $1 - \alpha/2$) and degree of freedom equal to 19 (corresponding to $m_0 - 1$).

On the basis of the values of the sample mean $\overline{T}_A$ and the sample standard deviation S_{T_A} of the transportation time T_A computed by (1.1) and (1.2), respectively, the confidence interval at $(1 - \alpha)$ level of T_A (in hours) is

$$\left[7.67 - 2.0930 \times 1.07/\sqrt{20}, 7.67 + 2.0930 \times 1.07/\sqrt{20}\right] = [7.17, 8.17].$$

Similarly, considering the values of the sample mean $\overline{T}_B$ and the sample standard deviation S_{T_B} of the transportation time T_B given by (1.3) and (1.4), respectively, the confidence interval at $(1 - \alpha)$ level of T_B (in hours) is

$$\left[7.31 - 2.0930 \times 2.31/\sqrt{20}, 7.31 + 2.0930 \times 2.31/\sqrt{20}\right] = [6.23, 8.39].$$

The alternative to be selected among the n available alternatives would be i^*, which corresponds to the best sample mean $\overline{Z}_{i^*}$ of the outcome, with a confidence level equal to $(1 - \alpha)$. However, it could happen that a second alternative $\hat{i}$ has a sample mean $\overline{Z}_{\hat{i}}$ very close to $\overline{Z}_{i^*}$, and therefore the difference between $\overline{Z}_{\hat{i}}$ and $\overline{Z}_{i^*}$ is so small that one would be indifferent between alternatives $\hat{i}$ and i^*. As a consequence, the confidence intervals corresponding to alternatives i^* and $\hat{i}$ overlap significantly (as happens, for example, in the Lugan problem for providers A and B). The correct selection of the best alternative (corresponding to the one with the best expected outcome which, as noted above, is not known), using the sample data could therefore fail; it could in fact happen that $\overline{Z}_{i^*} < \overline{Z}_{\hat{i}}$, while $E_\Theta[Z_{i^*}] \geq E_\Theta[Z_{\hat{i}}]$.

This motivates the introduction of *indifference zone methods*, such as the RINOTT procedure described below. In these methods, the decision maker must define an indifference parameter δ, which corresponds to the smallest difference between two

---

1: **procedure** RINOTT $(\alpha, n, m_0, \delta, \mathbf{m}, i^*)$

2:     # α is a value chosen in the $[0, 1]$ interval to define the confidence level $(1 - \alpha)$;

3:     # n is the number of alternatives to evaluate;

4:     # m_0 is the number of samples initially available for each alternative $i = 1, \dots, n$;

5:     # δ is the indifference parameter;

6:     # $\mathbf{m}$ is the vector, returned by the procedure, of n components, each of which corresponding to the number of samples used for each alternative;

7:     # i^* is the best alternative returned by the procedure;

8:     Set the Rinott's constant r corresponding to α, m_0 and n by using `Rinott_constant.py`;

9:     **for** $i = 1, \dots, n$ **do**

10:         # Compute the sample mean $\overline{Z}_i$ and the sample standard deviation S_i based on the m_0 observations available for alternative i;

11:         $\overline{Z}_i = \frac{1}{m_0} \sum_{j=1}^{m_0} z_{ij}$;

12:         $S_i = \sqrt{\dfrac{\sum_{j=1}^{m_0} (z_{ij} - \overline{Z}_i)^2}{m_0 - 1}}$;

13:         # μ is the number of samples to draw a conclusion on the additional observations possibly required for alternative i;

14:         $\mu = \left\lceil \left(\frac{r S_i}{\delta} \right)^2 \right\rceil$;

15:         **if** $\mu > m_0$ **then**

16:             Take $\mu - m_0$ more observations for alternative i;

17:             $m_i = \mu$;

18:             $\overline{Z}_i = \frac{1}{m_i} \sum_{j=1}^{m_i} z_{ij}$;

19:             $S_i = \sqrt{\dfrac{\sum_{j=1}^{m_i} (z_{ij} - \overline{Z}_i)^2}{m_i - 1}}$;

20:         **else**

21:             # No additional samples are required for alternative i;

22:             $m_i = m_0$;

23:         **end if**

24:     **end for**

25:     Select the alternative i^* to which corresponds the best value of $\overline{Z}_i$, $i = 1, \dots, n$;

26:     **return m,** i^*;

27: **end procedure**

---

expected outcomes considered to be significant. In other words, the decision maker cannot discriminate between two alternatives whose expected outcomes differ by less than δ. Observe that the probability of selecting the best alternative increases when the sample size for each alternative increases. Based on this, the RINOTT procedure, depending on the parameter of indifference δ chosen, is based on determining, for each alternative $i = 1, \dots, n$, the number of additional observations eventually necessary, starting from the number m_0 of observations initially available for each of them. This number is such that, with probability not lower than $(1 - \alpha)$, the alternative corresponding to the best sample mean (i.e., the best expected outcome) is selected.

[`Lugan.xlsx`, `Rinott.py`] With reference to the Lugan problem, suppose that the company wants to establish which transportation provider between A and B is the fastest on average. As illustrated before, $m_0 = 20$. On the basis of the sample means and sample standard deviations of the transportation times T_A and T_B computed by using (1.1)–(1.4), the company estimates the number of observations needed to achieve a confidence level equal to $(1 - \alpha) = 0.95$ with an indifference zone of 30 minutes (0.5 hour). In order to apply the $\mathtt{RINOTT}$ procedure, the proper Rinott's constant $r = 2.452$ is returned by the procedure implemented by $\mathtt{Rinott\_constant.py}$ with $\alpha = 0.05, m_0 = 20$ and $n = 2$. For provider A, the value of $\delta = 0.5$ leads to

$$\mu = \left\lceil \left(\frac{2.452 \times 1.07}{0.5} \right)^2 \right\rceil = 28.$$

Since $\mu > m_0$, the company needs to collect $28 - 20 = 8$ more observations of the transportation time for provider A (i.e., $m_A = \mu = 28$). Similarly, for provider B,

$$\mu = \left\lceil \left(\frac{2.452 \times 2.31}{0.5} \right)^2 \right\rceil = 129,$$

hence, additional $129 - 20 = 109$ observations are required ($m_B = 129$). The sample means and sample standard deviations, computed by using again the $\mathtt{Excel}$ functions AVERAGE and STDEV.S, based on $m_A = 28$ and $m_B = 129$ observations are (see $\mathtt{Lugan.xlsx}$):

$$\overline{T}_A = 7.67 \text{ hours, } S_{T_A} = 1.14 \text{ hours,}$$
$$\overline{T}_B = 6.96 \text{ hours, } S_{T_B} = 1.91 \text{ hours.}$$

Since $\overline{T}_B < \overline{T}_A$, provider B is finally selected as the best.

The confidence interval at $(1 - \alpha) = 0.95$ level of T_A (in hours) is updated as

$$\left[7.67 - 2.0518 \times 1.14 / \sqrt{28}, 7.67 + 2.0518 \times 1.14 / \sqrt{28} \right] = [7.23, 8.12],$$

since $t_{0.025,27} = \mathtt{T.INV(0.975;27)} = 2.0518$, and the updated confidence interval at $(1 - \alpha) = 0.95$ level of T_B (in hours) is

$$\left[6.96 - 1.9787 \times 1.91 / \sqrt{129}, 6.96 + 1.9787 \times 1.91 / \sqrt{129} \right] = [6.62, 7.29],$$

since $t_{0.025,128} = \mathtt{T.INV(0.975;128)} = 1.9787$. The two updated confidence intervals are still slightly overlapping (but definitely not as much as in the previous case), but this does not influence the choice of the best provider.

Numerical Optimization

In most logistics applications, the planning alternatives are not known explicitly. Rather, the decision is represented by a vector **x** of n continuous or discrete *decision variables*. For example, **x** may describe the configuration of a logistics system (such as the allocation of retailers to RDCs) or the parameters of a policy (such as the reorder point in inventory management, see Section 5.12). The outcome z is expressed as a function $z = f(\mathbf{x}, \mathbf{c})$ of $\mathbf{x}$ and a vector $\mathbf{c}$ of deterministic input parameters. In the simplest case, $f(\mathbf{x}, \mathbf{c})$ is known in *closed form*, as, for example, in the following quadratic function $x_1^2 + 2x_2$ (with $\mathbf{x} = [x_1; x_2]^T$ and $\mathbf{c} = [1; 2]^T$). Then, the decision making process can be cast as an *optimization problem*

$$\text{Minimize}_{\mathbf{x}} \, z = f(\mathbf{x}, \mathbf{c}) \tag{1.5}$$

subject to

$$\mathbf{g}(\mathbf{x}, \mathbf{b}) \le \mathbf{0}, \tag{1.6}$$

where the decision variables are constrained by (possibly non-linear) equalities and inequalities that can be always lead to the form $\mathbf{g}(\mathbf{x}, \mathbf{b}) \le \mathbf{0}$, where $\mathbf{g}(\mathbf{x}, \mathbf{b})$ is a set of m functions $g_i(\mathbf{x}, \mathbf{b})$, $i = 1, \dots, m$, and $\mathbf{b}$ is a vector of known parameters. Depending on the problem structure and its size, as well as on the available computing time, a decision maker may use:

- an off-the-shelf general purpose solver (from the simpler `Solver` available in `Excel` to more sophisticated optimization software packages such as CPLEX, Gurobi, and Minos). Such solvers are typically based on exact iterative algorithms (simplex algorithm, branch-and-bound procedure, constraint programming, etc.) which for some specified classes of problems converge to an optimal solution, or on heuristics that may provide approximate solutions to some problems, although their iterates need not necessarily converge. Most programming languages offer a modelling interface that hooks up to solvers: this is the case of the `Pulp` library in `Python`;
- an ad hoc exact algorithm, in which an available algorithmic paradigm (dynamic programming, branch-and-bound, branch-and-cut, branch-and-price, Benders decomposition, constraint programming, etc.) is tailored to make the search more efficient for the problem at hand;
- an ad hoc heuristic, based on a constructive procedure (such as a greedy method) or metaheuristic framework (such as simulated annealing, tabu search, genetic algorithms, GRASP, etc.).

When a heuristic is used, the outcome $\bar{z}$ of the best feasible solution $\bar{\mathbf{x}}$ (if any) is no better than the optimal outcome z^*, that is, $\bar{z} \ge z^*$. Heuristic algorithms are generally faster and more efficient than exact algorithms. They are often employed when the computation of exact solutions is computationally too expensive and it is far better to try computing a suboptimal solution within a limited available time.

Coping with Uncertainty in Planning

When the input parameters $\mathbf{c}$ and $\mathbf{b}$ in (1.5)–(1.6) are unknown at the time of decision making, they can be formally replaced by vectors $\boldsymbol{\gamma}$ and $\boldsymbol{\beta}$ of random variables. In logistics planning, this is the case of market demand, cargo ship travel times, assembly time in a manufacturing site on a given day, etc. In this context, model (1.5)–(1.6) makes no sense any more. In some application settings, a sensible approach is to replace random vector variables $\boldsymbol{\gamma}$ and $\boldsymbol{\beta}$ with their expected values $E[\boldsymbol{\gamma}]$ and $E[\boldsymbol{\beta}]$, if they are known.

The resulting optimization model is

$$\underset{\mathbf{x}}{\text{Minimize}}\, z = f(\mathbf{x}, E[\boldsymbol{\gamma}]) \tag{1.7}$$

subject to

$$\mathbf{g}(\mathbf{x}, E[\boldsymbol{\beta}]) \leq \mathbf{0}. \tag{1.8}$$

Of course, if $E[\boldsymbol{\gamma}]$ and $E[\boldsymbol{\beta}]$ are not known, they can be substituted by their sample means. Anyway, when using this approach, a decision maker should be advised that a solution to (1.7)–(1.8) may violate some constraint for some realizations of the random vector $\boldsymbol{\beta}$. This may be acceptable in some applications (and indeed this approach is used almost always in the remainder of this book), but in others it is unacceptable. Then, two approaches can be used if the probability distribution of $\boldsymbol{\gamma}$ and $\boldsymbol{\beta}$ is known: *chance-constrained optimization* and *stochastic programming with recourse*.

Chance-constrained Optimization

In chance-constrained optimization, the decision maker optimizes the expected outcome while ensuring that the probability of satisfying constraints $\mathbf{g}(\mathbf{x}, \boldsymbol{\beta}) \leq \mathbf{0}$ is above a certain level $p \in [0, 1]$:

$$\underset{\mathbf{x}}{\text{Minimize}}\, z = E_{\boldsymbol{\gamma}}[f(\mathbf{x}, \boldsymbol{\gamma})] \tag{1.9}$$

subject to

$$Pr\{\mathbf{g}(\mathbf{x}, \boldsymbol{\beta}) \leq \mathbf{0}\} \geq p. \tag{1.10}$$

Expression (1.10) ensures that *all* the constraints are satisfied with the prescribed minimum probability level p (this explain the reason why (1.10) is referred to as *joint chance constraint*). Since formulation (1.9)–(1.10) is very difficult to solve, even numerically, it is more common to impose that the probability of satisfying each constraint $j = 1, \ldots, m$, is above p, that is,

$$Pr\{g_j(\mathbf{x}, \boldsymbol{\beta}) \leq 0\} \geq p, \quad j = 1, \ldots, m. \tag{1.11}$$

The resulting formulation (1.9), (1.11) is usually much easier to solve than (1.9)–(1.10).

[Moderan.xlsx, Moderan.py] Moderan is a Croatian fashion apparel manufacturer serving the national market. In a few days, the logistician will define the production and transportation capacity for each of the main product groups for the next spring. For AlpinSkies backpacks, the company can use the production lines of three plants, located in Ljubljana, Zadar, and Sibenik, referred to as plants 1, 2, and 3, respectively, in the following. The market has been divided into three sales districts, Slavonia-Central Croatia, Istria-Kvarner, and Dalmatia (referred to in the following as districts 1, 2, and 3, respectively). The demand (in thousands of units) for AlpinSkies backpacks $\beta_j, j = 1, 2, 3$, in each of the three areas is assumed to be normally distributed with expected values μ_j and standard deviations σ_j reported in Table 1.14. The unit production and transportation costs (in €) from plant $i = 1, 2, 3$, to sales district $j = 1, 2, 3$, are random variables

γ_{ij} whose expected values are reported in Table 1.15. In each plant, the whole production capacity is shared among several product groups and the AlpinSkies backpacks may be assigned the capacity (in thousands of units) shown in the second column of Table 1.15.

A chance-constrained approach is used. The expected outcome to be minimized, corresponding to the overall production and transportation cost, is

$$z = E_\gamma \left[\sum_{i=1}^{3} \sum_{j=1}^{3} \gamma_{ij} x_{ij} \right] = \sum_{i=1}^{3} \sum_{j=1}^{3} E_\gamma[\gamma_{ij}] x_{ij},$$

where x_{ij} is the transportation capacity allocated, at the start of the season, to the replenishment of sales district $j = 1, 2, 3$, from plant $i = 1, 2, 3$.

Table 1.14 Expected value and standard deviation of the spring demand (in thousands of units) in the three districts of the Moderan problem.

Sales district (j)	Expected spring demand (μ_j)	Standard deviation of spring demand (σ_j)
1	90	5
2	120	15
3	110	10

Table 1.15 Production capacity, and overall expected unit production and transportation costs for each plant–district combination in the Moderan problem.

| Production plant (i) | Capacity [thousands of units] | Expected unit cost | | |
		$j = 1$ [€]	$j = 2$ [€]	$j = 3$ [€]
1	140	20	30	10
2	120	50	40	80
3	130	50	60	80

The following inequalities impose that the demand of each sales district is satisfied with a minimum probability level p:

$$Pr\left(\sum_{i=1}^{3} x_{ij} \geq \beta_j \right) \geq p, \qquad j = 1, 2, 3,$$

which can be equivalently written as

$$Pr\left(\frac{\beta_j - \mu_j}{\sigma_j} \leq \frac{\sum_{i=1}^{3} x_{ij} - \mu_j}{\sigma_j} \right) \geq p, \qquad j = 1, 2, 3,$$

or

$$Pr\left(Z \le \frac{\sum_{i=1}^{3} x_{ij} - \mu_j}{\sigma_j}\right) \ge p, \quad j = 1, 2, 3, \tag{1.12}$$

where $\frac{\beta_j - \mu_j}{\sigma_j}, j = 1, 2, 3$, is the standard normal random variable Z.

Let z_α be the value such that $Pr(Z \ge z_\alpha) = \alpha$ ($\alpha \in [0,1]$). Equivalently, z_α corresponds to $Pr(Z \le z_\alpha) = 1 - \alpha$ (see Figure 1.25), i.e., the value under which a standard normal random variable falls with probability $(1 - \alpha)$ (the quantile of order $(1 - \alpha)$ of the normal standard distribution). For example, if $\alpha = 0.05$, $z_\alpha = 1.645$ and corresponds to the quantile of order 0.95 of the normal standard distribution. In `Excel` it can be obtained by using the `NORM.S.INV` function, i.e., $z_{0.05} = $ `NORM.S.INV(0.95)` $ = 1.645$.

Hence, the satisfaction of the chance constraints (1.12) implies that

$$\frac{\sum_{i=1}^{3} x_{ij} - \mu_j}{\sigma_j} \ge z_{(1-p)}, \quad j = 1, 2, 3,$$

or, equivalently

$$\sum_{i=1}^{3} x_{ij} \ge \mu_j + z_{(1-p)}\sigma_j, \quad j = 1, 2, 3.$$

For $p = 0.95$ (and hence $z_{0.05} = 1.645$), the Moderan problem can be cast as follows:

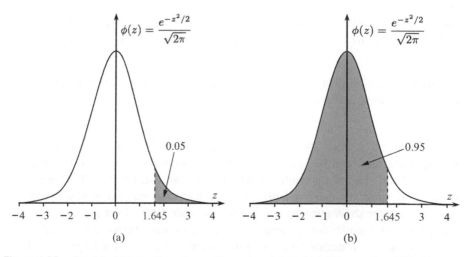

Figure 1.25 Graphical illustration of quantile $z_{0.05} = 1.645$: (a) $Pr(Z \ge z_{0.05}) = 0.05$; (b) $Pr(Z \le z_{0.05}) = 0.95$.

Minimize $20x_{11} + 30x_{12} + 10x_{13}$

$\qquad + \; 50x_{21} + 40x_{22} + 80x_{23}$

$\qquad + \; 50x_{31} + 60x_{32} + 80x_{33}$

subject to

$$x_{11} + x_{21} + x_{31} = 90 + 1.645 \times 5$$

$$x_{12} + x_{22} + x_{32} = 120 + 1.645 \times 15$$

$$x_{13} + x_{23} + x_{33} = 110 + 1.645 \times 10$$

$$x_{11} + x_{12} + x_{13} \leq 140$$

$$x_{21} + x_{22} + x_{23} \leq 120$$

$$x_{31} + x_{32} + x_{33} \leq 130$$

$$x_{11}, x_{12}, x_{13}.x_{21}, x_{22}, x_{23}, x_{31}, x_{32}, x_{33} \geq 0.$$

The optimal solution, $x_{11}^* = 0.000$, $x_{12}^* = 13.551$, $x_{13}^* = 126.449$, $x_{21}^* = 0.000$, $x_{22}^* = 120.000$, $x_{23}^* = 0.000$, $x_{31}^* = 98.224$, $x_{32}^* = 11.121$ and $x_{33}^* = 0.000$, corresponds to an expected cost $z^* = €\,12\,049.52$.

Stochastic Programming with Recourse

In stochastic programming with recourse, the decision maker may take corrective actions after a random event has taken place. In the simplest case (*two-stage recourse*),

- a first-stage decision is made (represented by a vector **x** of decision variables), producing some outcome;
- some random events happen; realizations of γ and β, indicated collectively as a *scenario s*, become known;
- some recourse actions (represented by a second-stage vector $\mathbf{y}^{(s)}$ of decision variables, one for each scenario s), are made to correct what may have been wrong in each scenario s.

The objective is to minimize the first-stage outcome plus the expected outcome of the second-stage recourse actions.

[HealthyLife.xlsx, HealthyLife.py] HealthyLife imports dry grain, cereals, vegetables, and fruit juice to Europe. For a particular product, the buying price at the start of the winter season is a known value $c_0 = 100$ €/tonne. On the other hand, the demand β (in tonnes) and the buying price γ (in €/tonne) during the winter show a strong uncertainty. In Table 1.16 a joint probability p_s characterizes the realizations β_s and γ_s of β and γ for each scenario $s = 1, 2, 3$. The company has to decide how much inventory x to buy at the beginning of the season (holding costs are neglected for the sake of simplicity) and how much inventory $y^{(s)}$ to buy, as a recourse action, in each scenario $s = 1, 2, 3$. The selling price can be assumed to be constant and unsold goods have no value. The model is formulated as follows:

$$\text{Minimize } z = 100x + 0.5 \times 80 y^{(1)} + 0.4 \times 160 y^{(2)} + 0.1 \times 180 y^{(3)}$$

subject to

$$x + y^{(1)} \geq 100$$
$$x + y^{(2)} \geq 120$$
$$x + y^{(3)} \geq 150$$
$$x, y^{(1)}, y^{(2)}, y^{(3)} \geq 0,$$

where the expected recourse cost is $p_1 \gamma_1 y^{(1)} + p_2 \gamma_2 y^{(2)} + p_3 \gamma_3 y^{(3)}$. The optimal solution is found in Excel by using the Solver tool. The total expected cost is equal to € 12 180 with $x^* = 100$, $y^{*(1)} = 0$, $y^{*(2)} = 20$, $y^{*(3)} = 50$. Therefore, if scenario 1 occurs, no additional amount of product should be purchased during the season. If scenario 2 occurs, the company needs 20 further tonnes of product, whereas, in case of occurrence of scenario 3, the amount of product to be purchased during the winter is 50 tonnes.

Table 1.16   Demand, buying price, and probability values for each scenario $s = 1, 2, 3$, of the HealthyLife problem.

Scenario (s)	Demand [tonnes] (β_s)	Buying price [€/tonne] (γ_s)	Probability (p_s)
1 (low demand)	100	80	0.5
2 (medium demand)	120	160	0.4
3 (high demand)	150	180	0.1

Simulation

In some complex logistics systems, the outcome z of a decision is represented by a function of variables $\mathbf{x}$ that is not known in *closed form*. In this case, the computation (or estimation) of z can be done by *simulation* for any given $\mathbf{x}$. Simulation means formalizing and reproducing cause–effect relations within a system, under given conditions, by using a *simulator*. In *computer simulation*, each part of the system is represented by a subroutine which reproduces its input–output behaviour. In this way, the logistician can obtain the relevant information about the outcome of the decision.

The Coca-Cola company, world leader in the production and distribution of soft drinks and syrup concentrates, has recently decided to reorganize the logistics activities carried out at the DC in Taguatinga (Brazil). A simulation model has been specifically developed by a consulting firm to reproduce how the different DC zones (see Section 5.1.2) and all the processes within them work, in order to identify operational limits in the current configuration and to suggest the best improvements to layout, storage, and flow within the DC (for further details in the warehouse design see Chapter 5). After several simulation experiments, the consulting firm has identified the greatest inefficiency in the picking and staging zones, and, hence, a new warehouse configuration has been proposed. In particular, the new DC layout configuration has been based on the introduction of a larger staging zone for palletized unit loads, a reduction of the total loading bays (from 16 to 8) and the complete redesign of the picking zone. Simulation has led to several improvements in the warehouse performance:

- with the new shipping zone, all the trucks can be loaded before 6:00 while in the current configuration around 46% of the trucks are late;
- with the new product allocation strategy, the productivity of the picking zone increases, with a reduction of 1.5 minutes on average for preparing a complete pallet;
- the vehicle loading time is reduced by 26% and the storage capacity is increased by 20%.

Simulation can be *deterministic* or *stochastic*. If the system does not show any degree of uncertainty, its simulation depends only on known (deterministic) input parameters.

[LogMe.py] LogMe is a French company operating a logistics platform near Nantes. The platform is an area where different activities of warehousing and transshipment are carried out by various operators, both for national and international transit. The current packing zone includes one packaging station. On a given day, at time $t_0 = 10{:}05$, a packaging station is expected to process a list of 10 parcels. Each parcel $i = 1, \ldots, 10$, is characterized by an arrival time a_i and has to be loaded onto a pallet $p_i \in \{A, B\}$ (see Table 1.17). The station needs $c = 2$ minutes to process any parcel. The truck on which pallet A has to be shipped must leave at time $t_f = 10{:}20$. The platform queue is managed according to a *first-in, first-out* (FIFO) policy. The objective is the minimization of the total completion time z for loading and packaging the parcels onto pallets A and B. Since no closed-form expression is known for the completion times of both A and B pallets, the logistics manager makes use of a simulation model (see PACKING_STATION_SIMULATION procedure), in order to both check the feasibility of the solution (that is, $g \leq t_f$, where g is the completion time for loading pallet A) and assess the outcome z. It turns out that the completion time for pallet

A is 10:20 (hence satisfying the company's operational requirement) and the total completion time *z* for loading the 10 parcels was 10:28.

Table 1.17 List of 10 parcels in the LogMe problem.

Parcel (i)	Arrival time (a_i)	Pallet (p_i)
1	10:05	A
2	10:07	B
3	10:08	A
4	10:09	A
5	10:10	B
6	10:11	A
7	10:18	A
8	10:19	B
9	10:20	B
10	10:26	B

If the system is stochastic, its behaviour is affected by uncertain parameters that can be treated as random variables or stochastic processes. The most common approach to analyse the influence of uncertain inputs on output variables relies on *Monte Carlo methods*. These are a broad class of computational algorithms in which repeated random sampling (from *pseudo-random number generators*, RNGs) is used to make numerical estimations of the unknown input parameters. In Monte Carlo methods, multiple simulation runs may be needed in order to draw a conclusion about the value of the outcome to be estimated.

[LogMe2.py] In LogMe problem, assume that packaging time *c* is a discrete random variable with probability distribution

$$Pr(c = 1) = 0.3, Pr(c = 2) = 0.4 \text{ and } Pr(c = 3) = 0.3. \tag{1.13}$$

A Monte Carlo method is used by the logistics manager to estimate the expected overall completion time, the expected completion time for loading pallet *A* and the probability that the company's operational constraint is violated. A single simulation run of the packaging station can be obtained from PACKING_STATION_SIMULATION procedure by substituting *c* with a call to a RNG, reproducing probability distribution (1.13). For $N = 1000$ simulation runs, the sample means and sample standard deviations of *z* and *g* are:

$\bar{z} = 28.12$ minutes;

$S_z = 0.93$ minutes;

$\bar{g} = 20.38$ minutes;

$S_g = 1.14$ minutes,

where $\bar{z}$ and $\bar{g}$ are expressed in minutes (in decimal notation) past 10:00 and, consequently, corresponding to times of 10:28:07 and of 10:20:23, respectively. Moreover, in $N' = 239$ simulation runs (out of $N = 1000$) the loading time g of pallet A exceeds $t_f = 10:20$. The confidence interval at $(1 - \alpha) = 0.95$ level of the expected outcome is (recall that $z_{0.05} = 1.645$)

$$\left[28.12 - 1.645 \times 0.93/\sqrt{1000}, 28.12 + 1.645 \times 0.93/\sqrt{1000} \right],$$

that is, a time interval given by [10:28:04, 10:28:10]. Similarly, the confidence interval of the expected completion time of loading pallet A is

$$\left[20.38 - 1.645 \times 1.14/\sqrt{1000}, 20.38 + 1.645 \times 1.14/\sqrt{1000} \right],$$

that is, a time interval corresponding to [10:20:19, 10:20:26]. Finally, the probability of violating the company's operational constraint is $N'/N = 0.24$.

```
 1: procedure PACKING_STATION_SIMULATION (n, L, a, t₀, c, z, g)
 2:     # n is the number of parcels;
 3:     # L is the list of n parcels;
 4:     # a is the vector of the arrival times of the parcels;
 5:     # t₀ is the starting time of the packing station;
 6:     # c is the time required by the packing station to load a parcel;
 7:     # z is the total completion time returned by the procedure;
 8:     # g is the completion time for pallet A, returned by the procedure;
 9:     t_now = t₀;
10:     while L ≠ ∅ do
11:         Extract the parcel i of least arrival time a_i from L;
12:         if t_now < a_i then
13:             t_now = a_i + c;
14:         else
15:             t_now = t_now + c;
16:         end if
17:         if p_i = A then
18:             g = t_now;
19:         end if
20:     end while
21:     Set z = t_now;
22:     return z, g;
23: end procedure
```

While in the LogMe problem the simulator was written in a general-purpose programming language (Python), it is sometimes easier to make use of special-purpose simulation languages and visual interactive modelling systems (VIMSs), such as Arena and Simio, to ease this task.

What-if Analysis

What-if analysis is a predictive technique whose objective is to evaluate how a change in the problem data (or the deletion of a constraint, and so on) affects the KPIs and performance measures (cost, profit, losses, efficiency, etc.) with respect to a baseline solution (current configuration of the logistics system, current policy, etc.). Each combination of such variations is called a *scenario* and may require the solution of a new optimization problem. *What-if* analysis is a very popular tool to support managers in making strategic and tactical decisions.

[FCT.xlsx] FCT, a German company specializes in the production and distribution of automotive spare parts at competitive prices, actually manages a logistics system composed of two production plants (nodes 1 and 2 in Figure 1.26), a CDC (node 3) and three RDCs (nodes 4, 5, and 6). Groups of retailers geographically close to each other are referred to as nodes 7 and 8 (for simplicity, a single node is considered for each group of retailers). The current monthly volumes of spare parts moved along the arcs and the unit transportation costs are reported in Table 1.18. The DCs 3, 4, 5, and 6 have a limited handling capacity per month equal to 600, 200, 200, and 200 pallets, respectively, and monthly running costs (including storage costs) equal to € 14 000, € 9500, € 7400 and € 8600, respectively. In the current configuration (see FCT.xlsx for details), the monthly total

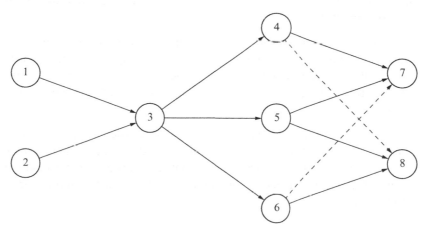

Figure 1.26 Representation of the FCT logistics system. Dashed arcs correspond to transportation links not activated in the current configuration.

Table 1.18 Monthly volume and unit transportation cost in the FCT problem.

Arc	Volume [pallets]	Unit cost [€/pallet]	Arc	Volume [pallets]	Unit cost [€/pallet]
(1,3)	200	30.0	(4,7)	150	13.0
(2,3)	200	30.0	(4,8)	0	16.5
(3,4)	150	14.0	(5,7)	100	14.0
(3,5)	200	16.0	(5,8)	100	14.5
(3,6)	150	15.0	(6,7)	0	15.0
			(6,8)	150	10.5

cost is equal to € 68 425, derived from the sum of € 39 500 (running costs of the DCs) and of € 28 925 (transportation costs along overall the arcs used). Considering an expected increase in sales of 20% for each retailer, the FCT director of the logistics function has proposed to review the current organization within the logistics system, making the distribution phase more efficient, if possible. To this end, the manager decides to perform a what-if analysis to evaluate three different possible scenarios: A, B, and C. The idea is to investigate how the divestment of one or two RDCs within the logistics system (scenarios B and C, respectively) will influence the total costs, compared with the alternative of maintaining the actual configuration (scenario A). The analysis has to simultaneously guarantee the satisfaction of the demand constraint. The three possible scenarios are described in the following.

A. Reorganization of the freight flows within the current logistics system for reducing the overall cost as much as possible. In this configuration, we allow an increase of 20% of the capacity of each RDC and, consequently, of the running costs.

B. Closure of one of the three RDCs, increasing the capacity of those remaining by 50%, against an increase of the operating costs by up to 50%. Consider that for the RDC candidate for closure it is necessary to pay a monthly quote of the total disposal cost equal to € 1200 for node 4, € 1300 for node 5 and € 1100 for node 6.

C. Use of only one of the three RDCs available, providing an appropriate increase of its capacity. In this configuration, the RDC monthly running costs are assumed equal to three times the current value. In addition, the monthly quote of the disposal cost is the same of that reported in the previous alternative.

It is assumed that there are no vehicle capacity constraints and the increase in sales is equally distributed for both groups of retailers. As far as the three alternatives are concerned, the hypothesis of the sales increase equally distributed among retailers leads to a monthly demand of 300 pallets for each of the nodes 7 and 8.

In the case of scenario A, the capacity of each RDC is increased to 240 pallets (20% more than 200 pallets). With the aim of optimizing the transportation costs, a least-cost path for moving 300 pallets from CDC (node 3) to the destination node 7 is determined. There exist three different feasible solutions: path $(3, 4, 7)$, with a unit cost of € $(14.0 + 13.0) = $ € 27.0; path $(3, 5, 7)$, with a unit cost of € $(16.0 + 14.0) = $ € 30.0; path $(3, 6, 7)$, with a unit cost of € $(15.0 + 15.0) = $ € 30.0. This implies that the maximum fraction allowed of the demand (240 pallets) is moved through the least-cost path $(3, 4, 7)$, saturating the capacity of node 4, and the remaining part (60 pallets) is transported through the second least-cost path $(3, 5, 7)$. Similarly, for moving freight from node 3 to node 8 there are only two feasible solutions: path $(3, 5, 8)$, with a unit cost of € $(16.0 + 14.5) = $ € 30.5 and path $(3, 6, 8)$, with a unit cost of € $(15.0 + 10.5) = $ € 25.5. Note that path $(3, 4, 8)$ is not considered since node 4 has no residual capacity. In this case, 240 pallets are moved through the least-cost path $(3, 6, 8)$ and 60 pallets along the path $(3, 5, 8)$. The monthly total cost of the logistics system corresponding to scenario A is of € $78\,830$, as reported in the third column of Table 1.19, generated in `Excel` by using the `What-If Analysis` → `Scenarios` tool, under `Data` tab (see `FCT.xlsx` for more details).

Scenario B corresponds to three different subscenarios: closure of node 4 (scenario B1), closure of node 5 (scenario B2) and closure of node 6 (scenario B3). Note that when a node is closed, the capacity of the remaining others increases up to 300 pallets each. A similar approach to that used in scenario A to optimize the transportation costs is considered. The results, in terms of monthly cost of the logistics system, obtained by using the `What-If Analysis` tool in `Excel` for the three subscenarios are reported in columns 5, 7 and 9 of Table 1.19.

Also scenario C leads to three different subscenarios: the use of only RDC 4, 5, and 6, respectively (corresponding to scenarios C1, C2, and C3, respectively). Note that when only one RDC is used, its capacity increases to 600 pallets. The monthly total cost of the logistics system corresponding to scenarios C1, C2, and C3 are summarized in columns 11, 13, and 15 of Table 1.19.

The conclusion driven from the what-if analysis is that the closure of the first RDC (node 4), corresponding to scenario B1, leads to the minimum monthly total cost of the FCT logistics system.

Table 1.19 Monthly costs of different scenarios in the FCT problem.

Arc/Node	Scenario A Volume [pallets]	Cost [k€]	Scenario B1 Volume [pallets]	Cost [k€]	Scenario B2 Volume [pallets]	Cost [k€]	Scenario B3 Volume [pallets]	Cost [k€]	Scenario C1 Volume [pallets]	Cost [k€]	Scenario C2 Volume [pallets]	Cost [k€]	Scenario C3 Volume [pallets]	Cost [k€]
(1,3)	300	9.00	300	9.00	300	9.00	300	9.00	300	9.00	300	9.00	300	9.00
(2,3)	300	9.00	300	9.00	300	9.00	300	9.00	300	9.00	300	9.00	300	9.00
(3,4)	240	3.36	0	0.00	300	4.20	300	4.20	600	8.40	0	0.00	0	0.00
(3,5)	120	1.92	300	4.80	0	0.00	300	4.80	0	0.00	600	9.60	0	0.00
(3,6)	240	3.60	300	4.50	300	4.50	0	0.00	0	0.00	0	0.00	600	9.00
(4,7)	240	3.12	0	0.00	300	3.90	300	3.90	300	3.90	0	0.00	0	0.00
(4,8)	0	0.00	0	0.00	0	0.00	0	0.00	300	4.95	0	0.00	0	0.00
(5,7)	60	0.84	300	4.20	0	0.00	0	0.00	0	0.00	300	4.20	0	0.00
(5,8)	60	0.87	0	0.00	0	0.00	300	4.35	0	0.00	300	4.35	300	4.50
(6,7)	0	0.00	0	0.00	0	0.00	0	0.00	0	0.00	0	0.00	300	4.50
(6,8)	240	2.52	300	3.15	300	3.15	0	0.00	0	0.00	0	0.00	300	3.15
3	–	14.00	–	14.00	–	14.00	–	14.00	–	14.00	–	14.00	–	14.00
4	–	11.40	–	1.20	–	14.25	–	14.25	–	28.50	–	1.20	–	1.20
5	–	8.88	–	11.10	–	1.30	–	11.10	–	1.30	–	22.20	–	1.30
6	–	10.32	–	12.90	–	12.90	–	1.10	–	1.10	–	1.10	–	25.80
Total cost		78.83		73.85		76.20		75.70		80.15		74.65		76.95

1.11   Segmentation Analysis

Segmentation analysis is a methodology that allows partitioning products and market in different parts (*segments*) that share common features. The ultimate goal is to manage each individual segment with specific procedures. In particular, as far as SCM is concerned, segmentation amounts to identifying which combinations of products and markets have to be serviced by the same supply chain (or with the same policy) in order to best pursue the logistics objectives (see Section 1.8). In other words, supply chain segmentation allows managers to recognize the logistics services required by each segment, to define the supply chain capabilities and implement solutions to improve performance. The most common segmentation techniques used in logistics are *customer* and *product* segmentations.

1.11.1   Customer Segmentation

In SCM, customer segmentation is aimed at partitioning customers into groups of individuals (or companies) expecting the same level of logistics service. Customer segmentation can be based on the following four criteria.

1. *Demographic.* In B2C markets, demographic customer segmentation relies on the customers' gender, age, level of education, profession, income, and marital situation. In B2B markets, it is based on the companies' annual turnover, number of employees, corporate sector, etc.
2. *Geographic.* In geographic market segmentation, users are subdivided with respect to their geolocation; this is the simplest criterion to create groups of customers with similar culture, climatic conditions, and habits.
3. *Psychographic.* In psychographic customer segmentation, customers are grouped with respect to their lifestyle, hobbies, values, etc.
4. *Behavioural.* In behavioural market segmentation, segments are generated by considering the past buying behaviour exhibited by customers towards the company's products and services.

Procter and Gamble is an American multinational consumer goods corporation, producing and marketing a wide range of cleaning and personal care products in over 180 countries. The main market segmentation operated by the company is realized by considering a geographical criterion, based on six different *selling and market areas* (SMAs): Europe, Asia–Pacific area, Greater China, Middle East, Africa and India, Latin America and North America. Inside each segment, the distribution channels, different for size and typology, are represented by a large variety of retailers: grocery stores, drug stores, hyper and super markets, distributors, baby stores, e-commerce, high-frequency stores, pharmacies, and nanostores. For this reason, in each SMA a local market access plan is specifically developed, through which dedicated retail customers, sales channels, and country-specific teams are clearly identified, focusing on efficiency and effectiveness in terms of SCM.

1.11.2 Product Segmentation

Product segmentation is a technique to deal with multiple products in supply chains.

ABC Classification

ABC classification is the most popular product segmentation technique and allows products to be subdivided into three classes, called *A*, *B*, and *C*, on the basis of their "value" (typically, the revenue they generate in a reference period, for example, a year). Class *A* is defined as the set of products achieving a given high percentage of the overall annual revenue (e.g., 80%). Class *B* is made up of the articles that constitute an additional 15% of the revenue, whereas class *C* is made up of the remaining articles. The classification is achieved by ordering the list of products in non-increasing fashion, and successively selecting the articles in that order, up to a predetermined cumulated value.

It is empirically observed that class *A* will most likely account for a modest fraction of the products. This property is often referred to as the *80–20 principle*, or *Pareto principle*, based on the observation made by the scholar V. Pareto who observed that, in the nineteenth century, 20% of the Italian population owned 80% of the national wealth. In contrast, class *C*, which accounts for a small percentage of total value, is usually made up of a large number of products. This observation suggests different logistics policies for the three product segments. For example, it is worth using several CDCs and RDCs, with high inventory levels, for class *A* products. On the other hand, the distribution of class *C* products can be done by using a single CDC and reducing the stocked quantity of products to a minimum.

[Blucker.xlsx] Blucker is the owner of an Irish plant manufacturing building materials. The Cork warehouse is used to store and distribute products in the water-based dispersion adhesives category to wholesalers. There is a total of 15 products. The annual revenue, sales volume, and average annual inventory value of the Blucker products are provided in Table 1.20. The *ABC* (80–15–5) classification of the products by annual revenue can be derived from Table 1.21, in which the products are sorted in non-increasing order with respect to revenue. Class *A* products make up 79.95% of the annual revenue, whereas they represent only 40.66% of the overall amount sold in the year. Class *B* products represent 14.45% of the annual revenue and generate 32.22% of the annual sales volume, whereas class *C* products make up only 5.60% of the annual revenue; the weight of these products is equal to 27.12% of the overall amount sold in the year. The cumulative percentages of the annual amounts sold and of the annual revenue for each of the 15 products are plotted in Figure 1.27. The same figure exhibits the 80–20 curve of equation $y = [(1 + \alpha)x] / (\alpha + x)$ which best fits the plotted values (y is the cumulative percentage of the annual revenue, x is the cumulative percentage of the amounts sold, $\alpha = 0.238$ is obtained by using the least-squares method, which involves in Excel the use of the Solver tool under the Data tab, see Blucker.xlsx for more details).

Table 1.20 Annual revenue, average annual inventory value and amounts sold of the Blucker products.

ID	Article	Revenue [€]	Amounts sold [kg]	Average inventory [€]
1	FIL12	424 764	38 614	109 000
2	BG1	126 000	33 452	26 000
3	BG2	959 800	24 522	401 400
4	BG3	84 540	25 545	53 000
5	P	441 280	24 767	23 800
6	TX	356 984	19 768	30 260
7	K0	762 250	32 234	157 000
8	K1	128 150	17 669	41 000
9	K2	51 206	22 600	9 900
10	K3	80 596	32 574	18 500
11	P-L1	144 625	30 578	33 900
12	P-L2	653 600	31 400	109 200
13	P-L3	35 608	33 560	127 000
14	P-L4	133 720	18 768	42 300
15	P-L5	118 300	35 287	27 000

Table 1.21 *ABC* classification of the Blucker products.

ID	Article	Annual revenue [€]	Annual amounts sold [kg]	Cumulated annual amounts sold [%]	Cumulated annual revenue [%]	Class
3	BG2	959 800	24 522	5.82	21.32	A
7	K0	762 250	32 234	13.47	38.26	A
12	P-L2	653 600	31 400	20.92	52.78	A
5	P	441 280	24 767	26.80	62.58	A
1	FIL12	424 764	38 614	35.97	72.01	A
6	TX	356 984	19 768	40.66	79.95	A
11	P-L1	144 625	30 578	47.91	83.16	B
14	P-L4	133 720	18 768	52.37	86.13	B
8	K1	128 150	17 669	56.56	88.98	B
2	BG1	126 000	33 452	64.50	91.77	B
15	P-L5	118 300	35 287	72.88	94.40	B
4	BG3	86 540	25 545	78.94	96.28	C
10	K3	80 596	32 574	86.67	98.07	C
9	K2	51 206	22 600	92.03	99.21	C
13	P-L3	35 608	33 560	100.00	100.00	C
	Total	4 501 423	421 338			

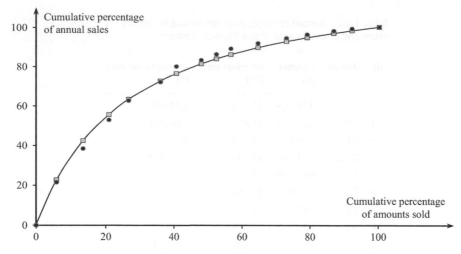

Figure 1.27 80–20 curve for the Blucker products. In black, the Blucker data. In grey, the corresponding values on the 80–20 curve.

ABC **Cross Analysis**

ABC cross analysis consists of performing simultaneously two independent *ABC* classifications with respect to two product values, for example, the revenue generated and the average inventory value in a reference period. This procedure allows identification of nine subclasses (*AA*, *AB*, *AC*, *BA*, *BB*, *BC*, *CA*, *CB* , and *CC*) that are reported in a double entry table (see Table 1.22). The nine classes give an in-depth representation of the current products' performance, allowing companies to evaluate possible corrective actions. On the main diagonal are the most virtuous product classes (*AA*, *BB*, and *CC*), characterized by a balanced and homogeneous logistics management. In particular, these classes are made up of products for which the high (medium or low) level of revenues corresponds to high (medium or low) value of stocks, respectively. The class *AA* is the most lucrative but also the most expensive in terms of inventory costs. A large number of products within this class could lead to a too conservative inventory management policy to avoid stockout; in this case, the company has to evaluate a stock reduction. Classes that are placed above the main diagonal (*AB*, *AC*, and *BC*) contain products with low inventory turnover (see Section 1.8), high stock level and low revenue. From a logistics viewpoint, these products are managed worse than the

Table 1.22 *ABC* cross analysis.

	Revenue		
Average inventory			
	Class A	**Class B**	**Class C**
Class *A*	*AA*	*AB*	*AC*
Class *B*	*BA*	*BB*	*BC*
Class *C*	*CA*	*CB*	*CC*

average. In particular, class AC usually contains obsolete products. In this case the company should reduce the stock level and revise its purchase policies. Classes that are placed under the main diagonal (BA, CA, and CB) are characterized by a high inventory turnover, a low inventory level, and a high revenue. This means that the products belonging to these classes are generally managed better than the average. However, they could be at risk of stockout because an unforeseen increase of the demand level may not be covered by inventories, especially for class CA.

[Blucker.xlsx] Blucker decided to perform an ABC cross analysis (with thresholds 80–15–5) with respect to the revenue generated and average inventory. The annual revenue, sales volume and average inventory of the Blucker products are provided in Table 1.20. Considering the average inventory criterion, class A is made up of the BG2, K0, PL-3, PL-2, FIL12, and BG3 products; class B contains the P-L4, K1, P-L1, TX, PL-5, and BG1 products; class C is composed of the P, K2, and K3 products. The resulting cross classification is summarized in Table 1.23. The analysis shows that the majority of the products fall along or under the main diagonal. Hence, they are correctly managed from a logistics point of view. Nevertheless, there are two obsolete products (PL-3 and BG3 in class AC) for which the company has to evaluate the opportunity of discontinuation. Finally, a product (P in class CA) is at high risk of stockout, which suggests increasing its stock level with the aim of moving the product to class BA.

Table 1.23   *ABC* cross analysis of the Blucker products.

Average inventory	Revenue		
	Class A	Class B	Class C
Class A	BG2, K0, P-L2, FIL12	–	P-L3, BG3
Class B	TX	P-L4, K1, P-L1, BG1, PL-5	–
Class C	P	–	K2, K3

1.12   Information Systems

Nowadays, the administrative processes and the managerial tasks of many organizations are supported by a variety of software applications. No unique classification of such systems exists, the functional separation between applications is blurred and even on the terminology there is no consensus in the scientific and business communities.

A *management information system* (MIS) is a software application which, in its basic form, collects, stores, and processes information from various sources, mainly for the purpose of reporting.

An *enterprise resource planning* (ERP) software is a category of MIS which uses data across various departments (manufacturing, distribution, procurement, sales, accounting, etc.) to monitor business resources (personnel, cash, inventories, production capacity, etc.) and the status of business commitments (orders from customers, purchase orders, payments, etc.), as well as interpret data, make forecasts, support organizational tasks (creating and keeping deadlines, assigning operational tasks, making schedules, etc.).

The modern versions of ERP are provided as a *cloud service*, so they can be accessed from anywhere using different devices (a Web browser and a basic internet connection are needed). In addition, they incorporate real-time analytics tools enabling discovery of hidden information that can be used to make critical decisions (see Section 1.10).

An ERP is composed of several modules which can be seen as a unique integrated software system or composed of independent parts that can interact each other through various *electronic data interchange* (EDI) mechanisms (see Problem 1.3). Figure 1.28 shows a schematic representation of the interaction of ERP modules with two specific logistics applications, namely the *warehouse management system* (WMS) and the *transportation management system* (TMS), briefly described in the subsequent chapters (in particular, WMS in Section 5.9 and TMS in Section 6.5).

The fourth generation of SAP Business Suite is called S/4HANA. S/4HANA is an ERP for large companies, also deployable as a cloud solution, whose strengths are, above all, its usability and flexibility. S/4HANA offers:

- a unified global SCM platform involving multiple countries, currencies, languages and subsidiaries;
- an e-commerce suite for Web stores;
- a retail software solution for supporting multichannel shoppers.

S/4HANA includes, among others, the following modules: finance and controlling, sales, sourcing and procurement, inventory management, production planning, quality management, plant maintenance, project management, and product life cycle management.

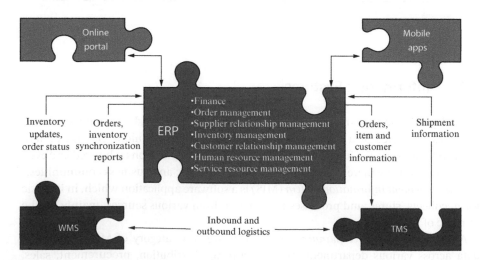

Figure 1.28 Interaction of ERP modules with WMS and TMS.

1.13 Questions and Problems

1.1 Adama is a French manufacturer of photovoltaic panels. The company has a production plant in Rennes, which supplies four warehouses located in Angers, Bourges, Clermont-Ferrand, and Montauban. The warehouses directly supply the installers, grouped, for this purpose, into 10 districts. The installers belonging to the same district are served from a single warehouse. Each installer returns defective photovoltaic panels to the corresponding warehouse; these panels are then sent to a repair centre located in Poitiers. Represent the logistics system of Adama as a directed graph. (Hint: assume that the installers of a district are concentrated in a single point.)

1.2 In 2010, Nokia and Yahoo! announced a partnership to let both companies broaden their global reach by offering messaging and navigation services on mobile devices and PCs. By making available the integration of its Ovi Maps, Nokia has become a global supplier for the navigation service and maps of Yahoo!. On the other hand, Yahoo! made its own messaging technology available to Nokia, becoming the official supplier of Ovi Mail and Ovi Chat. Which type of integration is implemented within the partnership described?

1.3 Barilla is an Italian multinational food company which "has believed in B2B e-commerce and, in particular, in EDI, especially in terms of Web-EDI, which supports the order–delivery–invoice cycle as well as in collaborative approaches, such as the *continuous replenishment program* (CRP) and the *collaborative forecasting and replenishment program* (CPFR)" [Mauro Viacava, CEO of Barilla]. What is presumably the role played by EDI, CRP, and CPFR in Barilla's integrated logistics system?

1.4 Search the website of a company operating in the beverage sector and sketch a representation of its logistics system.

1.5 Discuss logistics operations in a bank (including secure cash replenishment in *automated teller machines*, ATMs) compared with logistics in a manufacturing company such as a chemical producer.

1.6 Consider the case of a manufacturing company with a logistics network made up of two suppliers, a production plant, a CDC, two RDCs, and several retailers. Describe the structure of the information flow assuming that the supply chain is (a) MTS; (b) MTO; (c) ATO; (d) ETO.

1.7 Perform a Web search to look for examples of 1PL, 2PL, 3PL, 4PL, and 5PL providers and write a short report on their logistics operations.

1.8 A pneumatic refuse collection (PRC) system transports waste at high speed through underground tubes to a collection station where it is shredded and sealed in containers. Illustrate the difference in logistics activities of a PRC system and a traditional vehicle-operated door-to-door collection system in a densely populated urban area.

1.9 What are the key issues in relief supply management just after the occurrence of a environmental disaster? What are the information flows? What are the main challenges in logistics operations?

1.10 What is the job description for a position in a department of defence logistics?

1.11 Mercitalia Rail is the main rail cargo company in Italy and one of the most important in Europe. Find out about its logistics system and, in particular, about

Mercitalia Fast, the company's high speed service specialized in the transportation of roll containers.

1.12 Sustainable logistics aims at implementing a *closed-loop supply chain*. Perform a Web search to look for examples of such systems. Describe challenges and opportunities.

1.13 Deliveroo is one of the most popular online food delivery companies in Europe. Illustrate its service structure and identify the peculiar aspects of its crowdsourcing delivery approach.

1.14 Cargo bikes are becoming more and more popular in city logistics. Browse the Web to identify case studies where this freight transportation system is currently used. Moreover, explain how cargo bike hubs are designed and operated.

1.15 Belton is a US company producing wooden panels. Recently, its managers discovered some inefficiencies in the order information flow for both customers and suppliers. For this reason, the company invested in a new order processing software that allowed it to improve order registering, order processing, order preparation and receiving, shipping, billing collection, and payments. Furthermore, the company provided a unique contact person to coordinate the order approval process and manage exceptions. The result was a reduction in cost and a speed up of the entire ordering process. Which KPIs did the investment have an impact on?

1.16 Assume that, for a certain company, the estimated annual sales *versus* service level curve is $r(l) = 950\,000\,(l/100) - 328\,000\,(l/100)^2$, where l denotes the percentage of customers served within 24 hours and $r(l)$ is expressed in \$. The annual logistics costs (in \$) are estimated as 280 000, 320 000, 380 000, 410 000, 460 000, 510 000, with respect to the following service levels: 50%, 60%, 70%, 80%, 90% and 100%, respectively. Determine the service level at which the maximum estimated annual profit is achieved.

1.17 Five alternative configurations have been identified for a logistics system. Their costs and service levels are plotted in Figure 1.29. Determine which

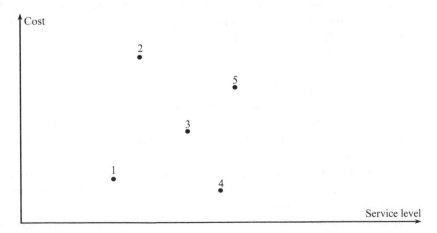

Figure 1.29 Costs and service levels of five alternatives for the design of the logistics system of Problem 1.17.

configurations should be taken into account as a possible solution and which ones should be discarded.

1.18 [Tranexpress.xlsx] Tranexpress is an international freight forwarder. Its OCT consists of two components: (a) shipment preparation, customs paperwork, customs inspections, warehousing, consolidation and container loading, dubbed additional services; (b) transportation. Some past observations are reported in Table 1.24. Characterize the OCT by computing the sample mean and the sample standard deviation of the corresponding random variable. Express service variability as a quantitative index.

1.19 Norsk is a Danish company which specializes in the production of food products for daily consumption. It has five associated subsidiaries in the European Union and a network of distributors in North America. Recently, the company decided to redesign its distribution network in Scandinavia where 140 DCs have been transformed into simple warehouses without administrative functions. The administrative functions have been concentrated in 14 regional logistics centres. Moreover, forecasting activities based on data analysis have been centralized in the company's headquarters. List and classify the decisions taken at a business management level during the reorganization phase of the logistics system.

1.20 Describe the functional structure of a petrochemical company and highlight the position of the logistics activities in the organizational chart. Compare the advantages and disadvantages of adopting a divisional organization.

1.21 [AjtSolar.xlsx] Ajt Solar is an Indian producer of solar-powered products. To control the supply chain, some critical logistics activities are monitored monthly. Three families of performance measures, related to order processing,

Table 1.24 Observed data related to additional services and transportation times (in days) of Tranexpress.

Additional services		Transportation	
Number of recorded data	Time	Number of recorded data	Time
1	25	20	159
3	26	29	160
4	27	56	161
9	28	66	162
18	29	52	163
35	30	26	164
42	31	16	165
22	32		
13	33		
2	34		
1	35		

Table 1.25 Performance measures used by Ajt Solar.

Family	Performance measure	Computing method
Order processing	Orders	Number of orders in a month/ number of weekdays in a month
	Complaints	Number of complaints in a month
Inventory	Stock value	Average inventory value on a monthly basis
Transportation	Deliveries	Average number of customers served in a vehicle trip
	Vehicle trips	Monthly number of trips carried out/ monthly number of planned trips

Table 1.26 Performance measure values for Ajt Solar in the last 15 months.

Month	Orders	Complaints	Stock value [$]	Deliveries	Vehicle trips
1	154.26	21	470 800	15.48	1.07
2	151.04	12	500 800	36.76	0.97
3	161.23	16	533 000	18.94	1.13
4	145.33	24	565 900	33.07	1.14
5	158.66	14	567 700	31.15	1.13
6	171.25	16	471 900	40.37	1.10
7	98.66	31	522 200	23.35	0.83
8	102.45	8	531 000	14.33	1.00
9	134.74	12	509 800	39.80	0.93
10	147.24	16	579 700	18.37	1.20
11	133.54	21	548 300	26.04	1.15
12	154.81	18	458 700	30.10	1.00
13	148.82	20	542 100	36.60	0.95
14	124.31	13	524 500	26.52	1.11
15	164.03	11	567 400	33.46	1.00

inventory management, and transportation are considered (see Table 1.25). The performance measures for the past 15 months are reported in Table 1.26. In the current month (with 23 working days), the company has recorded the following data: 3568 orders, 15 complaints, an inventory value of $ 560 400, 24.25 customers served on average in a trip, 4 950 planned vehicle trips and 4702 trips made. Build a control panel to monitor the three identified families of performance measures, according to the methodology illustrated in Section 1.9.3.

1.22 [Arka.xlsx] Arka is a Serbian manufacturer of diesel engines. The logistics manager in charge of designing the supply chain for product named WMF has identified four configurations (denoted as $i = 1, 2, 3, 4$ in the following). Based on the judgement of a panel of experts (see Section 2.2 for more details), the

Table 1.27 Profit (in k€) for each configuration-demand scenario pair in the Arka problem.

Configuration (i)	Demand in the next five years		
	low	medium	high
1	800	400	−400
2	500	250	100
3	400	650	100
4	−200	700	1200

demand in the next five years is expected to be high, medium, or low, with probabilities 0.5, 0.3, and 0.2, respectively. An in-depth analysis of the company's cost structure has allowed the manager to determine the profit of the four configurations for every demand scenario (see Table 1.27). Determine the optimal configuration with respect to the Bayes criterion. Also compute the expected value under perfect information.

1.23 [Lugan.xlsx] In the Lugan problem, assume that no additional travel time observations can be collected and a decision has been made on the basis of the initial 20 samples. Under this hypothesis, determine the *minimum* indifference zone width δ that allows the company to choose between providers A and B with a confidence level $(1 − \alpha) = 0.95$. Then identify the provider to be selected.

1.24 [Moderan.xlsx] In the Moderan problem, assume that $p = 0.99$ and determine the optimal solution of the corresponding chance-constrained model. Hence compare the solutions obtained for $p = 0.95$ and $p = 0.99$. Which is more robust? Which is more expensive?

1.25 [HealthyLife.xlsx] In the HealthyLife problem assume that the inventory $y^{(s)}$ to be bought, as a recourse action, in each scenario $s = 1, 2, 3$, cannot be greater than 20. Determine the optimal solution of the new version of the stochastic programming model with recourse. What is the company's new strategy?

1.26 A *supply chain digital twin* is a virtual supply chain replica that consists of simulated suppliers, plants, warehouses, etc. Perform a Web search and learn more about this paradigm. Which are the main differences between a digital twin and a traditional simulation model?

1.27 Garbi is a waste collection municipal agency operating in a small district near Bilbao (Spain). Every morning seven trucks arrive at the waste processing facility located at Ibaiondo at 8:42, 8:53, 8:55, 9:00, 9:10, 9:15, 9:25 to drop their content. Determine the average waiting time of a truck if vehicle unloading takes 10 minutes.

1.28 In the Garbi problem, assume that the time needed to unload a vehicle is normally distributed with an expected value of 10 minutes and a standard deviation of 2 minutes. Develop a Python code to simulate vehicle queueing at the waste processing facility. Execute 20 simulation runs and determine the confidence interval at 0.95 level for the expected vehicle waiting time.

1.29 In the second version of the Garbi problem, determine how many simulation runs are needed to obtain a confidence interval whose semi-width is 1% of the sample mean.

1.30 Eurlux is a Dutch company that has recently decided to start production and sale of a new energy-efficient light bulb. During the phase of the logistics system design, three alternatives are considered.

- Use the foreign manufacturing plant located in Tartu (Estonia), where the unit production cost is € 0.97 (the cost of raw materials purchase is included). The transportation cost to the CDC of Groningen is € 5 per box, where a box contains 100 units of the product. For simplicity, it is assumed that this cost includes also inventory costs at the CDC. The CDC of Groningen supplies two RDCs, situated in Delft and Eindhoven; their annual demands are 28 000 and 35 000 boxes, and the transportation costs per box are € 9 and € 10, respectively.

- Use the national manufacturing plant of Dordrecht, where the unit production cost is € 1.38. This facility supplies the RDC of Delft and Eindhoven and the unit transportation costs are, in this case, € 8.5 and € 7.0 per box, respectively.

- Use the national manufacturing plant of Dordrecht to satisfy the demand of the RDC of Delft; while the demand of RDC of Eindhoven is covered by a foreign manufacturing plant located in Antwerp (Belgium) with a unit production cost of € 1.18 and a transportation cost of € 5.6 per box.

Determine which alternative to select through the `Excel What-If Analysis` tool, considering the minimization of production and transportation costs as the logistics objective to pursue.

1.31 [`Florim.xlsx`] Florim is an Albanian company specializing in the manufacturing of ceramics. The company manufactures six products, whose sales are reported in Table 1.28. Let $y = [(1 + \alpha_1)x]/(\alpha_1 + x)$ be the equation of the 80–20 curve C_1 defined such that the first 21% of products sold corresponds to 68% of the annual sales; similarly, let $y = [(1 + \alpha_2)x]/(\alpha_2 + x)$ be the 80–20 curve C_2, obtained by assuming that the first 21% of products corresponds to 62% of the annual sales. Check which of the curves C_1 and C_2 is a better approximation of the actual cumulative percentage of the annual revenue with respect to the cumulative percentage of the annual quantity sold.

1.32 [`Zuick.xlsx`] Zuick is a German import–export company of household appliances. The company, whose headquarters are located in Hannover, distributes 15 products whose weekly sales volumes and revenues are reported in Table 1.29. The company is investing additional financial resources in two products, K-505 and K-506, for which the logistics manager proposes an aggressive distribution strategy, involving more CDCs and higher inventory levels. By using an *ABC* classification (20–30–50) with respect to weekly sales, verify whether the distribution strategy proposed by the logistics manager is correct and, if not, modify it accordingly.

Table 1.28 Annual revenue (in €) and quantity sold of the six Florim products.

Product	Revenue [€]	Quantity [quintals]
1	350 000	2700
2	160 000	2200
3	920 000	2500
4	125 000	1500
5	360 000	4200
6	160 000	1900

Table 1.29 Weekly quantity sold and corresponding revenue (in €) in the Zuick problem.

Product	Quantity [quintals]	Revenue [€]
K-501	155	119 806
K-502	64	31 448
K-503	70	25 607
K-504	66	24 406
K-505	61	15 196
K-506	58	13 112
K-507	60	10 106
K-508	197	11 395
K-509	154	9 489
K-510	56	8 664
K-511	74	13 955
K-512	208	16 283
K-513	164	14 085
K-514	71	12 984
K-515	163	123 935

1.33 [ElMa.xlsx] El.Ma is an American distributor of electrical equipment. The warehouse in Columbus (Ohio) has 18 products in stock. Monthly sales and average monthly stock values are reported in Table 1.30. Make an *ABC* classification (80–10–10) of the products with respect to monthly sales and monthly average stock values, respectively. Which inventory policy should El.Ma adopt for product "locking release 24V"?

Table 1.30 Monthly sales and average monthly stock values (in €) of the El.Ma products.

ID product	Description	Sales	Stock value
1	Digital starter	25 356	980
2	Differential block 4P	147 800	3667
3	Land trolley	10 450	1174
4	HCS cable	65 980	2030
5	Engine control unit 380V CA	17 654	652
6	Contacter 24-60V CC	27 580	1721
7	Control builder	19 768	558
8	Universal dimmer	46 225	1015
9	BRI interface	8 766	775
10	Circuit breaker 10KA	80 350	3159
11	Motoadaptor	13 746	1100
12	OPC server	57 558	3111
13	Spring relay	7 852	733
14	Electronic delayer	9 785	724
15	Sectioner	32 400	894
16	Locking release 24V	12 328	1020
17	TMA360	15 980	1058
18	Control unit for release	11 900	1062

2

Forecasting Logistics Data

2.1 Introduction

Forecasting is an attempt to determine in advance the future outcome of an uncertain variable. Planning and controlling logistics systems need predictions for those decisions that must be made before all the relevant data are known. Examples include the main phases of the logistics system planning process (in particular, facility location and capacity planning) as well as production scheduling, inventory management, and transportation planning. Logistics data to be predicted include customer demand, raw material prices, labour costs, and lead times. See Table 2.1 for a list of the main forecasts required by the various planning and control decisions.

Forecasting methods are equally relevant to every kind of logistics system, but they are truly vital for MTS supply chains (see Section 1.3.2), where inventory levels have to be set in advance to anticipate customer demand.

Forecasting is based on hypotheses. Each forecasting method is based on hypotheses. For instance, a common hypothesis is that some features (trend, seasonality, etc.) of sales past data will remain nearly the same in the future. Another example is the belief that future values of a variable depend to some extent on the past and present values of a set of observable variables. As a result, no forecasting method can be deemed to be superior to others in every respect and decision makers must only consider those methods relying on hypotheses they trust.

Timeline of forecasting. Forecasts can be classified on the basis of the time horizon they refer to. *Long-term* forecasts span a time horizon from one to five years. Predictions for longer periods are very unreliable, since political and technological issues come into play. Long-term forecasts are used for deciding whether a new product should be put on the market, or whether an old one should be withdrawn, as well as for designing a logistics system. Such forecasts are often generated for a group of products (or services) rather than for a single product (or service). Moreover, in the long term, sector forecasts are more common than corporate ones. *Medium-term* forecasts extend over a period ranging from a few months to one year. They are used for tactical logistics decisions, such as sales and marketing planning, production and distribution planning, inventory management, cash and capital budget planning. *Short-term* forecasts cover a time interval ranging from a few days to several weeks. They are utilized for day-to-day production, inventory, and distribution planning. As customer requests are received,

Introduction to Logistics Systems Management: With Microsoft® Excel® and Python® Examples, Third Edition. Gianpaolo Ghiani, Gilbert Laporte, and Roberto Musmanno.
© 2022 John Wiley & Sons Ltd. Published 2022 by John Wiley & Sons Ltd.

Table 2.1 Main forecasts required by logistics systems planning and control.

Decision area	Forecasts
Procurement	Price of raw materials, components, and semi-finished goods
	Availability of raw materials, components, and semi-finished goods
Facility location	Fixed facility costs
	Variable facility costs
	Aggregated demand
Warehousing	Demand of the sales points
	Inventory costs
Distribution	Travel times
	Demand of individual customers or retailers

there is less need for forecasts. Consequently, forecasts for a shorter time interval (a few hours or a single day) are quite uncommon, except for highly dynamic environments such as the e-commerce sector, urban express parcel delivery and warehouse picking, packing, and shipping.

Retail demand forecasting. In some settings, the variable to be predicted is not only merely observed but can be influenced by appropriate organizational initiatives. This is the case of retail demand which depends on marketing-mix decisions (see Section 1.8.3), in particular pricing and promotion (advertising campaigns, changes in display position, etc.). In such cases, forecasts are often carried out jointly by logisticians and marketing managers. Retail forecasting can be very intricate because of secondary impacts like *product cannibalization* which arises when, for example, a promotion on a product A leads to a decrease in sales for a similar product B. The complexity of such forecasting tasks, which are of the utmost importance when dealing with short shelf life products, may be successfully tackled with ML methods (see Section 2.7.8).

Forecasting techniques can be classified into two main categories: qualitative and quantitative methods.

2.2 Qualitative Methods

Qualitative methods are mainly based on expert judgement or on experimental approaches. They are usually employed for medium- and long-term forecasts when there are not enough data to use a quantitative approach. This is the case, e.g., when a new product or service is launched on the market, when a product packaging is changed, or when political changeovers or technological advances are expected to influence the variable to be predicted.

The most common qualitative methods are *expert judgement*, the *Delphi method*, and *market research*.

In the first approach, a forecast is developed on the basis of the opinions expressed by a group of people that have acquired a specific expertise in a technology field, in a product area or in an industry. These experts may be professional and technical organizations, stakeholders, customers as well as company employees. The latter include salespeople who, being in direct contact with customers, may be aware of shifts in their behaviour.

> In 2020 Experts of the German Ifo Institute expected that, due to the COVID-19 pandemic, the global logistics market will grow from 2734 billion dollars in 2020 to 3215 billion dollars in 2021.

The Delphi method overcomes some of the weaknesses of the judgement-based forecasting methodologies. A series of questionnaires is submitted to a panel of experts. Every time a group of questions is answered, new sets of information become available. Then a new questionnaire is prepared by a coordinator in such a way that every expert is faced with the new findings. This procedure eliminates the bandwagon effect of majority opinion. The Delphi method terminates as soon as all experts share the same viewpoint. This technique is mainly used to estimate the influence of political or macroeconomic changes on data patterns.

> The Delphi method has been used recently to estimate the demand of tourism and travel-related services in the Lazio coastal region in Italy. The group of experts was made up of 800 hotel managers and tour operators, coordinated by a team of 10 employees of the Lazio regional authority.

Market research is based on interviews with potential consumers or users. It is time consuming and requires a deep knowledge of *statistical sampling theory*. For these reasons, it is used only occasionally, e.g., when deciding whether a new product should be launched.

> Tienda is a Spanish company manufacturing and marketing aromatic oils. When launching a new lemon scented oil, obtained from citrus fruits and olives, the company asked a market research company to forecast its demand by means of a survey. A sample of 1455 customers were selected in 32 Spanish supermarkets in the area of Seville. Each questionnaire aimed to estimate the probability that the customer would buy the new product. On the basis of these data, a forecast of the new product's sales was generated.

Table 2.2 summarizes the features of the main qualitative forecasting methods.

2.3 Quantitative Methods

Quantitative methods can be used every time there are enough historical data. No matter where the data come from (relational databases, social media posts, log files, etc.), they are usually formatted as a *dataset* reporting the T past observations (realizations)

Table 2.2 Features of the main qualitative methods.

Feature	Forecasting method		
	Expert judgement	Delphi method	Market survey
Time horizon	Medium/long	Medium/long	Medium
Effort	Low, if based on company's experts	Moderate	Large
Cost	Low, if based on company's experts	Moderate	High
Accuracy	Low	Moderate	Moderate

$y_t, t = 1, \ldots, T$, of an *outcome* (or *dependent* or *target*) *variable y* as well as the corresponding past observations $x_{tj}, t = 1, \ldots, T, j = 1, \ldots, n$, of n correlated variables $x_j, j = 1, \ldots, n$, called *explanatory variables* (or *predictors*, *covariates* or *features*). An explanatory variable may take an integer or a real value (*numerical variable*) or one of a limited number of categories (*categorical variable*). In most cases, the past observations are indexed with respect to time, in such a way to form a sequence of data taken at successive equally spaced time periods (*time series*).

Elextronix is a German chain of consumer electronics retailers. Table 2.3 shows an extract of a dataset recently generated by the logistics manager. The extract reports, over a period of five months ($t = 1, \ldots, 5$), the company's sales y_t of a budget Oled TV set (in thousands of €), its unit product price x_{t1} (in €), the expenditure in television, radio, and social media advertising x_{t2} (in thousands of €), as well as the major direct marketing initiative x_{t3} of the month. The latter is a categorical variable that can take four values: TM (cell phone text messaging), EM (promotional email), FD (flier distribution) and CD (catalogue distribution).

Table 2.3 Extract of the Elextronix monthly sales dataset.

Month (t)	Oled TV sales [k€] (y_t)	Oled TV unit price [€] (x_{t1})	Advertising expenditure [k€] (x_{t2})	Direct marketing initiative (x_{t3})
1	5600	1400	150	FD
2	6900	1200	160	FD
3	6500	1200	120	EM
4	6700	1200	190	CD
5	7600	1150	250	TM

2.3.1 Explanatory Versus Extrapolation Methods

The choice of the most suitable quantitative forecasting method depends on the nature of the outcome variable as well as on the amount and features of the available data. These methods can be classified into two macrocategories: explanatory and extrapolation methods.

Explanatory methods are based on the hypothesis that the outcome variable y has a cause–effect relationship (or, at least, a correlation) with some explanatory variables $x_j, j = 1, \dots, n$.

> The demand for small cars is related to the business cycle of a country and can, therefore, be correlated to its *gross domestic product* (GDP).

The greatest advantage of explanatory methods lies in their capacity to anticipate the variations of the outcome variable. Unfortunately, it is sometimes difficult to identify explanatory variables having a high correlation with the outcome variable.

Extrapolation methods are of *univariate* type, since they rely solely on past observations of the outcome variable y. They assume that some features shown by y in the past will remain the same. Data patterns are first identified and then projected in the future. No other information is used. Extrapolation methods are accurate in the short–medium term, inexpensive and easily automated. As a result, they are widely used, especially for inventory, production, and distribution forecasts.

2.3.2 The Forecasting Process

The forecasting process is usually divided into four main steps.

1. *Exploratory data analysis* (EDA). EDA amounts to performing an initial investigation on data with the aim to create a short list of suitable forecasting methods; for instance, if EDA shows that a feature of the dataset is highly correlated to the outcome variable, the logistician may consider an explanatory model including that feature as a predictor (see Section 2.7); on the other hand, if no feature of the dataset shows a significant correlation, but the outcome variable has, for example, a constant trend and a remarkable seasonal component, the logistician may consider, among others, the Winters method (see Section 2.8.5).
2. *Data preprocessing.* Raw data (like those extracted from transactional databases) are seldom ready to be used to make a forecast. Data preprocessing is the step in which outliers, missing and inconsistent values are identified, and dealt with, in such a way they do not degrade the predictions. Moreover, data are possibly aggregated and transformed to improve forecasting accuracy.
3. *Choice of the forecasting method.* Here, the most suitable forecasting method is selected among a set of alternative techniques based on the accuracy they would have provided if used in the past. For parametrized methods, this step also includes the determination of the optimal value for each parameter.

4. *Evaluation of the forecasting accuracy.* Once the realizations of the outcome variable become known, errors can be computed retrospectively. Such errors are then combined in an aggregate measure to assess the accuracy of the method currently used. This measure can be used to finely tune the parameters of the method or even to select an alternative technique.

These four phases will be described in detail in the following sections.

2.4 Exploratory Data Analysis

The primary role of EDA is to help the logistician to make sense of data used in the forecasting process. In some settings, EDA can also be beneficial in providing a basis for further data collection.

There is not a predefined set of prescribed steps in EDA. Logisticians should follow their intuition in explaining the data. EDA methods can be *univariate* or *multivariate* (usually just *bivariate*). Univariate methods consider one variable at a time (whose past observations correspond to a column of the dataset), while multivariate methods look at two or more variables at the same time to explore relationships. It is customary to perform univariate EDA on each individual variable before performing a multivariate EDA. From another perspective, EDA tools can be classified into *numerical methods*, requiring the computation of summary statistics, and *graphical techniques*, summarizing the data in a diagrammatic or pictorial way. In the following, EDA is illustrated with reference to numerical variables, which are the most common in logistics management. The reader is referred to specialized texts for a general treatment of the topic, including the exploratory analysis of categorical data.

2.4.1 The Univariate Case

Univariate non-graphical EDA methods include the computation, among others, of *measures of location* and *measures of spread*. Measures of location provide information about where data lie, and include:

- the minimum value;
- the maximum value;
- the sample mean, i.e., the arithmetic average of all data values;
- the quartiles Q_1, Q_2, Q_3, i.e., the values under which 25%, 50%, and 75% of data (arranged in a non-decreasing order) fall, respectively;
- the median, defined as the midpoint of all observed data values when arranged in non-decreasing order, which is the same as Q_2.

Measures of spread describe the variability of the data and include the sample variance, the sample standard deviation and the length of the interquartile range $I = Q_3 - Q_1$.

In these definitions, the word "sample" emphasizes that the past observations are intended as realizations of a random variable.

[Mumbarai.csv, Mumbarai.py] Mumbarai is a producer of car spare parts with a plant located in Balikpapan (Indonesia). The finished parts are transported from the end of the production lines to the main warehouse by an AGV. The company's logistics manager has collected 1000 AGV travel times (in seconds) between the two points. They find out that data are characterized by the measures of location and spread reported in Table 2.4.

Table 2.4 Measures of location and spread (in seconds) for the AGV travel times in the Mumbarai problem.

Measures of location (L) and spread (S)	Value
Minimum value (L)	83.30
Maximum value (L)	108.90
Sample mean (L)	98.05
First quartile (L)	95.40
Second quartile, a.k.a. median (L)	98.15
Third quartile (L)	100.70
Sample standard deviation (S)	3.88

2.4.2 Histograms

A *histogram* is a graphical display of numerical data. It is made up of bars, associated with intervals of values, of different heights. Taller bars show that more data fall in the associated interval. The height of the bars can be the *absolute frequency* (number of observations that fall in the associated interval) or the *relative frequency* (absolute frequency divided by the overall number of observations) or the *probability density* (relative frequency divided by the width of the interval). In the latter case the histogram is referred to as a *density histogram*.

Histograms can be used to identify the *parent distribution* corresponding to a feature. To this end, the logistician estimates the parameters of the candidate parent distributions and plot their probability density function over the density histogram. A human observer looking at this representation can identify visually which distribution fits best. Several studies have shown that this approach has an accuracy comparable (or even superior) to goodness of fit tests (such as the *Chi-square* and the *Kolmogorov–Smirnov* tests).

[Mumbarai.csv, Mumbarai.py] In the Mumbarai problem, the logistics manager wants to verify visually whether the parent distribution of the AGV travel times is normal. They plot in the same diagram the density histogram of the available data and the probability density function of the normal distribution with the same mean and the standard deviation as the sample mean and the

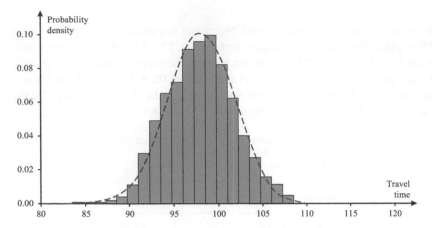

Figure 2.1 Density histogram of the AGV travel times along with the plot of probability density function of the normal distribution with the same mean and standard deviation, in the Mumbarai problem.

sample standard deviation of the data (see Table 2.4). The visual inspection of the resulting graphical representation (see Figure 2.1) shows that the intuition of the logistics manager is correct.

2.4.3 Boxplots

A *boxplot* (or *box-and-whisker plot*) is a graphical display depicting numerical data through their quartiles. In particular, the plot contains a *box* whose extreme sides represent quartiles Q_1 and Q_3. The box also includes an internal line representing median Q_2 and, possibly, a small triangle associated with the mean of the data. In addition, the representation includes two lines (called *whiskers*), extending from the box to the minimum and to the maximum data values which are not outliers, respectively. Finally, outliers (see Section 2.5.2) are plotted as individual points.

[Ravaioli.csv, Ravaioli.py] Ravaioli is an Italian producer of fresh pasta, renown for its ravioli. The sales of its spinach ravioli "Nonna Pina" are influenced heavily by the company's TV and social media advertising. Table 2.5 reports the latest 20 weekly expenditures (in k€) made by the company in TV and social media advertising, together with the sales (always in k€) realized in the same time periods. In order to devise sales forecast for inventory planning, the logistics manager performed preliminarily an EDA. This phase comprised the generation of the boxplots for the company's TV and social media advertising expenditures. The visual inspection of this representation (see Figure 2.2) allowed the manager to gain important insights into the statistical distribution of these two explanatory variables.

Table 2.5 Dataset used in the Ravaioli problem (all the data are expressed in k€).

Week (t)	TV advertising expenditure (x_{t1})	Social media advertising expenditure (x_{t2})	Sales (y_t)
1	2.96	4.45	371.88
2	5.25	0.64	484.55
3	0.90	5.01	85.47
4	6.77	5.07	613.85
5	5.72	3.27	531.31
6	5.69	6.01	628.30
7	11.76	6.45	1167.89
8	6.03	6.81	626.87
9	6.02	7.19	724.01
10	6.06	13.26	817.08
11	18.46	7.94	1325.71
12	6.52	21.38	1064.86
13	6.58	8.64	802.38
14	8.51	11.73	934.75
15	6.89	9.33	774.84
16	6.63	9.71	920.35
17	6.90	8.30	835.48
18	6.69	10.40	755.61
19	7.38	10.79	814.83
20	7.50	10.93	826.21

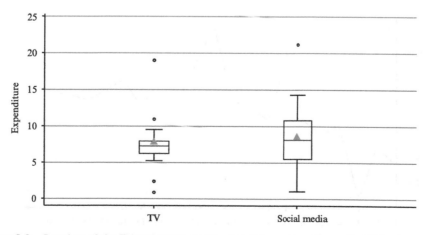

Figure 2.2 Boxplots of the TV and social media advertising expenditures (in k€) in the Ravaioli problem.

2.4.4 Time Series Plots

A *plot* of a time series $y_t, t = 1, \ldots, T$, is a Cartesian diagram (t, y_t) in which the horizontal axis shows graduations $t = 1, \ldots, T$ of time using an appropriate scale (weeks, months, quarters, years), while the vertical axis shows the corresponding numerical values y_t for each $t = 1, \ldots, T$. The visual analysis of the diagram can be used as a support to most complex methodologies (for example, to identify a linear trend in the data or to detect possible outliers) or even to visually analyse the data. In order to emphasize the main features of the time series (which is discrete in nature) the T dots corresponding to the past observations are often connected through a continuous line.

[Ravaioli.xlsx, Ravaioli.csv, Ravaioli.py] The EDA performed by the logistics manager of Ravaioli includes the generation of the plot of the weekly sales of spinach ravioli (see Figure 2.3).

2.4.5 The Bivariate Case

Bivariate non-graphical EDA aims to express the correlation between two variables x and y by computing the *sample covariance* from past observations $x_t, y_t, t = 1, \ldots, T$:

$$v_{x,y} = \frac{1}{T-1} \sum_{t=1}^{T} (x_t - \overline{x})(y_t - \overline{y}),$$

where $\overline{x}$ and $\overline{y}$ indicate the sample means of x and y, respectively. The sample covariance is positive if x and y are positively correlated, negative if negatively correlated and (close to) zero if uncorrelated. Since its magnitude is difficult to interpret, it is often computed by using its normalization

$$r_{x,y} = \frac{v_{x,y}}{S_x S_y}, \tag{2.1}$$

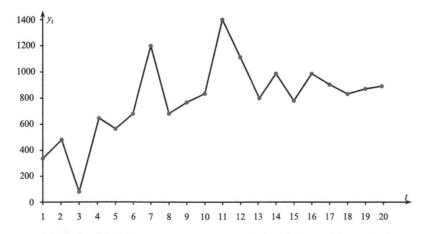

Figure 2.3 Plot of the weekly sales of spinach ravioli (in k€) in the Ravaioli problem.

where S_x and S_y are the sample standard deviations of x and y, respectively. The coefficient $r_{x,y}$ given by equation (2.1) is known as the *Pearson correlation coefficient* and ranges in the $[-1, 1]$ interval. In particular, it is equal to -1 or $+1$ if x and y are linearly dependent, i.e., $x = ay + b$, or, equivalently, $y = mx + q$, where a and b (m and q) are two parameters with $a < 0$ ($m < 0$) and $a > 0$ ($m > 0$), respectively.

[Ravaioli.xlsx, Ravaioli.csv, Ravaioli.py] The logistics manager of Ravaioli is interested in evaluating the correlation between the weekly sales y of spinach ravioli "Nonna Pina" and the company's weekly expenditure in TV and social media (x_1 and x_2, respectively). The Pearson correlation indices $r_{y,x_1} = 0.86$ and $r_{y,x_2} = 0.55$, obtained in Excel by using the PEARSON function, show that x_1 is a good predictor for y while x_2 is worse. Moreover, $r_{x_1,x_2} = 0.16$ indicates that explanatory variables x_1 and x_2 are substantially uncorrelated which would suggest that a regression model with both features might be superior to a model with x_1 alone.

2.4.6 Scatterplots

Another widely used graphical analysis amounts to drawing a (2D) *scatterplot* which is a diagram using Cartesian coordinates to display corresponding values for two numerical variables of a dataset. Scatterplots can be used to verify whether an outcome variable y is correlated to each individual explanatory variable x_j, $j = 1, \dots, n$ (in which case x_j is a good predictor for y). It can also be used to check whether two explanatory variables are strongly correlated (in which case one of them is redundant and can be removed). A *3D scatterplot* is a plot on the three axes in the attempt to show the relationship between three variables (i.e., the outcome variable and the other two explanatory variables).

[Ravaioli.csv, Ravaioli.py] The findings of the Ravaioli's logistics manager about the quality of x_1 and x_2 as predictors are confirmed by 2D scatterplots illustrated in Figures 2.4 and 2.5, as well as by the 3D scatterplot in Figure 2.6.

2.5 Data Preprocessing

Data entries are often characterized by missing data, errors, and outliers as well as inconsistencies. For these reasons, data usually need to be preprocessed by performing a cleaning, an interpolation, an aggregation, or a data transformation before they can be used to make forecasts. The main preprocessing operations on a (univariate) time series are illustrated below. The reader is referred to specialized textbooks for a general treatment of the topic, including preprocessing for multivariate analysis.

2.5.1 Insertion of Missing Data

In the simplest case, a missing observation can be replaced with the average of the previous and subsequent observations.

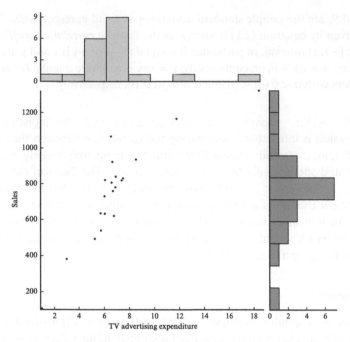

Figure 2.4 A 2D scatterplot of sales of spinach ravioli versus TV advertising expenditures (all in k€) in the Ravaioli problem. The plot also contains a histogram for each of the two variables.

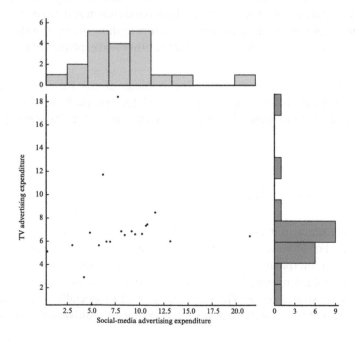

Figure 2.5 A 2D scatterplot of social media advertising expenditures versus TV advertising expenditures (all in k€) in the Ravaioli problem. The plot also contains a histogram for each of the two variables.

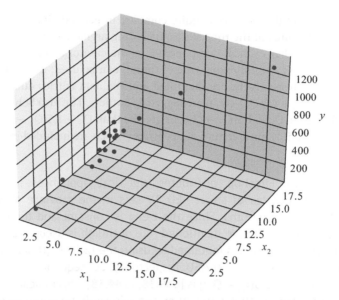

Figure 2.6 A 3D scatterplot of sales of spinach ravioli, TV and social media advertising expenditures (all in k€) in the Ravaioli problem.

Table 2.6 shows the number of cars sold in Argentina during the last 12 months. The missing observation (month 6) can be set as the average of sales in months 5 and 7,

$$y_6 = (y_5 + y_7)/2 = (38\,521 + 41\,345)/2 = 39\,933.$$

Table 2.6 Number of cars sold monthly in Argentina during last year.

Month	Sales	Month	Sales
1	61 945	7	41 345
2	35 379	8	43 866
3	43 535	9	49 379
4	39 317	10	39 533
5	38 521	11	41 611
6	–	12	52 458

2.5.2 Detection of Outliers

An outlier is an observation that appears to deviate significantly from the other members of the time series. Outliers can originate from an error in data transmission or recording, an instrument error or simply natural deviations (which is often the case of *heavy tailed* distributions). Except for the latter case, one may wish to discard the outliers that otherwise might mislead most forecasting methods. Then, the removed observations can be treated as missing data (see Section 2.5.1).

The identification of the outliers is quite complex in general since it has to consider a number of features of the time series, including an increasing or a decreasing trend, a seasonal or other cyclical components. In the case of a constant trend and no cyclical components, the following rule of thumb can be used: the first and third quartile, Q_1 and Q_3, respectively, are identified and the data entries outside interval $[Q_1 - 1.5(Q_3 - Q_1), Q_3 + 1.5(Q_3 - Q_1)]$ are tagged as outliers. The idea behind this rule is that entries less than $Q_1 - 1.5(Q_3 - Q_1)$, or greater than $Q_3 + 1.5(Q_3 - Q_1)$, deviate significantly from the 50% most central data entries (which are included in $[Q_1, Q_3]$).

[Elleshop.xlsx] Elleshop distributes electrical appliances in Austria. Table 2.7 reports its sales of smart LED TV sets in the province of Klagenfurt during the last 12 months. Since the trend is constant and there are no cyclical components, the above mentioned rule of thumb is used. The first and third quartile of the time series (obtained by using the Excel QUARTILE function) are $Q_1 = 866.25$, $Q_3 = 977.50$, respectively. Consequently, the interval $[Q_1 - 1.5(Q_3 - Q_1), Q_3 + 1.5(Q_3 - Q_1)]$ is $[699.38, 1144.38]$. This indicates that the observation related to month 8 is an outlier. The value y_8 is, therefore, eliminated and replaced by 825, the average of the sales volumes in months 7 and 9.

Table 2.7 Number of smart LED TV sets delivered during the last 12 months by Elleshop.

Month	Sales	Month	Sales
1	975	7	770
2	1025	8	200
3	895	9	880
4	1055	10	870
5	925	11	915
6	985	12	855

2.5.3 Data Aggregation

This consists of merging disaggregated data from multiple sources (e.g., sales from individual retailers in a given sales district) into a single time series (e.g., the overall sales in the district). Forecasts made on aggregated data are usually much more accurate than those made on disaggregated data. This phenomenon can be explained as follows. Let $X_1, \ldots, X_n$ be n random variables modelling the disaggregated data which are assumed to be independent and identically distributed (and, in particular, with the same mean μ_X and variance σ_X^2). Moreover, let Y be the random variable modelling the aggregated data

$$Y = X_1 + \ldots + X_n.$$

Then, the mean μ_Y and the variance σ_Y^2 of Y are

$$\mu_Y = n\mu_X;$$
$$\sigma_Y^2 = n\sigma_X^2.$$

Hence, the coefficient of variation of Y (representing the relative dispersion of Y around its mean μ_Y)

$$\sigma_Y/\mu_Y = (1/\sqrt{n})(\sigma_X/\mu_X)$$

is less than the coefficient of variation of each variable $X_1, \dots, X_n$ around their common mean μ_X. Therefore, the forecasts on Y are expected to be more accurate than those on each $X_1, \dots, X_n$ variable.

The demand for champagne (in thousand of bottles) in three regions of France (Burgundy, Alsace, and Provence) over the next 12 months can be modelled as three independent random variables $X_1, X_2,$ and X_3 whose mean, variance and coefficient of variation are reported in Table 2.8. The aggregated demand Y is characterized by the following statistics:

$$\mu_Y = \mu_{X_1} + \mu_{X_2} + \mu_{X_3} = 6800;$$
$$\sigma_Y^2 = \sigma_{X_1}^2 + \sigma_{X_2}^2 + \sigma_{X_3}^2 = 600;$$
$$\sigma_Y/\mu_Y = 0.0036.$$

As expected, the relative dispersion of the aggregated demand is less than that in each individual region.

Table 2.8   Mean (in thousands of bottles), variance and coefficient of variation of champagne demand in three regions of France over the next 12 months.

Geographic area (j)	Mean (μ_{X_j})	Variance ($\sigma_{X_j}^2$)	Coefficient of variation (σ_{X_j}/μ_{X_j})
Burgundy	2500	300	0.0069
Alsace	1800	200	0.0079
Provence	2500	100	0.0040

Data aggregation can be important when dealing with intermittent time series since the presence of zero values in the time series could cause errors in most forecasting methods. In order to overcome this problem, the available data can be grouped by product type, by different geographic areas, or by time periods.

The daily sales of Sidol75 aluminium polishing packages in each supermarket of the Suomen chain in the Lahti province, Finland, are lumpy (see, for example, the past week time series for a sample supermarket shown in Table 2.9). In order to

Table 2.9 Number of Sidol75 packages daily sold in the last week in a sample Suomen supermarket.

Day	Sales
1	4
2	0
3	0
4	3
5	8
6	7
7	0

Table 2.10 Number of Sidol75 packages weekly sold in all the Suomen supermarkets in the province of Lahti during the last 20 weeks.

Week	Sales	Week	Sales
1	254	11	263
2	262	12	265
3	260	13	271
4	264	14	256
5	255	15	269
6	258	16	262
7	262	17	258
8	267	18	262
9	256	19	264
10	259	20	265

carry out an accurate forecast, the sales manager decided to group the sales data of all supermarkets in the province of Lahti and to consider weekly sales figures instead of the daily data. The resulting time series is shown in Table 2.10. Based on these data, the demand for Sidol75 in the Lahti province for the following week was estimated at 264 packages.

2.5.4 Removing Calendar Variations

Time series representing a cumulative amount over a given period (e.g., monthly sales of a product) may contain *calendar effects* due to the variable month length, day-of-the-week effects, and holidays. Ignoring such calendar effects will lead to substantial forecasting errors. The simplest way to identify and filter out calendar effects is to replace each past observation $y_t, t = 1, \dots, T$, with an adjusted value $y'_t = w_t y_t$, where w_t is a suitably determined coefficient. If, for example, y_t is a monthly time series, coefficient $w_t, t = 1, \dots, T$, can be calculated as

$$w_t = \overline{n}/n_t, t = 1, \dots, T,$$

where $\overline{n}$ is the average number of workdays in a month, while n_t is the number of workdays during month t.

[Sotam.xlsx] Sotam is a Tunisian producer of orangeade. Table 2.11 shows the amount of fresh oranges (in quintals) used by its main plant over the last 10 weeks. The same table also indicates the workdays on which the plant was operating. The average number $\overline{n}$ of workdays in the 10 weeks was 4.80. Each weight $w_t, t = 1, \dots, 10$, can therefore be determined as $w_t = \overline{n}/n_t$. Once the weights are known, the modified values of the time series are easily determined, as shown in Table 2.12.

Table 2.11 Fresh orange consumption (in quintals) and number of workdays per week in the Sotam problem.

Week (t)	Sales (y_t)	Number of workdays	Week (t)	Sales (y_t)	Number of workdays
1	34 500	4	6	36 090	5
2	36 080	5	7	35 820	4
3	36 380	5	8	36 050	5
4	36 150	5	9	36 240	5
5	36 120	5	10	36 150	5

Table 2.12 Modified time series obtained by removing calendar variations in the Sotam problem.

t	w_t	y'_t	t	w_t	y'_t
1	1.20	41 400.0	6	0.96	34 646.4
2	0.96	34 636.8	7	1.20	42 984.0
3	0.96	34 924.8	8	0.96	34 608.0
4	0.96	34 704.0	9	0.96	34 790.4
5	0.96	34 675.2	10	0.96	34 704.0

2.5.5 Deflating Monetary Time Series

Inflation is a general increase in prices in an economy over a given time period. It causes the decline of purchasing power of a given currency over time. Inflation is a significant component of the apparent growth in any monetary time series, such as the sales of a finished product or the prices of a raw material. *Inflation adjustment* (or *deflation*) is the process of removing the effect of price inflation from monetary data. Let r_k be the average *inflation rate* in time period k. Inflation adjustment is accomplished by dividing each observation y_t of the monetary time series by a *price index* w_t which measures the variation in prices from time period 1 to time period t. The deflated series $y'_t = y_t/w_t$ is said to be measured in *constant currency* (dollars, euros, etc.), whereas the original series is measured in *nominal* or *current currency* (dollars, euros, etc.). Because of the capitalization effect, the price index is expressed as

$$w_t = \prod_{k=1}^{t-1}(1 + r_k), t = 2, \dots, T.$$
(2.2)

and $w_1 = 1$.

[Cavis.xlsx] Cavis is a wine-making company that sells its products almost exclusively in France. The annual sales (in M€) over the last 10 years are reported in Table 2.13. The same table also shows the annual rate of inflation recorded in the decade. The deflated data, obtained with price index formula (2.2), are reported in Table 2.14.

Table 2.13 Annual Cavis sales and the corresponding annual inflation rates.

Year	Sales [M€]	Inflation rate [%]	Year	Sales [M€]	Inflation rate [%]
1	1.03	2.80	6	1.31	2.10
2	1.13	2.50	7	1.36	1.80
3	1.20	2.70	8	1.29	3.30
4	1.24	2.20	9	1.33	0.80
5	1.26	2.00	10	1.37	1.50

Table 2.14 Price index and annual deflated sales in the Cavis problem.

Year	Price index [%]	Deflated sales [M€]	Year	Price index [%]	Deflated sales [M€]
1	100.00	1.03	6	112.81	1.16
2	102.80	1.10	7	115.18	1.18
3	105.37	1.14	8	117.25	1.10
4	108.21	1.15	9	121.12	1.10
5	110.60	1.14	10	122.09	1.12

2.5.6 Adjusting for Population Variations

When forecasting some economic variables such as sales in a certain geographic area, demographic variations need to be taken into account. Let a_t be the population of a given market in time period $t = 1, \ldots, T$ and let $y_t, t = 1, \ldots, T$, be the time series. Then, forecasts are devised on $y'_t = y_t/w_t$, where w_t is

$$w_t = a_t/a_1, t = 1, \ldots, T.$$

[Salus.xlsx] Salus is a private company providing home care services for the elderly in the Lombardy region, Italy. The annual number of customers over the past decade is shown in Table 2.15. Considering the annual population of Lombardy (second and sixth columns of Table 2.16) over the same 10 years, the modified time series is obtained as shown in Table 2.16 (fourth and eighth columns).

Table 2.15 Annual number of Salus customers.

Year	Number of customers	Year	Number of customers
1	1435	6	5056
2	2887	7	5432
3	3450	8	5382
4	4578	9	5920
5	4935	10	6003

Table 2.16 Adjustment of the number of Salus customers for variation of the population.

t	a_t	w_t	y'_t	t	a_t	w_t	y'_t
1	9 121 714	1.000	1435	6	9 475 202	1.039	4867
2	9 033 602	0.990	2915	7	9 545 441	1.046	5191
3	9 108 645	0.999	3455	8	9 642 406	1.057	5091
4	9 246 796	1.014	4516	9	9 742 676	1.068	5543
5	9 393 092	1.030	4792	10	9 826 141	1.077	5573

2.5.7 Data Normalization

The normalization of a time series $y_t, t = 1, \ldots, T$, is obtained by transforming it into a time series $y'_t, t = 1, \ldots, T$, whose observations belong to a given interval $[min, max]$ (where, of course, $min < max$). Each entry $y_t, t = 1, \ldots, T$ will correspond to

$$y'_t = \frac{y_t - a}{b - a}(max - min) + min,$$

where a and b represent the minimum and maximum values of y_t, $t = 1, ..., T$, respectively. The most common normalization is the $[0, 1]$-normalization.

[Sotam.xlsx] The time series of the Sotam problem shown in Table 2.12 (third and sixth columns), when normalized in interval $[2000, 4000]$, will result in Table 2.17.

Table 2.17 Sotam time series normalized in interval $[2000, 4000]$.

t	y'_t	t	y'_t
1	3621.78	6	2009.17
2	2006.88	7	4000.00
3	2075.64	8	2000.00
4	2022.92	9	2043.55
5	2016.05	10	2022.92

2.6 Classification of Time Series

If the percentage of zero values exceeds a given threshold (usually 30%), then the time series is called *intermittent* (or *sporadic*) (see Figure 2.7). Otherwise, it is called *continuous* (see Figure 2.8). Typical intermittent time series are those representing the sales of low-demand products. If zero values alternate regularly with non-zero observations, then the intermittent time series is called *periodic*, otherwise it is called *random*.

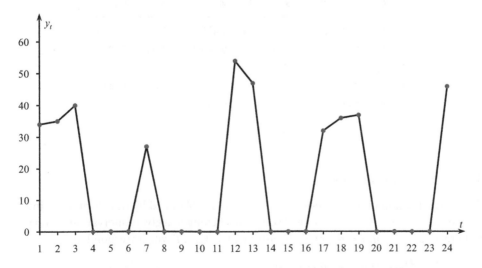

Figure 2.7 Plot of an intermittent time series.

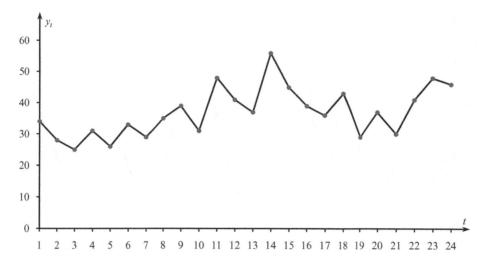

Figure 2.8 Plot of a continuous time series.

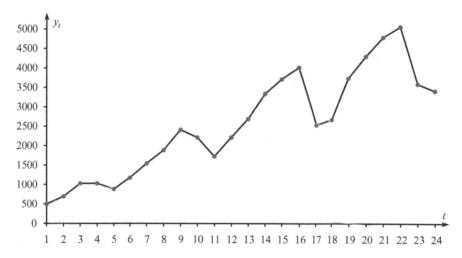

Figure 2.9 Plot of a regular time series.

Moreover, a time series is said to be *regular* if it can be decomposed into four main components: *trend*, *cyclical component*, *seasonal component* and *random component* (see Figure 2.9). Otherwise, it is defined as *irregular* (see Figure 2.10).

- *Trend.* The trend is the long-term modification of data pattern over time; it may depend on changes in population and on the product- (or service-) life cycle (see Figure 2.11).
- *Cyclical component.* This represents long-term fluctuations due to the so-called *business cycle*, which depends on external macroeconomic issues. A cyclical component usually extends over time periods from 3 to 10 years, with an average of 58.4 months since 1945. It is composed of four phases: prosperity (or boom), recession, depression, and recovery.

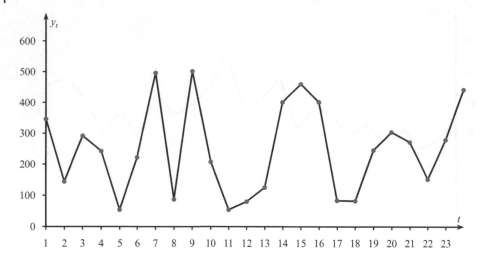

Figure 2.10 Plot of an irregular time series.

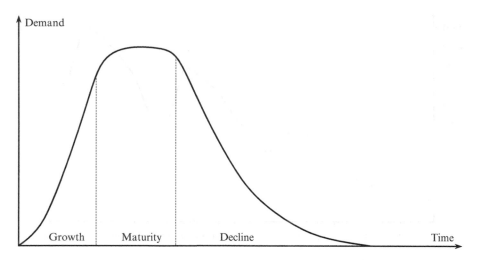

Figure 2.11 Life cycle of a product or service.

- *Seasonal component*. This represents repeating occurrences in certain time periods in a cyclical manner and is caused by the periodicity of several human activities. Typical examples include the sharp rise in most retail sales around December in response to the Christmas period, or the increase in ice cream consumption in the summer due to the warmer weather. This type of effect can also be observed on a weekly cycle, some product sales being higher on weekends than on workdays. In Figure 2.9 the length of the seasonal cycle (*periodicity*) is equal to six time periods.
- *Random component*. Also called *residual component* or *noise*, it is the irregular component of the historical data that cannot be interpreted as any of the three previous components. It is often the result of numerous causes, each of which are unpredictable and have a small impact.

2.7 Explanatory Methods

As anticipated in Section 2.3.1, explanatory methods rely on the hypothesis that the outcome variable y is characterized by a correlation with the explanatory variables $x_j, j = 1, \dots, n$.

Explanatory methods include *regression* and *classification* techniques, the latter being part of ML. In regression methods, the outcome variable is numerical while in classification methods it is categorical. In both methods, explanatory variables $x_1, \dots, x_n$ may be numerical or categorical. In the following, explanatory variables are initially considered as numerical; then, in Section 2.7.3, the main results will be extended to categorical predictors.

In explanatory methods, the relationship between the outcome variable and the explanatory variables can be expressed *ideally* by using a function $f(x_1, \dots, x_n; w_1, \dots, w_n)$ parametrized in $w_1, \dots, w_n$. Once the values of the weights $w_1, \dots, w_n$ are known it is possible to determine the forecast $\hat{y}$ of the outcome variable. By letting $\hat{x}_j, j = 1, \dots, n$, be the value of the jth explanatory variable that can be observed in advance (or can be set) by the decision maker, $\hat{y}$ is given by

$$\hat{y} = f(\hat{x}_1, \dots, \hat{x}_n; w_1^*, \dots, w_n^*), \tag{2.3}$$

where $w_1^*, \dots, w_n^*$ correspond to the optimal values for parameters $w_j, j = 1, \dots, n$ (determined as shown below).

2.7.1 Forecasting with Regression

A quite general regression model can be formulated as follows

$$Y = f(x_1, \dots, x_n; w_1, \dots, w_n) + \epsilon, \tag{2.4}$$

where the outcome (random) variable is represented by Y and ϵ is a random variable accounting for the uncertain effects.

In the linear regression case, function $f(\cdot)$ is linear and model (2.4) takes the form:

$$Y = w_0 + w_1 x_1 + \dots + w_n x_n + \epsilon, \tag{2.5}$$

where the explanatory variables are also called *regressors*.

Linear regression is fundamental since the hypothesis of linearity often holds in logistics forecasting. It is also relevant because a number of non-linear regression models (including polynomial and exponential models) can be transformed into a linear model, as shown in the following.

Equation (2.5) can be conveniently reformulated in the following vectorial form

$$Y = \mathbf{w}^{\mathsf{T}} \mathbf{x} + \epsilon,$$

where $\mathbf{x} = [x_0; x_1; \dots; x_n]^{\mathsf{T}}$ is the (column) vector of the explanatory variables (including a *dummy variable* x_0 whose value is always equal to 1) and $\mathbf{w} = [w_0; w_1; \dots; w_n]^{\mathsf{T}}$ is the (column) parameter vector.

Recall that, for each explanatory variable and for the outcome variable, T past observations are available. They are used as a *training set* in order to determine the optimal parameter vector $\mathbf{w}$, as shown in the following.

Let $\mathbf{X}$ be the $T \times (n+1)$ matrix made up of the columns of the dataset corresponding to the explanatory variables (including a dummy column made up of T 1s associated with x_0) and let $\mathbf{y}$ be the column of the dataset corresponding to the outcome variable.

According to the linear regression hypothesis, $w_0 + w_1 x_{t1} + \cdots + w_n x_{tn}$ can be seen as the forecast when the explanatory variables are equal to $x_{t1}, \ldots, x_{tn}$, while the corresponding observed value is y_t, $t = 1, \ldots, T$. Consequently, the forecasting errors made on the training set are

$$e_t = y_t - (w_0 + w_1 x_{t1} + \cdots + w_n x_{tn}), t = 1, \ldots, T. \tag{2.6}$$

In vectorial form, they can be expressed collectively as:

$$\mathbf{e} = \mathbf{y} - \mathbf{X}\mathbf{w}.$$

The optimal parameter vector $\mathbf{w}^*$ is determined by using the *least-squares method*, which consists of minimizing the *sum of squared errors (SSE)* made on the training set:

$$\underset{\mathbf{w}}{\text{Minimize}}\ SSE = \sum_t e_t^2 = \mathbf{e}^{\mathrm{T}}\mathbf{e}$$

$$= (\mathbf{y} - \mathbf{X}\mathbf{w})^{\mathrm{T}}(\mathbf{y} - \mathbf{X}\mathbf{w}) = \mathbf{y}^{\mathrm{T}}\mathbf{y} - 2\mathbf{y}^{\mathrm{T}}\mathbf{X}\mathbf{w} + \mathbf{w}^{\mathrm{T}}\mathbf{X}^{\mathrm{T}}\mathbf{X}\mathbf{w}.$$

A necessary condition for the *SSE* to be minimum is that the gradient of the *SSE*, with respect to $\mathbf{w}$, is zero:

$$\nabla_{\mathbf{w}} SSE = -2\mathbf{X}^{\mathrm{T}}\mathbf{y} + 2\mathbf{X}^{\mathrm{T}}\mathbf{X}\mathbf{w} = \mathbf{0}.$$

The solution of this system of linear equations (*normal equations*) is unique if square matrix $\mathbf{X}^{\mathrm{T}}\mathbf{X}$ is non-singular (i.e., its determinant $|\mathbf{X}^{\mathrm{T}}\mathbf{X}| \neq 0$), in which case

$$\mathbf{w}^* = (\mathbf{X}^{\mathrm{T}}\mathbf{X})^{-1}\mathbf{X}^{\mathrm{T}}\mathbf{y}. \tag{2.7}$$

Hence, in the linear regression case, the forecast $\hat{y}$ given by equation (2.3) corresponds to

$$\hat{y} = w_1^* \hat{x}_1 + \ldots + w_n^* \hat{x}_n. \tag{2.8}$$

[Ravaioli.xlsx, Ravaioli.csv, Ravaioli.py] In order to plan the inventories of spinach ravioli for the next month, the logistics manager of Ravaioli wants to make a one-period-ahead sales forecast. They know that the company plans to invest $\hat{x}_1 = 10$ k€ in TV advertising and $\hat{x}_2 = 10$ k€ in social media advertising for the next month. As a first step, the manager determines the optimal parameters of a linear regression model using formula (2.7). In Python, it corresponds to

$$w_0^* = 85.25, w_1^* = 64.09, w_2^* = 26.77.$$

Hence, by using formula (2.8),

$$\hat{y} = 85.25 + 64.09\hat{x}_1 + 26.77\hat{x}_2,$$

they derive a sales forecast equal to $(85.25 + 64.09 \times 10 + 26.77 \times 10) = 993.85$ k€. The optimal regression plane, as well as the errors made by the least-squares method on the training set, are illustrated in Figure 2.12.

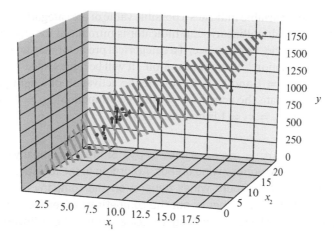

Figure 2.12 3D scatterplot of the Ravaioli problem with the optimal regression plane. For each error on the training set, the plot shows a segment joining the observed and the predicted values.

2.7.2 Multicollinearity

If matrix $\mathbf{X}^T\mathbf{X}$ is singular, no inverse $(\mathbf{X}^T\mathbf{X})^{-1}$ exists and it can be proved that the columns of the dataset corresponding to two (or more) predictors $x_0, x_1, \ldots, x_n$ are linearly dependent (*perfect multicollinearity*). In this case one of them can be removed. Finally, if the determinant $|\mathbf{X}^T\mathbf{X}|$ is close to zero, $(\mathbf{X}^T\mathbf{X})^{-1}$ is *ill-conditioned*, which can cause $\mathbf{w}^*$ to be highly sensitive to slight variations in the data (due to rounding error or slight variations in the training set).

[Ravaioli2.xlsx, Ravaioli2.csv, Ravaioli2.py] In the Ravaioli problem, assume that the past observations $x_{t2}, \ldots, x_{20,n}$ of the predictor x_2 are redefined as in the files `Ravaioli2.xlsx` or `Ravaioli2.csv`. The Pearson correlation coefficient between the two predictors becomes $r_{x_1,x_2} = 0.99$, indicating that the two variables are highly correlated. Based on formula (2.7), the linear regression model (2.5) becomes

$$Y = 266.76 - 15.99x_1 + 86.33x_2 + \epsilon.$$

The outcome variable is highly sensitive to slight variations in the data. For instance, if the first observation y_1 of the outcome variable were increased by just 30% (from 371.880 to 483.444), and all the other data remained the same, formula (2.5) would become

$$Y = 287.03 - 8.66x_1 + 76.87x_2 + \epsilon.$$

2.7.3 Categorical Predictors

The least-squares method presented in the previous sections can be extended to categorical predictors by introducing binary variables referred to as *indicator* (or *dummy*) variables. Each dummy variable indicates a possible value of the categorical variable. In order to avoid perfect multicollinearity, an indicator is introduced for any value of the categorical variable but one. For instance, if a categorical predictor may take values "no promotion", "light promotion", "heavy promotion", it will be represented by two binary indicators x_1 and x_2 ($x_1 = 1$ and $x_2 = 0$ indicate the activation of a light promotion; $x_1 = 0$ and $x_2 = 1$ indicate the activation of a heavy promotion; $x_1 = 0$ and $x_2 = 0$ indicate no promotion; $x_1 = x_2 = 1$ is meaningless).

[Sintra.xlsx, Sintra.csv, Sintra.py] Sintra is a company distributing a wide variety of Portuguese wine brands. According to the logistics manager, the sales of the Port and Douro wines depend on the company's advertising expenditure as well as on whether or not the competitors advertise their wines. After an explorative analysis, the manager decides to use the following linear regression model (2.5)

$$Y = w_0 + w_1 x_1 + w_2 x_2 + \epsilon, \tag{2.9}$$

where Y is the outcome variable representing the sales (in k€) of the Port and Douro wines, x_1 is the explanatory numerical variable corresponding to the company's advertising expenditure (in k€) and x_2 ($\in \{0, 1\}$) is an explanatory dummy variable indicating whether or not the competitors advertise their wines. The relevant past $T = 86$ observations were converted in such a way $x_{t2}, t = 1, \dots, 86$, correspond to binary values. The formula (2.7) provides $w_0^* = 928.95$, $w_1^* = 35.96$ and $w_2^* = -77.89$. Hence, assuming $\hat{x}_1 = 15.50$ k€ as the company's advertising expenditure planned for the next period and $\hat{x}_2 = 1$, which indicates that the competitors will not plan to invest in advertising their wines in the next period, the sales forecast of the Port and Douro wines for the next period is $\hat{y} = (928.95 + 35.96 \times 15.50 - 77.89 \times 1) = 1408.44$ k€.

2.7.4 Coefficient of Determination

A measure of how well the outcome variable is explained by the regression model is given by the *coefficient of determination* (or *R-squared*) R^2, defined as

$$R^2 = 1 - \frac{\sum_{t=1}^{T} e_t^2}{\sum_{t=1}^{T} (y_t - \bar{y})^2}, \tag{2.10}$$

where $e_t, t = 1, \dots, T$, given by formula (2.6), are the forecasting errors made by the regression model and $\bar{y}$ is the sample mean of $y_t, t = 1, \dots, T$, i.e., $\bar{y} = \frac{1}{T} \sum_{t=1}^{T} y_t$.

If parameters $w_0, w_1, \dots, w_n$ in formula (2.6) are determined by using formula (2.7), it can be proved that R^2 varies in the $[0, 1]$ interval. Of course, the higher the *R*-squared, the better the model fits the data. Ideally, if $e_t = 0, t = 1, \dots, T, \sum_{t=1}^{T} e_t^2 = 0$, then $R^2 = 1$. In contrast, when the regression model always predicts $\bar{y}$, it turns out that $e_t = y_t - \bar{y}$, $t = 1, \dots, T$, and, consequently, $R^2 = 0$. It can also be shown that in case of a single regressor, fitted by least squares, R^2 is the square of the Pearson correlation coefficient between the outcome variable and the unique predictor.

Expression (2.10) has a major drawback: R^2 increases as the number of predictors in the model goes up. Hence, one might be tempted to include some irrelevant factors because R^2 will never decrease as new explanatory variables are added. This practice (*kitchen-sink regression*) can be avoided by using the *adjusted R-squared* defined as

$$\hat{R}^2 = 1 - \frac{T-1}{T-n-1}(1 - R^2),$$

which penalizes the statistic when extra explanatory variables are included in the regression model (of course, the number of predictors n cannot be anyway equal or exceed the number T of observations minus one). $\hat{R}^2$ can be negative, and it will always be less than or equal to R^2. Unlike R^2, $\hat{R}^2$ increases only when the increase in R^2 due to the inclusion of a new predictor improves the forecasting accuracy.

[Sintra.xlsx, Sintra.csv, Sintra.py] In the Sintra problem, $R^2 = 0.45$, computed in Excel by using the RSQ function. Consequently, the adjusted R-squared is equal to 0.61. The logistics manager also evaluated the possibility to adopt the following linear regression model

$$Y = w'_0 + w'_1 x_1 + \epsilon, \tag{2.11}$$

without considering regressor x_2. In this case, formula (2.7) leads to $w'^*_0 = 886.77$ and $w'^*_1 = 36.57$ in equation (2.11). The corresponding adjusted R-squared is 0.43, showing that model (2.9) is more accurate than model (2.11).

2.7.5 Polynomial Regression

Sometimes the data are better explained by a non-linear model, e.g., by a simple quadratic regression model

$$Y = w_0 + w_1 x_1 + w_2 x_1^2 + \epsilon.$$

By setting $x_2 = x_1^2$ (and then adding the corresponding column to the dataset), the model becomes linear

$$Y = w_0 + w_1 x_1 + w_2 x_2 + \epsilon$$

and can be trained by using the least-squares method previously described. Similar transformations can be used for higher degree polynomial models, both simple and multiple.

[Alinef.xlsx, Alinef.csv, Alinef.py] Alinef manufactures heavy-duty process pumps in its plant located in Houston, USA. The company needs to predict the total production and logistics cost (in thousands of $) on the basis of the production volume (expressed in items). The data collected over 11 months are reported in Table 2.18. On the basis of the scatterplot in Figure 2.13, the logistics manager is convinced that the future costs can be forecast by using a quadratic regression model. Hence, they added a column given by

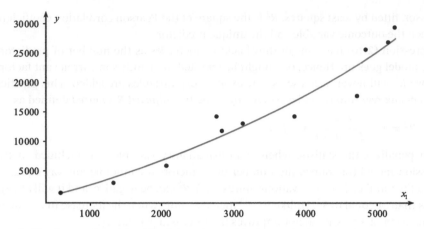

Figure 2.13 Scatterplot of the Alinef data along with the quadratic regression curve.

$x_{t2} = x_{t1}^2, t = 1, \ldots, 11$, to Table 2.18 and used the new (11×3) dataset to train the model

$$Y = w_0 + w_1 x_1 + w_2 x_2 + \epsilon.$$

The least-squares method, implemented in a `Python` library, provided parameter values $w_0^* = 2241.55$, $w_1^* = 2.48$ and $w_2^* = 0.000\,42$, which led to the quadratic regression model (see Figure 2.13)

$$y = 2241.55 + 2.48x_1 + 0.000\,42x_1^2.$$

Since next month the company plans to produce $\hat{x}_1 = 4200$ pumps, the total cost is expected to be $\hat{y} = 20\,066.35$ k$.

Table 2.18 Total production and logistics costs and production volume in the Alinef problem.

Month (t)	Costs [k$] ($y_t$)	Production volume [items] (x_{t1})
1	3 033	306
2	15 168	2628
3	4 650	1094
4	29 520	5294
5	25 998	4864
6	12 861	2702
7	18 459	4730
8	7 380	1886
9	14 055	3024
10	15 189	3788
11	26 904	5212

2.7.6 Linear–log, Log–linear and Log–log Regression Models

Other common non-linear regression models are the following

- linear–log model: $Y = w_0 + w_1 \ln x_1 + \epsilon$;
- log–linear model: $\ln Y = w_0 + w_1 x_1 + \epsilon$;
- log–log model: $\ln Y = w_0 + w_1 \ln x_1 + \epsilon$.

In all the three cases, a simple transformation allows the regression model to be linear and then trains it by using the least-squares method. For instance, for the linear–log regression model, by setting $x'_1 = \ln x_1$, the model becomes

$$Y = w_0 + w_1 x'_1 + \epsilon.$$

[UnitedPioneers.xlsx, UnitedPioneers.csv, UnitedPioneers-.py] United Pioneers is a venture capital company based in Palo Alto, USA. In 2021, one of its partners was evaluating a start-up company manufacturing an innovative drone specifically designed to distribute goods in urban areas. In order to assess the profitability of this investment, the partner needed to estimate the world's urban population in 2030. They would like to exploit the relationship between the share of population living in urban areas (y) and the GDP per capita $(x_1$, in $)$. The dataset is composed of 162 observations and, as shown in Figure 2.14, the two variables y and x_1 are not linearly correlated. However, by taking the logarithm of the GDP per capita, the plot exhibits a linear trend (see Figure 2.15). This led to the least-squares method being adopted. The resulting linear equation, depicted in Figure 2.15, is

$$y = -70.70 + 14.17 \ln x_1.$$

As a result, assuming a GDP annual average growth rate around 3%, a country with a GDP per capita equal to \$ 10 000 would have a GDP equal to \$ $[10\,000 \times (1 + 0.03)^{(2030-2021)}] = \$\,13\,048$ in 2031. Then, the partner estimated the future share of urban population in the country as $-70.70 + 14.17 \times \ln 13\,048 = 63.55\%$ (while, in 2021, it was 59.81%). The logarithmic regression curve is reported in Figure 2.16.

2.7.7 Underfitting and Overfitting

When a regression model is "too simple", it does not capture the essence of the relationship between the outcome variable and the predictors (*underfitting*). This gives rise to systematic errors, and predictions suffer a large *bias*. On the other hand, if the model is "too complex" (i.e., it has too many parameters with respect to the size of the training set), data are interpolated. From another perspective, the model memorizes the training data and is not able to perform good predictions for unseen instances. This produces predictions with a large variance (*overfitting*).

The choice of the "right" regression model can be done by randomly splitting the available data into a *training set* and a *validation set*. The former is used to train each candidate model while the latter is used to assess the accuracy of the forecasts by using

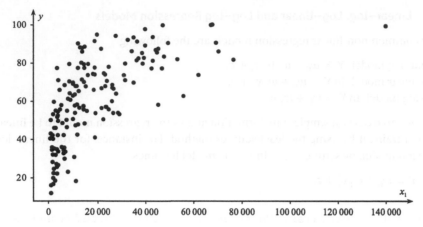

Figure 2.14 Scatterplot of the share of population living in urban areas versus the GDP per capita (in $) in the United Pioneers problem.

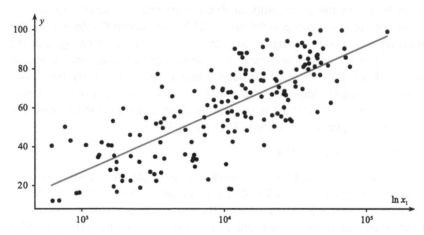

Figure 2.15 Scatterplot of the share of population living in urban areas versus the GDP per capita (in $, logarithmic scale) in the United Pioneers problem. The picture also reports the corresponding linear regression line.

the accuracy measures described in Section 2.9. According to a rule of thumb, an 80/20 split between training and validation sets is advisable.

[Alinef.xlsx, Alinef.csv, Alinef.py] In the Alinef problem, a polynomial regression model of degree six,

$$Y = w_0 + w_1 x_1 + w_2 x_1^2 + w_3 x_1^3 + w_4 x_1^4 + w_5 x_1^5 + w_6 x_1^6 + \epsilon,$$

would surely lead to overfitting since the number of parameters (seven) would be comparable to the number of observations in the training set (11). Indeed, a plot of the corresponding regression curve (see Figure 2.17), obtained by using the least-squares method, shows that the model tries to interpolate the data.

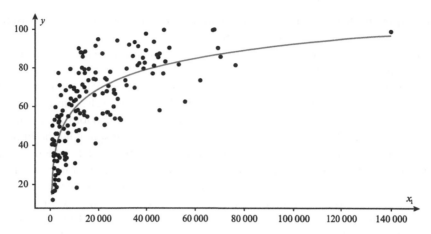

Figure 2.16 Scatterplot of the share of population living in urban areas versus the GDP per capita (in $) in the United Pioneers problem. The picture also shows the corresponding logarithmic regression curve.

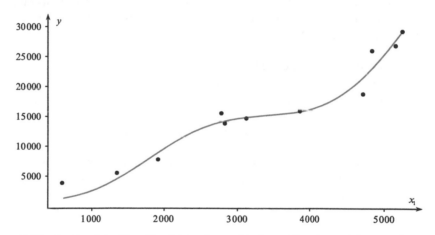

Figure 2.17 Scatterplot of the Alinef data along with the sextic polynomial regression curve.

2.7.8 Forecasting with Machine Learning

ML is a subfield of *artificial intelligence* that deals with the development of algorithms that *learn* from input data and improve their performance without being explicitly programmed. Formally, given a dataset D, a task T and a performance measure M, a computer program is said to learn from D to perform T if, *after a training stage*, the system's performance on T improves as measured by M. In other words, the program performs T better as compared to no learning.

ML has being used for speech and image recognition, medical diagnoses, credit checks, fraud detections, and making predictions.

In *supervised learning*, ML algorithms are trained on a *labelled* dataset (i.e., a dataset for which the output data corresponding to the input data are made available). Based on this dataset, the algorithms try to infer the relationship between the input data and the corresponding output data. In *unsupervised learning*, the data are unlabelled and the

ML algorithms try to extract features and patterns on their own. Supervised learning includes regression and classification methods.

ML include a wide range of methods such as the *naïve Bayes method, classification and regression trees* (CARTs), *random forests, support vectors machines* (SVMs) and *artificial neural networks* (ANNs). Some of them can be used as both regression and classification methods while others are more naturally inclined to classification.

In this section, the focus is on ANNs (or, simply, *neural networks*) for their ability to model problems of different types. ANNs (whose structure is inspired by the human brain, and, in particular, by biological neurons) can deal with any data which can be made numerical and are suitable for both regression and classification. Another feature is that, once trained, an ANN can make predictions pretty quickly. In contrast, the forecasts provided by ANNs are difficult to interpret since it is not straightforward to figure out how much each independent variable influences the outcome variable. Moreover, ANNs work best with large datasets and are time consuming to train with traditional CPUs.

An ANN is a layered architecture of *artificial neurons* (or, simply, *neurons* or *nodes*), each of which has the capability to process its inputs $x_1, \ldots, x_n$ and forward its output $y = \varphi(w_0 + w_1 x_1 + \ldots + w_n x_n)$ to other neurons or externally. The basic model of an artificial neuron (see Figure 2.18) consists of three units:

- *synaptic links* (*arcs*), characterized by weights $w_0, w_1, \ldots, w_n$ associated with the $x_1, \ldots, x_n$ inputs as well as to a dummy input x_0 constantly equal to 1 (*bias*);
- an *adder*, computing *potential* $u = w_0 + w_1 x_1 + \ldots + w_n x_n$ by summing weighted inputs;
- an *activation function* $\varphi(u)$, determining the output $y = \varphi(u)$.

The four more common activation functions are shown in Figures 2.19 and 2.20. If the linear activation function is selected (see Figure 2.19(a)), the neuron reduces to a linear regression model.

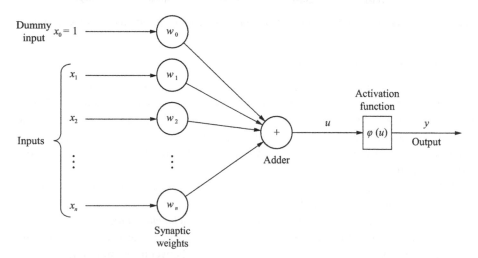

Figure 2.18 An artificial neuron.

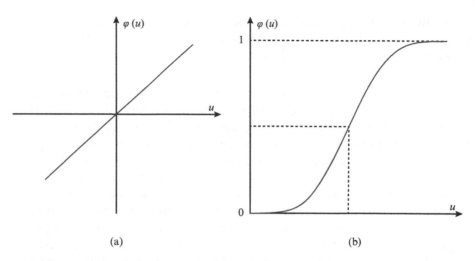

Figure 2.19 Two activation functions of an artificial neuron: (a) linear function $\varphi(u) = u$; (b) sigmoid function $\varphi(u) = \frac{1}{1+e^{-au}}$.

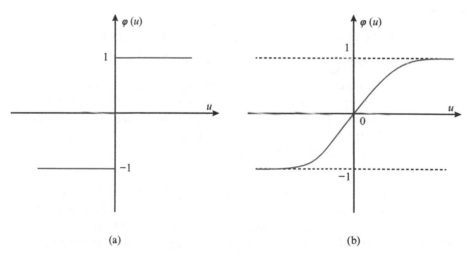

Figure 2.20 Other activation functions of an artificial neuron: (a) threshold function $\varphi(u) = sign(u)$ ($sign(u) = 1$, if $u \geq 0$, $sign(u) = -1$, otherwise); (b) hyperbolic tangent function $\varphi(u) = \tanh(\gamma u) = \frac{1-e^{-2\gamma u}}{1+e^{-2\gamma u}}$.

Neural networks can be divided into *feedforward neural networks*, in which connections between neurons do not form cycles, and *recurrent neural networks* which contain one or more feedback loops. In particular, *single-layer feedforward networks* consist of a single layer of *hidden* neurons connecting the input neurons to the output neurons, while *multi-layer feedforward networks* contain more layers of hidden neurons. If the input neurons do not modify their inputs then they may be omitted.

As for linear regression, the training problem aims to determine the weights vector **w** of the network in such a way that training data are best fitted. If the least-squares method (see Section 2.7.1) is used, the training problem amounts to minimizing the *SSE* (whose square is defined as *square loss* in this context) on the training set.

To this end, several algorithms are available, including general purpose nonlinear algorithms (the *gradient descent method*, the *Newton method*, the *conjugate gradient method*, *quasi-Newton methods* including the *Broyden–Fletcher–Goldfarb–*

Shanno (BFGS) *method*, etc.) and tailored methods such as the *backpropagation algorithm*. A description of such methods goes beyond the scope of this book. The interested reader is referred to specialized literature.

[Lmart.csv, Lmart.py] Lmart is a grocery retail chain with hundreds of sales point in the Netherlands, Belgium, and Germany. The logistics manager needs to predict the sales of blood oranges (imported from Sicily, Italy) as a function of the price discount offered to clients owning a fidelity card. The company has computed the percentage increase Δq in sales of blood oranges and the corresponding percentage discount Δp_1 for 50 weeks. The data (see Figure 2.21) show a strong non-linearity. After a preliminary study, the manager made use of a single-layer feedforward neural network with a single input (Δp_1), three neurons in the hidden layer and a single output neuron, whose output is Δq (see Figure 2.22). The activation function of the hidden neurons was the hyperbolic tangent (see Figure 2.20(b)) while the activation function of the output neuron is linear.

The network was trained with the BFGS algorithm which provided the following parameters:

$$w_1^* = 16.38, w_2^* = 0.09, w_3^* = -1.22, w_4^* = 0.45, w_5^* = -2.38;$$

$$w_6^* = -0.99, w_7^* = -78.34, w_8^* = 76.40, w_9^* = -78.04, w_{10}^* = 77.88.$$

The square loss equals 191.7, indicating a *SSE* equal to 13.85 on the training set. The sales of oranges predicted by the ANN as a function of the price discount are reported in Figure 2.21 (*response curve*). In particular, the neural network predicted $\Delta q = 8.8\%$ for $\Delta p_1 = 10\%$, $\Delta q = 38.7\%$ for $\Delta p_1 = 20\%$, $\Delta q = 142.6\%$ for $\Delta p_1 = 40\%$ and $\Delta q = 153.7\%$ for $\Delta p_1 = 60\%$.

[Lmart.csv, Lmart2.py] After a while, the logistics manager realized that the sales of blood oranges were related not only to the price discount on the product itself Δp_1, but also to the discount Δp_2 on the price of tangerines. They then built an enhanced ANN including both Δp_1 and Δp_2 as inputs. After a preliminary study, the manager made use of a single-layer feedforward neural network with five neurons in the hidden layer and a single output neuron, whose output is Δq (see Figure 2.23). Again, the activation function of the hidden neurons was the hyperbolic tangent (see Figure 2.20(b)) while the activation function of the output neuron is linear. The square loss equal to 88.9, indicating a *SSE* equal to 9.4 on the training set, less than the *SSE* characterizing the first ANN with a single input (Δp_1). The sales of blood oranges predicted by the ANN as a function of the price discount are reported in Figure 2.24 (*response surface*). In particular, for a discount $\Delta p_1 = 30\%$ on blood oranges, the neural network predicted $\Delta q = 135.6\%$ for $\Delta p_2 = 0\%$, $\Delta q = 120.8\%$ for $\Delta p_2 = 15\%$, $\Delta q = 99.6\%$ for $\Delta p_2 = 40\%$ and $\Delta q = 74.5\%$ for $\Delta p_2 = 45\%$, indicating that the sales of blood oranges decrease as the discount on tangerines increases.

It is worth noting that the selected network, characterized by five hidden neurons, contains $10 + 6 = 16$ weights. A network with 20 hidden neurons would have $40 + 21 = 61$ weights, more than the number of training observations (50), which would lead inevitably to overfitting.

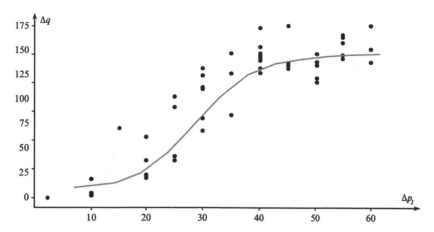

Figure 2.21 Scatterplot of the Lmart problem data along with the *response curve* provided the single-layer feedforward neural network in Figure 2.22 trained with the BFGS algorithm.

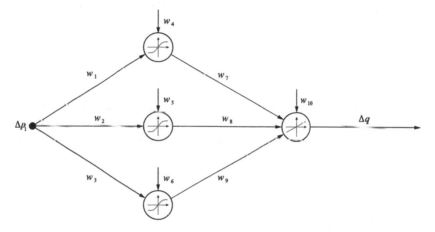

Figure 2.22 Single-layer feedforward neural network for the Lmart problem. Note the absence of the input neuron, since input Δp_1 does not need to be processed before being sent to the three neurons of the hidden layer.

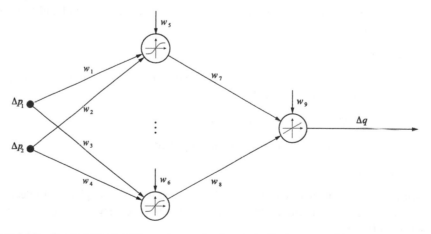

Figure 2.23 Single-layer feedforward neural network for the Lmart problem.

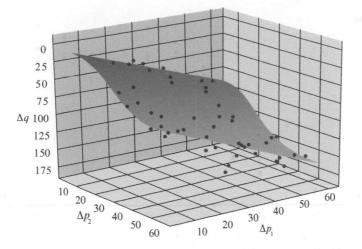

Figure 2.24 3D scatterplot of the Lmart problem data along with the *response surface* provided by the single-layer feedforward neural network in Figure 2.23 trained with the BFGS algorithm.

2.8 Extrapolation Methods

As indicated in Section 2.3.1, extrapolation methods are customarily used when no explanatory variable, having a high correlation with the outcome variable, can be observed in advance or set by the decision maker. This section describes the most common extrapolation methods in logistics.

2.8.1 Notation

In this section the following notation is used. Let

$$p_t(\tau), \tau = 1, \dots,$$

be the τ-period-ahead forecast made at time period t, i.e., the forecast for time period $t + \tau$ made at time period t. For the sake of simplicity, the one-period-ahead forecast ($\tau = 1$) is denoted as

$$p_t(1) = p_{t+1}.$$

Once y_t becomes known at time period t, it can be compared with the forecast $p_i(\tau)$ (with $i + \tau = t$) made τ time periods before, in order to compute the error

$$e_i(\tau) = y_t - p_i(\tau), \text{with } i + \tau = t.$$

Again, if $\tau = 1 (i = t - 1)$, the notation can be simplified to

$$e_{t-1}(1) = e_t = y_t - p_t.$$

Forecasting errors themselves constitute a time series. As better explained in Section 2.9, the errors can be used cumulatively to evaluate *retrospectively* the accuracy of the forecasting methods.

2.8.2   Decomposition Method

The decomposition method can be used for continuous and regular time series, which can be decomposed into trend, cyclical, seasonal, and random components (see Section 2.6). These components can be combined in a *multiplicative* model:

$$y_t = q_t v_t s_t r_t, t = 1, \dots, T, \tag{2.12}$$

where q_t represents the trend at time period t, v_t is an index representing the business cycle at time period t, s_t is the seasonal index at time period t while r_t is the index corresponding to the random component at time period t.

In model (2.12), q_t is expressed in the same unit of measurement as $y_t, t = 1, \dots, T$, while indices v_t, s_t, and $r_t, t = 1, \dots, T$, are dimensionless and non-negative. Of course, if one of the components is absent then the corresponding index is assumed to be equal to 1. Let M be the periodicity of the seasonal cycle. Then the average of the seasonal indices in M consecutive time periods should be 1:

$$\frac{\sum_{t=j+1}^{j+M} s_t}{M} = 1, j = 0, \dots, T - M.$$

Alternatively, the four components can be combined as follows

- additive model:

$$y_t = q_t + v_t + s_t + r_t, t = 1, \dots, T;$$

- mixed model:

$$y_t = (q_t + v_t)s_t r_t, t = 1, \dots, T;$$

- logarithmic model:

$$\ln y_t = \ln q_t + \ln v_t + \ln s_t + \ln r_t, t = 1, \dots, T.$$

The logarithmic model can be seen as the logarithmic transformation of the multiplicative model, used to transform the multiplicative model into an additive one.

As a rule, if the cyclical, seasonal, and random components do not depend on the trend, then the most suitable model is the additive one. Otherwise, a multiplicative model is preferable, which is often the case of most economic time series. In the following, we will refer to the multiplicative model. The decomposition method is divided into the following three steps.

1. time series $y_t, t = 1, \dots, T$, is broken down into its four components $q_t, v_t, s_t, r_t, t = 1, \dots, T$;
2. q, v and s are projected one or more time periods ahead;
3. these projections are combined in the following way

$$p_T(\tau) = q_T(\tau)v_T(\tau)s_T(\tau), \tau = 1, \dots, \tag{2.13}$$

to produce forecasts for the required future time periods $T + \tau, \tau = 1, \dots$.

Phase 1 is made up of the following steps.

Evaluating the Combined Trend-cyclical Component

The combined trend-cyclical component is obtained by removing from $y_t, t = 1, \dots, T$, the seasonal and the random components. This can be done by observing that the average value of M consecutive observations does not include any seasonal effect. Moreover, the influence of the random component is quite low, especially if M is sufficiently large (e.g., $M = 12$). Thus, the time series

$$\frac{y_1 + \cdots + y_M}{M}, \dots, \frac{y_{T-M+1} + \cdots + y_T}{M}, \tag{2.14}$$

includes only the trend and the cyclical components.

If M is an odd number, each element of time series (2.14) corresponds to a time period which is the median of the time periods at the numerator. In particular, the first element of (2.14) corresponds to time period $t = \lceil \frac{M}{2} \rceil$. Hence, the qv series is

$$(qv)_{\lceil \frac{M}{2} \rceil} = \frac{y_1 + \dots + y_M}{M},$$

$$\dots$$

$$(qv)_{T-\lceil \frac{M}{2} \rceil+1} = \frac{y_{T-M+1} + \dots + y_T}{M}.$$

If M is an even number, the elements of (2.14) do not refer to a well-defined time period. For instance, if $M = 12$, the first element of the time series (2.14) would correspond to a time period included between the sixth and the seventh time periods. This drawback can be overcome by weighting the observations $y_{t-M/2}$ and $y_{t+M/2}$ with a coefficient equal to $1/2$, so the following time series can be obtained

$$(qv)_t = \frac{\frac{1}{2}y_{t-M/2} + y_{t-M/2+1} + \dots + y_{t+M/2-1} + \frac{1}{2}y_{t+M/2}}{M}, \tag{2.15}$$

$$t = \frac{M}{2} + 1, \dots, T - \frac{M}{2}.$$

Separating the Trend and the Cyclical Component

In most cases, the trend can be described by a simple functional form, such as a linear or quadratic function. It can, therefore, be obtained by a simple regression model applied to the time series $(qv)_t, t = j, \dots, J$, where $j = \lceil \frac{M}{2} \rceil$ and $J = T - \lceil \frac{M}{2} \rceil + 1$, if M is odd; $j = \frac{M}{2} + 1$, and $J = T - \frac{M}{2}$, if M is even. For example, if the trend is linear then

$$q_t = a + bt, t = j, \dots, J,$$

where coefficients a and b can be obtained by using the least-squares method (see Section 2.7.1). Known $q_t, t = j, \dots, J$, the cyclical component $v_t, t = j, \dots, J$, can be calculated for every $t = j, \dots, J$ as

$$v_t = \frac{(qv)_t}{q_t}. \tag{2.16}$$

Evaluating the Combined Seasonal-random Component

The combined seasonal-random component $(sr)_t, t = j, \dots, J$, can be obtained as

$$(sr)_t = \frac{y_t}{(qv)_t}. \tag{2.17}$$

Separating the Seasonal and the Random Components

The seasonal component can be expressed by means of M indices $\bar{s}_1, \dots, \bar{s}_M$, satisfying

$$s_{kM+t} = \bar{s}_t, t = 1, \dots, M, k = 0, 1, \dots$$

Each index $\bar{s}_t, t = 1, \dots, M$, can be computed as the average of the elements of the time series $(sr)_t, t = j, \dots, J$, corresponding to *homologous* time periods. Time periods are homologous when they are equidistant from each other by M time periods, so that $\bar{s}_t, t = 1, \dots, M$, is the arithmetic mean of $(sr)_t, (sr)_{M+t}, \dots$. As observed previously, the arithmetic mean reduces the random effect noticeably. Hence,

$$\frac{\sum_{t=1}^{M} \bar{s}_t}{M} = 1. \tag{2.18}$$

Whenever relationship (2.18) is not satisfied, instead of $\bar{s}_t$, the following normalized index can be used

$$\tilde{s}_t = \frac{M\bar{s}_t}{\sum_{t=1}^{M} \bar{s}_t}, t = 1, \dots, M.$$

Of course,

$$\frac{\sum_{t=1}^{M} \tilde{s}_t}{M} = 1.$$

Finally, the random component $r_t, t = j, \dots, J$, can be obtained by dividing each element of the time series $(sr)_t, t = j, \dots, J$, by the corresponding seasonal index $s_t, t = j, \dots, J$:

$$r_t = \frac{(sr)_t}{s_t}.$$

If the decomposition has been done correctly, the mean value of $r_t, t = j, \dots, J$, is close to 1.

Phase 2 leads to project q, v and s over one or more future time periods. The way in which time series q_t and $s_t, t = j, \dots, J$, were computed makes this particularly easy to do. The cyclical component is estimated in a qualitative way, on the basis of information about the business cycle, or more simply by letting

$$v_t(\tau) = v_J, \tau = 1, \dots,$$

that is, setting $v_t(\tau), \tau = 1, \dots$, equal to the current value of the cyclical component. This assumption is particularly suitable for short-term forecasts, since variations in the business cycle can be appreciated only after several months or even a few years.

Finally, forecasts can be obtained at the end of Phase 3 of the decomposition method by combining, according to (2.13), the extrapolation of the trend, cyclical, and seasonal components.

[P&A.xlsx, P&A.py] P&A is a consulting firm, headquartered in Hannover, Germany, which has been asked to estimate the demand for electromedical equipment in the Niedersachsen region for the next six months. By examining the historical data, the company was able to assess the monthly sales during the last 163 months. These data (in k€), already preprocessed (see P&A.xlsx), are shown in Figure 2.25. The periodicity *M* of the seasonal cycle is equal to 12. The time series $(qv)_t$ (see Table 2.19 and Figure 2.26) was obtained using formula (2.15) for $t = 7, \dots, 157$.

Table 2.19 Combined trend-cyclical component $(qv)_t, t = 7, \dots, 157$, in the P&A problem.

t	$(qv)_t$	t	$(qv)_t$
7	627.70	...	...
8	630.23	152	864.65
9	632.95	153	871.73
10	636.50	154	879.05
11	639.01	155	885.91
12	638.14	156	890.95
...	...	157	894.86

An examination of Figure 2.26 allowed the firm to assume that the trend is linear, that is,

$$q_t = a + bt, t = 7, \dots, 157,$$

where the coefficients *a* and *b*, determined by using the Excel functions INDEX and LINEST (see P&A.xlsx), are

$$a = 638.51, b = 1.43.$$

The corresponding regression line is depicted in Figure 2.27. The availability of $q_t, t = 1, \dots, 157$, allowed the computation of $v_t, t = 1, \dots, 157$, through (2.16) (see Table 2.20 and Figure 2.28). The combined seasonal-random indices $(sr)_t$, $t = 1, \dots, 157$, computed by using (2.17), are reported in Table 2.21 and depicted in Figure 2.29.

Table 2.20 Trend $q_t, t = 7, \dots, 157$, and cyclical index $v_t, t = 7, \dots, 157$, in the P&A problem.

t	$(qv)_t$	q_t	v_t	t	$(qv)_t$	q_t	v_t
7	627.70	648.52	0.97	...	...	...	...
8	630.23	649.96	0.97	152	864.65	856.02	1.01
9	632.95	651.39	0.97	153	871.73	857.45	1.02
10	636.50	652.82	0.98	154	879.05	858.88	1.02
11	639.01	654.25	0.98	155	885.91	860.31	1.03
12	638.14	655.68	0.97	156	891.95	861.74	1.03
...	...	...	...	157	894.86	863.17	1.04

Table 2.21 Combined seasonal-random index $(sr)_t, t = 7, \ldots, 157$, in the P&A problem.

t	y_t	$(qv)_t$	$(sr)_t$	t	y_t	$(qv)_t$	$(sr)_t$
7	676.90	627.70	1.08	$\cdots$	$\cdots$	$\cdots$	$\cdots$
8	661.50	630.23	1.05	152	891.10	864.60	1.03
9	611.80	632.95	0.97	153	882.00	871.70	1.01
10	640.50	636.50	1.01	154	887.60	879.10	1.01
11	611.10	639.01	0.96	155	840.00	885.90	0.95
12	697.20	638.14	1.09	156	935.90	891.00	1.05
$\cdots$	$\cdots$	$\cdots$	$\cdots$	157	763.70	894.90	0.85

The indices $\bar{s}_1, \ldots, \bar{s}_{12}$ (see Table 2.22 and Figure 2.30) verify condition (2.18), while the mean value of the random component is approximately 1 (see Table 2.23 and Figure 2.31). Therefore, the hypothesis on which the decomposition was carried out can be considered correct.

The forecasts of the demand for the six subsequent time periods (see Table 2.24 and Figure 2.33) were obtained by combining the projections of the trend as well

Table 2.22 Seasonal index $\bar{s}_t, t = 1, \ldots, 12$, in the P&A problem.

t	$\bar{s}_t$	t	$\bar{s}_t$
1	0.82	7	1.07
2	0.76	8	1.02
3	0.92	9	0.98
4	1.05	10	1.02
5	1.15	11	0.99
6	1.11	12	1.08

Table 2.23 Random index $r_t, t = 7, \ldots, 157$, in the P&A problem.

t	$(sr)_t$	s_t	r_t	t	$(sr)_t$	s_t	r_t
7	1.08	1.07	1.01	$\cdots$	$\cdots$	$\cdots$	$\cdots$
8	1.05	1.02	1.02	152	1.03	1.02	1.01
9	0.97	0.98	0.98	153	1.01	0.98	1.03
10	1.01	1.02	0.98	154	1.01	1.02	0.99
11	0.96	0.99	0.97	155	0.95	0.99	0.96
12	1.09	1.08	1.01	156	1.05	1.08	0.98
$\cdots$	$\cdots$	$\cdots$	$\cdots$	157	0.85	0.82	1.04

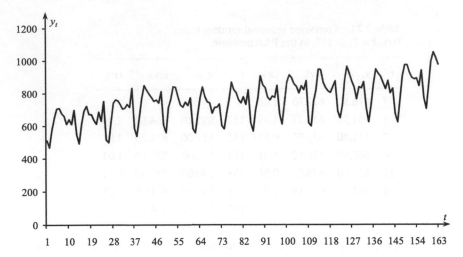

Figure 2.25 Plot of the P&A monthly sales of electromedical equipment (in k€).

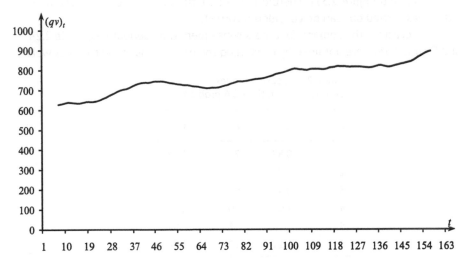

Figure 2.26 Plot of the combined trend-cyclical component $qv_t, t = 7, \dots, 157$, in the P&A problem.

Table 2.24 Demand forecast $p_T(\tau), T = 163, \tau = 1, \dots, 6$, in the P&A problem.

$T + \tau$	$q_T(\tau)$	$v_T(\tau)$	$s_T(\tau)$	$p_T(\tau)$
164	873.19	1.04	1.02	933.84
165	874.62	1.04	0.98	895.28
166	876.05	1.04	1.02	932.52
167	877.48	1.04	0.99	898.46
168	878.91	1.03	1.08	976.03
169	880.35	1.03	0.82	743.76

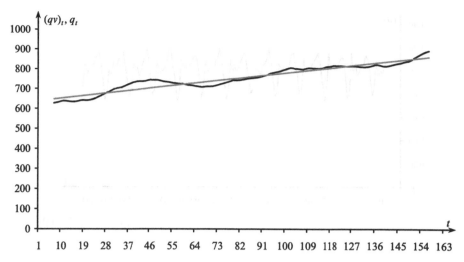

Figure 2.27 Linear trend (in grey) $q_t, t = 7, \ldots, 157$, in the P&A problem.

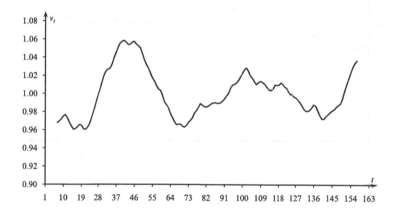

Figure 2.28 Plot of the cyclical component $v_t, t = 7, \ldots, 157$, in the P&A problem.

as the cyclical and seasonal components. The cyclical component was estimated (see Figure 2.32) by using a quadratic regression model, constructed on the basis of the values of $v_t, t = 150, \ldots, 157$:

$$v_t = a(t - 149)^2 + b(t - 149) + c, t = 150, \ldots$$

The parameters a, b, and c were obtained by using the least-squares method (through the `Excel Solver` tool under the `Data` tab) in time periods $t = 150, \ldots, 157$:

$$a = -0.0004, b = 0.0094, c = 0.9856.$$

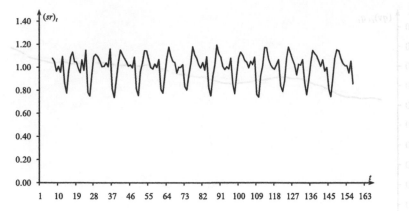

Figure 2.29 Plot of the combined seasonal-random component $(sr)_t$, $t = 7, \ldots, 157$, in the P&A problem.

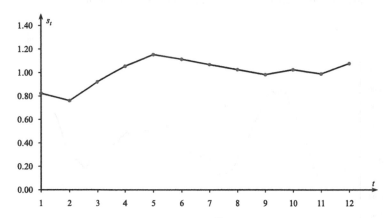

Figure 2.30 Plot of the seasonal component $s_t = \bar{s}_t$, $t = 1, \ldots, 12$, in the P&A problem.

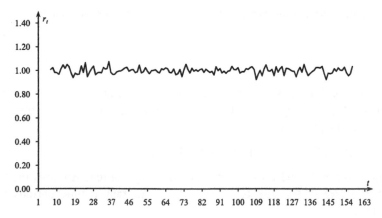

Figure 2.31 Plot of the random component r_t, $t = 7, \ldots, 157$, in the P&A problem.

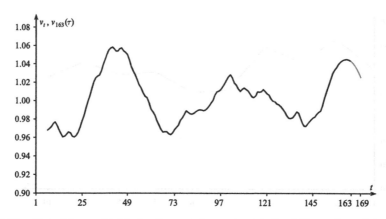

Figure 2.32 Plot of the cyclical-index forecast (in grey) $v_T(\tau)$, $T = 163$, $\tau = 1, \ldots, 6$, in the P&A problem.

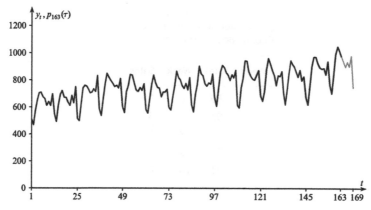

Figure 2.33 Plot of the P&A monthly demand forecast (in grey) of electromedical equipment in the next six months.

2.8.3 Further Extrapolation Methods: the Constant-trend Case

This section analyzes further extrapolation methods applicable when the time series does not show any relevant cyclical and seasonal components, and the trend is constant. First, the one-period-ahead forecast is considered.

Elementary Method

The forecast for the one-period-ahead is simply given by:

$$p_{T+1} = y_T.$$

Forecasts reproduce the data pattern with one-period delay. Consequently, this method usually produces rather poor predictions.

[RegensBooks.xlsx] Regens Books is an online bookseller with a DC located in Patton, USA. The sales of the last 10 weeks (in number of books) are reported in Table 2.25. The data plot (see Figure 2.34) shows that there is a constant trend. By using the elementary method, the logistics manager obtains

$$p_{11} = y_{10} = 86.$$

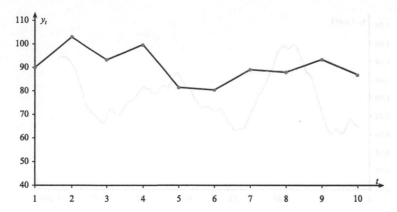

Figure 2.34 Plot of the weekly sales (in number of books) in the Regens Books problem.

Table 2.25 Weekly sales (in number of books) in the Regens Books problem.

Week	Sales	Week	Sales
1	89	6	80
2	101	7	88
3	92	8	87
4	98	9	92
5	81	10	86

Simple Moving Average Method

The simple moving average method uses the average of the $r \; (\geq 1)$ most recent data entries as the one-period-ahead forecast

$$p_{T+1} = \sum_{k=0}^{r-1} \frac{y_{T-k}}{r}.$$

Of course, if $r = 1$ this method reduces to the elementary method. A key aspect of the simple moving average method is the choice of parameter r. A small value of r allows a rapid adjustment of the forecast to data fluctuations but, at the same time, increases the influence of random perturbations. In contrast, a large value of r effectively filters the random components, but produces a slow adaptation to data variations. The optimal choice of r will be discussed in Section 2.9.

[RegensBooks.xlsx] Using the simple moving average method for solving the Regens Books problem, the following forecasts are obtained

$$p_{11} = \frac{y_{10} + y_9}{2} = 89,$$

and

$$p_{11} = \frac{y_{10} + y_9 + y_8}{3} = 88.33,$$

with $r = 2$ and $r = 3$, respectively. Of course, the value of p_{11} obtained with $r = 3$ should be rounded to the nearest integer (i.e., 88), since it refers to a number of books, but for the sake of simplicity, the forecast values are reported not rounded. This rule will be followed in the remainder of the chapter.

Weighted Moving Average Method

This method overcomes the limitations of the simple moving average method, which requires, as described above, that only the most recent $r \geq 1$ data (each with an equal weight of $1/r$) are used for the one-period-ahead forecast. In the weighted moving average method, the one-period-ahead forecast involves using all available data with a weight that decreases linearly as the data become more obsolete:

$$p_{T+1} = \frac{\sum_{t=1}^{T} t y_t}{\sum_{t=1}^{T} t}.$$

[RegensBooks.xlsx] By using the weighted moving average method to solve the Regens Books problem, the logistics manager obtains

$$p_{11} = \frac{y_1 + 2y_2 + 3y_3 + 4y_4 + 5y_5 + 6y_6 + 7y_7 + 8y_8 + 9y_9 + 10y_{10}}{1 + 2 + 3 + 4 + 5 + 6 + 7 + 8 + 9 + 10}$$

$$= \frac{4844}{55} = 88.07.$$

Exponential Smoothing Method

The exponential smoothing method (also called the *Brown method*) can be seen as an evolution of the simpler weighted moving average method. The one-period-ahead forecast is given by

$$p_{T+1} = \alpha y_T + (1 - \alpha) p_T, \tag{2.19}$$

where $\alpha \in (0, 1)$ is a *smoothing constant*. Here, p_T represents the forecast for time period T made at time period $T - 1$.

If $\alpha = 0.2$, $y_T = 1020$ and $p_T = 975$, the forecast for time period $T + 1$ is

$$p_{T+1} = 0.2 \times 1020 + (1 - 0.2) \times 975 = 984.$$

Rewriting equation (2.19) as

$$p_{T+1} = p_T + \alpha(y_T - p_T) = p_T + \alpha e_T,$$

allows the following interpretation to be obtained. The forecast for the time period $T + 1$ is equal to the latest forecast (i.e., the forecast made at time period $T - 1$ for time period T) *plus* a fraction α of the forecasting error made at time period T. Hence, if p_T is overestimated with respect to y_T, forecast p_{T+1} is lower than p_T; vice versa, when p_T is an underestimate of y_T, then p_{T+1} is increased with respect to p_T.

All the historical data are implicitly embedded into p_T. By applying equation (2.19) recursively, they can appear explicitly

$$p_T = \alpha y_{T-1} + (1 - \alpha)p_{T-1}.$$

From equation (2.19), it turns out that

$$p_{T+1} = \alpha y_T + (1 - \alpha)[\alpha y_{T-1} + (1 - \alpha)p_{T-1}]. \tag{2.20}$$

In equation (2.20), p_{T-1} can be substituted by rewriting equation (2.19) for p_{T-1}, and so on, backward to p_2, whose value can be set equal to y_1, as occurs for the elementary method and the simple and weighted moving average methods. In this way, past data entries $y_T, \ldots, y_2$ are multiplied by exponentially decreasing weights (this is why the method is called exponential smoothing)

$$p_{T+1} = \alpha \sum_{k=0}^{T-2} (1 - \alpha)^k y_{T-k} + (1 - \alpha)^{T-1} y_1. \tag{2.21}$$

It is worth noting that the sum of all weights in equation (2.21) is equal to 1. Moreover, the weight of y_1 is less than the weight of y_2, and less than the weight of all the other remaining data entries, if and only if $\alpha \in (1/2, 1)$ (see Problem 2.22).

[RegensBooks.xlsx] If the exponential smoothing method (with $\alpha = 0.1$) is applied to the Regens Books problem, the following forecasts are obtained

$$p_2 = y_1 = 89,$$

$$p_3 = \alpha y_2 + (1 - \alpha)p_2 = 90.20.$$

By iterating, the procedure generates the forecasts reported in Table 2.26 and, in particular, $p_{11} = 88.81$.

Table 2.26 Forecasts of the weekly sales provided by the exponential smoothing method (with $\alpha = 0.1$) in the Regens Books problem.

t	p_t	t	p_t
2	89.00	7	89.12
3	90.20	8	89.00
4	90.38	9	88.80
5	91.14	10	89.12
6	90.13	11	88.81

The choice of α plays a fundamental role in the exponential smoothing method. Large values of α involve a greater weight for more recent historical data and, therefore, a more outstanding capacity to follow rapidly the data variations observed in the time series with the passing of time. However, this corresponds to less filtering of the random fluctuations of the time series. In contrast, low values of α yield forecasts less subject to random components, but, at the same time, the most recent data variations progressively available are incorporated in the forecasts with a longer delay. In practice, the optimal value of α can be chosen on the basis of the same considerations that can be used to determine the optimal value of r in the simple moving average method, with the data-driven procedure illustrated in Section 2.9.

The forecasts for the subsequent time periods

So far the methods illustrated in this subsection only provided one-period-ahead forecasts. In order to make predictions for the subsequent time periods, it is sufficient to recall that the trend is assumed to be constant. Consequently,

$$p_T(\tau) = p_{T+1}, \tau = 2, \dots,$$

where the forecast p_{T+1} is obtained with any of methods previously described.

> [RegensBooks.xlsx] Figure 2.35 plots the Regens Books' weekly forecasts of online book sales for the subsequent four weeks (i.e., $p_{10}(\tau), \tau = 1, \dots, 4$), obtained by using the elementary method, the simple moving average method (with $r = 3$), the weighted moving average method and the exponential moving average method (with $\alpha = 0.1$).

Of course, once a new data entry becomes available, the time horizon shifts one-period-ahead and new updated forecasts can be computed, replacing the oldest ones. In such a context, the forecasting time horizon is said to be *rolling*.

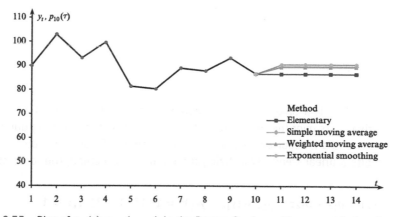

Figure 2.35 Plot of $p_{10}(\tau), \tau = 1, \dots, 4$, in the Regens Books problem, considering the elementary method, the simple moving average method (with $r = 3$), the weighted moving average method and the exponential moving average method (with $\alpha = 0.1$). Observe that, with the elementary method being the only exception, the forecasts $p_{10}(\tau), \tau = 1, \dots, 4$, of the other three methods almost overlap.

[RegensBooks.xlsx] In the Regens Books problem, forecasts for the next three months are needed. At time period $t = 10$, the rolling forecasting horizon includes time periods $t = 11, 12, 13$. By using the simple moving average method with $r = 2$, the forecasts are

$$p_{11}[= p_{10}(1)] = p_{10}(2) = p_{10}(3) = \frac{y_9 + y_{10}}{2} = 89.$$

If (at time period $t = 11$) it becomes known that $y_{11} = 90$, new forecasts are generated over the rolling forecasting horizon $t = 12, 13, 14$

$$p_{11}(1) = p_{11}(2) = p_{11}(3) = \frac{y_{10} + y_{11}}{2} = 88.$$

Thus, for $t = 12$, the new updated forecast ($p_{11}(1) = 88$) substitutes the previous one ($p_{10}(2) = 89$). Similar considerations are valid for $t = 13$ whereas, for time period $t = 14$, a first forecast ($p_{11}(3) = 88$) is determined.

2.8.4 Further Extrapolation Methods: the Linear-trend Case

If the trend is linear, and no cyclical and seasonal components are displayed by the time series, the forecasting methods are based on the following computational scheme

$$p_T(\tau) = a_T + b_T\,\tau, \tau = 1, \dots \tag{2.22}$$

The estimation of parameters a_T and b_T can be done with the methods illustrated below.

Elementary Method
This is the simplest forecasting method, on the basis of which

$$a_T = y_T;$$
$$b_T = y_T - y_{T-1}.$$

[BFT.xlsx] BFT is a Greek telecom company which needs to forecast overall administrative costs over the near future. The historical cost series (in €) during the past 14 months is reported in Table 2.27. As shown in Figure 2.36, the time series exhibits a linear trend. By utilizing the elementary method, the forecasts are

$$p_{14}(\tau) = a_{14} + b_{14}\tau = y_{14} + (y_{14} - y_{13})\tau = 194\,561 - 217\tau, \tau = 1, \dots$$

In particular, for the one-period-ahead ($\tau = 1$) the forecast is

$$p_{15} = €\,194\,344.$$

Table 2.27   Monthly administrative cost (in €) at BFT.

Month	Cost	Month	Cost
1	189 676	8	193 159
2	190 105	9	192 494
3	190 511	10	193 467
4	190 986	11	193 620
5	192 230	12	193 838
6	192 407	13	194 778
7	192 277	14	194 561

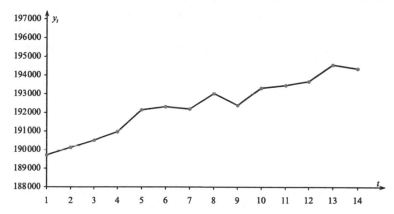

Figure 2.36   Plot of the monthly administrative cost (in €) at BFT.

Linear Regression Method

In order to estimate a_T and b_T, this method makes use of linear regression to fit the r (≥ 2) most recent data entries $y_{T-r+1}, \ldots, y_T$. Again, in order to determine the optimal r value, the reader is referred to Section 2.9.

[BFT.xlsx] By using the Excel functions INDEX and LINEST (with $r = 4$) in the BFT problem, the following estimates are obtained

$$b_{14} = 376.30;$$

$$a_{14} = 194\,763.70.$$

Hence,

$$p_{14}(\tau) = a_{14} + b_{14}\tau = 194\,763.70 + 376.30\tau, \tau = 1, \ldots$$

and, in particular, the one-period-ahead forecast ($\tau = 1$) is

$$p_{15} = €\,195\,140.$$

Double Moving Average Method

The method is an extension of the simple moving average method previously illustrated. Let r (≥ 2) be a double moving average parameter. The method applies the simple moving average method twice: once to the original data $y_t, t = 1, \ldots, T$, giving rise to time series

$$\gamma_t = \sum_{k=0}^{r-1} \frac{y_{t-k}}{r}, t = r, \ldots, T \tag{2.23}$$

and then to the resulting time series $\gamma_t, t = r, \ldots, T$, giving rise to time series

$$\eta_t = \sum_{k=0}^{r-1} \frac{\gamma_{t-k}}{r}, t = 2r - 1, \ldots, T. \tag{2.24}$$

Each moving average operation filters more random fluctuations, but accumulates additional delay with respect to the original time series. In particular, γ_T refers to time period $T - (r - 1)/2$ while η_T refers to time period $T - (r - 1) = T - r + 1$ (see, e.g., Figure 2.37). Hence, there is a time lag of $(r - 1)/2$ time periods between y_T and γ_T and an additional time lag of $(r - 1)/2$ time periods between γ_T and η_T. The coefficients of forecasting model (2.22) can be estimated as

$$a_T = \gamma_T + (\gamma_T - \eta_T) = 2\gamma_T - \eta_T \tag{2.25}$$

and

$$b_T = \frac{\gamma_T - \eta_T}{\frac{r-1}{2}} = \frac{2}{r-1}(\gamma_T - \eta_T). \tag{2.26}$$

In formula (2.25) the difference $\gamma_T - \eta_T$ is added to γ_T to compensate the time delay of γ_T with respect to y_T. Formula (2.26) expresses b_T as the ratio between the difference

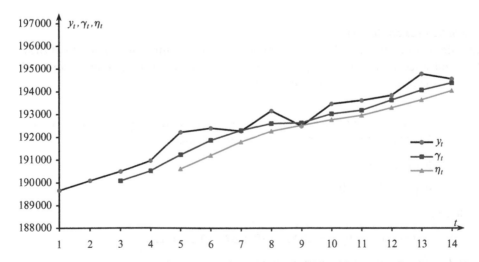

Figure 2.37 Plot of $y_t, t = 1, \ldots, 14$, $\gamma_t, t = 3, \ldots, 14$ (computed by using formula (2.23) with $r = 3$), and $\eta_t, t = 5, \ldots, 14$ (computed by using formula (2.24) with $r = 3$). Note that γ_{14} refers to the 13th time period, whereas η_{14} to the 12th time period.

of ordinates $(\gamma_T - \eta_T)$ and the difference of the corresponding abscissae $((r-1)/2)$ in the relevant Cartesian plane.

It is worth observing that, to simplify the calculations of γ_T and η_T, the following recursive relations may be used:

$$\gamma_T = \gamma_{T-1} + \frac{y_T - y_{T-r}}{r}$$

and

$$\eta_T = \eta_{T-1} + \frac{\gamma_T - \gamma_{T-r}}{r}.$$

Whenever r past observations are not available ($T < r$), the computation of γ_T and η_T can be executed along the guidelines illustrated for the simple moving average method. Again, the reader is referred to Section 2.9 to get an insight into the determination of the optimal value of r.

[BFT.xlsx] To apply the double moving average method (with $r = 3$) to the BFT problem, the following parameters are computed:

$$\gamma_{14} = \frac{y_{14} + y_{13} + y_{12}}{3} = 194\,392.33,$$

$$\eta_{14} = \frac{\gamma_{14} + \gamma_{13} + \gamma_{12}}{3} = 194\,037.56,$$

since

$$\gamma_{13} = \frac{y_{13} + y_{12} + y_{11}}{3} = 194\,078.67$$

and

$$\gamma_{12} = \frac{y_{12} + y_{11} + y_{10}}{3} = 193\,641.67.$$

Hence,

$$a_{14} = 2\gamma_{14} - \eta_{14} = 194\,747.11;$$

$$b_{14} = \gamma_{14} - \eta_{14} = 354.78;$$

$$p_{14}(\tau) = a_{14} + b_{14}\tau = 194\,747.11 + 354.78\tau, \tau = 1,\dots.$$

In particular, the one-period-ahead forecast ($\tau = 1$) is

$$p_{15} = €\ 195\,101.89.$$

Holt Method

The exponential smoothing method, introduced in Section 2.8.3, is unable to deal with a linear trend. The Holt method is a modification of the exponential smoothing method and is based on the following two relations

$$a_T = \alpha y_T + (1-\alpha)(a_{T-1} + b_{T-1}); \tag{2.27}$$

$$b_T = \beta(a_T - a_{T-1}) + (1-\beta)b_{T-1}. \tag{2.28}$$

In equation (2.27), the last data entry y_T is smoothed with the forecast $(a_{T-1} + b_{T-1})$ of y_T made at time period $T - 1$. Formula (2.28) updates the slope b_T, by smoothing the slope observed in the latest time period $(a_T - a_{T-1})$ with the previous estimation of the slope b_{T-1}.

By applying recursively equations (2.27) and (2.28), it is possible to express a_T and b_T as a function of the past data entries $y_t, t = 1, \dots, T$. In order to initialize the procedure, a_1 and b_1 must be specified. They can be chosen as $a_1 = y_1$ and $b_1 = 0$. Hence, the first forecast is $p_2 = p_1(1) = a_1 + b_1 = y_1$, as in the exponential smoothing method. The choice of the smoothing constants $\alpha, \beta \in (0, 1)$ is carried out according to the same criteria illustrated for the exponential smoothing method.

[BFT.xlsx] The Holt method (with $\alpha = \beta = 0.3$) is applied to the BFT problem. By recursively applying (2.27) and (2.28), a_t and b_t are computed for $t = 1, \dots, 14$, as reported in Table 2.28, Hence, the forecasts are

$$p_{14}(\tau) = a_{14} + b_{14}\tau = 194\,825.32 + 360.43\tau, \tau = 1, \dots.$$

In particular, the one-period-ahead forecast ($\tau = 1$) is

$$p_{15} = € 195\,185.75.$$

Figure 2.38 plots the forecasts $p_{14}(\tau), \tau = 1, \dots, 4$, obtained for the BFT problem, by using the elementary method, the linear regression method (with $r = 4$), the double moving average method (with $r = 3$) and the Holt method (with $\alpha = \beta = 0.3$).

Table 2.28 Parameters a_t and b_t, $t = 1, \dots, 14$, in the BFT problem.

t	a_t	b_t
1	189 676.00	0.00
2	189 804.70	38.61
3	190 043.62	98.70
4	190 395.42	174.63
5	191 068.04	324.03
6	191 696.55	415.37
7	192 161.44	430.23
8	192 761.87	481.29
9	193 018.41	413.86
10	193 442.69	416.99
11	193 787.78	395.42
12	194 079.64	364.35
13	194 544.19	394.41
14	194 825.32	360.43

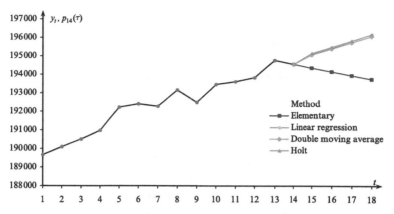

Figure 2.38 Plot of $p_{14}(\tau), \tau = 1, \dots, 4$, in the BFT problem, considering the elementary method, the linear regression method (with $r = 4$), the double moving average method (with $r = 3$) and, finally, the Holt method (with $\alpha = \beta = 0.3$). Observe that, with the only exception of the elementary method, the forecasts $p_{14}(\tau), \tau = 1, \dots, 4$, of the other three methods almost overlap.

2.8.5 Further Extrapolation Methods: the Seasonality Case

This section describes the main forecasting methods when the time series displays a constant or linear trend as well as a seasonal component with periodicity M.

Elementary Method

If the trend is constant, then

$$p_T(\tau) = y_{T+\tau-M}, \tau = 1, \dots, M. \tag{2.29}$$

On the basis of equation (2.29), the forecast related to time period $T + \tau$ corresponds to the data entry M time periods back. More generally, if the forecasting horizon is greater than the periodicity M (i.e., greater than on cycle), the forecast is

$$p_T(kM + \tau) = y_{T+\tau-M}, \tau = 1, \dots, M, k = 1, \dots$$

[Elna.xlsx] Elna distributes and installs air conditioning systems in the Nayarit region of Mexico. The number of systems sold monthly in the past two years is reported in Table 2.29. As shown in Figure 2.39, the time series shows a remarkable seasonal component with periodicity $M = 12$. By using the elementary method, the logistics manager gets

$$p_{24}(\tau) = y_{24+\tau-12}, \tau = 1, \dots, 12.$$

For instance, the one- and two-period-ahead forecasts are

$$p_{25} = y_{25-12} = y_{13} = 815;$$

$$p_{24}(2) = y_{26-12} = y_{14} = 1015.$$

Table 2.29 Monthly air conditioning systems sold by Elna.

Month	Quantity	Month	Quantity
1	915	13	815
2	815	14	1015
3	1015	15	915
4	1115	16	1315
5	1415	17	1215
6	1615	18	1615
7	1515	19	1315
8	1415	20	1115
9	815	21	1115
10	615	22	915
11	315	23	715
12	815	24	615

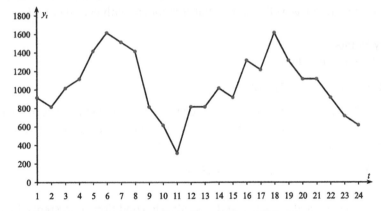

Figure 2.39 Plot of the monthly air conditioning systems sold by Elna.

Revised Exponential Smoothing Method

This method can be used whenever the trend is constant. It is based on the following forecasting model

$$p_T(\tau) = a_T s_{T+\tau}, \tau = 1, \dots, M,$$

where a_T takes into account the hypothesis of constant trend (and can be interpreted as the forecast without the seasonal component), whereas $s_{T+\tau}$ (≥ 0) is the seasonal index for time period $T + \tau$. More generally, for a forecasting horizon greater than M, the forecasting model is

$$p_T(kM + \tau) = a_T s_{T+\tau}, \tau = 1, \dots, M, k = 1, 2, \dots$$

Assuming, without loss of generality, that the available historical data are sufficient to cover an integer number $K = T/M$ of seasonal cycles, parameters a_T and $s_{T+\tau}$, $\tau = 1, \ldots, M$, can be computed with the following relations

$$a_T = \alpha \frac{y_T}{s_T} + (1 - \alpha)a_{T-1}, \tag{2.30}$$

$$s_{T+\tau} = s_{kM+\tau} = \gamma \frac{y_{(K-1)M+\tau}}{a_{(K-1)M+\tau}} + (1 - \gamma)s_{(K-1)M+\tau}, \tau = 1, \ldots, M, \tag{2.31}$$

where α and γ are smoothing constants such that $\alpha, \gamma \in (0, 1)$. Equation (2.30) expresses a_T as the weighted sum of two components: the former, y_T/s_T, represents the time series value at time period T without seasonal component, while the latter represents the forecast, without the seasonal component, at time period $T - 1$. A similar interpretation applies to equation (2.31). However, in this case, it is necessary to take into account the periodicity of the seasonal component. It is possible to develop recursively equations (2.30) and (2.31), so as to have all data entries $y_1, \ldots, y_T$ appear explicitly in the formulae of a_T and $s_{T+\tau}$. To initialize the procedure, a_0 is set equal to

$$a_0 = \frac{\overline{y}_{(1)} + \overline{y}_{(2)} \cdots + \overline{y}_{(K)}}{K} = \frac{\sum\limits_{t=1}^{T} y_t}{T}, \tag{2.32}$$

corresponding to the average value of all the historical data available, whereas the initial estimate of the seasonal indices is

$$s_t = \frac{y_t/\overline{y}_{(1)} + y_{t+M}/\overline{y}_{(2)} + \cdots + y_{t+(k-1)M}/\overline{y}_{(K)}}{K}, t = 1, \ldots, M. \tag{2.33}$$

In formula (2.33), the numerator is the sum of the data entries of the tth time period of each seasonal cycle, each divided by the average value of the corresponding seasonal cycle. Hence,

$$\sum_{t=1}^{M} s_t = \frac{T}{K} = M,$$

that is, the average seasonal index for the first seasonal cycle is equal to 1. However, this condition may not be satisfied for the subsequent seasonal cycles and for this reason it is necessary to normalize the indices $s_t, t = (k - 1)M + 1, \ldots, kM, k = 2, \ldots$.

The reader is referred to Section 2.9 to get an insight into the determination of optimal values for α and γ.

> [Elna.xlsx] The revised exponential smoothing method (with parameters $\alpha = \gamma = 0.05$) is applied to the Elna problem. Firstly, the average sales in the first and second seasonal cycles are computed

$$\bar{y}_{(1)} = \frac{y_1 + \dots + y_{12}}{12} = 1031.67,$$

$$\bar{y}_{(2)} = \frac{y_{13} + \dots + y_{24}}{12} = 1056.67,$$

which allows the calculation

$$a_0 = \frac{\bar{y}_{(1)} + \bar{y}_{(2)}}{2} = 1044.17.$$

Then, seasonal indices $s_t, t = 1, \dots, 12$ are computed by using formula (2.33) (see Table 2.30). It is worth noting that

$$\bar{s} = \frac{\sum_{t=1}^{12} s_t}{12} = 1.$$

By using the relationships (2.30) and (2.31), Tables 2.31 and 2.32 are generated, where the seasonal indices $s_t, t = 13, \dots, 36$, have already been normalized. These parameters allow computation of, among others, the one- and two-period-ahead forecasts

$$p_{25} = p_{24}(1) = 874.84;$$

$$p_{24}(2) = 924.17.$$

Figure 2.40 shows both the time series and its extrapolation over the next 12 months.

Table 2.30   Time series $s_t, t = 1, \dots, 12$, for the Elna air conditioning systems forecasting problem.

t	s_t	t	s_t
1	0.83	7	1.36
2	0.88	8	1.21
3	0.92	9	0.92
4	1.16	10	0.73
5	1.26	11	0.49
6	1.55	12	0.69

Table 2.31 Time series $a_t, t = 1, \ldots, 12$, for the Elna air conditioning systems forecasting problem.

t	a_t	t	a_t
1	1047.14	13	1021.65
2	1041.34	14	1028.83
3	1044.14	15	1026.70
4	1039.89	16	1032.11
5	1044.01	17	1028.49
6	1044.01	18	1029.24
7	1047.65	19	1026.06
8	1053.58	20	1020.43
9	1045.07	21	1030.28
10	1034.88	22	1041.91
11	1015.21	23	1063.95
12	1023.86	24	1055.19

Table 2.32 Times series $s_t, t = 13, \ldots, 36$, for the Elna air conditioning systems forecasting problem.

t	s_t	t	s_t
13	0.83	25	0.83
14	0.87	26	0.88
15	0.93	27	0.92
16	1.16	28	1.16
17	1.27	29	1.26
18	1.55	30	1.55
19	1.36	31	1.36
20	1.22	32	1.21
21	0.92	33	0.92
22	0.72	34	0.73
23	0.48	35	0.49
24	0.69	36	0.69

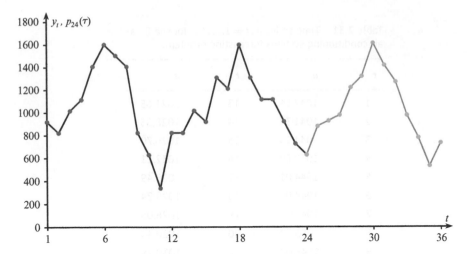

Figure 2.40 Plot of the monthly sales series during the past two years and its projection over the next 12 months (in grey) for the Elna air conditioning systems forecasting problem.

Winters Method

The Winters method can be used whenever there is a linear trend and a seasonal component of periodicity M. The forecasting model is

$$p_T(\tau) = [a_T + b_T(\tau)]s_{T+\tau}, \tau = 1, \ldots, M$$

and, more generally,

$$p_T(kM + \tau) = [a_T + b_T(kM + \tau)]s_{T+\tau}, \tau = 1, \ldots, M, k = 0, \ldots$$

As in the revised exponential smoothing method, it is assumed that the historical data available are enough to have an integer number $K = T/M$ of seasonal cycles ($K \geq 2$). The parameters of the Winters method are computed as follows

$$a_T = \alpha\left(\frac{y_T}{s_T}\right) + (1 - \alpha)(a_{T-1} + b_{T-1}), \tag{2.34}$$

$$b_T = \beta(a_T - a_{T-1}) + (1 - \beta)b_{T-1}, \tag{2.35}$$

$$s_{T+\tau} = s_{KM+\tau} = \gamma\left(\frac{y_{(K-1)M+\tau}}{a_{(K-1)M+\tau}}\right) + (1 - \gamma)s_{(K-1)M+\tau}, \tau = 1, \ldots, M, \tag{2.36}$$

where α, β and γ are smoothing constants chosen in the interval $(0, 1)$. The reader is referred to Section 2.9 for an insight into the determination of their optimal values.

Equations (2.34) and (2.35) are derived from equations (2.27) and (2.28) of the Holt method, except that y_T is deseasonalized (similarly to the revised exponential smoothing method). Moreover, relationship (2.36) is the same as (2.31) of the revised exponential smoothing method.

The Winters method can be initialized as follows

$$a_0 = \frac{\overline{y}_{(1)} + \overline{y}_{(2)} \cdots + \overline{y}_{(K)}}{K} = \frac{\sum_{t=1}^{T} y_t}{T};\tag{2.37}$$

$$b_0 = \frac{1}{K-1}\left(\frac{\overline{y}_{(2)} - \overline{y}_{(1)}}{M} + \cdots + \frac{\overline{y}_{(K)} - \overline{y}_{(K-1)}}{M}\right) = \frac{\overline{y}_{(K)} - \overline{y}_{(1)}}{(K-1)M}$$

$$= \frac{\overline{y}_{(K)} - \overline{y}_{(1)}}{T-M};\tag{2.38}$$

$$s_t = \frac{y_t/\overline{y}_{(1)} + y_{t+M}/\overline{y}_{(2)} + \cdots + y_{t+(k-1)M}/\overline{y}_K}{K}, t = 1, \dots, M.\tag{2.39}$$

Formulae (2.37) and (2.39) are the same as (2.32) and (2.33) used in the revised exponential smoothing method. In (2.38) the terms in brackets are $K - 1$. The quantity $\frac{\overline{y}_{(K)} - \overline{y}_{(1)}}{T-M}$ expresses the variation, in $T - M$ time periods, of the average of the historical data, from the first to the last seasonal cycle. This is because the average of the historical data of the first seasonal cycle is centered in time period $t = (M+1)/2$, whereas the average of the historical data of the last seasonal cycle is centered in time period $t = (K-1)M + (M+1)/2$. The values determined according to (2.39) are already normalized. For this reason, a normalization procedure should be executed only for seasonal indices determined by (2.36) and corresponding to the seasonal cycles subsequent to the first.

[Elna2.xlsx] Elna also distributes microwave ovens in the Nayarit region. The number of microwave ovens sold monthly over the past two years is reported in Table 2.33. As shown in Figure 2.41, the time series is characterized by a seasonal component with periodicity $M = 12$ as well as by a linear trend. To forecast the sales over the next 12 months with the Winters method, the logistics manager first computes the average sales in the first $K = 2$ seasonal cycles

$$\overline{y}_{(1)} = \frac{y_1 + \cdots + y_{12}}{12} = 2089.58,$$

$$\overline{y}_{(2)} = \frac{y_{13} + \cdots + y_{24}}{12} = 3674.50.$$

Table 2.33 Elna monthly sales of microwave ovens during the past 24 months.

Month	Quantity	Month	Quantity
1	682	13	416
2	416	14	1746
3	1613	15	2411
4	1613	16	2544
5	1746	17	4140
6	2677	18	4539
7	4672	19	7997
8	5603	20	8263
9	3741	21	7465
10	1480	22	3209
11	682	23	1081
12	150	24	283

Then, they calculate

$$a_0 = \frac{\bar{y}_{(1)} + \bar{y}_{(2)}}{2} = 2882.04,$$

$$b_0 = \frac{\bar{y}_{(2)} - \bar{y}_{(1)}}{24 - 12} = 132.08,$$

Table 2.34 Time series $s_t, t = 1, \ldots, 12,$ for the Elna forecasting problem.

t	s_t	t	s_t
1	0.22	7	2.21
2	0.34	8	2.47
3	0.71	9	1.91
4	0.73	10	0.79
5	0.98	11	0.31
6	1.26	12	0.07

as well as the seasonal indices $s_t, t = 1, \ldots, M$, as reported in Table 2.34. Then, by using formulae (2.34)–(2.36) with $\alpha = 0.2$, $\gamma = 0.1$ and $\beta = 0.3$, they get the parameters in Tables 2.35 and 2.36 (where $s_t, t = 13, \ldots, 36$, are normalized). Then, the forecasts for the next 12 months are computed (see Table 2.37 and Figure 2.41).

Table 2.35 Parameters a_t and b_t, $t = 1, \ldots, 24$, for the Elna microwave ovens forecasting problem.

t	a_t	b_t	t	a_t	b_t
1	3031.87	137.40	13	1814.88	−53.99
2	2782.21	21.28	14	2500.19	167.80
3	2694.59	−11.39	15	2817.44	212.63
4	2587.20	−40.19	16	3126.02	241.42
5	2393.52	−86.24	17	3555.85	297.94
6	2271.36	−97.01	18	3805.57	283.48
7	2163.03	−100.41	19	3994.18	255.02
8	2104.69	−87.79	20	4061.19	198.61
9	2005.06	−91.34	21	4187.14	176.81
10	1905.28	−93.87	22	4300.20	157.69
11	1888.72	−70.68	23	4248.50	94.87
12	1857.65	−58.79	24	4225.30	59.45

Table 2.36 Time series of normalized $s_t, t = 13, \ldots, 36$, for the Elna microwave ovens forecasting problem.

t	s_t	t	s_t
13	0.22	25	0.22
14	0.32	26	0.36
15	0.71	27	0.72
16	0.72	28	0.73
17	0.96	29	0.98
18	1.26	30	1.25
19	2.21	31	2.19
20	2.50	32	2.45
21	1.92	33	1.90
22	0.79	34	0.79
23	0.32	35	0.31
24	0.08	36	0.07

Table 2.37 Forecasts over the next 12 months for the Elna microwave ovens forecasting problem.

τ	$p_{24}(\tau)$	τ	$p_{24}(\tau)$
1	953.27	7	10 184.04
2	1557.05	8	11 535.31
3	3179.26	9	9 068.49
4	3279.20	10	3 805.78
5	4443.11	11	1 517.74
6	5735.14	12	368.72

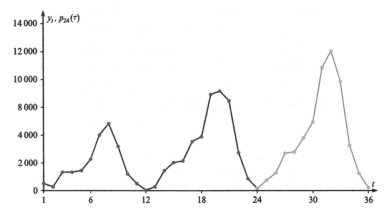

Figure 2.41 Plot of monthly sales of microwave ovens at Elna and their projection over the next 12 months (in grey).

2.8.6 Further Extrapolation Methods: the Irregular Time Series Case

The simplest forecasting method for the irregular time series is the *shift method* which is based on the assumption that the time series will be repeated in the future, possibly scaled up or down. The forecasting model is

$$p_T(\tau) = \alpha_{T+\tau} y_{T+\tau-L}, \tau = 1, \dots,$$

where $\alpha_{T+\tau}$ is the *scaling factor* at time period $T + \tau$. This means that at time period $T + \tau$ the forecast will be equal to $\alpha_{T+\tau}$ times the data entry at time period $T + \tau - L$. It is worth noting that L is not necessarily related to a possible seasonality of the time series. Its optimal value can be determined by following the guidelines illustrated in Section 2.9.

The scaling factor belongs to the interval $[0, 2]$ and can be computed on the basis of the data of the last two homologous time periods (that is, $T + \tau - L$ and $T + \tau - 2L$)

$$\alpha_{T+\tau} = 1 + \frac{y_{T+\tau-L} - y_{T+\tau-2L}}{\max\{y_{T+\tau-L}, y_{T+\tau-2L}\}}, \tag{2.40}$$

where it is assumed that $\alpha_{T+\tau} = 1$ whenever the fraction in equation (2.40) turns out to be 0/0. The choice of $\alpha_{T+\tau}$ according to (2.40) ensures that the forecast time series is intermittent if the time series is intermittent as well.

[MCE.xlsx] MCE is an international retail chain headquartered in New Zealand with more than 500 shops around the world. The daily sales (in NZ$) of a lotion at the Wellington centre during the past 24 days are shown in Table 2.38. As illustrated by Figure 2.42, the time series is irregular. In order to forecast the sales during the next six work days, the shift method can be used. By setting $L = 6$, from formula (2.40) the values of $\alpha_{T+\tau}$ and $p_T(\tau), \tau = 1, \ldots, 6$ are computed, as reported in Table 2.39. Figure 2.42 also illustrates the required sales forecasts.

Table 2.38 Daily sales (in NZ$) of a lotion at the Wellington MCE centre.

Day	Sales	Day	Sales
1	215	13	2 366
2	1 768	14	1 119
3	10 331	15	9 991
4	287	16	10 252
5	10 689	17	8 974
6	4 003	18	9 499
7	2 801	19	500
8	4 056	20	340
9	10 989	21	6 991
10	6 520	22	11 732
11	5 790	23	8 414
12	9 685	24	9 514

Table 2.39 Time series $\alpha_{T+\tau}$ and $p_T(\tau), \tau = 1, \ldots, 6$, for the MCE problem.

τ	$T+\tau$	$\alpha_{T+\tau}$	$p_T(\tau)$
1	25	0.21	105.66
2	26	0.30	103.31
3	27	0.70	4 891.81
4	28	1.13	13 212.00
5	29	0.94	7 888.95
6	30	1.00	9 529.00

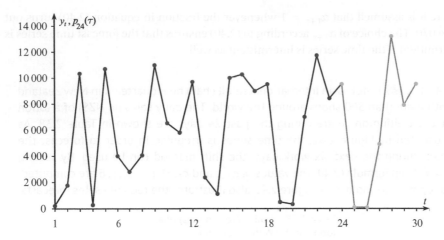

Figure 2.42 Plot of the daily sales and forecasts (in grey) at the Wellington MCE centre.

2.8.7 Further Extrapolation Methods: the Intermittent Time Series Case

If the time series is intermittent, two cases may occur:

- if the time series is random then the *Croston method*, described in the following, can be used; alternatively, if the time series is also irregular, the shift method can be utilized;
- if the time series is periodic, any method described in the previous section can be adapted.

Croston Method

The method is made up of three steps:

1. Remove all the zero data entries from the time series $y_t, t = 1, \ldots, T$. Let y'_t, $t = 1, \ldots, T'$, be the corresponding compact time series. Make the one-period-ahead forecast $p_{T'+1}$ through, for example, the exponential smoothing method.
2. Assign the previous forecast to a time period $T + \tau$, where τ is determined as follows. Let $z_k, k = 1, \ldots, K$, be the time series of the interarrival times between non-zero data entries in $y_t, t = 1, \ldots, T$. Time lag τ can be determined by applying a suitable forecasting method (such as the exponential smoothing method) to the time series $z_k, k = 1, \ldots, K$.
3. Once computed $p_T(\tau)$, the forecasts for the subsequent time periods are zero, except for time periods $T + 2\tau, T + 3\tau, \ldots$ when forecasts are set equal to $p_T(\tau)$.

It is worth noting that the forecasts defined above are periodic even if the time series $y_t, t = 1, \ldots, T$, is not periodic.

[Ipergent.xlsx] The logistics manager of Ipergent, a Belgian supermarket chain, wants to determine the sales of a 0.5 kg dog breed food package of a renowned brand at the sales point in Louvain. The sales recorded during the past 21 days are reported in Table 2.40 and in Figure 2.43. Since the time series is both intermittent and random, the Croston method is used. Firstly, the manager computes the time series $y'_t, t = 1, \dots, T'$ (see Table 2.41), where $T' = 9$. The forecast of the first non-zero entry p_{T+1} is calculated by applying the exponential smoothing method with $\alpha = 0.2$ to $y'_t, t = 1, \dots, T'$

$$p_{10} = 11.11.$$

Then, the manager determines the auxiliary time series $z_k, k = 1, \dots, K$ (as shown in Table 2.42) which is then used to forecast τ with the exponential smoothing method with $\alpha = 0.2$

$$\tau = 2.54.$$

The time lag τ is then approximated to three days. Hence,

$$p_{21}(1) = p_{21}(2) = 0,$$

and

$$p_{21}(3) = 11.11.$$

Table 2.40 Daily sales of dog breed food (in 0.5 kg packages) at the Ipergent supermarket in Louvain.

Day	Quantity	Day	Quantity	Day	Quantity
1	10	8	12	15	11
2	0	9	11	16	0
3	0	10	9	17	0
4	14	11	0	18	0
5	0	12	0	19	12
6	0	13	0	20	0
7	0	14	14	21	10

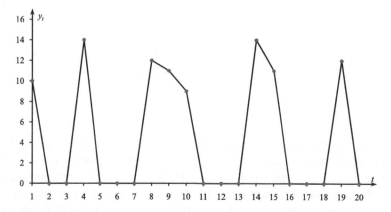

Figure 2.43 Plot of the daily sales of dog breed food (in 0.5 kg packages) at the Ipergent supermarket in Louvain.

Table 2.41 Time series y'_t, $t = 1, ..., T'$, in the Ipergent problem.

t	y'_t	t	y'_t
1	10	6	14
2	14	7	11
3	12	8	12
4	11	9	10
5	9		

Table 2.42 Auxiliary time series z_k, $k = 1, ..., 6$, in the Ipergent problem.

k	z_k	k	z_k
1	1	4	4
2	3	5	4
3	4	6	2

Periodic Time Series

Let M be the periodicity of the time series $y_t, t = 1, ..., T$. On the basis of the previous subsection, the following forecasting framework can be used.

1. The zero data entries are removed from $y_t, t = 1, ..., T$, and the corresponding compact time series $y'_t, t = 1, ..., T'$, is obtained.
2. The forecasts for one or more time periods ahead are computed by applying to $y'_t, t = 1, ..., T'$, one of the methods devised for regular time series (or the shift method in case y'_t is irregular).
3. Then, the one-, two-, ... period-ahead forecasts of y'_t are the forecasts of y_t in time periods $T + M, T + 2M, ...$, while the forecasts associated with the other time periods are set equal to 0.

[Belem.xlsx] Belem is a pastry producer located in Lisbon. The sales of the 1 kg package of its cake called "Torta de Viana" during the past three years are reported in Table 2.43. As shown in Figure 2.44, the time series is intermittent and periodic. In particular, the intervals with non-zero and zero data entries are five and seven time periods long, respectively. The compact time series y'_t (Table 2.44) shows a seasonal component with periodicity $M = 5$ as illustrated in Figure 2.45.

Table 2.43 Monthly sales of the "Torta de Viana" cake (in 1 kg packages) in the Belem problem.

Month	Quantity	Month	Quantity	Month	Quantity
1	200	13	189	25	187
2	11	14	7	26	9
3	0	15	0	27	0
4	0	16	0	28	0
5	0	17	0	29	0
6	0	18	0	30	0
7	0	19	0	31	0
8	0	20	0	32	0
9	0	21	0	33	0
10	14	22	18	34	17
11	121	23	118	35	114
12	450	24	489	36	498

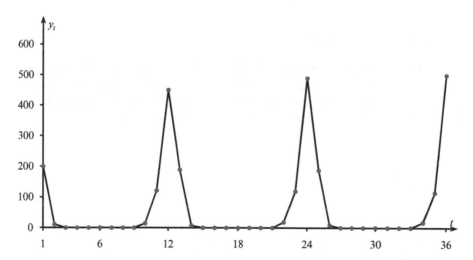

Figure 2.44 Monthly sales of the "Torta de Viana" cake (in 1 kg packages) in the Belem problem.

To estimate the sales during the next five months, the company uses the Winters method, considering that the number K of seasonal cycles is equal to three, and selecting the parameters $\alpha = \gamma = 0.1$ and $\beta = 0.3$. By taking into account that

$$\bar{y}_{(1)} = \frac{y_1 + \dots + y_5}{5} = 159.20;$$

Table 2.44 Time series $y'_t, t = 1, \ldots, 15$, for the Belem problem.

t	y'_t	t	y'_t	t	y'_t
1	200	6	189	11	187
2	11	7	7	12	9
3	14	8	18	13	17
4	121	9	118	14	114
5	450	10	489	15	498

$$\bar{y}_{(2)} = \frac{y_6 + \ldots + y_{10}}{5} = 164.20;$$

$$\bar{y}_{(3)} = \frac{y_{11} + \ldots + y_{15}}{5} = 165,$$

a_0 and b_0 are computed as follows

$$a_0 = \frac{\bar{y}_{(1)} + \bar{y}_{(2)} + \bar{y}_{(3)}}{3} = 162.80;$$

$$b_0 = \frac{\bar{y}_{(3)} - \bar{y}_{(1)}}{10} = 0.58,$$

while Table 2.45 reports the seasonal indices $s_t, t = 1, \ldots, 5$. Table 2.46 reports parameters a_t e $b_t, t = 1, \ldots, 15$, while Table 2.47 shows the normalized indices $s_t, t = 6, \ldots, 20$. Then, Table 2.48 shows the forecasts for the next five time periods for y'_t while Figure 2.46 illustrates the corresponding forecasts.

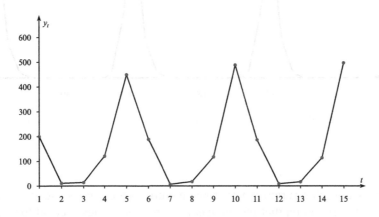

Figure 2.45 Plot of the time series $y'_t, t = 1, \ldots, 15$, for the Belem problem.

Table 2.45 Seasonal indices $s_t, t = 1, \dots, 5$, for the Belem problem.

t	s_t
1	1.18
2	0.06
3	0.10
4	0.72
5	2.94

Table 2.46 Parameters a_t and b_t, $t = 1, \dots, 15$, for the Belem problem.

t	a_t	b_t
1	163.99	0.64
2	168.01	0.98
3	166.07	0.69
4	166.81	0.69
5	166.05	0.55
6	165.61	0.45
7	161.29	−0.03
8	163.78	0.22
9	163.71	0.19
10	164.34	0.24
11	163.87	0.17
12	164.18	0.18
13	164.81	0.23
14	164.17	0.14
15	164.88	0.20

Table 2.47 Normalized seasonal indices s_t, $t = 6, \dots, 20$, for the Belem problem.

t	s_t	t	s_t	t	s_t
6	1.21	11	1.19	16	1.17
7	0.06	12	0.05	17	0.05
8	0.10	13	0.10	18	0.10
9	0.73	14	0.73	19	0.72
10	2.91	15	2.93	20	2.95

Table 2.48 Forecasts of the sales of the "Torta de Viana" cake (in 1 kg packages) for the next five months in the Belem problem.

Month	Forecast
1	193.52
2	9.00
3	16.77
4	118.98
5	489.92

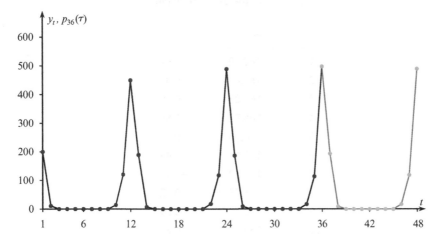

Figure 2.46 Plot of the monthly sales of the "Torta de Viana" cake (in 1 kg packages) and forecasts (in grey) in the Belem problem.

2.9 Accuracy Measures

Obviously, one cannot expect that forecasts are always 100% correct. This is because of the inherent random nature of the variables to be predicted. In addition to *random errors*, there may be *systematic errors*, i.e., errors caused by the fact that the hypotheses on which the forecasts are based no longer hold. This being said, it is crucial to determine by *how much* the forecasts generated by a given method can be wrong. To this end, suitable *accuracy measures* are calculated on the basis of the forecasting errors made in the past. Such measures can be employed to both evaluate the operational impact of these errors, and to select the most precise forecasting method. Moreover, for periodic predictions (like those required in inventory management), forecasting errors are *monitored* in order to adjust parameters *on the fly*, if needed. For the sake of brevity, these issues are examined for the case of a regular time series where an one-period-ahead forecast has to be generated.

To evaluate the accuracy of a forecasting method, the errors that would have been made in the past have to be computed. Then, an aggregate (cumulative) error is

Table 2.49 Evaluation of the quality of a forecasting method by means of the mean absolute percentage error.

Mean absolute percentage error	Quality of forecasting
$\leq 10\%$	Very good
$> 10\%, \leq 20\%$	Good
$> 20\%, \leq 30\%$	Moderate
$> 30\%$	Poor

calculated and used as an *accuracy measure*. The most common are the *mean absolute error (MAE)*, the *mean absolute percentage error (MAPE)* and the *mean squared error (MSE)* defined as follows at time period T:

$$MAE_T = \frac{\sum_{t=2}^{T} |e_t|}{T-1};\qquad(2.41)$$

$$MAPE_T = \frac{\sum_{t=2}^{T} \frac{|e_t|}{y_t}}{T-1};\qquad(2.42)$$

$$MSE_T = \frac{\sum_{t=2}^{T} e_t^2}{T-2}.\qquad(2.43)$$

where $T > 1$ for equations (2.41) and (2.42), and $T > 2$ for equation (2.43). These three accuracy measures can be used at time period T to establish a comparison between different forecasting methods. In addition, $MAPE_T$ is suitable for expressing concisely the quality of a forecasting method (see Table 2.49).

2.9.1 Calibration of the Parametrized Forecasting Methods

The accuracy measures can be used to tune the forecasting methods depending on one or more parameters, like exponential smoothing and the Winters method. The basic idea is to assign the parameters the values that would have optimized the accuracy of the forecasts in the past. For instance, by using the *MSE*, the most suitable parameter α^* of the exponential smoothing method at period T can be determined as the solution of the optimization problem

$$\underset{\alpha}{\text{Minimize}}\, MSE_T(\alpha)$$

subject to

$$\alpha \in (0, 1).$$

Since the number of parameters is usually small and the parameters themselves are usually bounded in nature, a good approximated solution can be found through a discretization. For the sake of simplicity, the FORECASTING_PARAMETER_SETTING procedure is illustrated below for a forecasting method with a single parameter δ and an accuracy measure indicated as $A(\delta)$.

1: **procedure** FORECASTING_PARAMETER_SETTING (Δ, δ_{min}, δ_{max}, $\overline{\delta}$)

2:      # Δ is the discretization step;

3:      # δ_{min} and δ_{max} are the minimum and the maximum feasible values of parameter δ, respectively;

4:      # $\overline{\delta}$ is the best approximated value of parameter δ returned by the procedure;

5:      $h = 1$;

6:      $\delta_h = \delta_{min}$;

7:      $min = \infty$;

8:      **while** $\delta_h \leq \delta_{max}$ **do**

9:          Determine the accuracy measure $A(\delta_h)$ corresponding to δ_h;

10:         **if** $A(\delta_h) < min$ **then**

11:             $min = A(\delta_h)$;

12:             $\overline{\delta} = \delta_h$;

13:         **end if**

14:         $\delta_{h+1} = \delta_h + \Delta$;

15:         $h = h + 1$;

16:     **end while**

17:     **return** $\overline{\delta}$;

18: **end procedure**

Of course, in case of multiple parameters there may be a different discretization step for each parameter.

[RegensBooks.xlsx] The logistics manager of Regens Books (see Section 2.8.3) wants to tune the parameter α of the exponential smoothing method, by using the mean absolute error as an accuracy measure. The application of the FORECASTING_PARAMETER_SETTING procedure for $\alpha_{min} = 0.1$, $\alpha_{max} = 0.9$ and $\Delta = 0.1$ returns $\overline{\alpha} = 0.2$, where $\overline{\alpha}$ is the most suitable value for α. The mean absolute errors computed by discretizing α during the execution of the procedure is reported in Table 2.50. It is worth observing that the best value of α computed by using the Excel Solver tool under the Data tab is equal to 0.25, and the corresponding *MAE* is 5.45.

Table 2.50 $MAE_{10}(\alpha)$ for the α values generated by the FORECASTING_PARAMETER_SETTING procedure in the Regens Books problem.

α	$MAE_{10}(\alpha)$	α	$MAE_{10}(\alpha)$
0.1	5.68	0.6	6.49
0.2	5.53	0.7	6.67
0.3	5.57	0.8	6.80
0.4	5.89	0.9	6.95
0.5	6.23		

2.9.2 Selection of the Most Accurate Forecasting Method

Once each parametrized forecasting method has been tuned, accuracy measures (2.41), (2.42), and (2.43) can be used to select the most accurate forecasting method.

[RegensBooks.xlsx] The logistics manager of Regens Books wants to assess the accuracy of four forecasting methods introduced in Section 2.8.3: the elementary method, the simple moving average method (with $r = 2$), the weighted moving average method and the exponential smoothing method (with $\alpha = 0.1$). Based on the forecasting errors summarized in Table 2.51, the values of the accuracy measures shown in Table 2.52 are determined. According to the *MAPE* values, all the four methods can be deemed to provide good quality forecasts. In particular, the most accurate forecasting method turns out to be the exponential smoothing method (with $\alpha = 0.1$).

Table 2.51   Forecasting errors in the Regens Books problem.

t	y_t	Elementary method		Simple moving average method ($r = 2$)		Weighted moving average method		Exponential smoothing method ($\alpha = 0.1$)	
		p_t	e_t	p_t	e_t	p_t	e_t	p_t	e_t
1	89	–	–	–	–	–	–	–	–
2	101	89	12.00	89.00	12.00	89.00	12.00	89.00	12.00
3	92	101	−9.00	95.00	−3.00	97.00	−5.00	90.20	1.80
4	98	92	6.00	96.50	1.50	94.50	3.50	90.83	7.62
5	81	98	−17.00	95.00	−14.00	95.90	−14.90	91.14	−10.14
6	80	81	−1.00	89.50	−9.50	90.93	−10.93	90.13	−10.13
7	88	80	8.00	80.50	7.50	87.81	0.19	89.12	−1.12
8	87	88	−1.00	84.00	3.00	87.86	−0.86	89.00	−2.00
9	92	87	5.00	87.50	4.50	87.67	4.33	88.80	3.20
10	86	92	−6.00	89.50	−3.50	88.53	−2.53	89.12	−3.12
11	–	86	–	89.00	–	88.07	–	88.81	–

Table 2.52   Accuracy measures of the forecasting methods used by Regens Books.

	Elementary method	Simple moving average method ($r = 2$)	Weighted moving average method	Exponential smoothing method ($\alpha = 0.1$)
MAE_{10}	7.22	6.50	6.03	5.68
$MAPE_{10}$	8.08%	7.42%	6.87%	6.39%
MSE_{10}	84.63	67.41	68.60	54.50

2.10 Forecasting Control

A forecasting method works correctly if the errors are random and not systematic. Typical systematic errors occur when the variable to be predicted is constantly underestimated or overestimated, or a seasonal component is not taken into account. Forecasting control aims at identifying systematic errors in periodic predictions (caused, e.g., by a shift in trend) in order to adjust parameters if needed. Forecasting control can be done through a *tracking signal* or a *control chart*.

2.10.1 Tracking Signal

The tracking signal K_T at time period T ($T > 1$) is defined as the ratio between the cumulative error at time period T and MAE_T, that is,

$$K_T = \frac{\sum_{t=2}^{T} e_t}{MAE_T}.$$

The tracking signal is greater than zero if the forecast systematically underestimates the data; in contrast, a negative value of K_T indicates a systematic overestimate of the data. For this reason, a forecast is assumed to be unbiased if the tracking signal falls in a range $\pm \overline{K}$. The value of $\overline{K}$ is established empirically, and usually falls between two and five. If the tracking signal is outside this interval, the parameters of the forecasting method should be modified or a different forecasting method should be selected.

[Plaza.xlsx] Months ago, the logistics manager of the retail chain Plaza, in Bolivia, was in charge of devising and monitoring monthly forecasts for the company's sportswear items. The sales of the previous 12 months (in k\$) are reported in Table 2.53. After a preliminary study of the time series, they decided to make use of the exponential smoothing method. The optimal smoothing parameter α was obtained by minimizing $MAE_{12}(\alpha)$ (in Excel with the Solver tool). The corresponding solution was $\alpha^* = 0.26$ which was associated with $MAE_{12}(\alpha^*) = 15.95$. On the basis of the forecasts $p_t, t = 2, \ldots, 12$, reported in the third and sixth columns of Table 2.53, the manager verified that the tracking signal K_{12} was in the range $[-4, 4]$ ($K_{12} = -1.65$). Hence, they devised the forecast for the next month, $p_{13} = 970.08$ k\$.

The subsequent month ($T = 13$), the sales turned out to be equal to $y_{13} = 1024$ k\$. The associated tracking signal value was $K_{13} = 1.44$, while the new forecast was $p_{14} = 984.22$ k\$.

Two months later ($T = 14$), the sales were known to be $y_{14} = 1035$ k\$; the tracking signal was $K_{14} = 3.63$ and the forecast was $p_{15} = 997.55$ k\$.

Three months later ($T = 15$), the sales were $y_{15} = 1047$ k\$ and the relative tracking signal was $K_{15} = 5.43$, out of the feasible range. Thus, the exponential smoothing parameter was replaced by $\alpha^* = 0.89$, associated with $MAE_{15}(\alpha^*) = 20.22$. The new forecasts are shown in Table 2.54. The new value of the tracking signal was $K_{15} = 3.81$ and the new forecast was $p_{16} = 1045.47$ k\$.

Table 2.53 Monthly sales of sportswear (in k$) during the first 12 months in the Plaza problem and the corresponding exponential smoothing forecasts with $\alpha = 0.26$.

Month	Sales	Forecast	Month	Sales	Forecast
1	977	–	7	948	968.00
2	958	977.00	8	996	962.76
3	989	972.01	9	955	971.48
4	966	976.47	10	959	967.15
5	956	973.72	11	960	965.02
6	965	969.07	12	988	963.70

Table 2.54 Monthly sales of sportswear (in k$) during the first 15 months in the Plaza problem and the corresponding exponential smoothing forecasts (with $\alpha = 0.89$).

Month	Sales	Forecast	Month	Sales	Forecast
1	977	–	9	955	990.84
2	958	977.00	10	959	959.00
3	989	960.12	11	960	959.00
4	966	985.78	12	988	959.89
5	956	968.21	13	1024	984.86
6	965	957.36	14	1035	1019.63
7	948	964.15	15	1047	1033.29
8	996	949.80			

2.10.2 Control Charts

Unlike tracking signals, control charts are based on the plot of single errors e_t, $t = 2, \ldots, T$. Let us assume that forecasting errors are due to a very large number of causes each of which has a very small impact. Then, the *central limit theorem* allows us to model $e_t, t = 2, \ldots, T$, as realizations of a normal random variable E with mean μ_E and standard deviation σ_E. A forecast is deemed to be effective if each error e_t is in the interval $\pm k\sigma_E$, where k is quite often set equal to three ("3-σ" control chart).

The standard deviation σ_E is usually substituted by the sample standard deviation, computed on the basis of the data available at time period T:

$$S_E = \sqrt{\frac{\sum\limits_{t=2}^{T} \left(e_t - \overline{E}\right)^2}{T - 2}}, \tag{2.44}$$

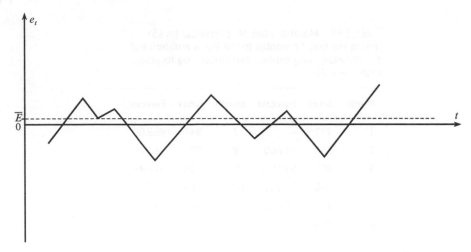

Figure 2.47 Visual examination of the control chart: errors with a sample mean $\bar{E}$ not equal to zero.

where $\bar{E}$ is the sample mean, close to zero in case there are no systematic errors, in which case (2.44) becomes

$$S_E \approx \sqrt{\frac{\sum\limits_{t=2}^{T} (e_t - 0)^2}{T - 2}} = \sqrt{MSE_T}. \tag{2.45}$$

In case $\bar{E}$ is significantly different from zero, the forecast is *biased* (see Figure 2.47), which would suggest updating the parameters of the forecasting method (or, eventually, changing it to a more appropriate one).

In addition to the previous analytical check, it is useful to verify visually the control chart to determine whether it is possible to improve the forecasts by introducing suitable modifications to the prediction method:

1. the time series of the errors shows a positive or a negative trend; in this case, the accuracy of the forecasting method is progressively diminishing;
2. the time series of the errors is periodic; this may happen if an existing seasonal component has not been identified.

[Softline.xlsx] Softline, a company manufacturing leather sofas, uses the simple moving average method (with $r = 2$) to forecast the monthly unit production cost (in €, assuming sales volume constant) of its top product called Trinity. To assess the forecasting process, the logistics manager uses a "3-σ" control chart. Table 2.55 reports the unit production costs as well as the forecasts and the errors made during the past two years. The corresponding sample mean error was

$$\bar{E} = € 1.40.$$

Table 2.55 Unit production costs, forecasts and errors (in €) in the Softline problem.

t	y_t	p_t	e_t	t	y_t	p_t	e_t
1	452.00	–	–	13	468.50	458.90	9.60
2	463.50	452.00	11.50	14	462.00	465.65	−3.65
3	478.00	457.75	20.25	15	470.00	465.25	4.75
4	466.50	470.75	−4.25	16	473.80	466.00	7.80
5	457.00	472.25	−15.25	17	472.00	471.90	0.10
6	465.20	461.75	3.45	18	478.00	472.90	5.10
7	460.00	461.10	−1.10	19	468.40	475.00	−6.60
8	457.50	462.60	−5.10	20	473.40	473.20	0.20
9	463.50	458.75	4.75	21	468.00	470.90	−2.90
10	456.00	460.50	−4.50	22	463.00	470.70	−7.70
11	455.00	459.75	−4.75	23	470.50	465.50	5.00
12	462.80	455.50	7.30	24	475.00	466.75	8.25
				25	–	472.75	–

Since this value is a small fraction of a typical monthly unit production cost, the forecast was deemed to be unbiased. The sample standard deviation of the error was approximated by using formula (2.45)

$$S_E = \sqrt{MSE_{24}} = €\ 7.89.$$

All the errors were in the range $\pm 3 \times €\ 7.89 = €\ \pm 23.67$ (a similar result can be obtained by considering the exact value of the sample standard deviation of the error, equal to 7.76, obtained in Excel by using the ST.DEV.S function. In this case, the confidence interval is $\pm 3 \times €\ 7.76 = €\ \pm 23.28$). Hence, the forecasting method was deemed to be under control. Moreover, the visual examination of the control chart in Figure 2.48 did not show any systematic forecasting error.

Figure 2.48 Plot of the control chart in the Softline problem.

2.11 Interval Forecasts

Previous sections were devoted to *point forecasts*, that is, forecasts expressed by a single possible outcome. This section deals with *interval forecasts*, defined as ranges of outcomes that the variable to be forecast is likely to take with a given confidence level (see Section 1.10.3). Here the focus will be on the one-period-ahead forecast case and it will be assumed that the forecasting errors e_t, $t = 2, \dots, T$, are independent realizations of a normal random variable E with mean $\mu_E = 0$ and standard deviation σ_E (usually unknown). It is worth noting that if μ_E were different from zero, there would be a systematic error and the forecasts could be improved by subtracting this constant offset μ_E. As a result, it can be assumed $\mu_E = 0$ without any loss of generality.

Given a confidence level $(1 - \alpha)$, if σ_E were unknown, normalized error $(E - \mu_E)/\sigma_E$ will satisfy relationship

$$Pr\left(-z_{\alpha/2} \leq \frac{E}{\sigma_E} \leq z_{\alpha/2}\right) = 1 - \alpha,$$

where $z_{\alpha/2}$ is the quantile of order $(1 - \alpha/2)$ of the standard normal distribution. If σ_E is unknown, as is often the case, its estimate S_E, given by (2.45), can be used to obtain

$$Pr\left(-t_{\alpha/2,T-2} \leq \frac{E}{\sqrt{MSE_T}} \leq t_{\alpha/2,T-2}\right) = 1 - \alpha, \tag{2.46}$$

where $t_{\alpha/2,T-2}$ is the quantile of order $(1-\alpha/2)$ of the Student's t-distribution with $T-2$ degrees of freedom. On the basis of (2.46), the following interval forecast is computed for the first-period ahead

$$\left[p_{T+1} - t_{\alpha/2,T-2}\sqrt{MSE_T}; p_{T+1} + t_{\alpha/2,T-2}\sqrt{MSE_T}\right]. \tag{2.47}$$

[Softline.xlsx] In the Softline problem, the logistics manager wants to devise, at time period $T = 24$, an interval forecast for the one-month-ahead unit production cost of the Trinity sofa, with a confidence level of $(1 - \alpha) = 0.90$. Since $\alpha/2 = 0.05$ and, consequently, $(1 - \alpha/2) = 0.95$, the quantile involved into the computation is $t_{\alpha/2,T-2} = t_{0.05,22} =$ T.INV(0.95;22) = 1.7171, where T.INV(0.95;22) is the Excel function which returns the left-tailed inverse of the the Student's t-distribution with probability equal to 0.95 and degree of freedom equal to 22. By utilizing (2.47) and taking into account that $p_{25} =$ € 472.75 and $\sqrt{MSE_{24}} =$ € 7.89, the interval forecast turns out to be

[€ (472.75 − 1.7171 × 7.89); € (472.75 + 1.7171 × 7.89)]

= [€ 459.20; € 486.30].

Hence, the unit production cost of a Trinity sofa in the 25th month will be included between € 459.20 and € 486.30 with a confidence level equal to 90%.

Table 2.56 Number of batteries sold by Shivoham between 2009 and 2018.

Year	Karnataka market sales [number of batteries]	Shivoham sales [number of batteries]	Shivoham market share [%]
2009	693 326	138 665	20
2010	803 666	152 696	19
2011	947 243	170 503	18
2012	1 136 433	193 192	17
2013	1 406 432	210 964	15
2014	1 666 011	233 241	14
2015	1 869 683	243 058	13
2016	2 136 463	256 375	12
2017	2 316 402	266 386	11
2018	2 507 929	275 872	11

2.12 Case Study: Sales Forecasting at Shivoham

Shivoham is an electromechanical company headquartered in Bangalore, India, manufacturing spare parts for cars for a large part of the Karnataka state. In 2018, the results of a survey showed that, although the company's car battery sales constantly increased during the previous decade, the company progressively lost market share (see Table 2.56). Until 2018, the company had traditionally based its production and marketing plans on sales forecasts provided by an extrapolation method. Applied to the data shown in Table 2.56, the method would result in the following expression

$$p_{10}(\tau) = 285\,875.06 + 15\,951.08\tau, \tau = 1, 2, \ldots,$$

which would provide the following demand forecasts: 301 826 units for 2019 ($\tau = 1$) (with a 9.4% increase with respect to 2018) and 317 777 units for 2020 ($\tau = 12$) (with a 15.2% increase with respect to 2018). However, the results of the survey convinced the company's managers that during the previous decade Shivoham had lost an opportunity to sell more. Based on this reasoning, it was decided to predict sales of batteries by first estimating the Karnataka's market demand and then taking appropriate marketing initiatives. First, it was decided to use a regression method (see Section 2.7.1) in which market sales of batteries were correlated to the number of cars sold two years before (see Table 2.57). This enabled to forecast the Karnataka's demand for the years 2019 and 2020, as 2 396 003 and 2 676 295 units, respectively. Then, the company's managers generated several scenarios based on different market shares. Assuming the firm maintained a market share equal to 11%, the company's demand would be equal to 263 560 units in 2019 (with a 4.5% increase with respect to 2018), and 294 392 units in 2020 (with a 6.7% increase with respect to 2018).

The two forecasting methods used by the company provided different results. Therefore, the company decided to better analyze the logic underlying the two approaches. Because the Karnataka's economy was undergoing a period of quick and remarkable change, the latter method was deemed to provide more accurate predictions than the

Table 2.57 Car sales (number) in Karnataka between 2007 and 2018.

Year	Sales	Year	Sales
2007	253 321	2013	886 297
2008	381 385	2014	1 014 975
2009	491 755	2015	1 162 246
2010	634 706	2016	1 167 614
2011	951 704	2077	1 217 929
2012	830 175	2018	1 363 594

former, which is more suitable when the past demand pattern is likely to be replicated in the future.

2.13 Case Study: Sales Forecasting at Orlea

Orlea is an Australian firm that produces and distributes bicycles. The company makes more than 150 models, divided into four classes: "Mountain Bike", "Racing", "City", and "Junior". Each class is further articulated into families (e.g., the class "City" consists of four families). The models also differ in colour and size, with more than 500 versions overall. The range of bicycles is renewed every sales period, which begins in September and ends in August of the subsequent year. The company is active on the Australian market (about 60% of the company's sales), with a network of more than 450 dealers across the country, and on some foreign markets, in collaboration with affiliate companies and distributors. In Australia, the supply chain is of MTS type (see Section 1.3.2), since dealers, due to the limited budget availability and to the difficulty in sales forecasting, work with low inventory levels and issue orders of low size which the company has to fulfil within a short time. Conversely, the affiliate companies and the foreign distributors require large lots of bicycles with a quite long delivery time (up to four months). In this case, the company works with an MTO-type supply chain. In order to keep inventory management costs low, in 2018 the company decided to adopt a sales forecasting system for the short term (a few months).

For products of the "Dream" family (which includes nine models and belongs to the "City" class), the time series related to the monthly sales in 2015/16, 2016/17 and 2017/18 (see Figure 2.49) are characterized by a strong seasonal component.

The sales forecasting method has been articulated into three phases. In the first, a sales forecast for the "Dream" family has been defined by using an extrapolation method. In the next phase, the data forecast in the previous step are disaggregated among the nine models of the family, by proportionally dividing the overall values and using specific coefficients which are updated monthly. Finally, the demand for each bicycle model has been estimated qualitatively by the marketing managers, based on their experience, starting from the corresponding data available at the end of the second phase. For the first phase, the following process was adopted.

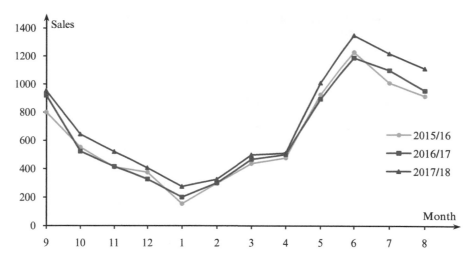

Figure 2.49 Plot of the monthly sales of bicycles (number) of the "Dream" family in 2015/16, 2016/17 and 2017/18.

- **Data preprocessing.** The available monthly sales of bicycles belonging to the "Dream" family were extracted from the company's information system and preprocessed; in particular, calendar variations and outliers were removed (see Sections 2.5.2 and 2.5.4).
- **Initialization.** Data from 2015/16 period were used to this end.
- **Parameter calibration.** The values of the parameters which characterize the adopted forecasting methods (like the Holt method and the Winters method) were determined by computing the *MSE* over 2016/17 period.
- **Choice of the forecasting method.** Once the optimal parameter values were determined for each of the adopted forecasting methods, these methods were compared by computing the *MSE* over 2017/18 period and the most suitable forecasting method was chosen.
- **Forecast.** The selected method was used to forecast the short-term demand of the bicycle model family.

The adoption of the described procedure has allowed the reduction of operative costs and the enhancement of the service level, increasing at the same time the level of company's awareness of market dynamics.

2.14 Questions and Problems

2.1 Groupe Danone is a French multinational dairy company, ranked third in Europe and seventh worldwide. In Italy, Danone has launched a number of probiotic products, such as Activia and Actimel. During the first three months of last year, the company carried out an opinion poll by launching a campaign called "Choose the Flavour" on the Facebook fanpage Activia Italia. The goal was to let consumers select the new yoghurt flavours. In this way, Facebook was used as a tool

Table 2.58 Number of Hot Spot heater installations and service requests.

	Heater installations			
Year	> 2 years ago	≤ 2 years ago	Total	Service requests
1	260 000	69 500	329 500	18 672
2	265 000	74 200	339 200	19 076
3	287 800	82 850	370 650	20 994
4	313 750	90 550	404 300	23 249
5	345 350	97 150	442 500	25 025
6	379 050	105 950	485 000	28 111
7	416 950	111 550	528 500	30 985
8	459 100	117 000	576 100	33 397
9	502 550	123 200	625 750	

to carry out a market survey. Discuss the limits and the opportunities offered by such a qualitative forecasting method.

2.2 In freight transportation, a typical logistics problem is the allocation of empty containers to depots, in order to satisfy future demands of transportation services. How can a carrier predict demands of containers for the transportation of a given category of products?

2.3 To what extent are the forecasting practices different in an MTS- and in an MTO-supply chain?

2.4 How would you predict the future demand of a new product?

2.5 [HotSpot.xlsx] Hot Spot is a firm based in the United States whose core business is the maintenance of home heaters. The company usually forecasts service requests on the basis of the number of installed heaters. The number of installed home heaters and the number of service requests received over past years in New Jersey are reported in Table 2.58.
Forecast the service request p_9 by using, to this purpose:

- a single regression model (service requests versus the total number of heaters installed);
- a multiple regression model (service requests versus the number of heaters installed in the last two years and more than two years ago).

Which forecasting method is the most accurate? Why?

2.6 [Sunshine.xlsx] Sunshine is one of the world's leading suppliers of fast-moving goods in household care and personal product categories. According to the company management, the sales of the facial-cleansing Blemish product mostly depend on promotion expenditure made by the company and its competitors. Table 2.59 reports facial soap sales (in millions CAN$) for the previous 10 quarters, and quarterly levels of sales in Canada for that detergent with respect to (a) Sunshine promotion expenditure, and (b) the competitors' promotion expenditure divided by the Sunshine promotion expenditure. Make a sales forecast for the next two time periods under the hypothesis that Sunshine will increase its

Table 2.59 Data for the forecasting problem of Sunshine.

Quarter	Sunshine promotion expenditure [millions CAN$]	Competitors' promotion expenditure /Sunshine promotion expenditure	Sales of Blemish in Canada [millions CAN$]
1	6.0	1.2	46.8
2	6.8	1.2	52.7
3	7.5	1.4	60.5
4	7.5	1.5	56.6
5	9.0	1.5	64.4
6	10.5	1.7	74.1
7	12.0	1.8	72.2
8	12.0	1.5	78.0
9	13.5	1.4	87.8
10	13.5	1.5	95.6

Table 2.60 Aqua-Floor sales (in thousands of m^2) realized by Carlinek in Poznan during the last nine months.

Month	Sales	Month	Sales
1	450	6	437
2	670	7	456
3	332	8	–
4	123	9	231
5	343		

promotion expenditure to CAN$ 14.5 million per quarter and the competitors' promotion expenditure will remain the same as in the current quarter.

2.7 [Carlinek.xlsx] Carlinek is a Polish company, based in Poznan, which retails Aqua-Floor, a well-known water-resistant laminated floor covering. The company needs to forecast the demand of this product, by using historical data on the sales of the latest nine months, reported in Table 2.60. After checking the presence of possible outliers and missing data in the dataset, use the weighted moving average method to determine the one-month-ahead sales forecast.

2.8 [Flanders.xlsx] Table 2.61 reports an estimate of the annual mean demand and of its standard deviation (in tonnes) of the fruit and vegetable products of Flanders, Belgium. The products are divided into five categories: fall–winter fresh vegetable, spring–summer fresh vegetables, exotic fruits, citrus fruits, and dried fruits.

Verify whether the aggregated forecast is more or less accurate than the estimated demand for each category.

Table 2.61 Estimate of the annual mean demand and of its standard deviation (in tonnes) of fruit and vegetable products of Flanders.

Category	Mean	Standard deviation
Fall–winter fresh vegetables	123 000	14 500
Spring–summer fresh vegetables	245 000	28 450
Exotic fruits	9 860	1 120
Citrus fruits	98 000	9 750
Dried fruits	2 450	230

2.9 [Ravaioli.csv] The optimal parameters of a large regression model are often determined by using a gradient descent method instead of the closed-form equation (2.7) (in which the inversion of matrix $\mathbf{X}$ can be time-consuming). Implement the gradient descent method in Python and use it to fit the linear regression model in the Ravaioli problem.

2.10 [Ravaioli.csv] Because of their universal function approximation properties, ANNs are sometimes used by practitioners without a preliminary detailed analysis of the available data. Following this approach, select and train an ANN model for the Ravaioli problem and compare its accuracy with that obtained with the linear regression model (see Section 2.7).

2.11 [UnitedPioneers.xlsx] In the United Pioneers problem, compute the coefficient of determination between y and $\ln x_1$.

2.12 [Lmart.csv] In the Lmart problem, train an ANN model with 10 hidden neurons. Then, split the data into a training set and a validation set, in order to compare such a model with the neural network described in Section 2.7.8.

2.13 [Lmart.csv] Does overfitting occur when using the predictive model developed in Problem 2.12? Why?

2.14 [VaalEngineering.xlsx] Vaal Engineering is a South African company that produces and distributes industrial slot machines. The number of 800xp machines sold during the last 60 months is reported in Table 2.62. After plotting the data, clean the time series from possible seasonal and random components and then determine the trend.

2.15 [Seasonality.xlsx] A time series $w_t, t = 1, \dots, 24$, is reported in Table 2.63. Can the time series correspond to seasonal indices, with periodicity $M = 12$, obtained through the application of a decomposition method (multiplicative model)? If so, explain the meaning of w_1, otherwise explain what it is possible to do.

2.16 The time series decomposition method (additive) into trend, cyclical, seasonal and random components is used to estimate the monthly sales (in kg) of a product. Characterize the seasonal component $s_{\bar{t}}$ for some $\bar{t}$, in terms of units, magnitude (can it be positive, negative, or zero? Explain its meaning) and the average value of M consecutive time periods (M is the seasonal periodicity).

2.17 [Mitsumishi.xlsx] Mitsumishi is a Korean company whose number of M5 light trucks sold during the last 42 months is reported in Table 2.64. The company

Table 2.62 Number of 800xp slot machines sold by Vaal Engineering during the last 60 months.

Month	Quantity	Month	Quantity	Month	Quantity	Month	Quantity
1	60	16	104	31	35	46	144
2	83	17	205	32	86	47	185
3	130	18	235	33	147	48	280
4	120	19	32	34	139	49	89
5	187	20	81	35	176	50	73
6	248	21	149	36	283	51	137
7	51	22	141	37	75	52	146
8	76	23	177	38	78	53	191
9	125	24	281	39	130	54	268
10	134	25	55	40	112	55	31
11	196	26	97	41	202	56	63
12	262	27	139	42	262	57	126
13	54	28	107	43	44	58	133
14	94	29	206	44	80	59	213
15	132	30	232	45	128	60	281

Table 2.63 Time series $w_t, t = 1, \ldots, 24$, used in Problem 2.15.

t	w_t	t	w_t	t	w_t
1	1.012	9	0.876	17	1.174
2	1.123	10	0.904	18	1.055
3	1.088	11	0.812	19	1.101
4	1.122	12	0.714	20	1.086
5	1.097	13	1.057	21	0.909
6	1.023	14	1.134	22	0.987
7	1.001	15	1.099	23	0.893
8	0.987	16	1.180	24	0.811

invested significant financial resources in promotion, and the M5 sales increased in some months, as shown in Table 2.65.

(a) Forecast sales for the next six months using an appropriate forecasting method.

(b) Plot a control chart. Are you able to detect any anomalies?

2.18 [Sit.xlsx] Sit is an American company which produces and sells high-quality printers. The number of XC2100 printers monthly sold in Berkeley, California, during the last 12 months is reported in Table 2.66. The company has always

Table 2.64 Number of M5 trucks sold by Mitsumishi during the last 42 months.

Month	Quantity	Month	Quantity	Month	Quantity
1	22 882	15	20 967	29	28 414
2	19 981	16	19 759	30	22 537
3	18 811	17	22 200	31	22 845
4	19 352	18	24 162	32	9 451
5	27 226	19	20 275	33	15 842
6	18 932	20	7 949	34	16 409
7	18 931	21	14 328	35	13 881
8	8 523	22	16 691	36	11 230
9	13 064	23	13 784	37	24 765
10	13 733	24	10 986	38	21 739
11	12 597	25	24 768	39	25 153
12	7 645	26	19 351	40	20 515
13	23 478	27	23 953	41	24 038
14	17 019	28	18 855	42	25 151

Table 2.65 M5 sales improvement (in %) achieved during some months by Mitsumishi.

Month	Sales improvement	Month	Sales improvement
5	30	19	8
6	13	29	30
7	3	30	12
17	20	31	5
18	15		

Table 2.66 Number of XC2100 printers sold during the last 12 months by Sit.

Month	Quantity	Month	Quantity
1	835	7	810
2	798	8	814
3	831	9	800
4	772	10	793
5	750	11	805
6	783	12	829

Table 2.67 Forecasting errors $e_t, t = 2, \ldots, 12$, generated by using the Q method.

Month	Error	Month	Error
1	–	7	16
2	24	8	x
3	−13	9	−8
4	19	10	41
5	−35	11	31
6	27	12	−28

Table 2.68 Monthly demand for lemons (in tonnes) and number of workdays in 10 months for the Artelemon problem.

Month	Demand	Number of workdays
1	4283	22
2	4187	20
3	4283	21
4	4141	20
5	4086	22
6	4121	20
7	4193	23
8	4246	23
9	4129	21
10	4476	23

forecast the monthly one-period-ahead demand by using the simple moving average method ($r = 2$), but the logistician is now evaluating the option of adopting a new forecasting method, called Q. Knowing the forecasting errors $e_t, t = 2, \ldots, 12$, made by method Q for one-period-ahead forecasts (and reported in Table 2.67), determine the value that e_8 should take to induce the logistician to replace the simple moving average method with Q.

2.19 [Artelemon.xlsx] Artelemon is a company which distributes citrus fruits in Spain. Consider the monthly demand for lemons (in tonnes) reported in Table 2.68. First, preprocess the data, specifying the type of task to be performed. Then use the weighted moving average method to make a one-month-ahead forecast, by considering that there will be 20 workdays in the eleventh month. Finally, extend the forecast to the twelfth month (which has 22 workdays).

2.20 [Juice.xlsx] Juice produces and markets fruit juice in Ireland. Table 2.69 reports the value of the company's sales (in €) over the last 12 years, and the corresponding inflation rates (in %). First, deflate the annual sales. Then, perform

Table 2.69 Annual Juice sales and the corresponding inflation rate.

Year	Sales [€]	Inflation rate [%]	Year	Sales [€]	Inflation rate [%]
1	132 000	2.7	7	155 000	2.2
2	140 000	2.1	8	153 000	1.7
3	143 000	2.4	9	158 000	2.6
4	140 000	1.9	10	163 000	2.1
5	150 000	3.1	11	165 000	2.5
6	152 000	2.3	12	170 000	2.4

Table 2.70 Aldes exports (in $) of tuna fish to Jordan during the last eight months.

Month	Sales	Month	Sales
1	100 000	5	92 000
2	98 000	6	87 500
3	70 500	7	95 000
4	90 000	8	99 500

an exploratory analysis to prove that the deflated time series presents a linear trend and no cyclical and seasonal components. Finally, provide the one- and two-ahead sales forecasts, using the double moving average method with $r = 3$.

2.21 [Aldes.xlsx] Aldes is a food and beverage company located in the Republic of Mauritius which specializes in the export of canned food, such as tuna fish, mainly sold in Africa, the Middle East, and Western Europe, in cartons of 48 cans of 185 g each. The monthly exportation value of tuna fish to Jordan during the last eight months is summarized in Table 2.70. Estimate the two-period-ahead exports of tuna fish by using the exponential smoothing method with $\alpha = 0.20$, $\alpha = 0.25$ and $\alpha = 0.35$. By comparing the obtained results, what is the best value of α? If the company decides to promote the product through a discount policy to increase the sales, how convenient is it to modify the choice of α? Is it possible to use the selected method for six-month-ahead forecasting?

2.22 Prove that the sum of weights in equation (2.21) is equal to 1. In addition, show that the weight of y_1 is less than the weight of y_2 if and only if $\alpha \in (1/2, 1)$.

2.23 A logistician decides to forecast the one-period-ahead demand of a product as follows

$$p_{T+1} = \gamma p_{T+1}^{(1)} + (1 - \gamma)p_{T+1}^{(2)}, \quad 0 \le \gamma \le 1, \tag{2.48}$$

where $p_{T+1}^{(1)}$ and $p_{T+1}^{(2)}$ are the forecasts provided, respectively, by the weighted moving average method and the exponential smoothing method. Demonstrate that the forecast (2.48) corresponds to

Table 2.71 Time series $y_t, t = 1, \ldots, 12$, used in Problem 2.24.

t	y_t	t	y_t
1	1100	7	920
2	935	8	985
3	1120	9	1070
4	1040	10	940
5	1060	11	1100
6	1100	12	970

Table 2.72 Annual sales and number of births in Campania during the last six years in the Babytoys problem.

Year	Sales [€]	Births
1	134 600	47 545
2	145 400	47 822
3	158 000	47 434
4	163 000	48 122
5	168 500	47 934
6	179 700	48 006

$$p_{T+1} = \sum_{t=1}^{T} w_t y_t,$$

where $\sum_{t=1}^{T} w_t = 1$.

2.24 [Brown.xlsx] Given the time series $y_t, t = 1, \ldots, 12$, reported in Table 2.71, use the exponential smoothing method to determine the forecasting value p_{13}, assuming a time-variable parameter $\alpha_t = \max_t \left\{ 0.24; \sqrt[3]{\frac{3}{5t^2}} \right\}, t = 2, \ldots, 12$. Modify equation (2.21) when the time-variable parameter is used and verify that the sum of weights still remains equal to 1. Determine whether the time-variable parameter implies a more accurate forecast than the value of $\alpha = 0.3$.

2.25 [Babytoys.xlsx] Babytoys is an Italian company producing hypoallergenic toys. Its logistics manager wants to forecast the overall company's demand in the Campania region in the next few years. The annual sales and the number of births registered in Campania over the last six years are shown in Table 2.72. Adjust the annual sales considering the number of births. Verify that the adjusted time series is characterized by a linear trend and has no cyclical and seasonal components. Estimate the one- and two-ahead yearly sales forecasts using the Holt method. Set $\alpha = 0.3$ and choose the most appropriate value of $\beta = 0.2$ or $\beta = 0.5$. Compute

Table 2.73 Quantity (in thousands of kg) of g/m² polyethylene and nylon film used by EuroPack in the last 30 days.

Day	Quantity	Day	Quantity
1	133	16	154
2	155	17	220
3	179	18	207
4	207	19	213
5	176	20	219
6	165	21	219
7	204	22	170
8	145	23	213
9	104	24	227
10	209	25	205
11	149	26	203
12	210	27	187
13	185	28	225
14	204	29	195
15	203	30	194

the forecast as a function of the estimated number of births for next year (e.g., 48 200).

2.26 [EuroPack.xlsx] EuroPack is a leading European firm of packaging products. The plant, located in Denmark, manufactures in a continuous cycle more than 150 products in 60 g/m² polyethylene and nylon film.

(a) Estimate the amount of polyethylene and nylon film needed to ensure daily production in the next week by using historical data related to the last 30 days (shown in Table 2.73).

(b) Determine the tracking signal (band ±4) at the current time period.

(c) Determine the one-day-ahead interval forecast of polyethylene and nylon film with a confidence level $(1 - \alpha) = 0.95$.

2.27 [Oasis.xlsx] Oasis is a pineapple soft drink sold in Germany. The monthly sales (in kl) over the past 44 months are reported in Table 2.74.

(a) Perform an exploratory analysis of the data. What important observations can be made about the demand pattern? Which data are relevant and should be used for forecasting purposes?

(b) Use two different forecasting methods to predict sales over the next four months.

(c) Compute *MAPE* of both forecasting methods using the data of the last six months. Which approach seems to work best?

(d) Select the most accurate method and then determine, with a confidence level $(1 - \alpha) = 0.90$, the interval forecast for the one-month-ahead sales at $T = 44$.

Table 2.74 Sales (in kl) of Oasis soft drink in Germany over the last 44 months.

Month	Quantity	Month	Quantity	Month	Quantity	Month	Quantity
1	9 050	13	10 000	25	9 150	37	8 900
2	8 050	14	9 750	26	8 750	38	9 450
3	7 000	15	9 300	27	8 800	39	8 750
4	6 120	16	10 100	28	9 400	40	9 500
5	8 250	17	10 400	29	12 000	41	13 400
6	11 450	18	15 650	30	13 450	42	14 000
7	10 900	19	16 350	31	14 900	43	16 850
8	12 850	20	17 000	32	18 760	44	21 000
9	10 650	21	13 600	33	12 250		
10	11 000	22	11 250	34	11 000		
11	9 200	23	9 500	35	9 600		
12	8 900	24	9 200	36	9 100		

Table 2.75 Number of vans daily used over the last five weeks by Hollaflowers.

	Week				
Day	1	2	3	4	5
Monday	3	24	21	25	24
Tuesday	8	16	25	23	16
Wednesday	10	12	12	19	8
Thursday	2	18	15	12	22
Friday	6	6	26	12	14

2.28 [Hollaflowers.xlsx] Hollaflowers is a Dutch cut-flowers company with its own distribution network that covers five European countries. In order to maintain product quality control in the last phase of the distribution system, the company uses its vans for timely delivery to customers. To this end, it is required to determine the number of vans needed to satisfy the future demand of customers. Based on daily data over the last five weeks, which are reported in Table 2.75, forecast the number of vans for the one-week-ahead deliveries.

2.29 [Neurozam.xlsx] The National Health Service of Belgium is responsible for the distribution of Neurozam to all national hospitals. Neurozam is a drug used to fight a rare neurological disease. Its high cost and perishability forced the National Health Service to have a small amounts of stocks, used to satisfy demand of no more than three days. Plan the supply of the drug of the National Health Service taking into account the number of packages of the drug (three vials each) distributed in the last 20 days (see Table 2.76).

Table 2.76 Number of packages (three vials each) of Neurozam distributed over the last 20 days from the Belgian National Health Service.

Day	Quantity	Day	Quantity
1	23	11	0
2	0	12	0
3	19	13	0
4	0	14	16
5	14	15	14
6	0	16	6
7	0	17	0
8	13	18	0
9	16	19	12
10	0	20	11

2.30 A company is planning to add extra capacity to a plant currently manufacturing 110 000 items per year. After an accurate sales forecast for the next few years (based on data entries of the last 12 years), the logistics manager is quite sure that the most likely value of the annual demand is 140 000 items and that the *MSE* is equal to 10^8 items. It is known that the company loses \$ 3 for each unit of unused capacity and \$ 7 for each unit of unsatisfied demand. How much capacity should the company buy? (Hint: suppose that the forecasting error can be assumed as a random variable normally distributed).

2.31 [Mitsumishi2.xlsx] Given the time series $y_t, t = 1, \dots, 42$, reported in Table 2.64, it is assumed that $y_t = 0$, for $t = 4, 5, 6, 16, 17, 18, 28, 29, 30, 40, 41, 42$. Determine $p_{42}(\tau), \tau = 1, \dots, 6$, using, if it is necessary, the same forecasting method selected in Problem 2.17.

2.32 [Plaza.xlsx] Monitor the forecasts for the monthly sales of sportswear of the Plaza supermarket (see Section 2.10) by using, as an accuracy measure, the *MAE* instead of the tracking signal.

2.33 [Aldes.xlsx] To control a one-period-ahead forecasting process determined with the exponential smoothing method ($\alpha = 0.35$) on the time series reported in Table 2.70 (see Problem 2.21), use a "3-σ" control chart. What you can say about the accuracy of the forecasting method? Determine the interval forecast for the one-period-ahead with a confidence level equal to 95%.

3

Designing the Logistics Network

3.1 Introduction

Logistics network design is a strategic decision that must ensure the entire logistics system works at the lowest cost while ensuring a proper customer service. It includes the determination of the number, type, size, location, and equipment of new facilities, as well as the divestment, displacement, or downsizing of existing ones. Even though the impact of logistics network design is generally over a long-term time horizon, it requires continuous monitoring and updating, in order to ensure the best fit to changes in demand, user expectations, and other environmental conditions. The most important factors that may trigger a redesign of the logistics network are:

- *Cost changes.* Logistics costs may grow incrementally over time for a number of reasons, in which case a more efficient logistics network is required.
- *Market changes.* A change in market demand, both in terms of volume and spatial distribution, can prompt an organization to review the configuration of its logistics network (e.g., by opening a facility to serve an emerging market in a timely fashion).
- *Technological changes.* Technological advances may lead to new products and services, and accelerate the obsolescence of old products and services. They may also increase productivity and reduce costs. As a consequence, a change in the logistics network may be desirable.
- *Economic changes.* Variations in taxes, tariffs, exchange rates, freight, and fuel costs may have a major impact on costs and profits. Hence, they may trigger a change in the configuration of the logistics network.
- *Political changes.* A change in the form of government or a legislative reform (e.g., the liberalization of a market) may push an organization to revise the structure of its logistics network.
- *Organizational changes.* A merger, acquisition, or divestment operation, as well as a corporate restructuring, can have a significant impact on the logistics network. For example, if a company acquires a competitor, their supply chains may be merged to reduce costs.

In the scientific literature, a logistics network design problem is referred to as a *facility location problem*. Indeed, such a problem comprises not only the choice of the location, type, size, and equipment of the facilities, but also the definition of their area

Introduction to Logistics Systems Management: With Microsoft® Excel® and Python® Examples, Third Edition. Gianpaolo Ghiani, Gilbert Laporte, and Roberto Musmanno.
© 2022 John Wiley & Sons Ltd. Published 2022 by John Wiley & Sons Ltd.

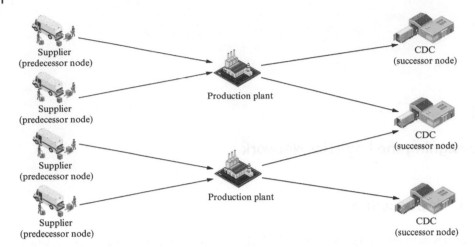

Figure 3.1 An example of a two-echelon single-type location problem.

boundaries (i.e., the allocation of demand to facilities). For example, in a two-echelon logistics system (see Figure 3.1), the opening of a new production plant must be accompanied by a redefinition of the allocation of DCs to production plants, and it may be necessary to modify the policies for the supply of raw materials and semi-finished products from suppliers to production plants. For this reason, logistics network design problems are sometimes referred to as *location-allocation problems*.

Some location decisions can be considered as tactical. This happens when changing facility locations or equipment does not involve a significant investment (e.g., if space and equipment are rented from a public warehouse and handling operations are subcontracted). See Section 3.16 for an example.

3.2 Classification of Logistics Network Design Problems

Facility location problems can be classified with respect to a number of criteria. The classification proposed in this chapter is logistics-oriented.

Key factors in decisions
In industrial logistics the main concern is to minimize the overall logistics cost while ensuring a reasonable customer service level. On the other hand, when locating facilities providing services of general interest (fire service units, ambulance parking areas, waste dump sites, police stations, etc.) it is of the utmost importance to seek equity among all the users.

Decision space
Location problems can be *continuous* if the candidate locations are points of a continuous space, represented by a Cartesian plane, or *discrete*, whenever the facilities have to be selected from a list of potential sites. Moreover, some location problems can be modelled on a graph (directed, undirected or mixed) on which facilities corresponding to vertices or points on arcs (or on edges) have to be located.

Although appearing at first glance to be of little practical interest, continuous location problems can be used to make a preliminary selection of the geographic area where potential sites may be considered. It is worth observing that the solution of such problems may correspond to infeasible locations (e.g., locations in the middle of the sea or on the top of a mountain) unless specific constraints are imposed to forbid the location of facilities in restricted areas (see Problem 3.10).

Single versus multi-commodity flows

In *single-commodity problems*, all the materials transported share the same characteristics (nature, density, cost of transportation per unit of flow and per unit of distance, etc.). Conversely, in *multi-commodity* problems, materials (or groups of materials) with different features are modelled individually, each with its own parameters. In the latter case, each commodity is associated with a specific flow pattern.

Facility typology

In *single-type* location problems, a single type of facility (e.g., only regional warehouses) must be located. In contrast, in *multi-type* problems several kinds of facilities (e.g., both central and regional warehouses) have to be located.

Interaction among facilities

In logistics systems there can be material flows between facilities of the same kind (e.g., component flows between production plants). In this case, optimal facility locations depend not only on the spatial distribution of finished product demand, but also on the mutual positions of the facilities (*location problems with interactions*).

Dominant flows

A further classification can be made on the basis of the relevance of the flows between the facilities to be located and those directly connected to them.

A theory of industrial location, formulated by A. Weber at the beginning of the last century, suggests locating a production plant where the total transportation cost of raw materials and finished product is minimum. In particular, two special cases can be identified:

- when the weight of a finished product is less than the weight of the raw materials used to produce it (*weight-losing case*), the optimal location is close to the sources of raw materials;
- in contrast, if a finished product is heavier than the raw materials used to produce it (because some ubiquitous raw material, such as water, is incorporated, *weight-gaining case*), then the optimal location is close to the markets.

Number of echelons

In a *one-echelon* location problem, facilities of a single type have to be located, and either the flow entering the facilities or the flow coming out from them is negligible. A typical case of a one-echelon problem arises in manufacturing plant location (e.g., in the steel industry), in which the weight of the finished products is significantly lower than that of the raw materials (iron and coal); under these assumptions, the transportation costs associated with flows incoming the manufacturing plants are dominant. Another example of a one-echelon problem is the location of the central warehouse of a retail company purchasing goods at pre-established prices (inclusive of the transportation cost up to the central warehouse). In this case, only the transportation cost from the warehouse to the sales points has to be considered when locating the warehouse.

On the other hand, in *two-echelon* location problems, both inbound and outbound flows are relevant. This is the case, e.g., of the location of production plants for which both the transportation of components from the suppliers to the plants, and the transportation of finished products from the plants to the CDs, are relevant (see Figure 3.1).

Divisibility of flows

In some systems, each logistics node has to be serviced by a single facility for administrative or book-keeping reasons. For example, in the logistics system depicted in Figure 3.1, each warehouse may have to be served by a single manufacturing plant. Then the corresponding location problem should consider this additional constraint. In other cases, the same successor node can be served by more facilities.

Influence of transportation policy on location decisions

Most facility location models do not consider the actual transportation policy. Rather, they simply rely on the hypothesis that there are direct transportation services. However, in a number of contexts, including *last-mile distribution*, where small loads have to be collected or delivered, the same vehicle typically serves several points (e.g., retailers in a sales district). In such cases, the transportation cost depends on the order in which destinations are served and this also impacts the location decisions. To illustrate this concept, consider Figure 3.2 where a distribution warehouse to be located should serve three sales points situated at the vertices of triangle *ABC*. Under the hypothesis that the facility costs are independent of where the warehouse is located, there can be two extreme cases:

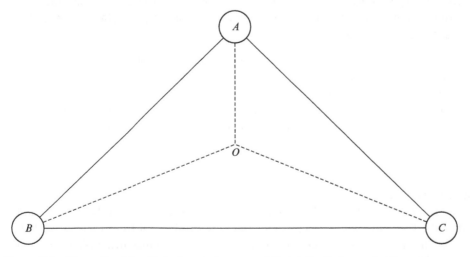

Figure 3.2 The optimal location of a warehouse could be at the Steiner point *O* or at any point of the triangle's *ABC* perimeter, depending on the transportation policy.

- each sales point requires a full load replenishment and, therefore, the optimal location of the warehouse corresponds to the Steiner point O (i.e., the internal point of the triangle forming three angles of $120°$ with the vertices);
- the sum of the demands of the three sales points is less than or equal to the vehicle capacity and, therefore, the warehouse can be located at any point on the triangle's *ABC* perimeter.

Location models taking into account explicitly the transportation policy are called *location-routing* models (see Section 3.15.2).

3.3 The Number of Facilities in a Logistics System

As pointed out in Section 3.1, the determination of the number of facilities to be located is a key issue in logistics network design. This section provides a detailed and comprehensive *qualitative* account of how this decision influences the trade-off between total logistics cost (which can be generally split into transportation, inventory, and facility costs) and service level (see Section 1.8).

The impact on transportation costs
Economies of scale in transportation make the use of intermediate facilities economically convenient, at least to some extent. To figure out why this happens, consider, for example, a logistics system composed of m origins (e.g., plants) and n destinations (e.g., retailers). If no intermediate facility (e.g., a warehouse or a crossdock, see Section 5.2.1) exists, direct transportation among all $m \times n$ origin–destination pairs is required (see Figure 3.3(a)). On the other hand, the presence of an intermediate facility reduces the number of direct shipments to only $m + n$ (see Figure 3.3(b)) while increasing the average material flow on each connection. As a result, in the former case shipments tend not to completely fill a truck, in which case they are called *less-than-truckload* (LTL), whereas in the latter case they are more likely to be full *truckloads* (TLs).

As the number of intermediate facilities increases, the total distance from the sources to these facilities goes up while the total distance from the intermediate facilities to the final destinations decreases. As a consequence, the decrease in outbound transportation costs is faster than the rise of inbound transportation costs, and the total transportation cost diminishes. This trend persists until the intermediate facilities are so numerous that inbound transportation is no longer a full load and, consequently, the cost reductions deriving from economies of scale are lost (see Figure 3.4).

The impact on inventory costs
If the facilities to be located are warehouses, inventory costs must be taken into account. Assuming that the order size is optimized according to the *economic order quantity* (EOQ) formula (see Section 5.12.1), it can be proven (see Problem 3.2) that the number of existing and planned facilities n_c and n_f, respectively, and the current and the planned inventory levels I_c and I_f, respectively, are related by the following *square root law*:

$$I_f = I_c \sqrt{\frac{n_f}{n_c}}.$$

(3.1)

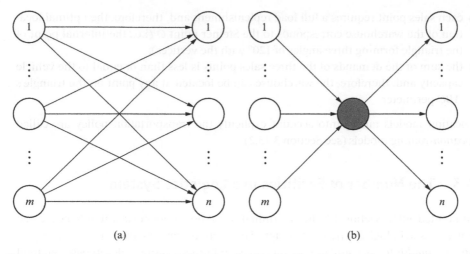

Figure 3.3 A logistics system (a) with no intermediate facility; (b) with an intermediate facility.

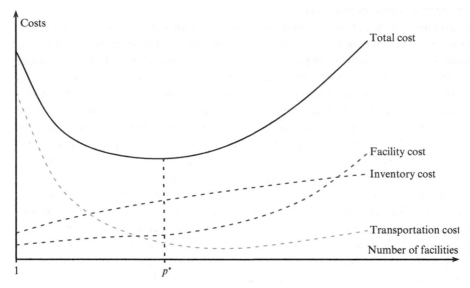

Figure 3.4 Transportation, inventory, facility, and total logistics cost as a function of the number of activated facilities.

Fruit Gourmand is a French company that produces fruit juice in packages composed of six one-litre bottles to be sold in supermarkets. Due to a reconfiguration of the distribution system, the company has decided to reduce its number of warehouses from six to three in order to lower the total average inventory, currently equal to $I_c = 570\,000$ packages. The new inventory level, computed with formula (3.1), is equal to $I_f = 570\,000 \times \sqrt{3/6} \approx 403\,050$ packages.

This implies that inventory costs can be assumed to increase proportionally to the square root of the number of facilities to be located (see Figure 3.4).

The impact on facility cost

Facility costs incorporate all the costs incurred in a facility, except for those classified as transportation or inventory costs. They are characterized by two main components: fixed costs and running costs. Fixed costs are related to facility purchase, designing, building, renting, equipment acquisition, installation, and disposal. Running costs correspond to energy consumption, labour, maintenance, staff, insurance, and taxes. These costs tend to be less than proportional the size of the facility because of economies of scale. Hence, it is more convenient to set up a small number of large facilities rather than many small ones. As a result, facility costs increase as the number of facilities grows.

The impact on total logistics cost

As the number of facilities increases, the total logistics cost (which is the sum of transportation, inventory, and facility costs), first decreases and then increases (see Figure 3.4). Hence, there is a number of facilities p^* that minimizes the total logistics cost.

Service level

Nowadays, the logistics service level is often seen by companies as a form of *competitive advantage*, i.e., a factor of differentiation with their competitors that is valued by customers. In such a context, logistics networks often include more (smaller) facilities near the customers. This explains why the number of facilities often lies beyond the value of p^* at which the total logistics cost is minimum (see Section 1.8.3 and, in particular the Ecopaper problem).

3.4 Qualitative Versus Quantitative Location Methods

Location problems can be solved by means of two distinct approaches: qualitative and quantitative methods. Qualitative methods are suitable whenever the number of candidate solutions is discrete (and fairly small) and the location decision is affected by some criteria that are difficult to assess in monetary terms (e.g., proximity to transportation infrastructures such as parking areas, motorways, railways, ports, and airports; proximity to shopping areas, competitors and so on; or the availability of a qualified workforce, a fall or a rise in population, etc.). The main qualitative methods are briefly illustrated in Sections 3.5 and 3.6. The second class is composed of quantitative methods based on the formulation of location problems as optimization models (see Sections 3.7 to 3.15).

3.5 The Weighted Scoring Method

The weighted scoring method is an intuitive and easy-to-apply form of multi-criteria decision analysis. It involves the identification of all monetary and non-monetary criteria relevant to the location decision; then, weights are assigned to each of them to reflect their relative importance; finally, experts are asked to assign a score to each criterion of the available alternatives. The result is a single weighted score for each alternative, which is then used to choose the most suitable solution. It is worth noting that, especially if the investment cost is large, it is of the utmost importance that the rationale for each weight and each score can be fully explained by the experts to the final decision maker.

Let m and n be the number of potential facilities and the number of location criteria, respectively. The WEIGHTED_SCORING procedure is as follows.

1: **procedure** WEIGHTED_SCORING (m, n, i^*)

2:      # i^* is the facility returned by the procedure;

3:      Assign weights $w_j \in (0, 1), j = 1, \dots, n$;

     # The weights are such that $\sum_{j=1}^{n} w_j = 1$;

4:      Assign scores $s_{ij}, i = 1, \dots, m, j = 1, \dots, n$;

     # Each score is typically from 0 to 10;

5:      **for** $i = 1, \dots, m$ **do**

6:          $r_i = \sum_{j=1}^{n} w_j s_{ij}$;

7:      **end for**

8:      Select facility $i^* = \arg \max_{i=1,\dots,m} \{r_i\}$;

9:      **return** i^*;

10: **end procedure**

[JetMarket.xlsx] Jet Market has to decide where to locate a retail outlet in Berne, Switzerland. External consultants have selected seven criteria which are considered the most important for the location decision. Each criterion has a weight from 0 to 1 (see Table 3.1). Three possible commercial areas (indicated as 1, 2, and 3) are evaluated by applying the WEIGHTED_SCORING procedure. Scores assigned to the location criteria for the three alternative locations vary between 0 and 10 and are reported in Table 3.2. The weighted scores for each site are 4.55, 5.30, and 4.50, respectively. On the basis of these results, the second site is selected.

Table 3.1   Weights associated with location criteria in the Jet Market problem.

j	Location criterion	Weight (w_j)
1	Renting cost	0.40
2	Availability of spaces requiring minor layout changes	0.20
3	Proximity to transportation infrastructure	0.05
4	Proximity to parking areas	0.10

(Continued)

Table 3.1 (Continued)

j	Location criterion	Weight (w_j)
5	Number of shop windows	0.05
6	Proximity of retail competitors	0.10
7	Proximity of complementary shops	0.10
	Total	1.00

Table 3.2 Scores received by the location criteria for the three alternative locations in the Jet Market problem.

Alternative location (i)	Location criterion						
	$j = 1$	$j = 2$	$j = 3$	$j = 4$	$j = 5$	$j = 6$	$j = 7$
1	5	3	6	4	5	3	7
2	7	5	4	2	4	5	4
3	4	5	5	5	7	5	3

3.6 The Analytical Hierarchy Process

The *analytical hierarchy process* (AHP) is a multi-criteria decision-making method relying on three main concepts: goal, criteria and alternatives. In AHP the decision is not treated as a whole, as in the weighted scoring method, but decomposed into a series of simpler pairwise comparisons (left to experts' or to decision maker's judgement), each of which expressing the relative importance of one criterion or alternative versus another. Several scales can be used in the pairwise comparison of two criteria or alternatives to convert the judgements into numerical values (*ratings*). The most popular is reported in Table 3.3. Additional ratings (8, 6, 4, and 2) are also possible if intermediate judgements (between any two reported in adjacent rows of Table 3.3) are decided to be used.

Table 3.3 Conversion scale of judgements into numerical values.

Judgement	Rating
Extremely preferred	9
Very strongly preferred	7
Strongly preferred	5
Moderately preferred	3
Equally preferred	1

AHP is based on the following steps:

- establish the relative importance of each criterion in meeting the goal by assigning weights to criteria;
- evaluate the extent to which each alternative contributes to each criterion; this leads to the computation of a score for each alternative;
- rank the alternatives to meet the goal.

Each step will be described below.

Computing the criteria weights

The computation of the weights of the n criteria involves the construction of an $n \times n$ pairwise comparison matrix $\mathbf{A}$ of real values. Each element $a_{jk}, j, k = 1, \dots, n$, represents a rating (from Table 3.3) expressing the comparison of criterion j to criterion k. For example, if $a_{jk} = 5$, criterion j is strongly preferred to criterion k.

To build matrix $\mathbf{A}$, only $[n \times (n - 1)]/2$ judgements have to be specified. This is because of the following two properties:

1. $a_{jk} = 1/a_{kj}, j, k = 1, \dots, n$ *(reciprocal ratings)*;
2. $a_{jj} = 1, j = 1, \dots, n$ (any criterion is equally preferred to itself).

Once matrix $\mathbf{A}$ has been defined, a vector $\mathbf{w}$ of criteria weights is computed by using the following CRITERIA_WEIGHTS procedure. It is easy to show (see Problem 3.4) that $\mathbf{w}$ is such that $\sum_{j=1}^{n} w_j = 1$.

1: **procedure** CRITERIA_WEIGHTS $(n, \mathbf{A}, \mathbf{w})$
2:     # n is the number of criteria;
3:     # $\mathbf{A}$ is the $n \times n$ pairwise comparison matrix;
4:     # $\mathbf{w}$ is a vector of n components (each of which indicates the weight assigned to the corresponding criterion, returned by the procedure;
5:     # Compute the normalized pairwise comparison matrix $\overline{\mathbf{A}}$ of $\mathbf{A}$ by imposing that the sum of the entries of each column is one;
6:     **for** $j = 1, \dots, n$ **do**
7:         **for** $k = 1, \dots, n$ **do**
8:             $\overline{a}_{jk} = a_{jk}/\sum_{h=1}^{n} a_{hk}$;
9:         **end for**
10:     **end for**
11:     # Compute the criteria weight vector $\mathbf{w}$ in such a way each weight is the average of the corresponding row of $\overline{\mathbf{A}}$;
12:     **for** $j = 1, \dots, n$ **do**
13:         $w_j = \sum_{k=1}^{n} \overline{a}_{jk}/n$;
14:     **end for**
15:     **return w**;
16: **end procedure**

[JetMarket.xlsx] In the Jet Market problem, a business expert has made pairwise comparisons among the location criteria reported in Table 3.1. This resulted in the following 7×7 matrix **A**:

$$
\mathbf{A} =
\begin{bmatrix}
1 & 3 & 7 & 5 & 7 & 3 & 5 \\
1/3 & 1 & 5 & 3 & 7 & 7 & 5 \\
1/7 & 1/5 & 1 & 1/3 & 1 & 1/3 & 1/5 \\
1/5 & 1/3 & 3 & 1 & 3 & 1/3 & 1/3 \\
1/7 & 1/7 & 1 & 1/3 & 1 & 1/5 & 1/5 \\
1/3 & 1/7 & 3 & 3 & 5 & 1 & 1 \\
1/5 & 1/5 & 5 & 3 & 5 & 1 & 1
\end{bmatrix}.
$$

In order to determine the weights to assign to each location criterion, the CRITERIA_WEIGHTS procedure is applied. The normalized pairwise comparison matrix $\overline{\mathbf{A}}$ of **A** corresponds to (code lines 6–10):

$$
\overline{\mathbf{A}} =
\begin{bmatrix}
0.425 & 0.598 & 0.280 & 0.319 & 0.241 & 0.233 & 0.393 \\
0.142 & 0.199 & 0.200 & 0.191 & 0.241 & 0.544 & 0.393 \\
0.061 & 0.040 & 0.040 & 0.021 & 0.034 & 0.026 & 0.016 \\
0.085 & 0.066 & 0.120 & 0.064 & 0.103 & 0.026 & 0.026 \\
0.061 & 0.028 & 0.040 & 0.021 & 0.034 & 0.016 & 0.016 \\
0.142 & 0.028 & 0.120 & 0.191 & 0.172 & 0.078 & 0.079 \\
0.085 & 0.040 & 0.200 & 0.191 & 0.172 & 0.078 & 0.079
\end{bmatrix},
$$

from which the following weight vector **w** is obtained (code lines 12–14):

$$
\mathbf{w} = [0.36; 0.27; 0.03; 0.07; 0.03; 0.12; 0.12]^{\mathsf{T}}.
$$

Note that $0.36 + 0.27 + 0.03 + 0.07 + 0.03 + 0.12 + 0.12 = 1$.

An important issue concerns the risk of inconsistent and contradictory judgements when a pairwise comparison matrix is built. This occurs, e.g., if criterion j is judged preferable to criterion k, criterion k preferable to criterion l, and, in turn, criterion l preferable to criterion j. Some inconsistency is acceptable, even unavoidable, as long as it remains within a reasonable threshold. If this threshold is exceeded, some judgements need to be corrected. AHP allows the consistency of a pairwise comparison matrix to be checked by means of a procedure, named CONSISTENCY_INDEX, which computes a *consistency index* (*CI*). If $CI = 0$, the pairwise comparison matrix is *consistent*. The rationale of the *CI* computation is the following. In an $n \times n$ consistent pairwise comparison matrix **A**, it can be proved that $\overline{b}$ (computed at code line 10 of the CONSISTENCY_INDEX procedure) represents the largest eigenvalue of **A**. Moreover, $\lambda_{\max}$ achieves its maximum (equal to n) for consistent matrices. Consequently, a consistence measure of a matrix **A** is defined as

$$
CI = (\lambda_{\max} - n)/(n - 1),
$$

which is zero when **A** is consistent. If $CI > 0$, this value is compared with a *random index* (*RI*) calculated from hundreds of randomly-generated pairwise comparison matrices of the same dimension (*RI* is reported in Table 3.4 for matrices of dimensions up to 10). As a rule of thumb, if $0 < CI < 0.1RI$ (*CI* is within a threshold equal to 10%

Table 3.4 *RI* values for different dimensions of a pairwise comparison matrix.

n	RI	n	RI	n	RI
2	0.00	5	1.12	8	1.41
3	0.58	6	1.24	9	1.45
4	0.90	7	1.32	10	1.51

of *RI*), the inconsistency of the pairwise comparison matrix is considered acceptable; otherwise a modification of the pairwise comparison matrix is recommended.

1: **procedure** CONSISTENCY_INDEX $(n, \mathbf{A}, \mathbf{w}, CI)$

2:   # $\mathbf{A}$ is an $n \times n$ pairwise comparison matrix;

3:   # $\mathbf{w}$ is the n-weight vector, computed by using the CRITERIA_WEIGHTS procedure;

4:   # CI is the consistency index returned by the procedure;

5:   # Compute vector $\mathbf{b}$ as the matrix-vector product $\mathbf{Aw}$;
$\mathbf{b} = \mathbf{Aw}$;

6:   # Divide each component of vector $\mathbf{b}$ by the corresponding weight;

7:   **for** $j = 1, \dots, n$ **do**

8:     $b_j = b_j / w_j$;

9:   **end for**

10:   # Determine the average of the components of the modified vector $\mathbf{b}$;
$$\bar{b} = \sum_{j=1}^{n} b_j / n;$$

11:   # Determine CI;
$CI = (\bar{b} - n)/(n - 1)$;

12:   **return** CI;

13: **end procedure**

[JetMarket.xlsx] In the Jet Market problem, *CI* can be determined as follows through the CONSISTENCY_INDEX procedure. The 7-component vector $\mathbf{b}$ (code lines 5–9) is:

$$\mathbf{b} = [8.240; 8.800; 7.542; 7.212; 7.296; 7.436; 7.430]^\mathsf{T},$$

whose average $\bar{b} = 7.708$. Hence, *CI* (code line 11) corresponds to

$$CI = (7.708 - 7)/6 = 0.118.$$

Considering that *RI* = 1.32 (see Table 3.4 for $n = 7$), it turns out that *CI* = 0.118 < 0.1 × 1.32 = 0.132. Consequently, the inconsistency of the pairwise comparison matrix $\mathbf{A}$ is acceptable.

Computing the scores of the alternatives

This step corresponds to creating an $m \times n$ matrix $\mathbf{S}$ of real values, where $s_{ij}, i = 1, \dots, m, j = 1, \dots, n$, is the score of alternative i with respect to criterion j. The score s_{ij} is a weight such that $\sum_{i=1}^{m} s_{ij} = 1$ for $j = 1, \dots, n$. To determine the score matrix $\mathbf{S}$, an $m \times m$ pairwise comparison matrix $\mathbf{B}^{(j)}$ is built for each criterion $j = 1, \dots, n$. Each element $b_{ih}^{(j)}, i, h = 1, \dots, m$, represents the rating (see Table 3.3) corresponding to the judgement of alternative i with respect to alternative h for criterion j. To each matrix $\mathbf{B}^{(j)}, j = 1, \dots, n$, is applied the same CRITERIA_WEIGHTS procedure described in the previous step for the pairwise comparison matrix $\mathbf{A}$. The procedure returns the vector S_j corresponding to the jth column of the score matrix $\mathbf{S}$. Of course, the same consistency check executed for matrix $\mathbf{A}$ can be repeated for each matrix $\mathbf{B}^{(j)}$, $j = 1, \dots, n$.

[JetMarket.xlsx] The three alternative commercial areas of the Jet Market problem are compared with respect to the seven location criteria of Table 3.1. As a result, seven 3×3 pairwise comparison matrices $\mathbf{B}^{(j)}, j = 1, \dots, 7$, are generated (see the JetMarket.xlsx file). For the sake of brevity, only the pairwise comparison matrix $\mathbf{B}^{(1)}$ corresponding to the first location criterion (i.e., renting cost) is reported:

$$\mathbf{B}^{(1)} = \begin{bmatrix} 1 & 1/3 & 3 \\ 3 & 1 & 5 \\ 1/3 & 1/5 & 1 \end{bmatrix},$$

with $CI = (3.039 - 3)/2 = 0.019$. Consequently, $CI = 0.019 < 0.1 \times 0.58 = 0.058$. The execution of the CRITERIA_WEIGHTS procedure on each of the matrices $\mathbf{B}^{(j)}, j = 1, \dots, 7$, allows the following 3×7 score matrix $\mathbf{S}$ to be determined:

$$\mathbf{S} = \begin{bmatrix} 0.26 & 0.11 & 0.63 & 0.41 & 0.19 & 0.11 & 0.75 \\ 0.63 & 0.41 & 0.11 & 0.11 & 0.08 & 0.41 & 0.18 \\ 0.11 & 0.48 & 0.26 & 0.48 & 0.72 & 0.48 & 0.07 \end{bmatrix}.$$

Ranking the alternatives

Once the criteria weight vector $\mathbf{w}$ and the score matrix $\mathbf{S}$ of the alternatives have been determined, a global score r_i for each alternative $i = 1, \dots, m$, is computed:

$$r_i = \sum_{j=1}^{n} w_j s_{ij}. \tag{3.2}$$

The best alternative i^* corresponds to

$$i^* = \arg \max_{i=1,\dots,m} \{r_i\}.$$

[JetMarket.xlsx]. The global scores of the three alternative commercial areas of the Jet Market problem, computed by using formula (3.2), are 0.28, 0.42, and 0.30, respectively. As for the weighted scoring method, the highest score is obtained at site 2.

The main drawback of AHP is the number of pairwise comparisons to make a decision which could be very large for problems with many criteria and alternatives. Fortunately, this number grows polynomially with the problem size (see Problem 3.6).

3.7 Single-commodity One-echelon Continuous Location Problems

This section describes how to optimally locate a single or multiple uncapacitated facilities in a Cartesian plane.

Locating a single facility

A facility has to supply a single commodity to a set of *successor* (or *sink*) *nodes* (e.g., a plant has to replenish a set of markets). The transportation cost between the facility to be located and every successor node is assumed to be proportional to the Euclidean distance between them.

To formulate the problem, let V be the set of successor nodes. Each successor node $j \in V$ is characterized by its Cartesian coordinates (x_j, y_j). Its demand of the single commodity d_{jt} over the planning horizon $t = 1, \dots, T$ is assumed to be constant ($d_{jt} = d_j, j \in V$, for each time period $t = 1, \dots, T$). Moreover, let (x, y) be the (unknown) Cartesian coordinates of the facility to be located. If the facility cost is independent of its location, the problem consists in finding the coordinates (x^*, y^*) which minimize the overall transportation cost in a single time period of the planning horizon:

$$\text{Minimize } f(x, y) = \sum_{j \in V} cd_j \left(\sqrt{(x_j - x)^2 + (y_j - y)^2} \right), \tag{3.3}$$

where c represents the transportation cost per unit of distance travelled and per unit of commodity transported from the facility to any successor node $j \in V$.

Since function (3.3) is convex, its minimizer (x^*, y^*) corresponds to a stationary point whose coordinates can be determined by imposing

$$\frac{\partial f(x, y)}{\partial x} = 0, \tag{3.4}$$

$$\frac{\partial f(x, y)}{\partial y} = 0. \tag{3.5}$$

By computing partial derivative (3.4), the following relation is obtained:

$$-\frac{1}{2} c \sum_{j \in V} \frac{d_j[2(x_j - x^*)]}{\sqrt{(x_j - x^*)^2 + (y_j - y^*)^2}} = 0.$$

After a few algebraic manipulations, the following equation is obtained:

$$x^* = \frac{\displaystyle\sum_{j \in V}\left[\frac{d_j x_j}{\sqrt{(x_j-x^*)^2+(y_j-y^*)^2}}\right]}{\displaystyle\sum_{j \in V}\left[\frac{d_j}{\sqrt{(x_j-x^*)^2+(y_j-y^*)^2}}\right]}. \tag{3.6}$$

Similarly, partial derivative (3.5) originates equation:

$$y^* = \frac{\displaystyle\sum_{j \in V}\left[\frac{d_j y_j}{\sqrt{(x_j-x^*)^2+(y_j-y^*)^2}}\right]}{\displaystyle\sum_{j \in V}\left[\frac{d_j}{\sqrt{(x_j-x^*)^2+(y_j-y^*)^2}}\right]}. \tag{3.7}$$

Relations (3.6) and (3.7) constitute a system of non-linear equations whose solution is not known in closed form. The following heuristic, proposed by E. Weiszfeld (see WEISZFELD procedure), can be used to find an approximate solution with a worst-case error within a tolerance ϵ. Setting $\epsilon = 0$, the described algorithm becomes exact. Indeed, it can be proved that the sequence $\{f(x^{(h)}, y^{(h)})\}_{h=0}^{\infty}$ generated by the algorithm is monotonically decreasing and converges to the optimal solution (x^*, y^*).

It is worth noting that the previous methodology can also be adapted to model and solve a variant of the problem in which the nodes in V represent the *predecessor* (or *source*) *nodes* of the facility to be located. This is the case, e.g., of locating a manufacturing plant supplied by a set of vendors. Under this hypothesis, each predecessor node is characterized by a commodity supply instead of a demand in each period of the planning horizon.

[`Karakum.xlsx`, `Karakum.py`] The Karakum Desert lies in Turkmenistan, in Central Asia. Nine small villages, located on the borders of the desert, are served by water tank trucks, for daily water supply. The trucks are supplied from a water reservoir and cover the distance between the water reservoir and each village on daily TL trips along a route whose length is very close to the Euclidean distance. Table 3.5 reports the coordinates (in km) of the villages with respect to a two-dimensional Cartesian system and the average daily number of trips required to the water tank trucks to guarantee the water supply of the inhabitants. The average kilometric transportation cost for each water tank truck is estimated to be € 0.35.

The position of the water reservoir (see Figure 3.5) was determined by solving the continuous location problem (3.3) in which $c = €\,0.7$ (to take into account the return truck trip). The optimal solution, obtained by using the `Excel Solver` tool under the `Data` tab, is the following

$$(x^*, y^*) = (17.282, 11.632);$$

$$f(x^*, y^*) = 1\,126.50 \text{ €/day.}$$

1: **procedure** WEISZFELD $(V, \mathbf{x}, \mathbf{y}, \mathbf{d}, c, \epsilon, \overline{x}, \overline{y})$

2:   # ϵ is a non-negative user-defined tolerance parameter;

3:   # $\overline{x}$ and $\overline{y}$ are the coordinates of the facility returned by the procedure;

4:   # $x^{(0)}$ and $y^{(0)}$ are the coordinates of the *centre of gravity*;

5:

$$x^{(0)} = \frac{\sum\limits_{j \in V} d_j x_j}{\sum\limits_{j \in V} d_j}; \tag{3.8}$$

6:

$$y^{(0)} = \frac{\sum\limits_{j \in V} d_j y_j}{\sum\limits_{j \in V} d_j}; \tag{3.9}$$

7:

$$f(x^{(0)}, y^{(0)}) = \sum_{j \in V} c d_j \left(\sqrt{(x_j - x^{(0)})^2 + (y_j - y^{(0)})^2} \right);$$

8:   $h = 1$;

9:   exit = FALSE;

10:   **while** exit = FALSE **do**

11:

$$x^{(h)} = \frac{\sum\limits_{j \in V} \left[\frac{d_j x_j}{\sqrt{(x_j - x^{(h-1)})^2 + (y_j - y^{(h-1)})^2}} \right]}{\sum\limits_{j \in V} \left[\frac{d_j}{\sqrt{(x_j - x^{(h-1)})^2 + (y_j - y^{(h-1)})^2}} \right]};$$

$$y^{(h)} = \frac{\sum\limits_{j \in V} \left[\frac{d_j y_j}{\sqrt{(x_j - x^{(h-1)})^2 + (y_j - y^{(h-1)})^2}} \right]}{\sum\limits_{j \in V} \left[\frac{d_j}{\sqrt{(x_j - x^{(h-1)})^2 + (y_j - y^{(h-1)})^2}} \right]};$$

$$f(x^{(h)}, y^{(h)}) = \sum_{j \in V} c d_j \left(\sqrt{(x_j - x^{(h)})^2 + (y_j - y^{(h)})^2} \right);$$

12:     **if** $f(x^{(h-1)}, y^{(h-1)}) - f(x^{(h)}, y^{(h)}) \leq \epsilon$ **then**

13:       # Stopping condition is reached;

14:       $\overline{x} = x^{(h)}$;

15:       $\overline{y} = y^{(h)}$;

16:       exit = TRUE;

17:     **else**

18:       $h = h + 1$;

19:     **end if**

20:   **end while**

21:   **return** $\overline{x}, \overline{y}$;

22: **end procedure**

Table 3.5 Cartesian coordinates and average daily number of trips to the villages on the borders of the Karakum Desert.

Village (j)	Abscissa [km] (x_j)	Ordinate [km] (y_j)	Average daily number of trips (d_j)
1	0.000	0.000	11
2	13.543	11.273	13
3	8.578	24.432	15
4	42.438	14.583	12
5	20.652	5.232	16
6	3.782	21.567	11
7	25.720	19.565	13
8	18.768	3.678	14
9	35.650	25.678	10

Using the WEISZFELD procedure with $\epsilon = 0.1€$/day, the sequence of solutions reported in Table 3.6 is obtained. When the procedure terminates, the point of coordinates $(17.325, 11.660)$, which is a good approximation of the optimal solution, is returned. It is also worth noting that the centre of gravity solution (3.8)–(3.9) is a good approximation of the optimal solution as well, its cost being only 1.01% worse than the cost of the solution provided by the WEISZFELD procedure.

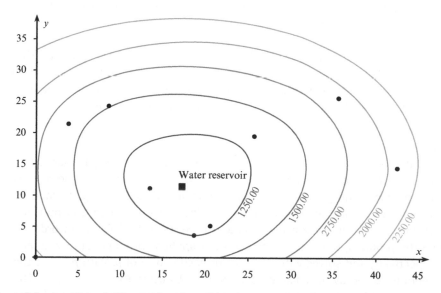

Figure 3.5 Location of villages, plot of some level curves of the objective function, and optimal location of the water reservoir to serve the villages on the borders of the Karakum Desert. The coordinates are expressed in km.

Table 3.6 Sequence of the water reservoir locations generated by the WEISZFELD procedure.

Iteration (h)	Abscissa [km] $(x^{(h)})$	Ordinate [km] $(y^{(h)})$	Object function [€/day]	Difference [€/day]
0	18.605	13.666	1138.04	
1	17.975	12.583	1129.42	8.62
2	17.629	12.053	1127.15	2.26
3	17.454	11.810	1126.64	0.52
4	17.368	11.704	1126.53	0.11
5	17.325	11.660	1126.51	0.02

Locating multiple facilities

In the case of p (> 1) facilities to be located, it is possible to extend the single-facility approach by using a two-phase procedure. First, successor nodes are divided into p clusters, each of which is intended to be served by a single facility. Then, facilities are located by solving p independent single-facility problems. The procedure, proposed by L. Cooper (see COOPER procedure) is described in the following.

Initially, each successor node is randomly assigned to a cluster $C_k, k = 1, \dots, p$ (code lines 3–5). In this way, a number of clusters equal to the number p of facilities being located is defined. Then, the coordinates $\bar{x}_{(k)}$ and $\bar{y}_{(k)}$ of the facility are determined for each cluster. Note that $\bar{x}_{(k)}$ and $\bar{y}_{(k)}$ can be an optimal solution found by an off-the-shelf general purpose solver (for example Excel Solver) or a feasible solution found by the WEISZFELD procedure described above. Each successor node is then reassigned to the closest facility (code line 15). If the composition of the clusters is modified, the two phases are iterated until the clusters do not change any more. Note that, even if the optimal coordinates of the facility for each cluster are found at each iteration, the COOPER procedure could lead to a local optimum instead of a global optimum. This is due to the fact that the final solution depends on the initial composition of the clusters. To overcome this problem, multiple runs of the procedure are advisable. The higher the number of runs, the more likely the optimal solution will be found.

[MelloDrink.xlsx] Mello Drink is a soft drink company headquartered in Louisville, Kentucky, USA. The logistics manager needs to locate up to four county DCs of soft drinks to serve a community of 19 towns. The towns, their coordinates in a Cartesian plane (in km) and the corresponding annual average demand (in hl) of soft drinks are reported in Table 3.7. The average cost to transport one hl of soft drinks for one km is estimated to be equal to $ 0.135. In order to solve the problem, the COOPER procedure is applied with p equal to 2, 3, and 4 respectively, and solved with 10 runs for each choice of p. With $p = 1$ the problem is solved by minimizing Equation (3.3) by using Excel Solver. The best solutions found, in terms of average annual transportation costs (in $), are 567 238, 343 666, 201 570, and 147 935, corresponding to p equal to 1, 2, 3, and 4, respectively.

1: **procedure** COOPER $(V, \mathbf{x}, \mathbf{y}, p, \overline{\mathbf{x}}, \overline{\mathbf{y}})$
2:    # $C_k, k = 1, \dots, p$, are clusters formed by the successor nodes in V such that $C_j \cap C_k = \varnothing, j, k = 1, \dots, p, j \neq k$, and $\bigcup_{k=1,\dots,p} C_k = V$;
   # $\overline{x}_{(k)}$ and $\overline{y}_{(k)}$ are the coordinates of the facility associated with the kth cluster, $k = 1, \dots, p$;
3:    **for** $i \in V$ **do**
4:       Assign each successor node i to a random cluster C_k;
5:    **end for**
6:    exit = FALSE;
7:    **while** exit = FALSE **do**
8:       **for** $k = 1, \dots, p$ **do**
9:          Determine $\overline{x}_{(k)}$ and $\overline{y}_{(k)}$ associated with C_k;
10:       **end for**
11:       exit = TRUE;
12:       **for** $i \in V$ **do**

$$j = \arg \min_{l=1,\dots,p} \left\{ \sqrt{(x_i - \overline{x}_{(l)})^2 + (y_i - \overline{y}_{(l)})^2} \right\};$$

13:          **if** $i \notin C_j$ **then**
14:             # A modification of the composition of cluster C_j is required;
15:             Move i in C_j;
16:             exit = FALSE;
17:          **end if**
18:       **end for**
19:    **end while**
20:    **return** $\overline{\mathbf{x}}, \overline{\mathbf{y}}$;
21: **end procedure**

Table 3.7 Cartesian coordinates and average annual demand of soft drinks of 19 towns of a Kentucky county to be served by Mello Drink.

Town	Abscissa [km]	Ordinate [km]	Annual demand [hl]
Bonnieville	4.77	26.36	400
Brodhead	137.19	29.41	1 800
Campbellsville	54.41	23.15	15 600
Cave City	0.00	0.00	3 200
Danville	105.67	56.42	14 400
Elizabethtown	8.08	62.77	19 700
Eubank	115.69	15.85	500
Greensburg	40.87	13.60	3 000
Hodgenville	19.43	48.15	4 700

(Continued)

Table 3.7 (Continued)

Town	Abscissa [km]	Ordinate [km]	Annual demand [hl]
Horse Cave	4.52	4.33	3 300
Lancaster	122.61	53.42	5 600
Lebanon	62.35	48.09	8 000
Liberty	91.35	20.41	3 100
Loretto	50.75	55.31	900
Munfordville	5.26	15.48	2 400
New Haven	32.73	57.93	1 200
Science Hill	117.61	4.23	1 000
Springfield	65.42	60.95	3 800
Stanford	115.41	43.84	5 300

The COOPER procedure resembles the *k-means* clustering algorithm used in ML (see Section 2.7.8) and does not take into account the fixed and variable costs connected to the activation of multiple facilities.

[MelloDrink.xlsx] Each county DC of the Mello Drink company, once activated, leads to an annual fixed cost estimated in $ 110 000 on average. In addition, the average inventory costs are estimated, according to the square root law, as $ (60 000 × $\sqrt{p}$), where p is the number of activated county DCs. The other logistics costs do not depend on the number of open facilities, so they do not affect the solutions found. Table 3.8 summarizes the total average annual logistics costs for different numbers (from one to four) of county DCs to activate. The best configuration is obtained when $p = 3$. The coordinates (in km) of the three county DCs are reported in Table 3.9. Figure 3.6 depicts their position and that of the 19 towns, clustered for each open facility. Note that each of the three county DCs is opened in the town of the cluster that corresponds to the largest annual demand.

Table 3.8 Annual logistics costs (in $) for different numbers (from one to four) of county DCs of the Mello Drink company.

Number of county DCs	Transportation costs	Fixed costs	Inventory costs	Total costs
1	567 238	110 000	60 000	737 238
2	343 666	220 000	84 853	648 519
3	201 570	330 000	103 923	635 493
4	147 935	440 000	120 000	707 935

Table 3.9 Cartesian coordinates (in km) of the three county DCs activated by Mello Drink.

City	Abscissa	Ordinate
Danville	105.67	56.42
Elisabethtown	8.08	62.77
Campbellsville	54.41	23.15

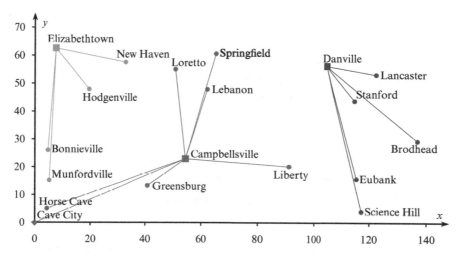

Figure 3.6 Location of the county DCs and that of the towns served by Mello Drink. The coordinates are expressed in km. The locations of the facilities coincide with three cities.

3.8 Single-commodity Two-echelon Continuous Location Problems

As observed in Section 3.1, when both the inbound and outbound transportation costs to the facility are relevant, it is necessary to use a two-echelon formulation.

Locating a single facility

The location problem of a single facility can be formulated as follows. Let V_1 and V_2 be the sets of predecessor and successor nodes, respectively. Denote by o_i, $i \in V_1$, the maximum amount of commodity available at predecessor node i in a time period, and by d_j, $j \in V_2$, the average demand of successor node j in a time period. Of course, it is assumed that $\sum_{i \in V_1} o_i \geq \sum_{j \in V_2} d_j$. Let c be the transportation cost per unit of distance travelled and per unit of commodity transported, independently of the origin–destination pair. Assuming that the facility cost is location independent, and denoting by (x, y) the

Cartesian coordinates of the facility to be located, the problem can be modelled as

$$\text{Minimize } f(x, y) = \sum_{i \in V_1} co_i \left(\sqrt{(x_i - x)^2 + (y_i - y)^2} \right) \tag{3.10}$$

$$+ \sum_{j \in V_2} cd_j \left(\sqrt{(x_j - x)^2 + (y_j - y)^2} \right).$$

Objective function (3.10) expresses the overall average transportation cost in a time period from the predecessor nodes to the facility to be located and from this facility to the successor nodes.

It is worth noting that problem (3.10) is equivalent to problem (3.3) in which predecessor and successor nodes form a unique and indistinct set V. More formally, the optimal coordinates (x^*, y^*) are such that

$$x^* = \frac{\sum_{i \in V_1} \left[\frac{o_i x_i}{\sqrt{(x_i - x^*)^2 + (y_i - y^*)^2}} \right] + \sum_{j \in V_2} \left[\frac{d_j x_j}{\sqrt{(x_j - x^*)^2 + (y_j - y^*)^2}} \right]}{\sum_{i \in V_1} \left[\frac{o_i}{\sqrt{(x_i - x^*)^2 + (y_i - y^*)^2}} \right] + \sum_{j \in V_2} \left[\frac{d_j}{\sqrt{(x_j - x^*)^2 + (y_j - y^*)^2}} \right]},$$

and

$$y^* = \frac{\sum_{i \in V_1} \left[\frac{o_i y_i}{\sqrt{(x_i - x^*)^2 + (y_i - y^*)^2}} \right] + \sum_{j \in V_2} \left[\frac{d_j y_j}{\sqrt{(x_j - x^*)^2 + (y_j - y^*)^2}} \right]}{\sum_{i \in V_1} \left[\frac{o_i}{\sqrt{(x_i - x^*)^2 + (y_i - y^*)^2}} \right] + \sum_{j \in V_2} \left[\frac{d_j}{\sqrt{(x_j - x^*)^2 + (y_j - y^*)^2}} \right]}.$$

[Avial.xlsx] Avial is a Dutch company that commercializes and distributes coffee in the Netherlands. The company has two depots located in the ports of Amsterdam and Rotterdam (more specifically, in the Europoort situated along the Calandkanaal). By considering a Cartesian coordinate system centred in the depot of the port of Rotterdam, the Cartesian coordinates (in km) of the Amsterdam depot are (45.70, 52.16). In these two depots the commodity is stocked in jute bags containing 60 kg of crude coffee beans each. Avial supplies several coffee roasting plants spread over a wide geographic area, including nine plants located in the Dutch cities reported in Table 3.10. The crude coffee supply for the nine plants occurs every week by using a crossdocking point (whose location must be determined), which is replenished from the two depots located in the ports of Amsterdam and Rotterdam. The average weekly demand of the nine coffee roasting plants is reported (in 60 kg jute bags) in Table 3.10. The maximum quantities of commodity available on average every week at the depots of Amsterdam and Rotterdam ports are 5000 and 8000 60 kg jute bags, respectively. The average cost to transport one 60 kg jute bag of crude coffee over one km is estimated to be equal to € 0.014. The location problem can be cast as model (3.10), with $|V_1| = 2$ (the depots) and $|V_2| = 9$ (the coffee roasting plants). The other data can be found in Table 3.10. The optimal solution of (3.10) can be determined by using Excel Solver and corresponds to coordinates $(x^*, y^*) = (38.94, 36.65)$.

Table 3.10 Cartesian coordinates of coffee roasting plants and their average weekly demand of crude coffee in the Avial problem.

Coffee roasting plant	Abscissa [km]	Ordinate [km]	Weekly crude coffee demand [60 kg jute bags]
Amersfoort	85.66	23.71	450
Amsterdam	49.73	47.11	6680
Arnhem	121.85	4.45	470
Delft	14.77	7.54	300
Den Haag	11.96	15.58	1590
Hoofddorp	37.73	40.38	220
Leiden	24.04	23.80	340
Rotterdam	18.82	0.74	1860
Utrecht	66.83	16.20	1040

The average weekly transportation cost is € 12 072. Figure 3.7 depicts the positions of the two depots, of the nine coffee roasting plants and of the crossdocking point.

It is worth observing that when the maximum quantity of commodity available per time period at all the predecessor nodes, $\sum_{i \in V_1} o_i$, is much greater than the average commodity demand of all the successor nodes in the considered time period, $\sum_{j \in V_2} d_j$, the position of the facility, for "gravity", is more attracted towards the predecessor nodes, even though this may be not really convenient, since the effective average outbound commodity flow can be lower than the value of o_i for some predecessor node $i \in V_1$ (see Problem 3.12).

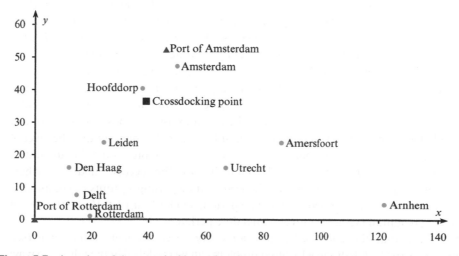

Figure 3.7 Location of the crossdocking point, depots (ports of Amsterdam and Rotterdam) and coffee roasting plants of the Avial problem. The coordinates are expressed in km.

Locating multiple facilities

The problem of locating multiple facilities can be tackled by using an *alternating heuristic* resembling the above mentioned COOPER procedure. Briefly, this approach consists of alternating between

- an *allocation phase*, in which some facilities have been selected, and both predecessor and successor nodes are allocated to them by solving a *minimum cost flow problem* (see Section 6.6.1);
- a *location phase*, where a single facility problem is solved for each cluster of predecessor and successor nodes.

The process terminates when no new improvement is achieved (see Problem 3.12 for more details).

If flows are divisible (see Section 3.1) some predecessor and successor nodes may be assigned to more facilities in the allocation phase. On the other hand, if flows are indivisible, integrality constraints must be imposed on the variables of the minimum cost flow problem and either the set of predecessor and successor nodes are partitioned into disjoint subsets.

3.9 Single-commodity One-echelon Discrete Location Problems

In *single-commodity one-echelon* (SCOE) discrete location problems, it is assumed that the facilities to be located are of the same type (e.g., they are all regional warehouses). For the sake of simplicity, the case where inbound flows are negligible is considered, although the same methodology can be applied without any change to the case where inbound flows are relevant and outbound flows are negligible (see Problem 3.18). The problem can be modelled through a bipartite complete directed graph $G = (V_1 \cup V_2, A)$, where vertices in V_1 represent the potential facilities to be located, vertices in V_2 are the successor nodes, and arcs in $A = V_1 \times V_2$ are associated with the commodity outbound flows from the potential facilities to the successor nodes (see Figure 3.8).

Let $t = 1, \dots, T$ be the time periods of the planning horizon. Let $d_{jt}, j \in V_2$, $t = 1, \dots, T$, be the demand estimate of successor node j at time period t and $q_{it}, i \in V_1$, $t = 1, \dots, T$, the throughput of potential facility i at time period t. Similarly to the continuous location cases, here $d_{jt}, j \in V_2, t = 1, \dots, T$, and $q_{it}, i \in V_1, t = 1, \dots, T$, do not vary over the planning horizon, i.e., $d_{jt} = d_j, j \in V_2, t = 1, \dots, T$, and $q_{it} = q_i, i \in V_1, t = 1, \dots, T$. The more general formulation is left to the reader as an exercise (see Problem 3.21). Under the previous assumptions, the problem can be modelled with the following decision variables: $u_i, i \in V_1$, representing the level of activity at facility i in a time period; $s_{ij}, i \in V_1, j \in V_2$, representing the amount of commodity delivered in a time period from facility i to the successor node j. Furthermore, let $C_{ij}(s_{ij}), i \in V_1, j \in V_2$, be the cost of transporting s_{ij} units of commodity from facility i to successor node j in a time period, and let $F_i(u_i), i \in V_1$, be the cost of operating potential facility i at level u_i in a time period.

Assuming that the commodity flow for every successor node is divisible (see Section 3.1), the SCOE discrete location problem can be modelled in the following way:

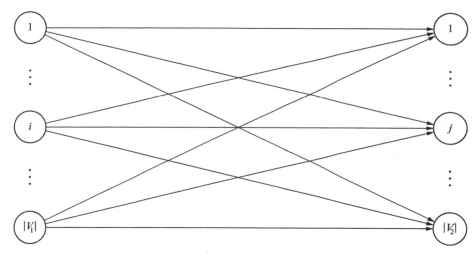

Figure 3.8 Representation of the SCOE discrete location problem on a bipartite complete directed graph.

$$\text{Minimize} \sum_{i \in V_1} \sum_{j \in V_2} C_{ij}(s_{ij}) + \sum_{i \in V_1} F_i(u_i) \tag{3.11}$$

subject to

$$\sum_{j \in V_2} s_{ij} = u_i, i \in V_1 \tag{3.12}$$

$$\sum_{i \in V_1} s_{ij} = d_j, j \in V_2 \tag{3.13}$$

$$\sum_{j \in V_2} s_{ij} \leq q_i, i \in V_1 \tag{3.14}$$

$$s_{ij} \geq 0, i \in V_1, j \in V_2 \tag{3.15}$$

$$u_i \geq 0, i \in V_1. \tag{3.16}$$

The values of the decision variables u_i, $i \in V_1$, implicitly define a location decision since a facility $l \in V_1$ is opened if and only if u_i is strictly positive. The values of the decision variables s_{ij}, $i \in V_1$, $j \in V_2$, determine the commodity flow's allocation to the successor nodes. Objective function (3.11) is the sum of the facility operating cost, plus the transportation cost between facilities and successor nodes in a time period. Constraints (3.12) state that the sum of the flows outgoing a facility to the successor nodes equals its activity level in a time period, while constraints (3.13) ensure the demand satisfaction of every successor node in a time period, while constraints (3.14) force the activity level of a facility not to exceed the corresponding throughput in a time period.

Model (3.11)–(3.16) is quite general and can be easily adapted to the case where, in order to have an acceptable service level, some arcs $(i, j) \in A$ having a travel time larger than a given threshold cannot be used (the corresponding decision variables s_{ij} are removed from the model).

The remainder of this section examines the SCOE location problem under the hypotheses that transportation costs per unit of commodity are constant and facility operating costs are piecewise linear and concave, that is,

$$C_{ij}(s_{ij}) = a_{ij}s_{ij}, i \in V_1, j \in V_2,$$

$$F_i(u_i) = \begin{cases} f_i + g_i u_i, & \text{if } u_i > 0 \\ 0, & \text{if } u_i = 0 \end{cases}, i \in V_1. \tag{3.17}$$

In equation (3.17), f_i, $i \in V_1$, is the average fixed cost of facility i in a time period (obtained by spreading fixed costs over its estimated useful life) and g_i, $i \in V_1$, is the running cost of facility i per unit of commodity in a time period.

Three years ago, Baja bought a warehouse in Debrecen (Hungary) from a competing company to distribute food products throughout the whole country. The asset cost 850 000 constant dollars (see Section 2.5.5). The annual fixed costs due to warehouse operations were 173 000, 168 000 and 125 000 constant dollars in the three subsequent years, respectively. Then, Baja decided to close the warehouse and sell it to another company. The selling price was 680 000 constant dollars. The warehouse operated for 48 weeks each year. Hence, the average weekly fixed cost of the warehouse was

$$(850\,000 + 173\,000 + 168\,000 + 125\,000 - 680\,000)/(48 \times 3)$$

$$= 4416.67 \text{ constant dollars.}$$

The facility cost function defined by (3.17) can be modelled by introducing a binary decision variable y_i for each $i \in V_1$, equal to 1 if potential facility i is opened, 0 otherwise. Constraints (3.12) allow avoiding the explicit use of continuous decision variable u_i for each $i \in V_1$. In particular, objective function (3.11) can be rewritten as

$$\sum_{i \in V_1} \sum_{j \in V_2} a_{ij}s_{ij} + \sum_{i \in V_1} \left(f_i y_i + g_i \sum_{j \in V_2} s_{ij} \right) = \sum_{i \in V_1} \sum_{j \in V_2} (a_{ij} + g_i) s_{ij} + \sum_{i \in V_1} f_i y_i,$$

where $a_{ij} + g_i = b_{ij}$, for each arc $(i, j) \in A$, represents the sum of the transportation cost per unit of flow from facility i to successor node j and of the running unit cost associated with the activity of facility i.

The SCOE model, therefore, becomes

$$\text{Minimize } \sum_{i \in V_1} \sum_{j \in V_2} b_{ij}s_{ij} + \sum_{i \in V_1} f_i y_i$$

subject to

$$\sum_{i \in V_1} s_{ij} = d_j, j \in V_2$$

$$\sum_{j \in V_2} s_{ij} \le q_i y_i, i \in V_1 \tag{3.18}$$

$$s_{ij} \ge 0, i \in V_1, j \in V_2 \tag{3.19}$$

$$y_i \in \{0, 1\}, i \in V_1. \tag{3.20}$$

Constraints (3.18) are facility throughput constraints and are used to express the relationship between continuous decision variables $s_{ij}, i \in V_1, j \in V_2$, and binary decision variables $y_i, i \in V_1$.

Alternatively, an equivalent model can be formulated by replacing decision variables s_{ij} with $x_{ij}, i \in V_1, j \in V_2$, according to the following relations:

$$s_{ij} = d_j x_{ij}, i \in V_1, j \in V_2.$$

Here, $x_{ij}, i \in V_1, j \in V_2$, represents the fraction of the demand of successor node j satisfied by facility i in a time period. The corresponding formulation is reported below:

$$\text{Minimize} \sum_{i \in V_1} \sum_{j \in V_2} c_{ij} x_{ij} + \sum_{i \in V_1} f_i y_i \qquad (3.21)$$

subject to

$$\sum_{i \in V_1} x_{ij} = 1, j \in V_2 \qquad (3.22)$$

$$\sum_{j \in V_2} d_j x_{ij} \leq q_i y_i, i \in V_1 \qquad (3.23)$$

$$x_{ij} \geq 0, i \in V_1, j \in V_2 \qquad (3.24)$$

$$y_i \in \{0, 1\}, i \in V_1, \qquad (3.25)$$

where

$$c_{ij} = b_{ij} d_j = a_{ij} d_j + g_i d_j, i \in V_1, j \in V_2.$$

SCOE discrete location problem (3.21)–(3.25) is known as the *capacitated plant location* (CPL) problem.

[Milatog.xlsx, Milatog.py] Milatog is a Russian company producing cattle forage. In the Volga region, there are seven farms which have an average daily forage demand (in quintals) equal to 36, 42, 34, 50, 27, 30, and 43, respectively.

Milatog intends to purchase some silos, to supply the seven farms. Six different potential sites in the area have been identified, with a daily forage throughput (expressed in quintals) equal to, respectively, 80, 90, 110, 120, 100, and 120. For the next four years, Milatog has estimated the following fixed costs (in €): 321 420, 350 640, 379 860, 401 775, 350 640, and 336 030, respectively. The daily average running cost (in €) per quintal of forage, for each potential site, is equal to 0.15, 0.18, 0.20, 0.18, 0.15, and 0.17, respectively.

The transportation cost per quintal of forage and per kilometre travelled is equal to € 0.06. The kilometric distances for each origin–destination pair are shown in Table 3.11. The daily transportation costs are computed by considering that every trip is made up of both an outward and a return journey.

Milatog is planning to keep the warehouses in operation for four years (corresponding to $365 \times 3 + 366 = 1461$ days). On a daily basis, the CPL model can be formulated as follows:

$V_1 = \{1, 2, 3, 4, 5, 6\}$ is the set of potential sites;

Table 3.11 Distances (in km) between each potential silo and each farm for the Milatog problem.

Potential silo (*i*)	Farm						
	j = 1	*j* = 2	*j* = 3	*j* = 4	*j* = 5	*j* = 6	*j* = 7
1	18	23	19	21	24	17	9
2	21	18	17	23	11	18	20
3	27	18	17	20	23	9	18
4	16	23	9	31	21	23	10
5	31	20	18	19	10	17	18
6	18	17	29	21	22	18	8

$V_2 = \{1, 2, 3, 4, 5, 6, 7\}$ is the set of farms;

$f_1 = 321\,420/1461 = €\ 220$.

Similarly, the other fixed costs f_i, $i = 2, \ldots, 6$, can be computed. Also

$c_{11} = 0.06 \times 2 \times 18 \times 36 + 0.15 \times 36 = €\ 83.16$.

A similar procedure is used to calculate the other costs c_{ij}, $i = 1, \ldots, 6$, $j = 1, \ldots, 7$. Furthermore, let y_i, $i = 1, \ldots, 6$, be the binary decision variable associated with potential site i (with value 1 if silo i is purchased by Milatog, 0 otherwise), and x_{ij}, $i = 1, \ldots, 6$, $j = 1, \ldots, 7$, be the decision variable that expresses the fraction of the average daily demand of farm j and satisfied by silo i.

Minimize $83.16x_{11} + 122.22x_{12} + 82.62x_{13} + 133.50x_{14} + 81.81x_{15}$

$+ 65.70x_{16} + 52.89x_{17} + 97.20x_{21} + 98.28x_{22} + 75.48x_{23}$

$+ 147.00x_{24} + 40.50x_{25} + 70.20x_{26} + 110.94x_{27} + 123.84x_{31}$

$+ 99.12x_{32} + 76.16x_{33} + 130.00x_{34} + 79.92x_{35} + 38.40x_{36}$

$+ 101.48x_{37} + 75.60x_{41} + 123.48x_{42} + 42.84x_{43} + 195.00x_{44}$

$+ 72.90x_{45} + 88.20x_{46} + 59.34x_{47} + 139.32x_{51} + 107.10x_{52}$

$+ 78.54x_{53} + 121.50x_{54} + 36.45x_{55} + 65.70x_{56} + 99.33x_{57}$

$+ 83.88x_{61} + 92.82x_{62} + 124.10x_{63} + 134.50x_{64} + 75.87x_{65}$

$+ 69.90x_{66} + 48.59x_{67} + 220y_1 + 240y_2 + 260y_3 + 275y_4$

$+ 240y_5 + 230y_6$

subject to

Table 3.12 Fraction of the daily forage demand of every farm satisfied by the purchased silos for the Milatog problem.

Purchased silo (i)	Farm						
	$j = 1$	$j = 2$	$j = 3$	$j = 4$	$j = 5$	$j = 6$	$j = 7$
1	1	0	11/34	0	0	1	0
5	0	0	23/34	1	1	0	0
6	0	1	0	0	0	0	1

$$x_{11} + x_{21} + x_{31} + x_{41} + x_{51} + x_{61} = 1$$

$$x_{12} + x_{22} + x_{32} + x_{42} + x_{52} + x_{62} = 1$$

$$x_{13} + x_{23} + x_{33} + x_{43} + x_{53} + x_{63} = 1$$

$$x_{14} + x_{24} + x_{34} + x_{44} + x_{54} + x_{64} = 1$$

$$x_{15} + x_{25} + x_{35} + x_{45} + x_{55} + x_{65} = 1$$

$$x_{16} + x_{26} + x_{36} + x_{46} + x_{56} + x_{66} = 1$$

$$x_{17} + x_{27} + x_{37} + x_{47} + x_{57} + x_{67} = 1$$

$$36x_{11} + 42x_{12} + 34x_{13} + 50x_{14} + 27x_{15} + 30x_{16} + 43x_{17} \leq 80y_1$$

$$36x_{21} + 42x_{22} + 34x_{23} + 50x_{24} + 27x_{25} + 30x_{26} + 43x_{27} \leq 90y_2$$

$$36x_{31} + 42x_{32} + 34x_{33} + 50x_{34} + 27x_{35} + 30x_{36} + 43x_{37} \leq 110y_3$$

$$36x_{41} + 42x_{42} + 34x_{43} + 50x_{44} + 27x_{45} + 30x_{46} + 43x_{47} \leq 120y_4$$

$$36x_{51} + 42x_{52} + 34x_{53} + 50x_{54} + 27x_{55} + 30x_{56} + 43x_{57} \leq 100y_5$$

$$36x_{61} + 42x_{62} + 34x_{63} + 50x_{64} + 27x_{65} + 30x_{66} + 43x_{67} \leq 120y_6$$

$$x_{ij} \geq 0, i = 1, \dots, 6, j = 1, \dots, 7$$

$$y_i \in \{0, 1\}, i = 1, \dots, 6.$$

The optimal solution of the problem (determined, e.g., by using the `Excel Solver` tool) includes silos 1, 5, and 6, at an overall daily cost equal to € 1218.08. The daily demand of the seven farms is satisfied as reported in Table 3.12.

The CPL model can be easily adapted to take into account further conditions of practical interest. For example, a potential facility cannot be run economically if its average level of activity in a time period is lower than a value q_i^- or higher than a threshold q_i^+. For intermediate values, the operating cost grows linearly. To take this condition into account, the CPL model can be modified by substituting the corresponding constraints (3.23) with the following pair of relations

$$\sum_{j \in V_2} d_j x_{ij} \leq q_i^+ y_i,$$

$$\sum_{j \in V_2} d_j x_{ij} \geq q_i^- y_i,$$

for all facilities $i \in V_1$ where these conditions apply.

[Milatog.xlsx] With reference to the Milatog location problem, it is now assumed that the sixth silo has a minimum daily level of activity equal to 90 quintals, whereas the maximum level of activity remains unchanged at 120. Consequently, the CPL problem illustrated in the previous box requires the following additional constraint:

$$36x_{61} + 42x_{62} + 34x_{63} + 50x_{64} + 27x_{65} + 30x_{66} + 43x_{67} \geq 90y_6.$$

Imposing this constraint implies that the optimal solution of the Milatog problem changes: the purchased silos will always be the first, fifth, and sixth, but the daily cost results are equal to € 1218.18 and the average daily demands of the seven farms are satisfied in a different way, as reported in Table 3.13.

Table 3.13   Fraction of the daily forage demand of each farm satisfied by the purchased silos for the modified Milatog problem.

Purchased silo (i)	Farm						
	$j = 1$	$j = 2$	$j = 3$	$j = 4$	$j = 5$	$j = 6$	$j = 7$
1	31/36	0	11/34	0	0	1	0
5	0	0	23/34	1	1	0	0
6	5/36	1	0	0	0	0	1

Another interesting case occurs when the cost $F_i(u_i)$ of a potential facility $i \in V_1$ can be represented through a concave piecewise linear function of its activity level because of economies of scale. In the simplest case, there are only two piecewise lines (see Figure 3.9). Then

$$F_i(u_i) = \begin{cases} 0, & \text{if } u_i = 0, \\ f_i' + g_i' u_i, & \text{if } 0 < u_i \leq \overline{u}_i \\ f_i'' + g_i'' u_i, & \text{if } \overline{u}_i < u_i \leq \hat{u}_i \end{cases}, i \in V_1 \tag{3.26}$$

where it results that $f_i' \leq f_i''$, $g_i' \geq g_i''$ and $f_i' + g_i'\overline{u}_i = f_i'' + g_i''\overline{u}_i$.

In order to model this problem, each potential facility is replaced by as many artificial facilities as the number of piecewise lines of its cost function. For instance, if equation (3.26) holds, facility $i \in V_1$ is replaced by two artificial facilities i' and i'' whose operating costs are characterized, respectively, by fixed costs equal to $f_{i'} = f_i'$ and $f_{i''} = f_i''$ and by running unit costs equal to $g_{i'} = g_i'$ and $g_{i''} = g_i''$. Moreover, it is necessary to add the constraint

$$y_{i'} + y_{i''} \leq 1, \tag{3.27}$$

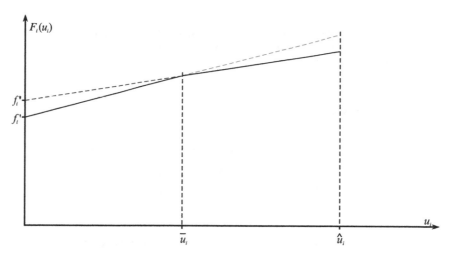

Figure 3.9 Plot of a concave two-piecewise linear cost function $F_i(u_i)$ of potential facility $i \in V_1$.

to ensure the non-simultaneous activation of the two artificial facilities i' and i'' (it is easy to verify that this constraint is redundant in case of $\hat{u}_i$ being sufficiently large), see Problem 3.14).

[Milatog.xlsx] With reference to the Milatog location problem, it is assumed that the daily average facility unit running cost of the sixth silo decreases by € 0.03 per quintal of forage and that the facility fixed costs are increased to € 339 536 when the level of daily activity is greater than 80 quintals. The above CPL formulation is modified by replacing facility 6 with two artificial facilities $6'$ and $6''$, characterized by fixed and unit running unit daily costs equal to, respectively, $f_{i'} = €\ 230$, $g_{i'} = €\ 0.17$, and $f_{i''} = €\ 232.4$, $g_{i''} = €\ 0.14$.

Both the artificial facilities are at the same distances to the seven farms as the original facility, so that the daily transportation and the facility unit running cost (in €) $c_{6'j}$ and $c_{6''j}$, $j = 1, \dots, 7$, are

$$c_{6'1} = 83.88; \quad c_{6''1} = 82.80;$$
$$c_{6'2} = 92.82; \quad c_{6''2} = 91.56;$$
$$c_{6'3} = 124.10; \quad c_{6''3} = 123.08;$$
$$c_{6'4} = 134.50; \quad c_{6''4} = 133.00;$$
$$c_{6'5} = 75.87; \quad c_{6''5} = 75.06;$$
$$c_{6'6} = 69.90; \quad c_{6''6} = 69.00;$$
$$c_{6'7} = 48.59; \quad c_{6''7} = 47.30.$$

The two artificial facilities are characterized by the following daily forage throughput (in quintals): $q_{6'} = 80$ and $q_{6''} = 120$.

Table 3.14 Fraction of the daily forage demand of every farm satisfied by the purchased silos for the Milatog problem for two artificial facilities.

Purchased silo (*i*)	Farm						
	j = 1	*j* = 2	*j* = 3	*j* = 4	*j* = 5	*j* = 6	*j* = 7
1	1/36	0	11/34	0	0	1	0
5	0	0	23/34	1	1	0	0
6	35/36	1	0	0	0	0	1

$$36x_{6'1} + 42x_{6'2} + 34x_{6'3} + 50x_{6'4} + 27x_{6'5} + 30x_{6'6} + 43x_{6'7} \leq 80y_{6'},$$

$$36x_{6''1} + 42x_{6''2} + 34x_{6''3} + 50x_{6''4} + 27x_{6''5} + 30x_{6''6} + 43x_{6''7} \leq 120y_{6''}.$$

The decision variable y_6 is replaced with $y_{6'}$ and $y_{6''}$ and the decision variables $x_{6j}, j = 1, \ldots, 7$, with $x_{6'j}$ and $x_{6''j}, j = 1, \ldots, 7$, respectively.

The throughput constraints (3.23) corresponding to the two artificial facilities become

It is also necessary to impose $y_{6'} + y_{6''} \leq 1$ to avoid the simultaneous activation of the two artificial facilities.

The optimal solution, obtained by using the Excel Solver tool, yields $y_1^* = y_5^* = y_{6''}^* = 1$ and $y_2^* = y_3^* = y_4^* = y_{6'}^* = 0$. Consequently, silos 1, 5 and 6 will be purchased by Milatog, in particular silo 6 with a daily throughput equal to 120 quintals of forage. The optimal daily cost is equal to € 1217.58. Table 3.14 shows how the daily forage demand of each farm is satisfied.

If the facilities to be activated have no throughput constraints, then constraints (3.23) should be replaced by constraints which can be used only to express the existing relationship between the decision variables $x_{ij}, i \in V_1, j \in V_2$, and $y_i, i \in V_1$, that is,

$$\sum_{j \in V_2} x_{ij} \leq |V_2| y_i, i \in V_1.$$

The resulting model corresponds to the *simple plant location* (SPL) model, reported below.

Minimize $\sum_{i \in V_1} \sum_{j \in V_2} c_{ij} x_{ij} + \sum_{i \in V_1} f_i y_i$

subject to

$$\sum_{i \in V_1} x_{ij} = 1, j \in V_2$$

$$\sum_{j \in V_2} x_{ij} \leq |V_2| y_i, i \in V_1$$

$$x_{ij} \geq 0, i \in V_1, j \in V_2$$

$$y_i \in \{0,1\}, i \in V_1.$$

The absence of throughput constraints means that an optimal solution of the SPL problem automatically leads to binary values of the x_{ij} variables. More specifically, there exists at least one optimal solution such as the demand of each successor node $j \in V_2$ is satisfied by a single activated facility $i \in V_1$ (*single assignment* property).

[Milatog.xlsx] With reference to the Milatog location problem, it is assumed that the silos have no throughput constraints (i.e., the average daily throughput for each silo is sufficient to satisfy the whole daily average forage demand for all the farms). The original throughput constraints are replaced with

$$x_{11} + x_{12} + x_{13} + x_{14} + x_{15} + x_{16} + x_{17} \leq 7y_1;$$

$$x_{21} + x_{22} + x_{23} + x_{24} + x_{25} + x_{26} + x_{27} \leq 7y_2;$$

$$x_{31} + x_{32} + x_{33} + x_{34} + x_{35} + x_{36} + x_{37} \leq 7y_3;$$

$$x_{41} + x_{42} + x_{43} + x_{44} + x_{45} + x_{46} + x_{47} \leq 7y_4;$$

$$x_{51} + x_{52} + x_{53} + x_{54} + x_{55} + x_{56} + x_{57} \leq 7y_5;$$

$$x_{61} + x_{62} + x_{63} + x_{64} + x_{65} + x_{66} + x_{67} \leq 7y_6.$$

The optimal solution of the problem, obtained again by using the Excel Solver tool, entails the purchase of just one silo (the first), which will supply all the farms, with an overall daily cost equal to € 841.90.

Other SCOE discrete location problems can be formulated by modifying the CPL model. In particular, if p facilities should be activated, then the following constraint has to be added to the CPL model

$$\sum_{i \in V_1} y_i = p. \tag{3.28}$$

If it is requested that a specific subset of facilities $V_1' \subseteq V_1$ be necessarily activated, then the following constraints can be imposed

$$y_i = 1, i \in V_1'.$$

Considering the SCOE discrete location formulation (3.21)–(3.25), (3.28) and assuming that

- the fixed costs are the same for each facility (i.e., $f_i = f$, for each $i \in V_1$; this means that in the objective function (3.21) the sum of the fixed costs of the activated facilities is equal to fp and, being a constant term, it can be omitted);
- $d_j = 1, j \in V_2$;
- $q_i = |V_2|, i \in V_1$,

the so-called *p-median* model is obtained,

$$\text{Minimize} \sum_{i \in V_1} \sum_{j \in V_2} c_{ij} x_{ij}$$

subject to

$$\sum_{i \in V_1} x_{ij} = 1, \ j \in V_2$$

$$\sum_{j \in V_2} x_{ij} \leq |V_2| y_i, \ i \in V_1$$

$$\sum_{i \in V_1} y_i = p$$

$$x_{ij} \geq 0, \ i \in V_1, j \in V_2$$

$$y_j \in \{0, 1\}, \ j \in V_2.$$

Similarly to the SPL model, the single assignment property holds also for the p-median problem, since the absence of throughput constraints, so there exists an optimal solution for which the x_{ij} decision variables assume binary values.

[UnitedBank.xlsx, UnitedBank.py] United Bank, a Bulgarian credit institute, has used the p-median model to determine the optimal location of two bank offices at Gabrovo. The city area was divided into eight different districts (which form the set V_2), whereas six different potential sites are identified (forming the set V_1) for the location of the bank offices. The six potential sites have identical facility fixed costs. The connection costs between the potential sites and the centroids of the eight districts are proportional to the corresponding kilometric distances, which are reported in Table 3.15. In the p-median problem (with $p = 2$) $y_i, i \in V_1$, is the binary decision variable assuming value 1 if a bank office is located at site i, 0 otherwise; $x_{ij}, i \in V_1, j \in V_2$, is a continuous decision variable that, according to the single assignment property, assumes a binary value, equal to 1 if the bank office located at site i serves district j, 0 otherwise; $c_{ij}, i \in V_1, j \in V_2$ (see Table 3.15), is the kilometric distance between site i and district j.

Note that the single assignment property of the p-median model allows the satisfaction of the binary constraints of the x_{ij} decision variables, $i \in V_1, j \in V_2$, if they are simply assumed continuous and non-negative.

The optimal United Bank problem solution, obtained by using the Excel Solver tool, leads to the location of the two bank offices at sites 5 and 6. Districts 1, 3, 6, and 7 are assigned to the bank office located at site 5, while districts 2, 4, 5, and 8 are assigned to the bank office located at site 6.

Table 3.15 Distances (in km) between the potential sites and the centroids of the eight districts of Gabrovo for the United Bank problem.

Potential site (i)	District							
	$j = 1$	$j = 2$	$j = 3$	$j = 4$	$j = 5$	$j = 6$	$j = 7$	$j = 8$
1	2.1	1.7	2.8	0.3	0.8	2.2	1.8	0.7
2	1.5	2.2	3.1	2.2	0.2	1.9	2.3	1.3
3	0.9	1.6	2.3	0.3	1.7	1.6	0.9	2.7
4	1.8	3.1	2.7	2.6	3.1	0.6	0.2	0.7
5	0.1	2.5	1.8	3.1	0.4	1.2	0.7	1.1
6	0.5	1.4	3.1	0.5	0.2	1.5	2.2	0.8

The CPL model is a location–allocation problem. If the set $\overline{V}_1 \subseteq V_1$ of open facilities is known, it is clear that an optimal allocation of the average quantities demanded by the successor nodes in a time period can be determined by solving the following *linear programming* (LP) problem:

$$\text{Minimize} \sum_{i \in \overline{V}_1} \sum_{j \in V_2} c_{ij} x_{ij} + \sum_{i \in \overline{V}_1} f_i \tag{3.29}$$

subject to

$$\sum_{i \in \overline{V}_1} x_{ij} = 1, j \in V_2 \tag{3.30}$$

$$\sum_{j \in V_2} d_j x_{ij} \leq q_i, \ i \in \overline{V}_1 \tag{3.31}$$

$$0 \leq x_{ij} \leq 1, i \in \overline{V}_1, \ j \in V_2. \tag{3.32}$$

[Milatog.xlsx, Milatog2.py] In the Milatog problem, if the set of purchased silos is known to be $\overline{V}_1 = \{1, 5, 6\}$ (whose daily total fixed costs are equal to € 690), the optimal fraction of daily forage demand satisfied by the purchased silos (shown in Table 3.12) can be obtained by solving the following LP problem:

$$\text{Minimize } 83.16x_{11} + 122.22x_{12} + 82.62x_{13} + 133.50x_{14} + 81.81x_{15}$$
$$+ 65.70x_{16} + 52.89x_{17} + 139.32x_{51} + 107.10x_{52} + 78.54x_{53}$$
$$+ 121.50x_{54} + 36.45x_{55} + 65.70x_{56} + 99.33x_{57} + 83.88x_{61}$$
$$+ 92.82x_{62} + 124.10x_{63} + 134.50x_{64} + 75.87x_{65} + 69.90x_{66}$$
$$+ 48.59x_{67} + 690$$

subject to

$$x_{11} + x_{51} + x_{61} = 1$$
$$x_{12} + x_{52} + x_{62} = 1$$
$$x_{13} + x_{53} + x_{63} = 1$$
$$x_{14} + x_{54} + x_{64} = 1$$
$$x_{15} + x_{55} + x_{65} = 1$$
$$x_{16} + x_{56} + x_{66} = 1$$
$$x_{17} + x_{57} + x_{67} = 1$$
$$36x_{11} + 42x_{12} + 34x_{13} + 50x_{14} + 27x_{15} + 30x_{16} + 43x_{17} \leq 80$$
$$36x_{51} + 42x_{52} + 34x_{53} + 50x_{54} + 27x_{55} + 30x_{56} + 43x_{57} \leq 100$$
$$36x_{61} + 42x_{62} + 34x_{63} + 50x_{64} + 27x_{65} + 30x_{66} + 43x_{67} \leq 120$$
$$x_{ij} \geq 0, i = 1, 5, 6, j = 1, \dots, 7.$$

It can happen that the optimal solution of the demand allocation problem (3.29)–(3.32) is such that the demand of a successor node $j \in V_2$ can be satisfied by more than one open facility $i \in V_1$ (i.e., some x_{ij}^* values may be fractional), because of throughput constraints. However, in the SCOE discrete location models without constraints (3.23) (as in the SPL and p-median models), the single assignment property holds, i.e., the demand of each successor node $j \in V_2$ is satisfied by a single facility $i \in V_1$. This solution can be obtained as follows. Let $i_j \in \overline{V}_1$ be a facility such that

$$i_j = \arg \min_{i \in \overline{V}_1} \{c_{ij}\}.$$

Then, the values of the decision variables can be found as follows:

$$x_{ij}^* = \begin{cases} 1, & \text{if } i = i_j \\ 0, & \text{otherwise.} \end{cases}$$

[Milatog.xlsx] To determine the optimal solution (already reported in a previous box) of the Milatog SPL problem, it is sufficient to execute the following steps.

1. Solve the problem assuming the purchase of a single silo. By exploiting the single assignment property, the silo $i^* \in V_1$ corresponds to

$$\min_{i \in V_1} \left\{ f_i + \sum_{j \in V_2} c_{ij} \right\}.$$

The minimum value is reached for the first silo, so that $f_1 + \sum_{j \in V_2} c_{1j} = €\,841.90$. Let $z^{(1)}$ be this cost.

2. Verify that a *lower bound* $LB^{(p)}$ on the optimal cost of the Milatog SPL problem is greater than or equal to $z^{(1)}$, supposing that p silos are purchased, with $p \geq 2$. The value of $LB^{(p)}$ can be easily determined in the following way. Since the single assignment property is valid, a lower bound on the daily cost of the demand allocation of each farm $j = 1, \ldots, 7$, can be determined. In particular, for farm 1 at least €\,75.60 per day has to be paid; for the other farms at least €\,92.82, €\,42.84, €\,121.50, €\,36.45, €\,38.40 and €\,48.59 per day has to be paid, respectively. A lower bound on the overall daily demand allocation costs is, therefore, equal to €\,456.20. In order to obtain $LB^{(p)}$, it is necessary to add to these costs the fixed costs for the p purchased silos, whose lower bound is easily determined by sorting fixed costs $f_i, i = 1, \ldots, 6$, in non-decreasing order and taking the first p sorted values. In this way, $LB^{(2)} = 456.20 + (220 + 230) = €\,906.20$, and similarly, $LB^{(3)} = 456.20 + (220 + 230 + 240) = €\,1146.20$. Consequently, $LB^{(p)} \geq LB^{(2)}$, for $p = 3, \ldots, 6$. Since $z^{(1)} < LB^{(2)}$, the optimal cost of the Milatog SPL problem corresponds to $z^{(1)}$.

A Lagrangian heuristic for the capacitated plant location problem

An SCOE discrete location problem is a *mixed integer programming* (MIP) problem, whose optimal solution can in principle be determined by means of a general purpose or a tailored branch-and-bound or a branch-and-cut algorithm. As a rule, capacitated problems are harder than uncapacitated ones. Nowadays, thanks to the development of such algorithms, it is possible to solve fairly large instances by using a suitable solver. Here, a heuristic capable of determining a good feasible solution within a reasonable amount of time is illustrated. Such a heuristic is valuable for more difficult location problems (i.e., multi-commodity and two-echelon problems) whose large instances are still out of reach for branch-and-bound or branch-and-cut algorithms, or require unacceptable computing times.

To evaluate whether a heuristic solution value is a tight *upper bound* (UB) on the optimal solution value, it is useful to determine a lower bound LB on the optimal solution value. This yields a ratio (UB − LB)/ LB (LB > 0) which represents an overestimate of the relative deviation of the heuristic solution value from the optimum. In the following, a Lagrangian heuristic is illustrated for the CPL problem, although this algorithm may be used to solve other SCOE discrete location problems.

The fundamental step of the heuristic is the determination of an LB, obtained by relaxing demand satisfaction constraints (3.22) in a Lagrangian fashion. Let $\lambda_j \in \mathfrak{R}$ be the Lagrangian multiplier associated with the jth constraint (3.22). Then the Lagrangian relaxed problem is

$$\text{Minimize} \sum_{i \in V_1} \sum_{j \in V_2} c_{ij} x_{ij} + \sum_{i \in V_1} f_i y_i + \sum_{j \in V_2} \lambda_j \left(\sum_{i \in V_1} x_{ij} - 1 \right)$$

$$= \sum_{i \in V_1} \sum_{j \in V_2} (c_{ij} + \lambda_j) x_{ij} + \sum_{i \in V_1} f_i y_i - \sum_{j \in V_2} \lambda_j \tag{3.33}$$

subject to

$$\sum_{j \in V_2} d_j x_{ij} \le q_i y_i, i, \in V_1 \tag{3.34}$$

$$0 \le x_{ij} \le 1, i \in V_1, j \in V_2 \tag{3.35}$$

$$y_i \in \{0, 1\}, i \in V_1, \tag{3.36}$$

whose optimal objective function value is denoted as $\text{LB}_{\text{CPL}}(\lambda)$. It should be observed that the elimination of constraints (3.22) imposes the introduction of constraints (3.35) which define an upper bound equal to 1 on decision variables $x_{ij}, i \in V_1, j \in V_2$.

In the Milatog problem, the Lagrangian relaxation corresponding to multipliers

$$\lambda = [-186; -170; -140; 60; -115; -166; -112]^{\text{T}}.$$

is

$$\text{Minimize} - 102.84 x_{11} - 47.78 x_{12} - 57.38 x_{13} + 193.50 x_{14} - 33.19 x_{15}$$

$$- 100.30 x_{16} - 59.11 x_{17} - 88.80 x_{21} - 71.72 x_{22} - 64.52 x_{23}$$

$$+ 207.00 x_{24} - 74.50 x_{25} - 95.80 x_{26} - 1.06 x_{27} - 62.16 x_{31}$$

$$- 70.88x_{32} - 63.84x_{33} + 190.00x_{34} - 35.08x_{35} - 127.60x_{36}$$
$$- 10.52x_{37} - 110.40x_{41} - 46.52x_{42} - 97.16x_{43} + 255.00x_{44}$$
$$- 42.10x_{45} - 77.80x_{46} - 52.66x_{47} - 46.68x_{51} - 62.90x_{52}$$
$$- 61.46x_{53} + 181.50x_{54} - 78.55x_{55} - 100.30x_{56} - 12.67x_{57}$$
$$- 102.12x_{61} - 77.18x_{62} - 15.90x_{63} + 194.50x_{64} - 39.13x_{65}$$
$$- 96.10x_{66} - 63.41x_{67} + 220y_1 + 240y_2 + 260y_3 + 275y_4$$
$$+ 240y_5 + 230y_6 + 829$$

subject to

$$36x_{11} + 42x_{12} + 34x_{13} + 50x_{14} + 27x_{15} + 30x_{16} + 43x_{17} \leq 80y_1$$
$$36x_{21} + 42x_{22} + 34x_{23} + 50x_{24} + 27x_{25} + 30x_{26} + 43x_{27} \leq 90y_2$$
$$36x_{31} + 42x_{32} + 34x_{33} + 50x_{34} + 27x_{35} + 30x_{36} + 43x_{37} \leq 110y_3$$
$$36x_{41} + 42x_{42} + 34x_{43} + 50x_{44} + 27x_{45} + 30x_{46} + 43x_{47} \leq 120y_4$$
$$36x_{51} + 42x_{52} + 34x_{53} + 50x_{54} + 27x_{55} + 30x_{56} + 43x_{57} \leq 100y_5$$
$$36x_{61} + 42x_{62} + 34x_{63} + 50x_{64} + 27x_{65} + 30x_{66} + 43x_{67} \leq 120y_6$$
$$0 \leq x_{ij} \leq 1, i = 1, \ldots, 6, j = 1, \ldots, 7$$
$$y_i \in \{0, 1\}, i = 1, \ldots, 6.$$

It is easy to show that problem (3.33)–(3.36) can be decomposed into $|V_1|$ subproblems, one for each potential facility $i \in V_1$, as follows:

$$\text{Minimize} \sum_{j \in V_2} (c_{ij} + \lambda_j) x_{ij} + f_i y_i \qquad (3.37)$$

subject to

$$\sum_{j \in V_2} d_j x_{ij} \leq q_i y_i \qquad (3.38)$$

$$0 \leq x_{ij} \leq 1, j \in V_2 \qquad (3.39)$$

$$y_i \in \{0, 1\}. \qquad (3.40)$$

Let $\text{LB}_{\text{CPL}}^{(i)}(\lambda), i \in V_1$, be the optimal objective function value of ith subproblem (3.37)–(3.40). It happens that

$$\text{LB}_{\text{CPL}}(\lambda) = \sum_{i \in V_1} \text{LB}_{\text{CPL}}^{(i)}(\lambda) - \sum_{j \in V_2} \lambda_j.$$

The optimal solution of subproblem (3.37)–(3.40) can be determined easily by inspection, by observing that:

- For $y_i = 0$, constraint (3.38) implies $x_{ij} = 0$, for each $j \in V_2$, and therefore $\text{LB}_{\text{CPL}}^{(i)}(\lambda)|_{y_i=0} = 0$.

- For $y_i = 1$, subproblem (3.37)–(3.40) is a *continuous knapsack* problem; it is well known that the optimal solution of this problem can be found in polynomial time by means of a greedy method, by sorting the decision variables $x_{ij}, i \in V_1, j \in V_2$, according to non-decreasing values of the ratios $(c_{ij} + \lambda_j)/d_j, j \in V_2$, and assigning, one by one, following the order in which the decision variables are sorted, the maximum possible value to each of them (compatibly with constraints satisfaction), whenever the corresponding Lagrangian cost $c_{ij} + \lambda_j$ is negative, and zero when the corresponding Lagrangian cost is non-negative (see the following example). It is worth noting that this case has to be taken into account only if there is at least one negative Lagrangian cost coefficient in objective function (3.37).

Consequently,

$$\text{LB}_{\text{CPL}}^{(i)}(\lambda) = \min\left\{0, \ \text{LB}_{\text{CPL}}^{(i)}(\lambda)|_{y_i=1}\right\},$$

that is,

$$\text{LB}_{\text{CPL}}^{(i)}(\lambda) \leq 0, \lambda \in \Re^{|V_2|}.$$

With reference to the Lagrangian relaxation of the Milatog problem reported in the previous box, subproblem $i = 1$ is

Minimize $- 102.84x_{11} - 47.78x_{12} - 57.38x_{13} + 193.50x_{14}$

$\qquad - 33.19x_{15} - 100.30x_{16} - 59.11x_{17} + 220y_1$

subject to

$\qquad 36x_{11} + 42x_{12} + 34x_{13} + 50x_{14} + 27x_{15} + 30x_{16} + 43x_{17} \leq 80y_1$

$\qquad 0 \leq x_{1j} \leq 1, j = 1, \dots, 7$

$\qquad y_1 \in \{0, 1\}.$

By setting $\bar{y}_1 = 1$, this subproblem becomes the following continuous knapsack problem:

Minimize $- 102.84x_{11} - 47.78x_{12} - 57.38x_{13} + 193.50x_{14} - 33.19x_{15}$

$\qquad - 100.30x_{16} - 59.11x_{17} + 220$

subject to

$\qquad 36x_{11} + 42x_{12} + 34x_{13} + 50x_{14} + 27x_{15} + 30x_{16} + 43x_{17} \leq 80$

$\qquad 0 \leq x_{1j} \leq 1, j = 1, \dots, 7,$

which can be solved in the following manner. Decision variables $x_{1j} = 1, j = 1, \dots, 7$, are sorted by non-decreasing values of ratios $\{-102.84/36, -47.78/42, -57.38/34, 193.50/50, -33.19/27, -100.30/30, -59.11/43\}$, that is,

$\qquad (x_{16}, x_{11}, x_{13}, x_{17}, x_{15}, x_{12}, x_{14}).$

Hence,

$\qquad \bar{x}_{16} = \min\{1, 80/30\} = 1$

is set. The throughput constraint becomes

$$36x_{11} + 42x_{12} + 34x_{13} + 50x_{14} + 27x_{15} + 43x_{17} \leq 50.$$

Then,

$$\bar{x}_{11} = \min\{1, 50/36\} = 1$$

is set. The throughput constraint becomes

$$42x_{12} + 34x_{13} + 50x_{14} + 27x_{15} + 43x_{17} \leq 14.$$

Then,

$$\bar{x}_{13} = \min\{1, 14/34\} = 7/17$$

is set. Since the residual throughput is reduced to 0,

$$\bar{x}_{12} = \bar{x}_{14} = \bar{x}_{15} = \bar{x}_{17} = 0;$$

$$LB_{CPL}^{(i)}(\lambda)|_{y_i=1} = -102.84 - 57.38 \times (7/17) - 100.30 + 220 = -6.77.$$

Since

$$LB_{CPL}^{(i)}(\lambda)|_{\bar{y}_i=1} < 0,$$

it follows that

$$LB_{CPL}^{(i)}(\lambda) = -6.77,$$

and, therefore,

$$y_1^* = 1, x_{11}^* = 1, x_{12}^* = 0, x_{13}^* = 7/17, x_{14}^* = 0, x_{15}^* = 0, x_{16}^* = 1, x_{17}^* = 0.$$

In a similar way, the other subproblems $i = 2, \ldots, 6$, can be solved, obtaining

$$LB_{CPL}^{(2)}(\lambda) = -11.70;$$

$$y_2^* = 1; x_{21}^* = 11/12; x_{22}^* = 0; x_{23}^* = 0; x_{24}^* = 0; x_{25}^* = 1; x_{26}^* = 1; x_{27}^* = 0;$$

$$LB_{CPL}^{(3)}(\lambda) = -10.48;$$

$$y_3^* = 1; x_{31}^* = 1; x_{32}^* = 5/21; x_{33}^* = 1; x_{34}^* = 0; x_{35}^* = 0; x_{36}^* = 1; x_{37}^* = 0;$$

$$LB_{CPL}^{(4)}(\lambda) = -41.55;$$

$$y_4^* = 1; x_{41}^* = 1; x_{42}^* = 0; x_{43}^* = 1; x_{44}^* = 0; x_{45}^* = 20/27; x_{46}^* = 1; x_{47}^* = 0;$$

$$LB_{CPL}^{(5)}(\lambda) = -13.79;$$

$$y_5^* = 1; x_{51}^* = 0; x_{52}^* = 3/14; x_{53}^* = 1; x_{54}^* = 0; x_{55}^* = 1; x_{56}^* = 1; x_{57}^* = 0;$$

$$LB_{CPL}^{(6)}(\lambda) = -63.10;$$

$$y_6^* = 1; x_{61}^* = 1; x_{62}^* = 1; x_{63}^* = 0; x_{64}^* = 0; x_{65}^* = 0; x_{66}^* = 1; x_{67}^* = 12/43.$$

The optimal cost of the Lagrangian relaxation corresponds to

$$LB_{CPL}(\lambda) = -147.39 + 829 = 681.61.$$

Starting from the optimal solution of the Lagrangian relaxation, it is possible to construct a CPL feasible solution as follows.

1. *Finding the facilities to be activated.* Let L be the list of potential facilities $i \in V_1$ sorted by non-decreasing values of $\mathrm{LB}_{\mathrm{CPL}}^{(i)}(\lambda)$, $i \in V_1$ (note that $\mathrm{LB}_{\mathrm{CPL}}^{(i)}(\lambda) \leq 0$, $i \in V_1, \lambda \in \mathfrak{R}^{|V_2|}$). Extract from L the minimum number of facilities capable of satisfying the total demand $\sum_{j \in V_2} d_j$ of the successor nodes. Let $\overline{V}_1$ be the set of facilities selected. Then $\overline{V}_1$ satisfies relation

$$\sum_{i \in \overline{V}_1} q_i \geq \sum_{j \in V_2} d_j.$$

2. *Customer allocation to the selected facilities.* Solve the demand allocation problem (3.29)–(3.32) considering $\overline{V}_1$ as the set of facilities to be opened. Let $\mathrm{UB}_{\mathrm{CPL}}(\lambda)$ be the cost (3.29) associated with the optimal allocation.

The heuristic first selects the facilities characterized by the smallest $\mathrm{LB}_{\mathrm{CPL}}^{(i)}(\lambda)$ values and then allocates optimally the demand to them.

With reference to the Lagrangian relaxation of the Milatog problem, the list L of facilities is:

$$L = (6, 4, 5, 2, 3, 1).$$

and the corresponding set $\overline{V}_1$ of silos to be purchased is

$$\overline{V}_1 = \{4, 5, 6\},$$

since $\sum_{i \in \overline{V}_1} q_i = 340 > \sum_{j \in V_2} d_j = 262$. The demand allocation problem is therefore the following:

Minimize $75.60x_{41} + 123.48x_{42} + 42.84x_{43} + 195.00x_{44} + 72.90x_{45}$

$+ 88.20x_{46} + 59.34x_{47} + 139.32x_{51} + 107.10x_{52} + 78.54x_{53}$

$+ 121.50x_{54} + 36.45x_{55} + 65.70x_{56} + 99.33x_{57} + 83.88x_{61}$

$+ 92.82x_{62} + 124.10x_{63} + 134.50x_{64} + 75.87x_{65} + 69.90x_{66}$

$+ 48.59x_{67} + 745$

subject to

$x_{41} + x_{51} + x_{61} = 1$

$x_{42} + x_{52} + x_{62} = 1$

$x_{43} + x_{53} + x_{63} = 1$

$x_{44} + x_{54} + x_{64} = 1$

$x_{45} + x_{55} + x_{65} = 1$

$x_{46} + x_{56} + x_{66} = 1$

$x_{47} + x_{57} + x_{67} = 1$

$$36x_{41} + 42x_{42} + 34x_{43} + 50x_{44} + 27x_{45} + 30x_{46} + 43x_{47} \leq 120$$

$$36x_{51} + 42x_{52} + 34x_{53} + 50x_{54} + 27x_{55} + 30x_{56} + 43x_{57} \leq 100$$

$$36x_{61} + 42x_{62} + 34x_{63} + 50x_{64} + 27x_{65} + 30x_{66} + 43x_{67} \leq 120$$

$$x_{ij} \geq 0, i = 4, 5, 6, j = 1, \ldots, 7,$$

with an optimal solution whose cost is

$$UB_{CPL}(\lambda) = €1229.48.$$

The daily forage demand of the seven farms is satisfied as reported in Table 3.16.

Table 3.16   Fraction of the daily forage demand of each farm satisfied by the purchased silos 4, 5, and 6 for the Milatog problem.

Purchased silo (*i*)	Farm						
	j = 1	*j* = 2	*j* = 3	*j* = 4	*j* = 5	*j* = 6	*j* = 7
4	1	0	1	0	0	0	0
5	0	0	0	1	1	23/30	0
6	0	1	0	0	0	7/30	1

Thus, for each set of Lagrangian multipliers $\lambda \in \mathfrak{R}^{|V_2|}$, the above procedure computes both an LB and a UB ($LB_{CPL}(\lambda)$ and $UB_{CPL}(\lambda)$, respectively). If these bounds coincide, an optimal solution has been found. Otherwise, in order to determine the Lagrangian multipliers λ^* corresponding to the maximum possible Lagrangian LB

$$\lambda^* = \arg \max_{\lambda} LB_{CPL}(\lambda)$$

(or at least a satisfactory approximation), the *subgradient method* can be used (the Lagrangian relaxed problem corresponding to λ^* is called *dual Lagrangian problem*). This procedure can also be used to generate, in many cases, better upper bounds, since the feasible solutions generated from improved lower bounds are generally less costly. Here is a schematic description of the CPL_SUBGRADIENT procedure implemented in the file CPL_subgradient.py.

At each iteration, this algorithm updates the Lagrangian multipliers of the previous iteration, thus generating new lower and upper bounds. Computational experiments have shown that the initial values of the Lagrangian multipliers do not significantly affect the behaviour of the heuristic. Hence, the Lagrangian multipliers are set equal to zero in code line 7 of the CPL_SUBGRADIENT procedure. The Lagrangian multipliers are updated by formula (3.41) which can be explained in the following way. If, at the hth iteration, the left-hand side of constraint (3.22) is higher than the right-hand side ($\sum_{i \in V_1} x_{ij}^{(h)} > 1$) for a certain $j \in V_2$, the subgradient $s_j^{(h)}$ is positive and the corresponding Lagrangian multiplier has to be increased in order to heavily penalize the constraint

1: **procedure** CPL_SUBGRADIENT $(\beta_{min}, \beta_{max}, \gamma, \overline{\mathbf{y}}, \overline{\mathbf{x}}, \overline{z})$

2:     # β_{min}, β_{max} and γ are non-negative user-defined parameters;

3:     # $\overline{\mathbf{y}}$ and $\overline{\mathbf{x}}$ constitute the CPL feasible solution returned by the procedure;

4:     # $\overline{z}$ is the cost corresponding to the CPL feasible solution;

5:     LB $= -\infty$;

6:     UB $= \infty$;

7:     $\lambda_j^{(1)} = 0, j \in V_2$;

8:     $\beta^{(1)} = \beta_{max}$;

9:     $h = 1$;

10:     **while** $(\beta^{(h)} > \beta_{min})$ **and** (LB < UB) **do**

11:         # Computation of a new LB;

12:         Solve Lagrangian relaxation (3.33)–(3.36) using $\lambda^{(h)} \in \mathfrak{R}^{|V_2|}$ as a vector of Lagrangian multipliers;

13:         **if** $\text{LB}_{\text{CPL}}(\lambda^{(h)}) > \text{LB}$ **then**

14:             $\text{LB} = \text{LB}_{\text{CPL}}(\lambda^{(h)})$;

15:         **end if**

16:         # Computation of a new UB;

17:         Determine the CPL feasible solution $\overline{\mathbf{y}}^{(h)}, \overline{\mathbf{x}}^{(h)}$ (with cost $\text{UB}_{\text{CPL}}(\lambda^{(h)})$) corresponding to $\text{LB}_{\text{CPL}}(\lambda^{(h)})$;

18:         **if** $\text{UB}_{\text{CPL}}(\lambda^{(h)}) < \text{UB}$ **then**

19:             $\text{UB} = \text{UB}_{\text{CPL}}(\lambda^{(h)})$;

20:             $\overline{\mathbf{y}} = \overline{\mathbf{y}}^{(h)}$;

21:             $\overline{\mathbf{x}} = \overline{\mathbf{x}}^{(h)}$;

22:             $\overline{z} = \text{UB}_{\text{CPL}}(\lambda^{(h)})$;

23:         **end if**

24:         **if** LB < UB **then**

25:             # Determine the subgradient of the relaxed constraints;

26:

$$s_j^{(h)} = \sum_{i \in V_1} x_{ij}^{(h)} - 1, \quad j \in V_2;$$

27:             # where $x_{ij}^{(h)}$ is the solution of Lagrangian relaxation (3.33)–(3.36) with multipliers $\lambda^{(h)} \in \mathfrak{R}^{|V_2|}$;

28:             # Update the Lagrangian multipliers;

29:

$$\lambda_j^{(h+1)} = \lambda_j^{(h)} + \beta^{(h)} s_j^{(h)}, j \in V_2; \qquad (3.41)$$

30:             # Compute $\beta^{(h+1)}$;

31:

$$\beta^{(h+1)} = \gamma \beta^{(h)}$$

32:             $h = h + 1$;

33:         **end if**

34:     **end while**

35:     **return** $\overline{\mathbf{y}}, \overline{\mathbf{x}}, \overline{z}$;

36: **end procedure**

violation. Conversely, if the left-hand side of constraint (3.22) is lower than the right-hand side ($\sum_{i \in V_1} x_{ij}^{(h)} < 1$) for a certain $j \in V_2$, the associated subgradient $s_j^{(h)}$ is negative and the value of the associated Lagrangian multiplier must be decreased to make the service of the unsatisfied demand fraction $1 - \sum_{i \in V_1} x_{ij}^{(h)}$ more attractive. Finally, if the jth constraint (3.22) is satisfied ($\sum_{i \in V_1} x_{ij}^{(h)} = 1$), the corresponding Lagrangian multiplier is unchanged.

It can be proved that the sequence of multiplier vectors $\lambda^{(1)}, \lambda^{(2)}, \ldots$ converges to vector λ^* if parameters $\beta^{(h)}, h = 1, \ldots, \infty$, satisfy the following two conditions:

$$\lim_{h \to \infty} \beta^{(h)} = 0; \tag{3.42}$$

$$\sum_{h=1}^{\infty} \beta^{(h)} = \infty. \tag{3.43}$$

In practice, it is required that multipliers tend to the zero, but not too fast.

Finally, it is worth noting that the $\{LB_{CPL}(\lambda^{(h)})\}$ sequence produced by the CPL_SUBGRADIENT procedure does not decrease monotonically. Therefore, there could exist iterations h for which $LB_{CPL}(\lambda^{(h)}) < LB_{CPL}(\lambda^{(h-1)})$ (this explains the LB update in code lines 13–15 of the CPL_SUBGRADIENT procedure). In practice, $LB_{CPL}(\lambda^{(h)})$ values exhibit a zigzagging pattern.

Computational experiments have shown that the CPL_SUBGRADIENT procedure generally requires a few thousand iterations for CPL problems with hundreds of vertices in V_1 and in V_2.

[Milatog.xlsx, Milatog3.py] In the Milatog problem, the CPL_SUBGRADIENT procedure is applied as follows:

1. the proportionality coefficient $\beta^{(h)}$ is initialized (i.e., $h = 1$) to $\beta_{max} = 100$ and decreased by 1% at each iteration ($\gamma = 0.999$) until it reaches $\beta_{min} = 0.1$;
2. when choosing the facilities to be inserted into list L, ties are broken according to the (minimum) f_i/q_i ratios, $i \in L$, if possible, or randomly, otherwise.

In what follows, the results of the first three iterations are reported.

- Initialization

   $\lambda^{(1)} = [0.00; 0.00; 0.00; 0.00; 0.00; 0.00; 0.00]^T$;

- Iteration 1

   $LB_{CPL}(\lambda^{(1)}) = 0.00$;

   $LB = 0.00$;

   $\bar{\mathbf{y}}^{(1)} = [0; 0; 1; 1; 0; 1]^T$;

   $UB_{CPL}(\lambda^{(1)}) = 1266.15$;

UB = 1266.15;

$\mathbf{s}^{(1)} = [-1.00; -1.00; -1.00; -1.00; -1.00; -1.00; -1.00]^{\mathsf{T}}$;

$\beta^{(1)} = 100.00$;

$\lambda^{(2)} = [-100.00; -100.00; -100.00; -100.00; -100.00; -100.00;$
$\qquad -100.00]^{\mathsf{T}}$;

- Iteration 2

  $LB_{CPL}(\lambda^{(2)}) = 700.00$;

  $LB = 700.00$;

  $\overline{\mathbf{y}}^{(2)} = [0; 0; 1; 1; 0; 1]^{\mathsf{T}}$;

  $UB_{CPL}(\lambda^{(2)}) = 1266.15$;

  $UB = 1266.15$;

  $\mathbf{s}^{(2)} = [-1.00; -1.00; -1.00; -1.00; -1.00; -1.00; -1.00]^{\mathsf{T}}$;

  $\beta^{(2)} = 99.90$;

  $\lambda^{(3)} = [-199.90; -199.90; -199.90; -199.90; -199.90; -199.90;$
  $\qquad -199.90]^{\mathsf{T}}$;

- Iteration 3

  $LB_{CPL}(\lambda^{(3)}) = 266.26$;

  $LB = 700.00$;

  $\overline{\mathbf{y}}^{(3)} = [0; 0; 0; 1; 1; 1]^{\mathsf{T}}$;

  $UB_{CPL}(\lambda^{(3)}) = 1229.48$;

  $UB = 1229.48$;

  $\mathbf{s}^{(3)} = [0.36; -0.55; 3.65; -1.00; 5.00; 5.00; 0.21]^{\mathsf{T}}$;

  $\beta^{(3)} = 99.80$;

  $\lambda^{(4)} = [-163.86; -254.55; 164.08; -299.70; 299.10; 299.10; -179.01]^{\mathsf{T}}$.

The procedure terminates after 6905 iterations with LB = 1079.14 and UB = 1218.08. It is worth noting that LB is greater than the continuous lower bound, equal to 1027.43, obtained by solving the CPL problem allowing that each binary decision variable y_i, $i \in V_1$, should have a continuous $[0, 1]$ domain instead of $\{0, 1\}$. Moreover, UB coincides with the optimal solution, previously determined. Finally, (UB−LB)/LB = 12 %. Figure 3.10 shows how UB and LB evolve during the algorithm's execution, while Figure 3.11 depicts $LB_{CPL}(\lambda^{(h)})$ and $UB_{CPL}(\lambda^{(h)})$, as well as UB and LB, for iterations 200 to 1500. As previously observed, both $LB_{CPL}(\lambda^{(h)})$ and $UB_{CPL}(\lambda^{(h)})$ do not decrease monotonically.

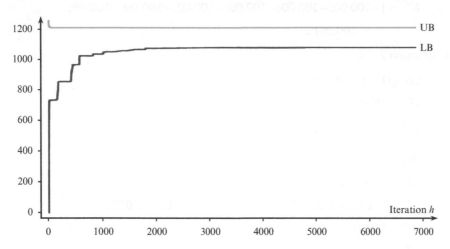

Figure 3.10 LB and UB values determined during iterations $h = 1, \ldots, 6905$ of the `CPL_SUBGRADIENT` procedure for the Milatog problem.

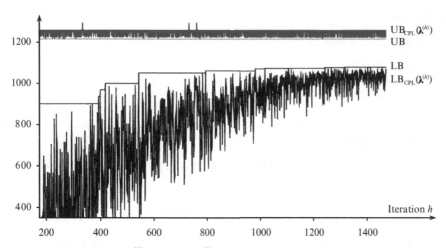

Figure 3.11 LB, UB, $LB_{CPL}\lambda^{(h)}$ and $UB_{CPL}\lambda^{(h)}$ values determined during iterations $h = 200, \ldots, 1500$ of the `CPL_SUBGRADIENT` procedure for the Milatog problem.

3.10 Single-commodity Two-echelon Discrete Location Problems

This section presents a *single-commodity two-echelon* (SCTE) discrete location model, based on the following assumptions. Let $G = (V_1 \cup V_2 \cup V_3, A_1 \cup A_2)$ be a complete directed tripartite graph in which the vertices in V_1 represent the predecessor nodes, the vertices in V_2 represent the potential facilities to be located and the vertices in V_3 represent the successor nodes; the arcs in $A_1 = V_1 \times V_2$ are associated with the commodity flows between the predecessor nodes and the potential facilities, while those in $A_2 = V_2 \times V_3$ correspond to the commodity flows between the potential facilities and the successor nodes.

Let $t = 1, \ldots, T$ be the time periods composing the planning horizon. Denote by o_{it}, $i \in V_1$, the maximum quantity of commodity available at the predecessor node i at time period t, and q_{jt}, $j \in V_2$, the throughput of potential facility j, at time period t, and d_{rt}, $r \in V_3$, the demand estimate of the successor node r at time period t. Similarly to the SCOE formulation illustrated in the previous Section 3.9, the focus is limited to the case in which o_{it}, $i \in V_1$, q_{jt}, $j \in V_2$ and d_{rt}, $r \in V_3$, are constant for each time period $t = 1, \ldots, T$, of the planning horizon, that is,

$$o_{it} = o_i, i \in V_1, t = 1, \ldots, T,$$

$$q_{jt} = q_j, j \in V_2, t = 1, \ldots, T,$$

$$d_{rt} = d_r, r \in V_3, t = 1, \ldots, T.$$

Under these assumptions, the SCTE discrete location problem can be associated with a generic time period, instead of the whole planning horizon. Let a_{ijr}, $i \in V_1$, $j \in V_2$, $r \in V_3$, be the unit transportation cost of the commodity from the predecessor node i to successor node r through facility j.

Finally, it is assumed that the facility cost of each potential facility $j \in V_2$, in a time period can be expressed in terms of a fixed cost f_j and a constant unit running cost g_j.

Supposing that the commodity flows are divisible, the SCTE discrete location problem can be formulated using the following decision variables: y_j, $j \in V_2$, is binary, equal to 1 if potential facility j is activated, 0 otherwise; $s_{ijr} \geq 0$, $i \in V_1$, $j \in V_2$, $r \in V_3$, which represents the average quantity of commodity transported, in each time period $t = 1, \ldots, T$, from the predecessor node i to the successor node r through facility j. The formulation is the following:

$$\text{Minimize} \sum_{i \in V_1} \sum_{j \in V_2} \sum_{r \in V_3} a_{ijr} s_{ijr} + \sum_{j \in V_2} \left(f_j y_j + g_j \sum_{i \in V_1} \sum_{r \in V_3} s_{ijr} \right) \tag{3.44}$$

subject to

$$\sum_{j \in V_2} \sum_{r \in V_3} s_{ijr} \leq o_i, i \in V_1 \tag{3.45}$$

$$\sum_{i \in V_1} \sum_{j \in V_2} s_{ijr} = d_r, r \in V_3 \tag{3.46}$$

$$\sum_{i \in V_1} \sum_{r \in V_3} s_{ijr} \leq q_j y_j, j \in V_2 \tag{3.47}$$

$$y_j \in \{0, 1\}, j \in V_2 \tag{3.48}$$

$$s_{ijr} \geq 0, i \in V_1, j \in V_2, r \in V_3. \tag{3.49}$$

The objective function (3.44) includes both the transportation and the facility costs. Constraints (3.45) impose an UB on the commodity flow from each predecessor node; equation (3.46) imposes the demand satisfaction for every successor node; relations (3.47) represent the throughput constraints on the facilities to be activated. They also impose that no flow can traverse a facility if it is not activated.

[Milatog.xlsx] With reference to the Milatog location problem, the presence of two production plants for the daily supply of the silos to be purchased is now considered.

These plants have a maximum daily quantity of forage (in quintals) to provide equal to 120 and 150, respectively. The kilometric distances from the manufacturing plants to the potential silos are shown in Table 3.17.

Let

$V_1 = \{1, 2\}$ be the set of the manufacturing plants,

$V_2 = \{1, 2, 3, 4, 5, 6\}$ be the set of the potential silos,

$V_3 = \{1, 2, 3, 4, 5, 6, 7\}$ be the set of farms.

The fixed costs associated with the silos and the average daily unit storage costs are those shown in the previous section; regarding the unit transportation costs, $a_{111} = 0.12 \times (18 + 20) = €\,4.56$, whereas the other costs $a_{ijr}, i \in V_1$, $j \in V_2, r \in V_3$, can be calculated in a similar way.

Indicating with $y_j, j = 1, \ldots, 6$, the binary decision variable associated with each potential site j, having value 1 if the silo j is purchased by Milatog, 0 otherwise, and with $s_{ijr}, i = 1, 2, j = 1, \ldots, 6, r = 1, \ldots, 7$, the decision variable expressing the average daily quantity of forage requested by farm r satisfied by silo j and coming from manufacturing plant i, the specific formulation of the SCTE discrete location problem (3.44)–(3.49) is the following:

$$\text{Minimize } 4.56s_{111} + 5.16s_{112} + \cdots + 3.48s_{117} + 5.52s_{121} + 5.16s_{122}$$

$$\cdots + 5.40s_{127} + \cdots + 4.68s_{261} + 4.56s_{262} + \cdots + 3.48s_{267}$$

$$+ 220y_1 + 240y_2 + 260y_3 + 275y_4 + 240y_5 + 230y_6$$

$$+ 0.15 \times (s_{111} + s_{112} + \cdots + s_{117} + s_{211} + s_{212} + \cdots + s_{217})$$

$$+ 0.18 \times (s_{121} + s_{122} + \cdots + s_{127} + s_{221} + s_{222} + \cdots + s_{227})$$

$$\cdots$$

$$+ 0.17 \times (s_{161} + s_{162} + \cdots + s_{167} + s_{261} + s_{262} + \cdots + s_{267})$$

Table 3.17 Distances (in km) between the manufacturing plants and the potential silos in the Milatog problem.

Manufacturing plant (*i*)	Silo					
	$j=1$	$j=2$	$j=3$	$j=4$	$j=5$	$j=6$
1	20	25	18	22	15	25
2	19	22	25	28	24	21

subject to

$$s_{111} + s_{112} + \cdots + s_{117} + s_{121} + s_{122} + \cdots + s_{127} + \cdots$$
$$+ s_{161} + s_{162} + \cdots + s_{167} \leq 120$$

$$s_{211} + s_{212} + \cdots + s_{217} + s_{221} + s_{222} + \cdots + s_{227} + \cdots$$
$$+ s_{261} + s_{262} + \cdots + s_{267} \leq 150$$

$$s_{111} + s_{121} + \cdots + s_{161} + s_{211} + s_{221} + \cdots + s_{261} = 36$$

$$s_{112} + s_{122} + \cdots + s_{162} + s_{212} + s_{222} + \cdots + s_{262} = 42$$

$$\cdots$$

$$s_{117} + s_{127} + \cdots + s_{167} + s_{217} + s_{227} + \cdots + s_{267} = 43$$

$$s_{111} + s_{112} + \cdots + s_{117} + s_{211} + s_{212} + \cdots + s_{217} \leq 80y_1$$

$$s_{121} + s_{122} + \cdots + s_{127} + s_{221} + s_{222} + \cdots + s_{227} \leq 90y_2$$

$$\cdots$$

$$s_{161} + s_{162} + \cdots + s_{167} + s_{261} + s_{262} + \cdots + s_{267} \leq 120y_6$$

$$y_j \in \{0, 1\}, j = 1, \ldots, 6$$

$$s_{ijr} \geq 0, i = 1, 2, j = 1, \ldots, 6, r = 1, \ldots, 7.$$

The optimal solution amounts to purchasing silos 1, 5, and 6, and corresponds to an overall daily cost equal to € 1788.86. Table 3.18 shows the average quantities of forage transported to every farm.

Table 3.18 Average daily quantity (in quintals) of forage transported from the manufacturing plants to the farms by means of the three silos purchased by Milatog.

Manufacturing plant (i)	Purchased silo (j)	Farm (r)						
		1	2	3	4	5	6	7
	1	0	0	9	0	0	0	3
1	5	0	0	23	50	27	0	0
	6	0	0	0	0	0	0	0
	1	36	0	2	0	0	30	0
2	5	0	0	0	0	0	0	0
	6	0	42	0	0	0	0	40

3.11 The Multi-commodity Case

Most logistics systems deal with multiple products having different features. The corresponding location–allocation problems are much more complex than those previously examined. In the following, the discussion will be restricted to the multi-commodity version of the SCTE discrete location model (3.44)–(3.49). The problem can be formulated on the complete directed tripartite graph G used for the SCTE discrete location problem. Let H be the set of commodities, whose flows are assumed to be expressed in the same units of measurement (i.e., palletized unit loads, see Section 5.4.4). Denoted by $o_{ih}, i \in V_1, h \in H$, the maximum quantity of commodity h available at the predecessor node i, assumed to be constant for every time period of the planning horizon; by $q_j, j \in V_2$, the throughput of potential facility j, assumed to be constant for every time period of the planning horizon; by $d_{rh}, r \in V_3, h \in H$, the demand estimate of commodity h requested by successor node r in a time period, and by $a_{ijrh}, i \in V_1, j \in V_2, r \in V_3, h \in H$, the unit transportation cost of commodity h from predecessor node i to successor node r through facility j. As in the SCTE model, it is assumed that the cost of each facility $j \in V_2$, in a time period can be expressed in terms of fixed costs f_j and running costs g_j.

The *multi-commodity two-echelon* (MCTE) discrete location problem can be formulated using the binary decision variables $y_j, j \in V_2$, equal to 1 if and only if potential facility j is selected, and the continuous decision variables $s_{ijrh}, i \in V_1, j \in V_2, r \in V_3, h \in H$, representing the average quantity of commodity h transported in a time period from predecessor node i to successor node r through facility j. The model is as follows:

$$\text{Minimize} \sum_{i \in V_1} \sum_{j \in V_2} \sum_{r \in V_3} \sum_{h \in H} a_{ijrh} s_{ijrh} + \sum_{j \in V_2} \left(f_j y_j + g_j \sum_{i \in V_1} \sum_{r \in V_3} \sum_{h \in H} s_{ijrh} \right)$$

subject to

$$\sum_{j \in V_2} \sum_{r \in V_3} s_{ijrh} \leq o_{ih}, i \in V_1, h \in H$$

$$\sum_{i \in V_1} \sum_{j \in V_2} s_{ijrh} = d_{rh}, r \in V_3, h \in H$$

$$\sum_{i \in V_1} \sum_{r \in V_3} \sum_{h \in H} s_{ijrh} \leq q_j y_j, j \in V_2$$

$$y_j \in \{0, 1\}, j \in V_2$$

$$s_{ijrh} \geq 0, i \in V_1, j \in V_2, r \in V_3, h \in H.$$

[K9.xlsx, K9.py] K9 is a German petrochemical company. The firm's management intends to renovate its production and distribution network, which is presently composed of two refining plants, two DCs and hundreds of sales points (gas pumps and liquefied gas retailers). After a series of meetings, it was decided to relocate the DCs, leaving the locations and features of the two production plants unchanged. The products of K9 are subdivided into two homogeneous commodities (represented by the indices $h = 1, 2$): fuel for motor transportation and liquefied gas (the latter sold in cylinders). There are four potential sites suited to receive a DC and their daily throughput q_j (expressed in hl) are, respectively, 1500, 1200, 2300 and 2500.

The sales points have been grouped into three districts ($r = 1, 2, 3$) characterized by the daily demands shown in Table 3.19. The annual fixed costs (in €) of the DCs $f_j, j \in V_2$, are the following: 960 000, 880 000, 1 540 000, 1 610 000. The daily unit storage facility costs are, respectively, 0.15, 0.14, 0.20, and 0.25 €/hl. The unit transportation costs $a_{ijrh}, i \in V_1$, $j \in V_2, r \in V_3, h = 1, 2$, are obtained by multiplying the cost per kilometre and per hectolitre (equal to € 0.0067 for $h = 1$ and to € 0.0082 for $h = 2$) by the return trip distances between the manufacturing plants $i \in V_1$ and the centroids of the sales districts $r \in V_3$ through DC $j \in V_2$ (see Table 3.20).

Finally, Table 3.21 shows the daily average quantities of the two commodities available at the two manufacturing plants. The unit transportation costs $a_{ijrh}, i = 1, 2, j = 1, ..., 4, r = 1, 2, 3, h = 1, 2$, can be computed from Tables 3.22 and 3.23. For example, cost a_{1111} is calculated as

$$a_{1111} = 0.0067 \times 2 \times 423 = 5.6682 \text{ €/hl.}$$

Table 3.19 Average daily demand (in hl) of the two commodities from the sales districts in the K9 problem.

Sales district (r)	Commodity	
	$h = 1$	$h = 2$
1	800	300
2	600	400
3	700	500

Table 3.20 Distances (in km) between the refining plants and the centroids of the sales districts through the potential DC in the K9 problem.

Refining plant (i)	Potential DC (j)	Sales district		
		$r = 1$	$r = 2$	$r = 3$
1	1	423	612	1108
	2	613	434	927
	3	1031	631	918
	4	1628	1236	954
2	1	826	1028	1531
	2	864	638	1158
	3	838	464	782
	4	1227	871	544

Table 3.21 Average daily amount (in hl) of the two commodities available at the refining plants in the K9 problem.

Refining plant (*i*)	Commodity	
	h = 1	*h* = 2
1	1200	500
2	1500	800

Table 3.22 Transportation costs (in €/hl) for the first commodity in the K9 problem.

Refining plant (*i*)	Potential DC (*j*)	Sales district		
		r = 1	*r* = 2	*r* = 3
1	1	5.6682	8.2008	14.8472
	2	8.2142	5.8156	12.4218
	3	13.8154	8.4554	12.3012
	4	21.8152	16.5624	12.7836
2	1	11.0684	13.7752	20.5154
	2	11.5776	8.5492	15.5172
	3	11.2292	6.2176	10.4788
	4	16.4418	11.6714	7.2896

Table 3.23 Transportation costs (in €/hl) for the second commodity in the K9 problem.

Refining plant (*i*)	Potential DC (*j*)	Sales district		
		r = 1	*r* = 2	*r* = 3
1	1	6.9372	10.0368	18.1712
	2	10.0532	7.1176	15.2028
	3	16.9084	10.3484	15.0552
	4	26.6992	20.2704	15.6456
2	1	13.5464	16.8592	25.1084
	2	14.1696	10.4632	18.9912
	3	13.7432	7.6096	12.8248
	4	20.1228	14.2844	8.9216

The fixed annual costs of the DCs are transformed into daily fixed costs by assuming that a year is composed of 220 workdays during which the distribution of the two commodities takes place. Therefore, the following fixed daily costs (in €) are obtained for each of the four potential DCs: 4363.64, 4000.00, 7000.00, and 7318.18. Indicating with $y_j, j \in V_2$, the binary decision variable associated with each potential DC j, having value 1 if the site j is used to host a K9 DC, 0 otherwise, and with $s_{ijrh}, i \in V_1, j \in V_2, r \in V_3$ and $h = 1, 2$, the decision variable expressing the average daily amount of commodity h, requested by sales district r, satisfied by DC j, and coming from refining plant i, the problem can be formulated as follows:

$$\text{Minimize } 5.6682s_{1111} + 8.2008s_{1112} + 14.8472s_{1121} + 6.9372s_{1122}$$

$$+ \cdots + 14.2844s_{2431} + 8.9216s_{2432}$$

$$+ 4\,363.64y_1 + 0.15 \times (s_{1111} + s_{1112} + \cdots + s_{2131} + s_{2132})$$

$$+ 4\,000.00y_2 + 0.14 \times (s_{1211} + s_{1212} + \cdots + s_{2231} + s_{2232})$$

$$\cdots$$

$$+ 7\,318.18y_4 + 0.25 \times (s_{1411} + s_{1412} + \cdots + s_{2431} + s_{2432})$$

subject to

$$s_{1111} + s_{1121} + s_{1131} + \cdots + s_{1411} + s_{1421} + s_{1431} \leq 1200$$

$$s_{1112} + s_{1122} + s_{1132} + \cdots + s_{1412} + s_{1422} + s_{1432} \leq 500$$

$$\cdots$$

$$s_{2112} + s_{2122} + s_{2132} + \cdots + s_{2412} + s_{2422} + s_{2432} \leq 800$$

$$s_{1111} + s_{1211} + \cdots + s_{2311} + s_{2411} = 800$$

$$s_{1112} + s_{1212} + \cdots + s_{2312} + s_{2412} = 300$$

$$\cdots$$

$$s_{1132} + s_{1232} + \cdots + s_{2332} + s_{2432} = 500$$

$$s_{1111} + s_{1112} + \cdots + s_{2131} + s_{2132} \leq 1500y_1$$

$$s_{1211} + s_{1212} + \cdots + s_{2231} + s_{2232} \leq 1200y_2$$

$$\cdots$$

$$s_{1411} + s_{1412} + \cdots + s_{2431} + s_{2432} \leq 2500y_4$$

$$y_j \in \{0, 1\}, j = 1, \ldots, 4$$

$$s_{ijrh} \geq 0, i = 1, 2, j = 1, \ldots, 4, r = 1, 2, 3, h = 1, 2.$$

The optimal solution is characterized by the location of two DCs (the first and the third). The optimal flows are

$$s^*_{1111} = 800; s^*_{1112} = 300; s^*_{1332} = 100; s^*_{2321} = 600;$$

$$s^*_{2322} = 400; s^*_{2331} = 700; s^*_{2332} = 400,$$

where, for the sake of conciseness, only the amounts different from zero are shown. Finally, the minimum daily cost is equal to € 39 329.36.

3.12 Location-covering Problems

In location-covering problems, the aim is to locate a least-cost set of service facilities in such a way that each customer can be reached within a limited travel time from the closest facility. The basic location-covering problem can be modelled on an undirected graph $G = (V_1 \cup V_2, E)$, where the vertices in V_1 represent the potential service facilities, those in V_2 describe the users to be reached and each edge $(i, j) \in E$ corresponds to a least-duration path between i and j. Let $f_i, i \in V_1$, be the fixed cost of potential facility i and $a_{ij}, i \in V_1, j \in V_2$, a binary constant equal to 1 if potential facility i is able to serve customer j, 0 otherwise. Given a user-defined time limit T, constant a_{ij} may be defined as 1 if the travel time t_{ij} between $i \in V_1$ and $j \in V_2$ does not exceed T, and as 0 otherwise. The decision variables are binary: $y_i, i \in V_1$, is equal to 1 if facility i is activated, 0 otherwise. The problem is modelled as follows:

$$\text{Minimize} \sum_{i \in V_1} f_i y_i \tag{3.50}$$

subject to

$$\sum_{i \in V_1} a_{ij} y_i \geq 1, j \in V_2 \tag{3.51}$$

$$y_i \in \{0, 1\}, i \in V_1, \tag{3.52}$$

and corresponds to a set covering (SC) problem that may be difficult to solve. A good feasible solution can be determined by using the following simple heuristic, originally proposed by V. Chvátal:

[Coimbra.xlsx, Coimbra.py] In Portugal, the municipal administration of Coimbra needs to locate emergency fire stations to cover all the seven residential districts of the city within a maximum time of 16 minutes. The minimum distances between the centroids of the seven residential districts are reported in Table 3.24. The stations' annual costs (in tens of thousands of €) are: 200, 160, 240, 220, 180, 180, and 220. The average travelling speed is assumed to be 65 km/h. The SC model is as follows:

Minimize $200y_1 + 160y_2 + 240y_3 + 220y_4 + 180y_5 + 180y_6 + 220y_7$

subject to

$y_1 + y_2 + y_7 \geq 1$

$y_1 + y_2 + y_5 + y_6 + y_7 \geq 1$

$y_3 + y_4 + y_5 + y_6 \geq 1$

```
1:  procedure CHVATAL (V₁, V₂, f, A, ȳ, z̄)
2:      # A is the |V₁| × |V₂| matrix formed by the aᵢⱼ binary values, i ∈ V₁, j ∈ V₂;
3:      # ȳ is the SC feasible solution returned by the procedure;
4:      # z̄ is the cost corresponding to the SC feasible solution;
5:      # nᵢ, i ∈ V₁, is the number of constraints in which variable yᵢ appears;
6:      while V₁ ≠ ∅ do
7:          for i ∈ V₁ do
8:              if aᵢⱼ = 0 for all j ∈ V₂ then
9:                  # Variable yᵢ does not appear in any constraint;
10:                 ȳᵢ = 0;
11:                 # Update set V₁, containing only the indices corresponding to variables
                    to be set;
12:                 V₁ = V₁\{i};
13:             end if
14:         end for
15:         if V₁ ≠ ∅ and V₂ ≠ ∅ then
16:             k = arg min{fᵢ/nᵢ};
                       i∈V₁
17:             ȳₖ = 1;
18:             # Update the set V₁;
19:             V₁ = V₁\{k};
20:             for j ∈ V₂ do
21:                 if aₖⱼ = 1 then
22:                     # Remove constraint j since it is satisfied;
23:                     V₂ = V₂\{j};
24:                     for i ∈ V₁ do
25:                         if aᵢⱼ = 1 then
26:                             nᵢ = nᵢ − 1;
27:                         end if
28:                     end for
29:                 end if
30:             end for
31:         else if V₂ = ∅ then
32:             ȳᵢ = 0, for all i ∈ V₁;
33:             V₁ = ∅;
34:         end if
35:     end while
36:     z̄ = Σ fᵢȳᵢ;
              i∈V₁
37:     return ȳ, z̄;
38: end procedure
```

$$y_3 + y_4 + y_7 \geq 1$$

$$y_2 + y_3 + y_5 + y_6 \geq 1$$

$$y_2 + y_3 + y_5 + y_6 \geq 1$$

$$y_1 + y_2 + y_4 + y_7 \geq 1$$

$$y_1, y_2, y_3, y_4, y_5, y_6, y_7 \in \{0, 1\}.$$

Table 3.24 Distances (in km) between the centroids of the seven residential districts of Coimbra.

Residential district (i)	Residential district (j)						
	1	2	3	4	5	6	7
1	0	8	24	18	30	20	16
2		0	30	120	14	4	6
3			0	16	12	10	18
4				0	18	20	6
5					0	4	100
6						0	54
7							0

The CHVATAL procedure is applied. At the beginning, all the decision variables appear in at least a constraint, so the first execution of code lines 7–14 of the procedure is void. Then, the following ratios are computed:

$$f_1/n_1 = 200/3 = 66.67;$$

$$f_2/n_2 = 160/5 = 32.00;$$

$$f_3/n_3 = 240/4 = 60.00;$$

$$f_4/n_4 = 220/3 = 73.33;$$

$$f_5/n_5 = 180/4 = 45.00;$$

$$f_6/n_6 = 180/4 = 45.00;$$

$$f_7/n_7 = 220/4 = 55.00.$$

The lowest ratio is obtained for $k = 2$ (code line 16); for this reason, $\bar{y}_2$ is set equal to 1 and the first two and the last three constraints are removed (code lines 20–30). The problem formulation reduces to

Minimize $200y_1 + 240y_3 + 220y_4 + 180y_5 + 180y_6 + 220y_7$

subject to

$$y_3 + y_4 + y_5 + y_6 \geq 1$$

$$y_3 + y_4 + y_7 \geq 1$$

$$y_1, y_3, y_4, y_5, y_6, y_7 \in \{0, 1\}.$$

Decision variable y_1 does not appear in any constraint, so that $\bar{y}_1$ can be set equal to 0 (code lines 7–14). The following ratios are computed:

$$f_3/n_3 = 240/2 = 120;$$

$$f_4/n_4 = 220/2 = 110;$$

$$f_5/n_5 = 180/1 = 180;$$
$$f_6/n_6 = 180/1 = 180;$$
$$f_7/n_7 = 220/1 = 220.$$

The lowest ratio is obtained for $k = 4$, so that $\bar{y}_4$ is set equal to 1 and the two constraints are removed because y_4 appears in both with coefficient equal to 1. The problem now reduces to the following formulation:

Minimize $240y_3 + 180y_5 + 180y_6 + 220y_7$

subject to

$$y_3, y_5, y_6, y_7 \in \{0, 1\},$$

whose solution is $\bar{y}_3 = \bar{y}_5 = \bar{y}_6 = \bar{y}_7 = 0$ (code lines 31–34). Hence, the feasible solution involves the location of the emergency fire stations in residential districts 2 and 4. The corresponding total annual cost amounts to € 3 800 000.

Several variants of the SC model can be used in practice. For example, if fixed costs f_i are identical for all potential facilities $i \in V_1$, it can be convenient to discriminate among all the solutions with the least number of open facilities the one corresponding to the least total travelling time, or to the most equitable demand distribution among the facilities. In the former case, let $x_{ij}, i \in V_1, j \in V_2$, be a binary decision variable equal to 1 if customer j is served by facility i, 0 otherwise. The problem can be modelled as follows

$$\text{Minimize} \sum_{i \in V_1} My_i + \sum_{i \in V_1} \sum_{j \in V_2} t_{ij} x_{ij} \tag{3.53}$$

subject to

$$\sum_{i \in V_1} a_{ij} x_{ij} \geq 1, j \in V_2 \tag{3.54}$$

$$\sum_{j \in V_2} x_{ij} \leq |V_2| y_i, i \in V_1 \tag{3.55}$$

$$y_i \in \{0, 1\}, i \in V_1 \tag{3.56}$$

$$x_{ij} \in \{0, 1\}, i \in V_1, j \in V_2. \tag{3.57}$$

In the objective function (3.53), M is an arbitrarily large positive constant chosen so that the number of facilities to be activated is always as small as possible; constraints (3.54) guarantee that all customers $j \in V_2$ are serviced, while constraints (3.55) ensure that if facility $i \in V_1$ is not set up ($y_i = 0$), then no customer $j \in V_2$ can be served by it.

[Cornwall.xlsx, Cornwall.py] In the English county of Cornwall, a consortium of 10 municipalities (Sennen Cove, Porth Curno, Trevilley, Botallack,

Morvah, Treen, Zennor, St. Ives, St. Erth, and Hayle) has decided to improve its fire fighting service. The person responsible for the project has established that each centre of the community must be reached within 10 minutes from the nearest fire station. Since the main aim is just to provide a first response in case of fire, the decision maker has decided to assign a single vehicle to each station. The annual cost of a station inclusive of the expenses of personnel is £ 123 000. To determine the optimal number and location of the fire stations, location-covering model (3.53)–(3.57) can be used with $V_1 = V_2 = \{$Sennen Cove, Porth Curno, Trevilley, Botallack, Morvah, Treen, Zennor, St. Ives, St. Erth, Hayle$\}$. See `Cornwall.xlsx` for travel times $t_{ij}, i \in V_1, j \in V_2$.

The coefficients $a_{ij}, i \in V_1, j \in V_2$, were obtained by imposing that $a_{ij} = 1$ if $t_{ij} \leq 10$ minutes, $a_{ij} = 0$ otherwise. The minimum number of fire stations, obtained by using `Excel Solver`, turned out to be two. The facilities were located in Trevilley and St. Ives. The fire station located in Trevilley serves Sennen Cove, Porth Curno, Botallack, Morvah, and Trevilley itself, and the remaining municipalities are served by the fire station located in St. Ives.

The presence of coefficients with different orders of magnitude in the objective function (3.53) can cause serious numerical difficulties to any solver. To overcome this problem, a different solution method can be adopted. First, the problem defined by (3.50)–(3.52) is solved with $f = f_i, i \in V$, in such a way to determine the number p^* of facilities to be activated. Then, a modified version of problem (3.53)–(3.57) is solved, in which objective function (3.53) is replaced by

$$\text{Minimize} \sum_{i \in V_1} \sum_{j \in V_2} t_{ij} x_{ij}$$

and the following additional constraint is considered

$$\sum_{i \in V_1} y_i = p^*.$$

3.13 *p*-centre Problems

In the *p*-centre problem, the aim is to locate p service facilities in such a way that the maximum travel time from a user to the closest facility is minimized. It finds its application when it is necessary to ensure equity in servicing users spread over a wide geographical area. The problem can be modelled on a directed, undirected, or mixed graph $G = (V, A, E)$, where V is a set of vertices representing both user sites and road intersections, while A and E (the set of arcs and edges, respectively) describe the road connections among the vertices. Exactly p facilities have to be located either on a vertex, or on an arc or edge.

If G is a directed graph, it can be easily shown that, in an optimal solution of the *p*-centre problem, every facility location is a vertex (*vertex location* property). If G is undirected or mixed, the optimal location of a facility may fall in a vertex or in an internal point of an edge. The remainder of this section illustrates the solution of the

1-centre problem. The reader is referred to the literature for a discussion of the more general case.

If G is directed, the 1-centre can be easily determined based on the vertex location property. Let $t_{ij}, i, j \in V$, be the travel time that corresponds to a least-duration path from i to j.

1: **procedure** 1-CENTRE_DIRECTED_GRAPH $(V, \mathbf{T}, i^*)$

2:    # i^* is the 1-centre returned by the procedure;

3:    # Determine, for each $i \in V$ the maximum travel time from i to every vertex $j \in V$;

4:    $T_i = \max_{j \in V}\{t_{ij}\}, i \in V$;

5:    # Determine the 1-centre i^* in correspondence of the minimum value of T_i, $i \in V$;

6:    $T_{i*} = \min_{i \in V}\{T_i\}$;

7:    **return** i^*;

8: **end procedure**

[Krosno.xlsx] To locate a first response mobile unit in the town of Krosno in Poland, the town centre was subdivided into 13 urban areas and the minimum distances (in metres) between their centroids were computed (see Table 3.25). Assuming that the unit can be located in any centroid, the farthest area from each centroid was determined by using the data reported in Table 3.25. Hence, applying the 1-CENTRE_DIRECTED_GRAPH procedure (code line 4), the vector $\mathbf{T}$ of the maximum travel times was computed $\mathbf{T} = [738; 841; 748; 707; 839; 779; 790; 732; 827; 847; 828; 742; 769]^T$. Consequently, the 1-centre fell in the fourth centroid, corresponding to the minimum value of $T_i, i = 1, \dots, 13$, equal to 707 m (code line 6). On the basis of this choice, the most disadvantaged urban area of the town was the tenth.

Table 3.25 Minimum distances (in m) between the centroids of the 13 urban areas of the town of Krosno.

(i)						(j)							
(i)	1	2	3	4	5	6	7	8	9	10	11	12	13
1	0	682	188	428	548	621	653	723	738	377	404	149	143
2	655	0	841	754	804	438	506	257	352	524	283	762	755
3	139	748	0	301	579	572	665	639	611	452	381	102	160
4	356	658	270	0	547	565	563	603	363	707	462	337	316
5	600	839	575	523	0	701	707	738	428	458	612	675	698
6	441	779	604	482	393	0	482	630	562	436	643	575	582

(Continued)

Table 3.25 (Continued)

(i)	1	2	3	4	5	6	7	8	9	10	11	12	13
7	625	314	594	587	790	457	0	269	309	624	437	696	737
8	406	369	585	705	710	323	555	0	497	732	483	520	512
9	786	609	745	529	827	568	364	505	0	663	622	823	657
10	652	847	701	486	391	746	747	822	709	0	259	638	528
11	437	616	613	733	793	582	692	810	828	644	0	519	489
12	108	720	156	318	480	576	706	742	637	478	285	0	185
13	141	596	191	294	435	544	769	727	650	471	408	208	0

The header spans columns labelled (j): 1 through 13.

In the case of undirected or mixed graph, the 1-centre could be a vertex or an internal point of an edge, as already observed. To simplify the discussion, the solution of the problem is illustrated for an undirected graph ($A = \varnothing$), although the procedure can be easily adapted to a mixed graph. For each $(i, j) \in E$, let a_{ij} be the traversal time of edge (i, j).

Furthermore, for each pair of vertices $i, j \in V$, denote by t_{ij} the travel time between i and j, corresponding to a least duration path between i and j. Note that, on the basis of the definition of travel time,

$$t_{ij} \leq a_{ij}, (i, j) \in E.$$

Finally, denote by $\tau_h(p_{hk})$ the travel time along edge $(h, k) \in E$ between vertex $h \in V$ and a point p_{hk} of the edge. In this way, the travel time $\tau_k(p_{hk})$ along the edge (h, k) between vertex $k \in V$ and p_{hk} is (see Figure 3.12):

$$\tau_k(p_{hk}) = a_{hk} - \tau_h(p_{hk}).$$

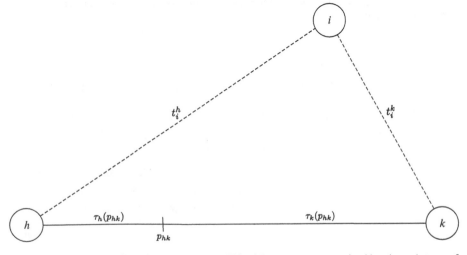

Figure 3.12 Computation of the travel time $T_i(p_{hk})$ between a vertex $i \in V$ and a point p_{hk} of the edge (h, k).

The 1-centre problem can be solved by using the following algorithm proposed by S. L. Hakimi.

1: **procedure** HAKIMI $(V, E, \mathbf{T}, p^*)$
2:       # $\mathbf{T}$ is the $|V| \times |V|$ matrix whose element t_{ij} is the travel time between $i \in V$ and $j \in V$;
3:       # p^* is the 1-centre returned by the procedure;
4:       # $\tau_h(p_{hk}), h \in V, (h,k) \in E$ is the travel time along edge (h,k) between vertex h and a point p_{hk} of the edge;
5:       # Computation of the travel time;
6:       **for** $(h,k) \in E$ **do**
7:           **for** $i \in V$ **do**
8:               # Determine the travel time $T_i(p_{hk})$ between $i \in V$ and a point p_{hk} of the edge (h,k) (see Figure 3.13);
9:

$$T_i(p_{hk}) = \min\{t_{ih} + \tau_h(p_{hk}), t_{ik} + \tau_k(p_{hk})\}; \tag{3.58}$$

10:           **end for**
11:       **end for**
12:       **for** $(h,k) \in E$ **do**
13:           # Determine the local centre p_{hk}^* as the point on (h,k) minimizing the travel time of the most disadvantaged vertex;
14:

$$p_{hk}^* = \arg \min_{p_{hk} \in (h,k)} \max_{i \in V} \{T_i(p_{hk})\};$$

15:           # $\max_{i \in V}\{T_i(p_{hk})\}$ is the superior envelope of functions $T_i(p_{hk})$, $i \in V$ (see Figure 3.14);
16:       **end for**
17:       # Determine the 1-centre as the best local centre p_{hk}^*, $(h,k) \in E$;
18:

$$(h^*, k^*) = \arg \min_{(h,k) \in E} \left\{ \min_{p_{hk} \in (h,k)} \max_{i \in V} \{T_i(p_{hk})\} \right\};$$

19:

$$p^* = p_{h^*k^*}^*;$$

20:       **return** p^*;
21: **end procedure**

In the La Mancha region of Spain (see Figure 3.15), a consortium of municipalities, situated in a rural area, decided to locate a parking place for ambulances. A preliminary examination of the problem revealed that the probability of receiving

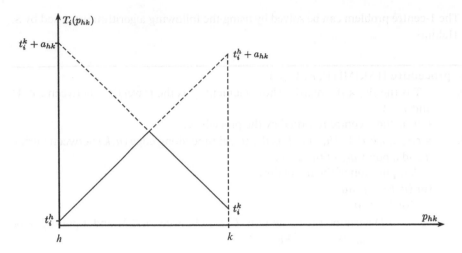

Figure 3.13 Travel time $T_i(p_{hk})$ as a function of the position of the point p_{hk} along edge (h, k).

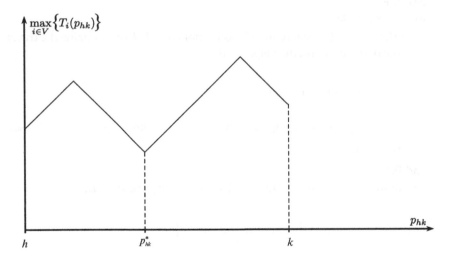

Figure 3.14 Determination of the local centre on edge $(h, k) \in E$.

a service request during the completion of a previous call was extremely low because of the small number of inhabitants in the zone. For this reason the team responsible for the service decided to use only one vehicle. In the light of this observation, the problem was modelled as a 1-centre problem on a road network G, characterized by $|V| = 11$ vertices (see Table 3.26) and $|E| = 19$ edges (all links are two-way streets), whose travel times (see Table 3.27) were calculated assuming an average vehicle speed of 90 km/h.

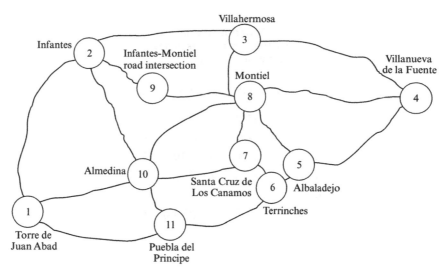

Figure 3.15   Location problem in the La Mancha region.

Table 3.26   Vertices of the La Mancha 1-centre problem.

Vertex	Locality
1	Torre de Juan Abad
2	Infantes
3	Villahermosa
4	Villanueva de la Fuente
5	Albaladejo
6	Terrinches
7	Santa Cruz de los Canamos
8	Montiel
9	Infantes-Montiel crossing
10	Almedina
11	Puebla del Principe

Travel times t_{ij}, $i, j \in V$, determined on the basis of the traversal times in Table 3.27, are reported in Table 3.28. The HAKIMI procedure was applied to solve the problem. For each edge $(h, k) \in E$ and for each vertex $i \in V$, the travel time $T_i(p_{hk})$ between vertex i and a point p_{hk} of the edge (h, k) can be defined through formula (3.58). Hence, for each edge $(h, k) \in E$, function $\max_{i \in V} \{T_i(p_{hk})\}$ was determined. Their minimum values correspond to local centres p_{hk}^*. Figure 3.16 depicts the function $\max_{i \in V} \{T_i(p_{23})\}$, and Table 3.29 gives, for each edge $(h, k) \in E$, both the position of p_{hk}^* and the value $\max_{i \in V} \{T_i(p_{hk}^*)\}$.

Table 3.27 Traversal time (in minutes) of the road network edges in the La Mancha problem.

(i,j)	a_{ij}	(i,j)	a_{ij}	(i,j)	a_{ij}
(1,2)	12	(3,9)	4	(6,11)	5
(1,10)	6	(4,5)	9	(7,8)	4
(1,11)	8	(4,8)	10	(7,10)	5
(2,3)	9	(5,6)	2	(8,9)	1
(2,9)	8	(5,8)	6	(8,10)	7
(2,10)	9	(6,7)	3	(10,11)	4
(3,4)	11				

Table 3.28 Travel times (in minutes) t_{ij} $i, j \in V$, in the La Mancha problem.

(i)	(j) 1	2	3	4	5	6	7	8	9	10	11
1	0	12	18	23	15	13	11	13	14	6	8
2		0	9	19	15	16	13	9	8	9	13
3			0	11	11	12	9	5	4	12	16
4				0	9	11	14	10	11	17	16
5					0	2	5	6	7	10	7
6						0	3	7	8	8	5
7							0	4	5	5	8
8								0	1	7	11
9									0	8	12
10										0	4
11											0

Consequently the 1-centre corresponds to the local centre of edge $(8, 10)$. Therefore, the optimal location of the ambulance parking should be on the road between Montiel and Almedina, at 2.25 km from the centre of Montiel. The most disadvantaged villages are Villanueva de la Fuente and Torre de Juan Abad, since the ambulance takes an average time of 11.5 minutes to reach them from the 1-centre.

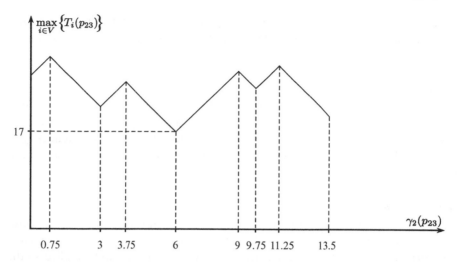

Figure 3.16 Time $\max_{i \in V} \{T_i(p_{23})\}$ when varying the position of p_{23} (indicated the distance $\gamma_2(p_{23})$ (in km) from node 2), in the La Mancha problem.

Table 3.29 Position of local centre p^*_{hk} (as a function of the distance $\gamma_h(p^*_{hk})$ (in km) from node h), and $\max_{i \in V} \{T_i(p^*_{hk})\}$ (in minutes), for each edge $(h, k) \in E$, in the La Mancha problem.

(h, k)	$\gamma_h(p^*_{hk})$	$\max_{i \in V} \{T_i(p^*_{hk})\}$	(h, k)	$\gamma_h(p^*_{hk})$	$\max_{i \in V} \{T_i(p^*_{hk})\}$
(1,2)	18.00	19.0	(1,11)	12.00	16.0
(2,3)	6.00	17.0	(2,9)	12.00	14.0
(3,4)	0.00	18.0	(2,10)	13.50	17.0
(4,5)	13.50	15.0	(3,9)	6.00	14.0
(5,6)	0.00	15.0	(4,8)	15.00	13.0
(6,7)	3.75	13.5	(5,8)	9.00	13.0
(7,8)	2.25	12.5	(6,11)	4.50	15.0
(8,9)	0.00	13.0	(7,10)	0.00	14.0
(8,10)	2.25	11.5	(1,10)	9.00	17.0
(10,11)	6.00	16.0			

3.14 Data Aggregation

Modeling and solving a location problem often requires a considerable amount of data. For instance, when designing the distribution network for a large food producer, the number of products may be in the order of tens of thousands and the demand points may exceed several hundreds. Therefore, in order to keep computation times and hardware requirements acceptable, it is often advisable to aggregate data. This can mainly be done in two ways.

Demand aggregation

Demand points are clustered in such a way that customers close to one another (e.g., customers having the same zip code) are aggregated into a single *customer zone*. Alternatively, customers are first clustered with respect to service level (or frequency of delivery), and then each cluster is further subdivided on the basis of a geographical criterion.

Item aggregation

Items are aggregated into a suitable number of product groups. This can be done according to their distribution pattern or their features (weight, volume, shape, cost, etc.). In the former case, products manufactured by the same plants and supplied to the same customers are treated as a single commodity. In the latter case, similar items (e.g., variants of the same basic product model) are aggregated.

Whatever the aggregation method, the optimization problem can be modelled with fewer decision variables and constraints. In what follows, a demand-aggregation method for the CPL model (see Section 3.9) is analysed. Let $S \subset V_2$ be a subset of successor nodes whose demand is aggregated and let s be the unique node replacing the aggregated nodes belonging to S. This means that

$$c_{is} = \sum_{j \in S} c_{ij}, \ i \in V_1 \tag{3.59}$$

and

$$d_s = \sum_{j \in S} d_j. \tag{3.60}$$

The number of decision variables is reduced since the variables x_{ij}, $j \in S$, for each $i \in V_1$, are replaced with the unique variable x_{is}. The following relations hold:

$$x_{ij} = x_{is}/|S|, \ i \in V_1, \ j \in S.$$

Consequently, each successor node $j \in S$ receives the same fraction of demand from each facility $i \in V_1$.

The reduced CPL model becomes

$$\text{Minimize} \sum_{i \in V_1} \sum_{j \in (V_2 \setminus S) \cup \{s\}} c_{ij}x_{ij} + \sum_{i \in V_1} f_i y_i$$

subject to

$$\sum_{i \in V_1} x_{ij} = 1, \ j \in (V_2 \setminus S) \cup \{s\}$$

$$\sum_{j \in (V_2 \setminus S) \cup \{s\}} d_j x_{ij} \le q_i y_i, \ i \in V_1$$

$$x_{ij} \ge 0, \ i \in V_1, \ j \in (V_2 \setminus S) \cup \{s\}$$

$$y_i \in \{0, 1\}, i \in V_1.$$

The optimal cost of the aggregated problem is not better than that of the original problem. It is useful to evaluate an upper bound on the additional cost paid in the aggregation procedure. Let z^*_{CPL} and $z^{(a)}_{\text{CPL}}$ be the costs of the optimal solutions of the original problem and of the aggregate problem, respectively. The following relation holds

$$z^*_{\text{CPL}} \leq z^{(a)}_{\text{CPL}} \leq z^*_{\text{CPL}} + \epsilon,$$

where

$$\epsilon = \sum_{j \in S} \max_{i \in V_1} \left\{ \frac{d_j \sum_{r \in S} c_{ir}}{\sum_{r \in S} d_r} - c_{ij} \right\}. \tag{3.61}$$

In equation (3.61), $\sum_{r \in S} c_{ir}$ represents the variable cost when the whole aggregated demand of customers in S is satisfied by facility $i \in V_1$, while $d_j / \sum_{r \in S} d_r$ represents the fraction of the demand of node j with respect to the aggregated demand. The difference $(d_j \sum_{r \in S} c_{ir} / \sum_{r \in S} d_r) - c_{ij}$ represents, therefore, the variation (positive, negative, or zero) of the cost of assigning the whole demand of successor node $j \in V_2$ to facility $i \in V_1$ when the nodes belonging to S are considered as a cluster. Therefore, equation (3.61) expresses the additional cost paid in the worst-case for the aggregation procedure. It can also be proved that bound (3.61) is *tight*, i.e., there are instances such that the cost increment is nearly equal to ϵ.

[Milatog.xlsx]. In the Milatog problem, assume that the fifth, sixth, and seventh farms are aggregated, and let s be the node replacing the aggregated farms. Therefore, the aggregated CPL problem is characterized by the first four original farms and farm zone s. According to relation (3.59), the costs $c_{is}, i \in V_1$, become $[200.40; 221.64; 219.80; 220.44; 201.48; 194.36]^{\mathsf{T}}$, whereas the cumulative average daily forage demand of the cluster equals $d_s = 100$ quintals (see formula (3.60)). Based on relation (3.61), $\epsilon = (19.34 + 27.54 + 35.45) = €82.33$. The optimal solution, found by using Excel Solver, corresponds to an optimal daily cost of €1261.71, which is €(1261.71 − 1218.08) = €43.63 worse than the optimal cost of the original CPL. Of course, this increase is not greater than ϵ.

The above aggregation procedure can obviously be extended to the case of K disjoint subsets $S_1, \ldots, S_K$ of customers. An a priori upper bound on the additional cost paid is given by

$$z^*_{\text{CPL}} \leq z^{(a)}_{\text{CPL}} \leq z^*_{\text{CPL}} + \sum_{k=1}^{K} \epsilon_k,$$

where

$$\epsilon_k = \sum_{j \in S_k} \max_{i \in V_1} \left\{ \frac{d_j \sum_{r \in S_k} c_{ir}}{\sum_{r \in S_k} d_r} - c_{ij} \right\}.$$

[Milatog.xlsx]. In the Milatog problem, assume that the first and the second farms are aggregated as well. Let s_1 be the corresponding cluster node, whereas the fifth, sixth, and seventh farms are associated with the cluster node s_2. The two upper bounds are $\epsilon_1 = (16.28 + 25.59) =$ € 41.87 and (as computed previously) $\epsilon_2 =$ € 82.33. Hence, an upper bound on the additional cost paid because of the aggregation procedure is € $(41.87 + 82.33) =$ € 124.20. The optimal daily cost of the resulting aggregated CPL problem, found by using Excel Solver, is € 1264.60. The cost increment with respect to the optimal cost of the original CPL is therefore € $(1264.60 - 1218.08) =$ € 46.52, not greater than $\epsilon_1 + \epsilon_2$.

3.15 Location Models Under Uncertainty

The facility location models discussed so far are deterministic, i.e., they are based on the hypothesis that the relevant parameters (customer demands, location, and transportation costs etc.) are known with certainty. However, very often this assumption proves not to be very realistic. A location decision taken on the basis of point estimates or expected parameter values can be very poor (see Section 1.10.3). There exist a very large variety of location models which take into account the parameter uncertainty in some way. The attention in this section will be limited to two cases, which seem to be of great practical interest. The reader is referred to the specialized literature for a more detailed discussion of the matter.

3.15.1 A Stochastic Location–allocation Model

As previously mentioned, facility location is interlinked with commodity flow allocation, so that the overall decision-making process has an intrinsic two-stage nature, corresponding to two distinct decision levels. At a strategic level, the location of the facilities must be established before knowing exactly the demands and costs. Subsequently, once the values of such uncertain parameters are disclosed, the commodity flow allocation decision is taken at a tactical level.

Regarding the representation of parameter uncertainty, the simplest way is to use discrete random variables, whose combinations of possible realizations constitute the scenarios, each characterized by a specific probability of occurrence. On the basis of these scenarios, the recourse decisions, which represent reactions to the observation of the evolution of the uncertain parameters, are defined.

In the following, a model is described for the SCOE discrete location problem with uncertain demands and costs and represented by means of a set of scenarios S. Each scenario $s \in S$ is characterized by a probability of occurrence p_s, a demand d_{js} for every successor node $j \in V_2$ and transportation costs $c_{ijs}, i \in V_1, j \in V_2$. The first-stage decision amounts to locating the facilities (i.e., to set binary decision variables $y_i, i \in V_1$), whereas the second-stage decision is to allocate commodity flows to successor nodes for each scenario $s \in S$. The SCOE discrete location model with uncertain parameters can then be formulated as follows:

$$\text{Minimize} \sum_{s \in S} p_s \sum_{i \in V_1} \sum_{j \in V_2} c_{ijs} x_{ijs} + \sum_{i \in V_1} f_i y_i \qquad (3.62)$$

subject to

$$\sum_{i \in V_1} x_{ijs} = 1, j \in V_2, s \in S \tag{3.63}$$

$$\sum_{j \in V_2} d_{js} x_{ijs} \leq q_i y_i, i \in V_1, s \in S \tag{3.64}$$

$$x_{ijs} \geq 0, i \in V_1, j \in V_2, s \in S \tag{3.65}$$

$$y_i \in \{0, 1\}, i \in V_1. \tag{3.66}$$

The objective function (3.62) represents the sum of facility costs and expected transportation cost over all scenarios, while constraints (3.63) and (3.64) extend constraints (3.22) and (3.23) of the deterministic model to the stochastic case.

Problem (3.62)–(3.66) belongs to the family of two-stage stochastic programming problems with some binary decision variables. As a rule, the need to use a large number of scenarios makes the size of this model fairly large.

[`Milatog.xlsx`] With reference to the Milatog location problem, it is supposed that the company would purchase one or more existing silos located at six sites, and which require renovations which will take about six months to be completed. At the end of the renovation, the daily throughput of a silo increases by 10%. The purchasing and renovation costs (in €) are estimated to be, on a four-year basis, 428 000, 456 900, 526 400, 558 000, 496 000, and 542 000 for the six potential silos, respectively. They affect forage distribution from the silos to the farms for the next three and a half years, i.e., for $365 \times 3 + 183 = 1268$ days.

The forage demand of the farms cannot be estimated with certainty. However, the company's logistics manager believes it plausible to imagine four ($|S| = 4$) equiprobable scenarios ($p_s = 0.25, s \in S$), identified by using a forecasting method. The scenarios are characterized by the daily average demands shown in Table 3.30. Costs $c_{ijs}, i \in V_1, j \in V_2, s \in S$, are: $c_{ijs} = c_{ij}, i \in V_1, j \in V_2, s \in S$. Model (3.62)–(3.66) can be formulated by considering that

- $V_1 = \{1, 2, 3, 4, 5, 6\}$ is the set of the potential silos to be purchased and renovated;
- $V_2 = \{1, 2, 3, 4, 5, 6, 7\}$ is the set of farms;
- $S = \{1, 2, 3, 4\}$ is the set of scenarios;
- $f_1 = 428\,000/1268 = €\,334.90$ (similarly, the other costs $f_i, i = 2, \ldots, 6$ can be obtained);
- the daily average forage demands $d_{ijs}, i \in V_1, j \in V_2, s \in S$, can be obtained from Table 3.30;
- $q_1 = 80 \times (1 + 0.1) = 88$ (an identical calculation is done for the other silo daily throughputs $q_i, i = 2, \ldots, 6$, which are all increased by 10%).

Denote by $y_i, i = 1, \ldots, 6$, the binary decision variable associated with each potential site i (equal to 1 if silo i is purchased and renovated by Milatog, 0 otherwise) and by $x_{ijs}, i = 1, \ldots, 6, j = 1, \ldots, 7, s = 1, \ldots, 4$, the decision variable corresponding to the fraction of the average daily forage demand requested by farm j and satisfied by silo i whenever scenario s occurs.

Table 3.30 Average daily demand scenarios (in quintals) for the Milatog problem.

Farm	Scenario			
(j)	$s = 1$	$s = 2$	$s = 3$	$s = 4$
1	36	38	36	34
2	42	44	40	42
3	34	36	30	32
4	50	50	54	54
5	27	25	25	27
6	30	28	30	34
7	43	41	51	39

The problem can hence be formulated as

Minimize $0.25 \times (83.16x_{111} + \cdots + 48.59x_{671}) + \ldots$

$\qquad + 0.25 \times (83.16x_{113} + \cdots + 48.59x_{673})$

$\qquad + 0.25 \times (83.16x_{114} + \cdots + 48.59x_{674})$

$\qquad + 334.90y_1 + 357.51y_2 + 411.89y_3 + 436.62y_4$

$\qquad + 388.11y_5 + 424.10y_6$

subject to

$x_{111} + x_{211} + x_{311} + x_{411} + x_{511} + x_{611} = 1$

$\ldots$

$x_{114} + x_{214} + x_{314} + x_{414} + x_{514} + x_{614} = 1$

$\ldots$

$x_{171} + x_{271} + x_{371} + x_{471} + x_{571} + x_{671} = 1$

$\ldots$

$x_{174} + x_{274} + x_{374} + x_{474} + x_{574} + x_{674} = 1$

$36x_{111} + 42x_{121} + 34x_{131} + 50x_{141}$

$\qquad + 27x_{151} + 30x_{161} + 43x_{171} \leq 88y_1$

$\ldots$

$36x_{114} + 44x_{124} + 32x_{134} + 48x_{144}$

$\qquad + 29x_{154} + 27x_{164} + 46x_{174} \leq 88y_1$

$\ldots$

$36x_{611} + 42x_{621} + 34x_{631} + 50x_{641}$

$$+ 27x_{651} + 30x_{661} + 43x_{671} \leq 132y_6$$

$$\ldots$$

$$36x_{614} + 44x_{624} + 32x_{634} + 48x_{644}$$

$$+ 29x_{654} + 27x_{664} + 46x_{674} \leq 132y_6$$

$$x_{ijs} \geq 0, i = 1, \ldots 6, j = 1, \ldots, 7, s = 1, \ldots, 4$$

$$y_i \in \{0, 1\}, i = 1, \ldots, 6.$$

The optimal solution, found by using `Excel Solver` under the `Data` tab, amounts to purchase and renovate silos 1, 2, and 5, with an expected overall cost equal to € 1613.98. The forage demand allocation from the silos to the farms depends on the scenario that will occur in the six months after the renovation of the facilities, as shown in Table 3.31.

As can be seen, changing the scenario leads to a different allocation of the forage flow between silos and farms. For example, if scenario 3 occurs (which corresponds to an average daily demand of 30 quintals of forage for farm 6), only one quintal would be supplied by silo 1 to farm 6, whereas, if scenario 4 occurred, five out of the 34 required by farm 6 would be supplied by silo 1.

Table 3.31 Demand allocation scenarios in the Milatog problem.

Scenario (s)	Activated silo (i)	Farm						
		$j = 1$	$j = 2$	$j = 3$	$j = 4$	$j = 5$	$j = 6$	$j = 7$
	1	1	0	0	0	0	9/30	1
1	2	0	1	1	0	0	0	0
	5	0	0	0	1	1	21/30	0
	1	1	0	0	0	0	0	1
2	2	0	1	1	0	0	0	0
	5	0	0	0	1	1	1	0
	1	1	0	0	0	0	1/30	1
3	2	0	1	1	0	0	0	0
	5	0	0	0	1	1	29/30	0
	1	1	0	0	0	0	5/34	1
4	2	0	1	1	0	0	0	0
	5	0	0	0	1	1	29/34	0

3.15.2 A Location-routing Model with Uncertain Demand

As mentioned in Section 3.1, in many location models it is assumed that shipments between logistics nodes are direct, but this is economically convenient only if vehicles travel with a full load. On the other hand, in the case of less-than-full-load shipments, the same vehicle usually serves multiple nodes; as a result, the transportation costs are dependent on the order in which the logistics nodes to be served are visited, and the optimal facility locations depend on the vehicle routes (location-routing problem). The location decision is usually made at a strategic level, while vehicle routes have to be defined at a tactical or operational level. However, location and routing decisions are interdependent and a number of studies carried out over the years suggest that the overall logistics cost can be significantly larger than the minimum if these two decisions are dealt with separately.

Although routing problems will be extensively illustrated in Chapter 6, it is worth anticipating the formulation of a routing model within a location-routing framework. The location-routing problem considered in this subsection is of a SCOE type and deals with a single facility to be selected from a list of n potential sites. The facility serves a set V of successor nodes, each of which is characterized by a demand $d_j, j \in V$ whose expected outcome is not known and can only be estimated from a sample of finite size. Let $\bar{d}_j$ the corresponding sample mean, $j \in V$.

For the sake of simplicity, assume that the costs, which are dependent on the location chosen, are only those related to vehicle routing. If $\bar{i}$ is the selected site, the corresponding routing costs can be determined as follows. Let K be the set of *plausible* routes, each of which starting from $\bar{i}$, serving a subset of successor nodes respecting all the deterministic constraints involved in the problem (i.e., time duration of the route). Let $a_{kj}, k \in K, j \in V$, be a binary constant, used to identify if a successor node j is served by route k (in this case $a_{kj} = 1$) or not ($a_{kj} = 0$). Let $c_k, k \in K$, be the cost associated with route k. Each route $k, k \in K$, is assigned to a vehicle of capacity q and there are m vehicles of this type available. A plausible route becomes feasible if the capacity constraint is satisfied for the sample values of the demand associated with the successor nodes served. If a successor node j cannot be served by the available fleet of vehicles, it is possible to outsource the service delivery to an LSP at a cost equal to f_j. Let $y_k, k \in K$, be a binary decision variable, equal to 1 if the corresponding route k is selected, 0 otherwise. Let $x_j, j \in V$, be a binary decision variable, equal to 1 if the corresponding successor node j is served by the LSP, 0 otherwise. By considering the sample mean $\bar{d}_j$ of the demand d_j for each successor node $j \in V$, it is possible to formulate a deterministic version of the problem in the following way:

$$\text{Minimize} \sum_{k \in K} c_k y_k + \sum_{j \in V} f_j x_j \tag{3.67}$$

subject to

$$\sum_{k \in K} a_{kj} y_k + x_j \geq 1, j \in V \tag{3.68}$$

$$\left(q - \sum_{j \in V} a_{kj} \bar{d}_j \right) y_k \geq 0, k \in K \tag{3.69}$$

$$\sum_{k \in K} y_k \leq m \tag{3.70}$$

$$y_k \in \{0, 1\}, k \in K \tag{3.71}$$

$$x_j \in \{0, 1\}, j \in V. \tag{3.72}$$

Constraints (3.68) ensure that each successor node is served. Constraints (3.69) ensures the feasibility of the routes in terms of vehicle capacity. In particular, if $(q - \sum_{j \in V} a_{kj}\overline{d}_j) < 0$ for some $k \in K$, then the vehicle capacity constraint is violated and, consequently, y_k should be 0. Constraint (3.70) limits to m the number of identical vehicles that can be used.

A possible drawback of formulation (3.67)–(3.72) is that the number of decision variables y_k, $k \in K$, may be very large especially for weakly-constrained problems. However, there exist applications in which $|K|$ is relatively small. This happens, e.g., in fuel distribution where the demand of a successor node (a gas pump) is customarily a significant fraction (usually a half or a third) of a vehicle capacity. Therefore, the successor nodes along a route can be at most three. As a consequence, $|K| = O\left(\binom{|V|}{3} + \binom{|V|}{2} + \binom{|V|}{1}\right) = O(|V|^3)$.

The optimal solution of formulation (3.67)–(3.72) corresponds to an average routing cost indicated by $\overline{z}$ associated with the choice of opening the facility in the site location $\overline{i}$. By solving the same model for each possible choice of the facility location within the set of potential sites, it is possible to determine the best site. However, the decision maker could be interested in computing a confidence interval on the expected outcome (see Section 1.10.3), since a point estimate of the routing cost does not necessarily coincide with the expected value. This amounts to solving an optimization problem obtained by replacing the sample mean $\overline{d}_j$ of the demand of each successor node $j \in V$ with the corresponding sample values in model (3.67)–(3.72). The procedure may lead to selecting a different site, as will be shown in the following example.

[Adival.xlsx, Rinott.py] Adival is a French award-winning dairy farm. The logistics department needs to solve the problem of best locating a cold storage warehouse, supplied directly by the main production plant, and used to serve daily 12 supermarkets of different sizes (see Table 3.32). The daily demand of the dairy products requested by each supermarket can be estimated by using historical data and the most probable values (measured in kg) are reported in the third column of Table 3.32, even though it is worth observing that the past data entries (see Adival.xlsx) show a constant trend with seasonality and a significant (±20%) random component (see Section 2.6). The logistics department has considered three different potential sites to host the cold storage warehouse located in the towns of Ennezat, Vic-le-Comte, and Saint Ours. The warehouse is intended to be also used as a depot for the fleet of four identical refrigerator vans, each with a capacity of 1100 kg.

Model (3.67)–(3.72) is solved for each of the three alternatives of the site potentially chosen to host the cold storage warehouse. A set of 298 "plausible" routes has been preliminarily identified. Each of them starts at the depot, serves at most three supermarkets and ends at the depot. Each route satisfies the van capacity constraint, with respect to the most probable values of the daily demand of dairy products associated with the supermarkets served by the route. The cost of each route is computed by considering a unit cost of 0.25 €/km multiplied by the length of the route. The company can use an LSP at a cost that comprises a fixed term of € 30, plus a variable part which is proportional to the distance

Table 3.32 Expected demand level (in kg) of dairy products for the supermarkets in the Adival problem.

Supermarket	City	Demand
1	Pontaumur	255
2	Thiers	351
3	Le Cendre	426
4	Ambert	429
5	Ménétrol	342
6	Clermont-Ferrand	441
7	Besse-et-Saint-Anastaise	366
8	Orcines	237
9	Royat	252
10	Courpière	240
11	Murol	237
12	Cusset	363

Table 3.33 Summary of the results obtained by considering the 480 scenarios of the Adival problem.

	City		
	Ennezat	Vic-le-Comte	Saint Ours
Sample mean [€/day]	164.92	162.36	177.90
Sample standard deviation [€/day]	37.80	39.05	40.32
Confidence interval [€/day]	(161.53; 168.31)	(158.85; 165.86)	(174.28; 181.51)
Thirdy-party provider activation [%]	43.54	43.54	42.92

between the potential site and the supermarket, with a unit cost equal to 1.23 €/km for trips longer than 50 km and to 1.43 €/km for trips less than or equal to 50 km. The average daily transportation cost (in €) corresponds to 131.93, 139.73, and 144.25 for the the cold storage warehouse located in the town of Ennezat, Vic-le-Comte, and Saint Ours, respectively. Note that the costs do not include the fixed and the variable facility costs which, regardless of its location, can be assumed to be constant. The best solution corresponds, therefore, to the site located in Ennezat. The daily variable transportation cost corresponds to the sum of the costs associated with the activated routes and no external services is selected.

However, the solution thus found does not capture the variability of the daily demand for dairy products. For this reason, the decision maker also computes a confidence interval on the expected daily cost, with a confidence level $(1 - \alpha) = 0.95$ and an indifference parameter of € 6 (see Section 1.10.3). The optimization model is solved 480 times for each potential town, on the basis of the different realizations on the daily demand. Because of the variability of the daily demand, the van capacity constraint is not satisfied for some realizations and the fleet of vans is not sufficient to serve all the supermarkets. The results of the runs are summarized in Table 3.33. It is worth observing that, considering $\alpha = 0.05$, $m_0 = 480$ and $n = 3$, the Rinott's constant r is equal to 2.7704. Consequently, running the RINOTT procedure with $\delta = €$ 6 leads to

$$m_1 = \lceil (\tfrac{rS_1}{\delta})^2 \rceil = \lceil (\tfrac{2.7704 \times 37.80}{6})^2 \rceil = 305, \; m_2 = \lceil (\tfrac{rS_2}{\delta})^2 \rceil = \lceil (\tfrac{2.7704 \times 39.05}{6})^2 \rceil = 326,$$

and $m_3 = \lceil (\tfrac{rS_3}{\delta})^2 \rceil = \lceil (\tfrac{2.7704 \times 40.32}{6})^2 \rceil = 347$; m_1, m_2 and m_3 are less than m_0, so no additional observations on the daily demand are required to draw a conclusion about the site selection.

This analysis identifies the best location as the one in Vic-le-Comte. The solution is in contrast to that obtained considering the average daily demands. Although Vic-le-Comte has the largest LSP activation percentage, it remains the most competitive solution in terms of daily transportation cost.

Finally, it is worth noting that the scenario analysis allows Adival's logistics manager to focus on a critical aspect of the distribution activity, that is, the necessity to outsource part of the delivery service. This aspect, not present in the average case analysis, allows greater awareness to be gained of the need to manage demand peaks in a more structured way and not under emergency circumstances.

3.16 Case Study: Intermodal Container Depot Location at Hardcastle

Hardcastle is a North European leader in intermodal transportion. In the last year, the company operated nearly 700 000 containers, with a revenue of about $ 4.8 billion. Like other intermodal transportation companies, Hardcastle manages both full and empty containers. When a customer places an order for freight transportation, Hardcastle sends a number of empty containers of the appropriate sizes and characteristics (refrigeration etc.) to the point of origin. The containers are then loaded and sent to their destination. At the arrival, the containers are emptied and sent back to the company unless there is an outbound load requiring the same kind of container. If feasible, this kind of *compensation* between demand and supply of empty containers allows operating costs to be reduced. Unfortunately, this is seldom possible for the following reasons:

- the origin–destination demand matrix of containers is strongly asymmetrical (some locations are mainly sources of materials while others are mainly points of consumption);

- at a given location, the demand and supply for empty containers do not usually occur at the same time;
- containers may have a large number of sizes and features; as a result, it may happen that the incoming containers at a customer facility are unsuitable for outbound goods.

To avoid surpluses or shortages of containers at the customer locations, the company has to implement a periodic redistribution of the empty containers. The way this allocation is performed is strongly affected by the cost structure. Fixed costs are usually low because public container depots (e.g., depots located inside port terminals, see Section 6.3.1) may be used to store the empty containers for short time periods. Consequently, the location decision is typically reversible in the short or medium term. Transportation cost is an increasing and concave function of the number of containers (both empty and full) along a route. Therefore, the company prefers to *consolidate* and ship together multiple loads whose origins and destinations are adjacent, instead of sending each single loads along an individual route (see Figure 3.17).

Empty containers are stored in regional depots, usually sited next to ports and railways (see Figure 3.18). The management of the empty containers is a complex decision-making process made up of two stages:

- at a tactical level, one has to determine, on the basis of forecast origin–destination transportation demands, the number and locations of the depots, as well as the expected container flows among depots;
- at an operational level, shipments are scheduled and vehicles are dispatched daily to fulfil the orders collected by the company.

Before its redesign, the company operated 87 depots (23 of which were close to a sea terminal) and the empty containers' movements accounted for nearly 40% of the total freight traffic. In order to redesign its logistics system, Hardcastle aggregated its

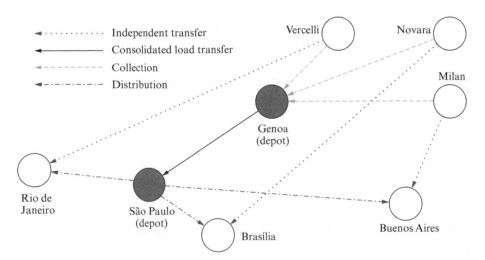

Figure 3.17 Freight consolidation at Hardcastle: containers originated in Vercelli, Novara, and Milan (Italy) are consolidated and shipped together to three destination in South America (Rio de Janeiro, Brasilia, and Buenos Aires) through the depots located in Genoa (Italy) and São Paulo (Brazil).

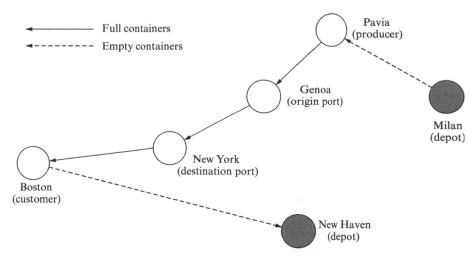

Figure 3.18 Transfer of containers at Hardcastle: empty containers are sent from the depot in Milan (Italy) to a manufacturing company in Pavia (Italy), where they are loaded and sent to their final destination in Boston (USA) along a route comprising the ports of Genoa (Italy) and New York (USA); then the containers are emptied and sent to a depot located in New Haven (USA).

customers into 300 demand points (defined in the following as customers). Let C be the set of customers, D the set of potential depots, P the set of different types of containers, $f_j, j \in D$, the fixed cost of depot j, $a_{ijp}, i \in C, j \in D, p \in P$, the transportation cost of a container of type p from customer i to depot j; $b_{ijp}, i \in C, j \in D, p \in P$, the transportation cost of a container of type p from depot j to customer i; $c_{jkp}, j \in D$, $k \in D, p \in P$, the transportation cost of an empty container of type p from depot j to depot k; $d_{ip}, i \in C, p \in P$, the number of containers of type p requested by customer i, and $o_{ip}, i \in C, p \in P$, the supply of containers of type p from customer i. Furthermore, let $y_j, j \in D$, be a binary decision variable equal to 1 if depot j is selected, and 0 otherwise; $x_{ijp}, i \in C, j \in D, p \in P$, the flow of empty containers of type p from customer i to depot j; $s_{ijp}, i \in C, j \in D, p \in P$, the flow of empty containers of type p from depot j to customer i, and $w_{jkp}, j \in D, k \in D, p \in P$, the flow of empty containers of type p from depot j to depot k. The problem was formulated as follows:

$$\text{Minimize} \sum_{j \in D} f_j y_j + \sum_{p \in P} \left[\sum_{i \in C} \sum_{j \in D} (a_{ijp} x_{ijp} + b_{ijp} s_{ijp}) + \sum_{j \in D} \sum_{k \in D} c_{jkp} w_{jkp} \right] \quad (3.73)$$

subject to

$$\sum_{j \in D} x_{ijp} = o_{ip}, \ i \in C, \ p \in P \quad (3.74)$$

$$\sum_{j \in D} s_{ijp} = d_{ip}, \ i \in C, \ p \in P \quad (3.75)$$

$$\sum_{i \in C} x_{ijp} + \sum_{k \in D} W_{kjp} - \sum_{i \in C} s_{ijp} - \sum_{k \in D} W_{jkp} = 0, \ j \in D, \ p \in P \quad (3.76)$$

$$\sum_{p \in P} \sum_{i \in C} (x_{ijp} + s_{ijp}) + \sum_{p \in P} \sum_{k \in D} (W_{jkp} + W_{kjp})$$

$$\leq y_j \sum_{p \in P} \sum_{i \in C} (o_{ip} + d_{ip} + 2M), \ j \in D \tag{3.77}$$

$$x_{ijp} \geq 0, \ i \in C, \ j \in D, \ p \in P \tag{3.78}$$

$$s_{ijp} \geq 0, \ i \in C, \ j \in D, \ p \in P \tag{3.79}$$

$$w_{jkp} \geq 0, \ j \in D, \ k \in D, \ p \in P \tag{3.80}$$

$$y_j \in \{0,1\}, \ j \in D, \tag{3.81}$$

where M in constraints (3.77) is an upper bound on the $W_{jkp}, j \in D, k \in D, p \in P$. The objective function (3.73) is the sum of depot fixed costs and empty container variable transportation costs (between customers and depots, and vice versa, as well as between pairs of depots). Constraints (3.74)–(3.76) impose empty container flow conservation. Constraints (3.77) state that if $y_j = 0, j \in D$, then the incoming and outbound flows from site j are equal to zero. Otherwise, constraints (3.77) are not binding since

$$x_{ijp} \leq o_{ip}, \ i \in C, \ j \in D, \ p \in P,$$

$$s_{ijp} \leq d_{ip}, \ i \in C, \ j \in D, \ p \in P,$$

$$w_{jkp} \leq M, \ j \in D, \ k \in D, \ p \in P.$$

The implementation of the optimal solution of model (3.73)–(3.81) yielded a reduction in the number of depots from 87 to 48, and a 47% reduction in transportation cost.

3.17 Case Study: Location–Allocation Decisions at the Italian National Transplant Centre

In Italy, lifesaving organ transplants are managed by the National Transplant Centre (CNT), under contract with the national Ministry of the Health. CNT ensures the coordination of all the hospital facilities where organs are explanted and transplanted. Italy is divided into three areas each of which is controlled by an inter-regional centre (CIR). In particular, the North Italy Transplant association covers five regions (Friuli-Venezia Giulia, Liguria, Lombardy, Marche, Veneto) as well as the independent Province of Trento; the Inter-Regional Association of Transplant (AIRT), covers five further regions (Piedmont, Aosta Valley, Tuscany, Emilia-Romagna, Apulia) and the independent Province of Bolzano; the Centre-South Organization of Transplant, covers nine more regions (Abruzzo, Basilicata, Umbria, Campania, Lazio, Calabria, Molise, Sardinia, and Sicily).

Transplant waiting lists are prioritized based on the degree of histological compatibility. When an organ becomes available within a region, it is first made available to the regional waiting list; if it is not required, it is proposed to the corresponding CIR waiting list, and finally, to the nation at large. For example, if there are no compatible patients in Apulia for an organ available in the same region, it is then offered to the patient having the highest priority on the AIRT waiting list. The main reason for this is that the chances of successful organ transplant surgery and subsequent recovery strongly depend on the organ cold-ischaemia time, defined as the maximum time lag between the chilling of an organ after its blood supply has been cut off and the time it is warmed by having its blood supply restored.

At some point, this organ allocation policy was deemed unsatisfactory since it produced unexpected and unwanted consequences on the patient selection, especially for heart, liver, and kidney transplants. Therefore, CNT carried out a more efficient reorganization of the whole organ-sharing system, based on the solution, for each organ, of a location–allocation model.

Let V_1 and V_2 be, respectively, the set of explantation centres and the set of potential transplant centres, p of which have to be opened. Each explantation centre will supply a single transplant centre. Moreover, let K be the set of "demand points", which aggregate potential patients of a specific geographic area. For each demand point $k \in K$ the annual demand d_k, that is, the number of patients on the waiting list, is assumed to be known in advance.

Fifty-two potential locations were selected for the transplant centres. Each transplant centre is assumed to have a circular covering area, with radius r which depends on the cold-ischaemia time of the organ.

Let a_{ij}, $i \in V_1$, $j \in V_2$, be the aerial distance between the explantation centre i and the transplant potential centre j (transportation of explanted organs is usually performed by means of an emergency helicopter) and b_{kj}, $k \in K$, $j \in V_2$, the terrestrial distance between demand point k and transplant potential centre j. Moreover, let T_k, $k \in K$ be the set of candidate sites that are within an acceptable distance from demand point k, i.e., $T_k = \{j \in V_2 : b_{kj} \leq r\}$.

Binary decision variables are the following: x_{ij}, $i \in V_1$, $j \in V_2$, equal to 1 if explantation centre i is assigned to transplant centre j, 0 otherwise; y_{kj}, $k \in K$, $j \in V_2$, equal to 1 if demand point k is assigned to transplant centre j, 0 otherwise; z_j, $j \in V_1$, equal to 1 if transplant potential centre j is activated, 0 otherwise.

The location–allocation model has been formulated as a multi-objective optimization model as follows:

$$\text{Minimize} \sum_{i \in V_1} \sum_{j \in V_2} a_{ij} x_{ij} \tag{3.82}$$

$$\text{Minimize} \sum_{k \in K} \sum_{j \in T_k} d_k b_{kj} y_{kj} \tag{3.83}$$

$$\text{Minimize } M \tag{3.84}$$

subject to

$$\sum_{j \in V_2} x_{ij} = 1, i \in V_1 \tag{3.85}$$

$$\sum_{j \in T_k} y_{kj} = 1, k \in K \tag{3.86}$$

$$x_{ij} \leq z_j, i \in V_1, j \in V_2 \tag{3.87}$$

$$y_{kj} \leq z_j, k \in K, j \in V_2 \tag{3.88}$$

$$\sum_{j \in V_2} z_j = p \tag{3.89}$$

$$M \geq \sum_{k \in K} d_k y_{kj}, j \in V_2 \tag{3.90}$$

$$x_{ij} \in \{0, 1\}, i \in V_1, j \in V_2 \tag{3.91}$$

$$y_{kj} \in \{0, 1\}, k \in K, j \in V_2 \tag{3.92}$$

$$z_j \in \{0, 1\}, j \in V_2. \tag{3.93}$$

The objective function (3.82) represents the total distance between explantation centres and transplant centres that have been activated; the objective function (3.83) is the total distance between demand points and transplant centres, weighted with the annual demand levels; the objective function (3.84) is the number of patients on the longest waiting list. The minimization of variable M aims at balancing the allocation of demand points to the transplant centres (as observed in Section 3.1, the location of facilities providing services of general interest must seek equity/fairness among all the users). Since the three objectives can be conflicting, a reasonable trade-off among them was imposed, by using weights proportional to the priority levels assigned to the single objectives.

Constraints (3.85) guarantee that each explantation centre $i \in V_1$ can be assigned to only one transplant centre $j \in V_2$, while constraints (3.86) impose that each demand point $k \in K$ can be associated with only one transplant centre $j \in V_2$. Conditions (3.87) and (3.88) are linking constraints among decision variables and, therefore, restrict the assignment of explantation centres and demand points to transplant centres that are really activated; equation (3.89) imposes the activation of exactly p transplant centres. Constraints (3.90), together with the minimization of objective function (3.84), ensure that M is equal to the number of patients on the longest waiting list.

The solution of model (3.82)–(3.93) has led to a more efficient and effective organ-sharing system. In particular, with respect to heart transplants, a new configuration has been generated by the model: the overall number of transplant centres has remained unchanged (and equal to 24); nine existing transplant centres have been closed and nine new centres have been opened in different locations. The new logistics configuration has allowed an 18% reduction in the overall distance between the explantation centres and the transplant centres to be achieved, as well as a 24% reduction in the distance between demand points and transplant centres. It has also yielded a better balance among the waiting lists of the different transplant centres.

3.18 Questions and Problems

3.1 Explain why crude oil refineries are customarily located near home heating and automotive fuel markets.

3.2 Let I_c and I_f be the average inventory levels in case a logistics system has n_c and n_f intermediate facilities, respectively. Prove that, under the hypotheses that the demand of the successor nodes is equally divided among the facilities and order size from the successor nodes is optimized according to the EOQ formula, the square root law $I_f = I_c \sqrt{\frac{n_f}{n_c}}$ holds.

3.3 [Conforge.xlsx] Conforge is a world leading casual shoe producer. Its authorized dealership in Switzerland has recently decided to open a retail outlet in Zurich. Four existing potential commercial areas have been selected. The location criteria considered are: potential market, size of the area, variety of the products sold, estimated position in the area, presence of competing stores, proximity of complementary shops, presence of residential area and workplaces within one km, pedestrian and vehicular traffic, and presence of free parking. The potential market is expressed as the shoes of all brands sold yearly to the resident population in the area. The position in the area is evaluated both qualitatively and quantitatively by considering whether the retailer outlet is located in: (a) a primary area (i.e., an area with a percentage of customers between 55% and 70%); (b) a secondary area (an area with a percentage of customers approximately equal to 15–20%) or (c) a marginal area (an area with a percentage of customers approximately equal to 5–10%). A group of seven experts has assigned a weight to each of the nine location criteria (see Table 3.34) and a score to the four potential sites for each location criterion (see Table 3.35). By applying the WEIGHTED_SCORING procedure, determine the commercial area to be selected.

3.4 Show that the vector **w** returned by the CRITERIA_WEIGHTS procedure illustrated in Section 3.6 is normalized.

3.5 [Consistency.xlsx] Evaluate the consistency of the following 4×4 pairwise comparison matrix **A**:

$$\mathbf{A} = \begin{bmatrix} 1 & 3 & 9 & 1/7 \\ 1/3 & 1 & 3 & 1/5 \\ 1/9 & 1/3 & 1 & 1/7 \\ 7 & 5 & 7 & 1 \end{bmatrix}.$$

3.6 [JetMarket.xlsx] Determine the number of pairwise comparisons required by AHP when applied to a facility location problem with m alternative sites and n location criteria. Then, compute this value for the Jet Market problem.

3.7 [Conforge.xlsx] Solve the Conforge problem (see Problem 3.3) by using AHP.

3.8 According to an article appearing in a newspaper, the decision to locate the major FedEx (see Section 1.6.6) air hub in Memphis (Tennessee) was taken by

Table 3.34 Weights associated with the location criteria in the Conforge problem.

ID	Location criterion	Weight
1	Potential market	0.30
2	Size of the area	0.15
3	Variety of the products sold	0.10
4	Position in the area	0.10
5	Presence of competing stores	0.10
6	Proximity of complementary shops	0.05
7	Presence of workplaces within 1 km	0.05
8	Pedestrian and vehicular traffic	0.10
9	Presence of free parking	0.05
	Total	1.00

Table 3.35 Scores of the location criteria for the four potential commercial areas in the Conforge problem.

Location criterion	Score			
	Site 1	Site 2	Site 3	Site 4
1	7	6	5	5
2	7	8	5	7
3	6	8	7	6
4	6	7	6	5
5	8	8	6	6
6	7	7	6	5
7	8	8	7	6
8	8	8	6	5
9	7	7	8	7

minimizing the sum of distances from US major cities, each weighted by the corresponding population. Perform a Web search to get a list of such cities as well as their latitude, longitude, and population. Then, use the WEISZFELD procedure to confirm or reject the location chosen by FedEx. Finally, identify other criteria (e.g., availability of skilled workforce) that may be relevant to the decision.

3.9 [Ring.xlsx] The Ring Offshore company has to connect 13 remote subsea wells in the North Sea (whose features are reported in Table 3.36) to a platform by flow lines. Determine the location of such a facility by using the WEISZFELD procedure with at least three iterations.

3.10 [Karakum.xlsx] Model and solve a variant of the SCOE continuous location problem of a water reservoir in the Karakum Desert in which the square area, whose vertices have the following coordinates (in km): $(17.000, 11.000)$, $(18.000, 11.000)$, $(18.000, 12.000)$, and $(17.000, 12.000)$, cannot accommodate the water reservoir.

3.11 [Monagas.xlsx] The collection of plastic waste produced in the 13 municipalities of the State of Monagas in Venezuela (see Table 3.37) is carried out weekly, using garbage trucks with a capacity of 11 tonnes. Plastic waste is collected from curbside bins and transported to two different recycling plants in the area (see Table 3.38), each with a capacity of 500 tonnes per week. However, to ensure proper operation of the two recycling plants, plastic waste is usually divided equally between the two plants. Table 3.37 also shows the Cartesian coordinates of the 13 municipalities and the amount of plastic waste (in tonnes) produced on average each week. The cost per km to transport one tonne of plastic waste is estimated to be VEF 60 000. In order to optimize transportation cost, an intermediate transfer station will be used. In that facility, plastic waste will be temporarily stored and then transferred to the two facilities using higher-capacity compactor trucks of 14 tonnes each, whose cost per km is 20% greater than that

Table 3.36 Cartesian coordinates and average flow rates of the remote subsea wells in the Ring Offshore problem.

Remote subsea well	Abscissa [km]	Ordinate [km]	Flow rate [quintals/day]
1	0.00	8.54	198
2	11.56	0.00	191
3	5.58	12.45	212
4	22.60	11.35	279
5	8.88	17.38	205
6	37.28	21.56	230
7	12.72	18.65	278
8	35.65	5.42	198
9	9.27	24.32	226
10	25.56	34.54	188
11	15.87	28.55	244
12	31.53	10.12	215
13	29.72	31.40	248

Table 3.37 Cartesian coordinates and plastic garbage production of the municipalities in the State of Monagas.

Municipality	Abscissa [km]	Ordinate [km]	Plastic garbage production [tonnes/week]
Acosta	0.00	159.68	18.2
Aguasay	38.62	96.94	11.8
Bolívar	67.69	158.99	38.6
Caripe	57.29	154.39	33.7
Cedeño	11.82	125.87	34.4
Ezequiel Zamora	12.10	113.28	62.7
Libertador	118.42	36.89	45.2
Maturin	59.12	119.06	302.3
Piar	25.63	143.30	46.6
Punceres	55.13	144.06	27.9
Santa Bárbara	12.36	103.26	9.8
Sotillo	145.59	0.00	24.2
Uracoa	150.22	35.62	9.6

of collection trucks. Formulate and solve the problem to locate the intermediate transfer station.

3.12 [Avial.xlsx] Assume that the maximum amount of 60 kg jute bags of crude coffee available every week at the ports of Amsterdam and Rotterdam is three times the value indicated in the Avial problem description. Solve the new location

Table 3.38 Cartesian coordinates of the two recycling plants in the State of Monagas problem.

Recycling plant	Abscissa [km]	Ordinate [km]
RC1	3.16	124.79
RC2	118.32	15.50

model and show how the crossdock moves in the Cartesian plane. Modify the problem formulation in order to balance the total maximum quantity of commodity available at the ports and the total average weekly quantity of commodity required by the coffee roasting plants. Finally, solve the continuous location problem of two crossdocks on the basis of the original data.

3.13 A company has to shut down 20 out of its 125 warehouses. Suppose that the SCOE hypotheses hold. Define V_1 and define the value of p.

3.14 Prove that constraint (3.27) (see Section 3.9) is redundant in the case where the value of $\hat{u}_i$ in Figure 3.9 is sufficiently large (determine a lower bound for this value).

3.15 Modify the CPL model to take into account the fact that a subset of already existing facilities $V_1' \subseteq V_1$ cannot be closed (but can be upgraded). To this end, indicate the current fixed cost and throughput of facility $i \in V_1'$ as f_i' and q_i', respectively. Moreover, denote the fixed cost and the throughput of a facility $i \in V_1'$ as, respectively, f_i'' and q_i'' if it is upgraded.

3.16 Borachera is a major Spanish wine wholesaler currently operating two central warehouses in Salamanca and Albacete, and a number of regional warehouses all over the Iberian peninsula. In order to reduce its overall logistics cost, the company aims to redesign its distribution logistics system by replacing its current regional warehouses with three (possibly new) facilities. Based on a preliminary qualitative analysis, a regional warehouse should be located in the Castilla-Leon region, in Valladolid, Burgos, or Soria. A second regional warehouse should be located in the Extremadura region, in Badajoz, Plasencia, or Caceres. Finally, the third regional warehouse should be located in the Argon region, in Barbastro, Saragossa, or Teruel. Transportation costs from regional warehouses to retailers are charged to retailers. Formulate the Borachera problem as an SCOE discrete location model.

3.17 As illustrated in Section 3.9, it is customary to remove excessively long transportation connections from the CLP model in order to allow a timely delivery to customers. How should the CPL Lagrangian heuristic be modified in this case?

3.18 Labro is a Portuguese producer of olive oil. It is interested in establishing two new tasting stands for its products in hypermarkets located in the Central Alentejo region. From a preliminary analysis, the hypermarkets available to host them are in the following locations (one tasting stand per hypermarket): Alandroal, Borba, Évora, Mourão, Portel, Redondo, Sousel, Vendas Novas, Viana do Alentejo, and Vila Viçosa. The oil mill plant has three warehouses, which are located in the towns of Borba, Mourão, and Redondowhich, respectively, and can be used for distributing the oil products to the two tasting stands. The following data are available:

- f_j costs (in €per year) for opening and managing tasting stand j;
- average daily amount of oil r_j (expressed as a number of packages of 12 l bottles) needed to supply the tasting stand if it is located at hypermarket j;
- average daily availability d_i of oil (in number of packages of 12 l bottles) at warehouse i.

Formulate the location problem by considering that: (a) each tasting stand will be supplied with a daily door-to-door connection, (b) the average daily transportation cost grows linearly with the number of packages of 12 l oil bottles transported and (c) 220 workdays should be considered per year.

3.19 [Coffa.xlsx] Coffa is a Sicilian distributor of canned tuna which is interested in opening a new warehouse in the province of Syracuse. Avola and Buccheri are the two potential locations identified by a preliminary analysis: in the former case, the annual total facility cost is € 127 500, whereas in the latter case, this cost is € 123 000. The new facility has to supply all the retailers located in the following towns and villages four days a week: Syracuse, Augusta, Avola, Buccheri, Ferla, Lentini, Melilli, Noto, Pachino, and Priolo (see Figure 3.19). Due to the limited capacity of the vehicles, transportation is carried out with direct shipments. The distances (in km) between the town and villages are listed in Table 3.39. Formulate and solve the location problem by assuming 200 workdays a year and a transportation cost per km equal to € 0.95. Alternatively, determine the warehouse location corresponding to the centre of gravity. To this end, use the Cartesian coordinates reported in the last two columns of Table 3.39.

Table 3.39 Distances (in km) between the 10 towns and villages and Avola and Buccheri as well as their Cartesian coordinates (in km), in the Coffa problem.

	Avola	Buccheri	Abscissa	Ordinate
Syracuse	25	55	7.8	7.7
Augusta	56	54	6.6	11.4
Avola	0	57	5.2	4.3
Buccheri	57	0	0.0	9.0
Ferla	52	12	1.7	8.9
Lentini	67	35	2.8	12.6
Melilli	45	36	5.0	10.2
Noto	9	45	4.0	3.9
Pachino	26	69	4.3	0.0
Priolo	38	59	6.1	9.6

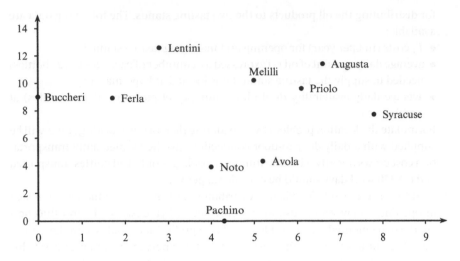

Figure 3.19 Location of the 10 towns and villages and Avola and Buccheri in the Coffa problem.

3.20 Consider the following SCOE discrete location problem:

$$\text{Minimize} \sum_{i \in V_1} \sum_{j \in V_2} c_{ij} x_{ij} + \sum_{i \in V_1} f_i y_i$$

subject to

$$\sum_{i \in V_1} x_{ij} = 1, j \in V_2$$

$$\sum_{j \in V_2} d_j x_{ij} \le q_i y_i, \ i \in V_1$$

$$\sum_{i \in V_1} y_i = 2$$

$$x_{ij} \ge 0, i \in V_1, \ j \in V_2$$

$$y_i \in \{0, 1\}, \ i \in V_1,$$

where $|V_1| = 3, |V_2| = 7, \mathbf{f} = [132; 138; 147]^T, \mathbf{q} = [1500; 1300; 2800]^T$, $\mathbf{d} = [52; 67; 88; 47; 91; 45; 68]^T$, and

$$\mathbf{C} = \begin{bmatrix} 13 & 11 & 17 & 16 & 19 & 27 & 13 \\ 21 & 13 & 19 & 11 & 15 & 17 & 22 \\ 16 & 10 & 21 & 13 & 22 & 18 & 18 \end{bmatrix}.$$

Determine its optimal solution (hint: observe that $q_i \ge \sum_{j \in V_2} d_j, i \in V_1$).

3.21 Extend the CPL model to the case of demand varying over the planning horizon. Assume that, once opened, a facility cannot be closed.

3.22 [CPL.xlsx] Consider a CPL problem defined by the following parameters: $|V_1| = 3, |V_2| = 7, \mathbf{f} = [44; 46; 21]^T, \mathbf{q} = [220; 100; 240]^T, \mathbf{d} = [72; 80; 68; 45; 58; 68; 60]^T$ and

Table 3.40 Distances (in km) between potential production plants and sales districts in the Goutte problem.

	Brossard	Granby	Sainte-Julie	Sherbrooke	Valleyfield	Verdun
Brossard	0.0	76.1	30.4	139.4	72.6	11.7
Granby	76.1	0.0	71.0	77.2	144.5	83.7
LaSalle	20.8	92.9	47.2	156.1	47.5	11.7
Mascouche	54.7	113.3	52.9	187.2	93.0	45.2
Montréal	13.5	85.5	28.0	148.7	67.3	9.3
Sainte-Julie	30.4	71.0	0.0	138.2	94.5	38.1
Sherbrooke	139.4	77.2	138.2	0.0	207.9	146.9
Terrebonne	47.8	106.5	46.2	180.2	86.7	38.9
Valleyfield	72.6	144.5	94.5	207.9	0.0	63.4
Verdun	11.7	83.7	38.1	146.9	63.4	0.0

Table 3.41 Daily fixed costs and daily throughput of the potential plants in the Goutte problem.

Potential plant	Daily fixed cost [CAN$]	Daily throughput [hl]
Brossard	2710	740
Granby	2790	800
LaSalle	2950	940
Mascouche	3030	990
Montréal	2630	670
Sainte-Julie	2870	870
Sherbrooke	2960	950
Terrebonne	3050	880
Valleyfield	2650	650
Verdun	2680	700

$$C = \begin{bmatrix} 12 & 13 & 11 & 13 & 19 & 15 & 11 \\ 16 & 15 & 17 & 13 & 9 & 15 & 19 \\ 9 & 13 & 11 & 9 & 11 & 13 & 13 \end{bmatrix}.$$

and the following Lagrangian multipliers:

$$\lambda^{(h)} = [-28; -43; -34; 12; -25; -38; -42]^T$$

at iteration $h = 2$ of the CPL_SUBGRADIENT procedure. Determine the optimal solution of the corresponding Lagrangian relaxed problem; then, formulate and solve the corresponding demand allocation problem.

3.23 Modify the CPL Lagrangian heuristic to account for the case where the demand is indivisible (i.e., the demand of any successor node must be satisfied by a single

Table 3.42 Demands (in hl per day) of the sales districts in the Goutte problem.

Sales district	Demand
Brossard	460
Granby	330
Sainte-Julie	270
Sherbrooke	400
Valleyfield	340
Verdun	300

facility). Is the modified heuristic still polynomial-time? How can the feasibility of the problem be established?

3.24 [Goutte.xlsx] Goutte is a Canadian company manufacturing and distributing soft drinks. The firm has recently achieved an unexpected increase in its sales mostly because of the launch of a new beverage which has become very popular with young consumers. Management is now considering the opportunity of opening one or more new plants. Considering that the main raw material (water) is available ubiquitously, the inbound transportation cost is negligible compared to finished product distribution costs. Table 3.40 provides the distances between potential plants and sales districts. The daily fixed costs and the daily throughput of the potential plants are shown in Table 3.41, while the daily demands of the sales districts are reported in Table 3.42. The trucks have a capacity of 150 hl and a cost of 0.92 CAN\$/km. Formulate and solve the corresponding SCOE discrete location problem.

3.25 Formulate a polling station location problem, taking into account the following binding constraints: (a) the number of polling stations is fixed for each municipality and calculated according to the number of resident voters; (b) the number of voters assigned to a polling station must not lie outside given lower and upper bounds; (c) the suitability of the potential sites is established by specific criteria (e.g., in some countries, only public buildings, such as schools, are eligible). A soft constraint imposes that almost the same number of voters has to be assigned to each polling station, with the aim of limiting long queues at the entrance. Another important requirement, generally handled as an objective to be optimized rather than a constraint to be imposed, is the minimization of the total distance covered by voters to reach their respective polling stations.

3.26 [Koster.xlsx] Koster Express is an American LTL express carrier operating in Oklahoma. The logistics system is made up of a distribution subsystem, a group of terminals and a long-haul transportation subsystem. The distribution subsystem uses a set of trucks, based at the terminals, where the outbound items are collected and consolidated on palletized unit loads, the inbound palletized unit loads are opened up, and their items are classified for the subsequent distribution phase. The firm has 12 terminals, located in Ardmore, Bartlesville, Dunkan, Enid, Lawton, Muskogee, Oklahoma City, Ponca City, Tulsa, Altus, Edmond,

Table 3.43 Traversal time (in minutes) of the edges of the 1-centre problem in Romania.

(i,j)	a_{ij}	(i,j)	a_{ij}	(i,j)	a_{ij}
(1,2)	5	(3,7)	4	(5,6)	4
(1,7)	7	(3,9)	7	(6,8)	12
(2,3)	4	(4,5)	5	(7,9)	8
(2,6)	11	(4,8)	5	(8,9)	7
(3,5)	6	(4,9)	3		

and Stillwater, of which the last three are newly opened. Their introduction into the logistics system has shown the need to reallocate flows, which were previously handled by two hubs (Dunkan and Tulsa) for the long-haul transportation subsystem. The hubs have the function of moving loads between the origin and destination terminals, and to the other hub, if necessary. In order to efficiently relocate the two hubs, the logistics team hired to carry out a preliminary analysis decided to consider only the transportation cost from each terminal to the hub and vice versa (neglecting, therefore, both the transportation cost between the two hubs and the cost, still considerable, for the possible divestment of the pre-existing hub). Under the hypothesis that each terminal can accommodate a hub, formulate and solve the corresponding p-median problem. To this end, since parcels daily entering and leaving each terminal can be transported using a TL service, the daily transportation cost (in $) between a pair of terminals i and j is given by $c_{ij} = 2 \times 0.74 \times l_{ij}$, where 0.74 is the transportation cost (in $ per mile), and l_{ij} is the distance (in miles) between the terminals, reported in the Koster.xlsx file.

3.27 [Coimbra.xlsx] In the location problem of Coimbra (see Section 3.12), assume that the average travel speed is 55 km/h and that the annual cost of the seventh urban area is € 1 700 000 instead of € 2 200 000. Solve the related SC problem by the CHVATAL procedure illustrated in Section 3.12. Compare the solution with the optimal solution obtained by using the Excel Solver. What about the quality of the heuristic solution? How does the solution change if an emergency fire station has to be located in the seventh urban area?

3.28 A consortium of municipalities in Romania has decided to locate a first response centre in a given area. Each intersection of the road network of the interested area is associated with a vertex of the set $V = \{1, 2, 3, 4, 5, 6, 7, 8, 9\}$ and each road segment is associated with an edge of the set $E = \{(1, 2), (1, 7), (2, 3), (2, 6), (3, 5), (3, 7), (3, 9), (4, 5), (4, 8), (4, 9), (5, 6), (6, 8), (7, 9), (8, 9)\}$. The traversal time a_{ij} for each edge $(i, j) \in E$ is shown in Table 3.43. The average speed of emergency vehicles on the road segments can be assumed to be equal to 60 km/h. Assuming that the first response centre should be located on the edge $(3, 9)$, solve the 1-centre problem by using the HAKIMI procedure.

3.29 [UnitedBank.xlsx] Consider the p-median problem of United Bank. Assume that the connection cost between each potential site and any district centroid is proportional to the product of the distance and the number of the customers in the district. Under the hypothesis that the number of customers in each district is not known in advance, but can assume three possible values, according to a given probability of occurrence, formulate a stochastic version of the p-median problem in which the decision variables $x_{ij}, i \in V_1, j \in V_2$, are scenario-dependent.

3.30 A variant of the location-covering models described in Section 3.12 arises when one must locate facilities to ensure double coverage of customers. Formulate the ambulance location problem when users are better protected if two ambulances are located in their vicinity. If one of the two ambulances has to answer a call, there will remain one ambulance to provide coverage.

4

Selecting the Suppliers

4.1 Introduction

Procurement is a fundamental part of any supply chain management. It amounts to:

- selecting the suppliers for every required product (raw material, component, sub-assembly, etc.) and service;
- negotiating and stipulating procurement contracts;
- periodically controlling that suppliers meet their contractual commitments and, if necessary, taking corrective actions.

The relevance of procurement relies on the fact that the consequences of the poor performance of even a single supplier may reflect on the entire organization and, in extreme cases, may cause severe long-term damage to their reputation.

In recent decades, procurement has evolved from an operational to a strategic perspective. More and more often, companies need to manage their operations in an integrated fashion with the suppliers, with whom logistics alliances are increasingly being established (see Section 1.1), in order to maximize efficiency, competitiveness, and sustainability of the logistics system. This strategic perspective was firstly introduced by P. Kraljic, who proposed a model to segment purchases into four categories: *strategic items*, *bottleneck items*, *leverage items*, and *non-critical items*, characterized by different combinations of *supply risk* and *profit impact*.

The supply risk is defined as the probability that the organization is unable to meet its customer expectations (or even causes threats to customer safety) because of an individual supplier failure to fulfil its obligations (in terms of quality standards, lead times, etc.). Any time an organization introduces a new supplier, a new element of risk is added. Supply risk is strictly connected to the complexity of purchases, the number of alternative suppliers, materials and service availability, bargaining power of suppliers, presence of entry barriers, etc.

The profit impact is a measure of the importance of the purchases in terms of value of the provision and their influence on profitability. As a general rule, it is worth establishing well-structured relationships with multiple suppliers (including potential suppliers) even for the same product or service, with the aim of monitoring their efficiency and deciding periodically whether to renew existing contracts.

Introduction to Logistics Systems Management: With Microsoft® Excel® and Python® Examples,
Third Edition. Gianpaolo Ghiani, Gilbert Laporte, and Roberto Musmanno.
© 2022 John Wiley & Sons Ltd. Published 2022 by John Wiley & Sons Ltd.

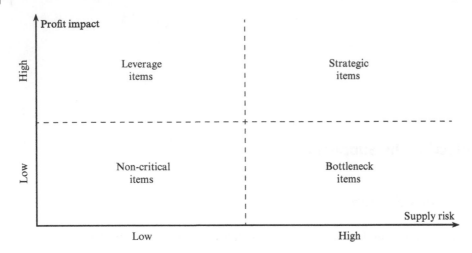

Figure 4.1 The Kraljic matrix, used to segment the purchases of an organization.

The Kraljic matrix (see Figure 4.1) can be used by organizations to identify differentiated supply policies for each category of products and services, on the basis of time considerations as well as on the cost and quality of the provisions. In this respect, there are three alternative procurement policies for each specific product or service: a *single-source* (SS) policy, a *dual-source* (DS) policy, and a *multiple-source* (MS) policy, depending on whether the number of suppliers is one, two, or more than two, respectively. The choice of an appropriate policy for each category of products and services is based on the following considerations:

- strategic items (rare and valuable materials, tailor-made products) are characterized by high profit impact and supply risk, as the number of possible suppliers is limited and the technical specifications make it difficult to identify substitutes. For this class an SS or a DS policy (characterized by long-term partnerships) is the most appropriate;
- bottleneck items (essential to ensure continuity to the company activities) are characterized by high supply risk, because of the scarcity of valid alternatives. As the company needs to ensure a necessary volume to avoid stock shortages, an SS or a DS policy is recommended through exclusive alliances and, in extreme cases, through vertical integration with the suppliers;
- for leverage items (products with high cost and direct impact on quality and customer perception) the company should exploit its contractual power and encourage competition between suppliers, in order to reduce procurement costs. In this case, an SS, a DS, or an MS (with a limited number of suppliers) policy could be equally suitable;
- non-critical items can be managed through an MS policy for reaching process efficiency; their procurement is indeed not risky because of the presence of numerous supply alternatives and substitute products.

The SS and DS policies have the advantage of encouraging suppliers to carry on targeted investments, thus contributing to the development of a useful cooperation and simplifying the integration of the logistics processes between the company and each supplier. On the other hand, this solution increases the risk of dependence on the suppliers and does not allow the company to make comparisons with alternative sources. The MS policy reduces the risk of dependence and allows benefits resulting from competition. However, coordinating several suppliers can be rather complex.

Supplier management includes the following steps:

1. definition of the set of potential suppliers;
2. definition of the selection criteria;
3. supplier selection.

These phases will be examined in detail in the following sections.

4.2 Definition of the Set of Potential Suppliers

The search for suppliers is performed either when a company does not yet have a portfolio of suppliers (e.g., when entering a new market) or when it is about to renew its current pool of suppliers. Various sources of information can be used when searching for suppliers:

1. *Specialized journals.* They allow the technological features and prices of alternative products and services to be compared by drawing information from the advertisements of the various suppliers and articles written by the editorial staff and industry experts.
2. *Search engines.* The Web contains plenty of information on alternative suppliers.
3. *Trade fairs.* These constitute a good opportunity to meet potential suppliers and analyse their offer of products and services.
4. *Organized meetings.* This amounts to meeting with suppliers' representatives, in order to negotiate prices, lead times and other relevant aspects.
5. *E-sourcing.* A set of Internet-based technological tools that systematically support the supplier research, identification, qualification, and contract management. It could be both a standalone software application or part of a *supplier relationship management* (SRM) system, and can sometimes include an online auction functionality.

Inspection visits to suppliers' plants can also be useful, although not very common. The aim is to collect directly detailed information about potential suppliers to assess their qualifications.

Eni is an Italian multi-national oil and gas company, considered one of the world's seven largest companies in this sector. In May 2020, the company introduced an e-sourcing platform, named *eniSpace*, in order to: provide news and information to companies who collaborate or want to collaborate with Eni; make it easier to select suppliers; promote new partnerships with suppliers, involving them into Eni's sustainable energy transition strategy. The platform is divided into four sections: *Business Opportunity*, where suppliers can apply spontaneously and stay updated about goods and services required by Eni; *Innovation Match*, a channel to encourage the development of innovation in the energy field (e.g., sustainability and digitalization); *Agorà*, a forum space for discussing sustainable best practices and *JUST* (Join Us in a Sustainable Transition), an environment for encourage new partnership to define sustainable energy projects. The digital platform helps the company to enhance the selection supplier process in terms of effectiveness, efficiency, and transparency, making it easier to pursue strategic sustainability goals.

4.3 Definition of the Selection Criteria

Defining the selection criteria is certainly the most critical phase of the decision-making process, because these and their relative weights determine the selected suppliers. Once identified, the selection criteria are recorded on the supplier file. This file generally contains the classical information records about the supplier (company name, average business volume, kind of activity, certifications etc.). The choice of selection criteria should correspond to the company's strategies and also depends on the kind of market in which the company operates. Thus, it is not possible to define a generally applicable list of selection criteria. However, the specialized literature contains several lists of criteria which may be refined for the specific application context. A study of particular interest is that conducted by G. W. Dickson, which identifies 23 different evaluation criteria sorted according to their importance (see Table 4.1). It is based on a survey carried out with 273 American managers belonging to the National Association of Purchasing Managers (NAPM). The factors that were considered the most critical are: (a) product quality, (b) contract terms for delivery, (c) performance history, and (d) guarantee terms.

Later on, T. Y. Choi and J. L. Hartley have studied supplier selection criteria in the auto industry. They have identified 23 criteria divided into eight main categories (see Table 4.2).

More recently, as awareness of the environmental impact of industrial processes has been increasing, managers have begun to recognize the importance of sustainability aspects in the criteria definition for supplier selection. In this way, the procurement environmental impact can be reduced, pursuing a broader strategical sustainable logistics development (see Section 1.7.1). In this respect, companies usually better evaluate suppliers with a high corporate environmental responsibility, which implement green solutions inside their operations (e.g., reducing the total life-cycle product impact, energy, and water saving within production process, reduction of pollutant emissions, etc.). Nowadays, a green procurement approach is usually compliant with customer values, but is also required to satisfy standards imposed by specific laws and regulations. In order to build a green supplier selection system, some criteria associated with sustainability have to be considered, in addition to those presented above. Among them, 13 seem to be more frequently used; they are shown in Table 4.3, divided into specific categories.

The Volkswagen Group, a German automotive company, introduced in 2019 a sustainability index (called *S-Rating*) to evaluate its suppliers. The company requires each supplier to complete a self-assessment questionnaire on its operation's sustainability, investigating the use of natural resources, waste management,

Table 4.1 Supplier selection criteria according to G. W. Dickson.

Classification	Criterion
1	Quality
2	Contract terms for delivery
3	Performance history
4	Guarantee terms
5	Structural and manufacturing capacity
6	Cost
7	Technical capacity
8	Financial position
9	Conformity to the procedures
10	Communication system
11	Reputation
12	Business attractiveness
13	Management and organization
14	Operative controls
15	Assistance service
16	Attitude
17	Impression
18	Packaging ability
19	Ended-works reports
20	Geographic position
21	Total of ended business
22	Training aids
23	Reciprocal agreements

pollution, and other key factors. If the answers are considered satisfactory, experts proceed with an inspection visit to the supplier, in order to complete the S-Rating evaluation. Approximately 12 500 candidate suppliers completed the questionnaire, but only 1331 were selected for an inspection visit. This selective process allows the company to rely only on suppliers who share its ethical, social, and environmental values.

Summarizing, it can be stated that the most relevant selection criteria are: (a) purchase price and quality, specifically related to the product or service considered; (b) delivery efficiency, flexibility, financial capabilities, and reputation, and (c) sustainability performances, related to an evaluation of the potential supplier rather than to the specific product or service to be purchased.

The presence of several selection criteria forces the decision maker to define their relative importance, with the aim of establishing their impact on the supplier selection.

Table 4.2 Selection criteria for suppliers (according to T. Y. Choi and J. L. Hartley) and their subdivision into categories.

Category	Criteria
Finances	Financial conditions
	Profitability
	Financial information availability
	Performance awards
Consistency	Product conformity
	Consistent delivery times
	Quality philosophy
	Response times
Relations capacity	Long-term relations
	Closeness in relations
	Openness in communications
	Reputation
Flexibility	Changes in production volumes
	Reduction of equipping times
	Reduction of delivery times
	Resolution of conflicts
Technological capabilities	Design capabilities
	Technical capabilities
Services	Post-sales assistance
	Sale representatives competences
Reliability	Incremental improvements
	Product reliability
Price	Initial price

In this respect, it is worth classifying the selection criteria in a hierarchical order. The highest level usually individuates the main criteria influencing the supplier selection (e.g., price, quality, sustainability, finance, etc.) and, successively, different subsequent levels can be added with increasing details. In particular, the selection factors influencing the main criteria are included in the second level and, in turn, they can be possibly split into other subcriteria. Although each company may consider any number of criteria levels, a rule of thumb suggests not exceeding a maximum of three. In fact, increasing the number of levels implies the need to adopt several and very specific criteria in the last level, for many of which sufficient information may not be always available.

[Suntech.xlsx] Suntech Solar is a Mexican company specializing in the assembly of monocrystalline photovoltaic panels. To evaluate its suppliers of monocrystalline photovoltaic cells, Suntech decided to adopt five main criteria:

Table 4.3 Green selection criteria for suppliers and their subdivision into categories.

Category	Criteria
Pollution level	Greenhouse gas emissions
	Water pollution
	Solid and water waste
	Energy consumption
	Use of harmful materials
Green image	Environmental regulations and standards
	Environmental certificates
	Social responsibility
Green product	Recycling
	Green packaging
Green competencies	Clean technology
	Reverse logistics
	Internal process control

product quality, product price, financial situation, geographic position, and sustainability. The purchase manager also decided to include a series of subcriteria to identify the factors mostly impacting on product quality, sustainability, and financial situation. No further specifications are required for product price and geographic position. The complete list of subcriteria is detailed in the second level of the hierarchical diagram of Figure 4.2.

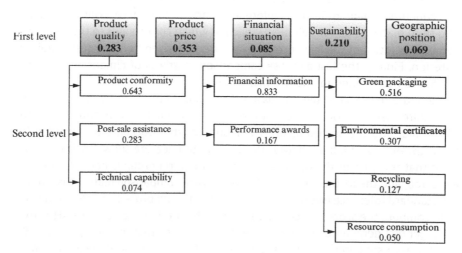

Figure 4.2 Hierarchical diagram of criteria and subcriteria and their weights (obtained by using AHP, see Section 4.4) in the Suntech Solar supplier evaluation.

Finally, it is worth observing that not all the criteria have the same weight for all companies. A company focusing on the quality of its products will give a higher weight to the quality criterion; on the other hand, a company interested in reducing its inventory costs will pay more attention to the geographic position of its suppliers.

4.4 Supplier Selection

A large variety of methods can be used for supplier selection, and an exhaustive presentation of them is outside the scope of this book. In the following, attention will be focused on three of the most widespread methods of practical interest. The first two are qualitative and have been already illustrated in Chapter 3 for solving discrete location problems:

- the weighted scoring method (see Section 3.5 for more details) is based on the availability of a score (from 0 to 10) assigned by experts to each supplier and of a weight (from 0 to 1) for each selection criterion to reflect its relative importance with respect to the other ones (for this reason, the sum of the weights of all the selection criteria is one);
- the AHP is also a multi-criteria decision-making method in which the supplier selection procedure is decomposed into a series of simpler pairwise comparisons, expressing the relative importance either of one criterion versus another or between alternative suppliers. A detailed description of the AHP is available in Section 3.6.

[Suntech.xlsx] In the Suntech Solar problem, the evaluation of five suppliers of monocrystalline photovoltaic cells has been conducted using the AHP. In order to determine the weight of each criterion (step 1 of the AHP), the manager of the purchase office first generated four pairwise comparison matrices. The first one is a 3×3 matrix used to compare the three subcriteria (product conformity, post-sales assistance, and technical capability) of the main criterion named product quality. The second one is a 2×2 matrix through which a comparison between the subcriteria named financial information and performance awards is made. The third matrix is of 4×4 elements, each of which corresponds to the comparison between two of the four second-level criteria derived from the sustainability criterion. Finally, the last pairwise comparison matrix is of dimension equal to 5×5 and was defined for the five main criteria.

Then, the manager of the purchase office adopted the CRITERIA_WEIGHTS procedure (see Section 3.6) to determine the weight for each criterion (both of the first and the second level). The results of the procedure are detailed in the Suntech.xlsx file, whereas the values obtained are shown in Figure 4.2. Note that the sum of the weights within each criteria group is equal to one, as expected, and the inconsistency of all the pairwise comparison matrices is within the standard tolerance (see the CONSISTENCY_INDEX procedure in Section 3.6).

To compute the score of each supplier for each criterion (step 2 of the AHP), the manager first generated nine 5×5 pairwise comparison matrices $\mathbf{B}^{(j)}, j = 1, \dots, 9$, one for each subcriterion in the second level of the hierarchical diagram shown in Figure 4.2. The details are in the Suntech.xlsx file. In the following, for the sake of brevity, only the first three are reported, associated with the three subcriteria of the main criterion named product quality.

$$\mathbf{B}^{(1)} = \begin{bmatrix} 1 & 3 & 5 & 3 & 5 \\ 1/3 & 1 & 3 & 1/3 & 3 \\ 1/5 & 1/3 & 1 & 1/3 & 1 \\ 1/3 & 3 & 3 & 1 & 3 \\ 1/5 & 1/3 & 1 & 1/3 & 1 \end{bmatrix};$$

$$\mathbf{B}^{(2)} = \begin{bmatrix} 1 & 1/3 & 3 & 1 & 1/3 \\ 3 & 1 & 3 & 1 & 1 \\ 1/3 & 1/3 & 1 & 1/5 & 1/3 \\ 1 & 1 & 5 & 1 & 3 \\ 3 & 1 & 3 & 1/3 & 1 \end{bmatrix};$$

$$\mathbf{B}^{(3)} = \begin{bmatrix} 1 & 1/5 & 5 & 1 & 7 \\ 5 & 1 & 7 & 5 & 9 \\ 1/5 & 1/7 & 1 & 1/5 & 3 \\ 1 & 1/5 & 5 & 1 & 5 \\ 1/7 & 1/9 & 1/3 & 1/5 & 1 \end{bmatrix}.$$

Then, the `CRITERIA_WEIGHTS` procedure has been applied to each $\mathbf{B}^{(j)}$ matrix, $j = 1, \dots, 9$. It returns a five-component score vector $\mathbf{F}_j, j = 1, \dots, 9$, that can be used to determine the final score matrix. For example, the first three vectors $\mathbf{F}_j$, $j = 1, \dots, 3$, associated with the matrices $\mathbf{B}^{(j)}, j = 1, \dots, 3$, define the following 5×3 score matrix of the suppliers with respect to product conformity, post-sale assistance and technical capability:

$$\begin{bmatrix} 0.449 & 0.151 & 0.189 \\ 0.164 & 0.285 & 0.541 \\ 0.072 & 0.063 & 0.063 \\ 0.243 & 0.308 & 0.173 \\ 0.072 & 0.221 & 0.034 \end{bmatrix}.$$

Multiplying this matrix with the corresponding three-component weight vector (reported in Figure 4.2),

$$\begin{bmatrix} 0.449 & 0.151 & 0.189 \\ 0.164 & 0.285 & 0.541 \\ 0.072 & 0.063 & 0.063 \\ 0.243 & 0.308 & 0.173 \\ 0.072 & 0.221 & 0.034 \end{bmatrix} \times \begin{bmatrix} 0.643 \\ 0.283 \\ 0.074 \end{bmatrix} = \begin{bmatrix} 0.345 \\ 0.218 \\ 0.069 \\ 0.256 \\ 0.111 \end{bmatrix} = \mathbf{S}_1,$$

the first column of the 5×5 score matrix $\mathbf{S}$ is obtained, associated with the product quality criterion. Proceeding in a similar way, the vectors $\mathbf{F}_j, j = 4, \dots, 5$, are used to determine the third column ($\mathbf{S}_3$) of the matrix $\mathbf{S}$, corresponding to the score assigned to each supplier for the financial situation criterion, and the vectors $\mathbf{F}_j, j = 6, \dots, 9$, allow the fourth column of $\mathbf{S}$ (i.e., $\mathbf{S}_4$ associated with the sustainability criterion).

By using a bottom-up logic, the manager then generated the remaining two 5×5 pairwise comparison matrices for the two main criteria (product price and geographic position) which are not split in subcriteria (see the `Suntech.xlsx` file). The application of the `CRITERIA_WEIGHTS` procedures allowed the second and the fifth columns (S_2 and S_5, respectively) of the score matrix S. Summarizing, the 5×5 score matrix S to be determined, at the end of the step 2 of the AHP, was

$$S = \begin{bmatrix} 0.345 & 0.372 & 0.240 & 0.324 & 0.349 \\ 0.218 & 0.295 & 0.063 & 0.285 & 0.184 \\ 0.069 & 0.073 & 0.076 & 0.102 & 0.349 \\ 0.256 & 0.199 & 0.202 & 0.187 & 0.067 \\ 0.111 & 0.062 & 0.419 & 0.102 & 0.051 \end{bmatrix}.$$

Multiplying S by the five-component weight vector (the values are also reported in Figure 4.2), the scores of the five suppliers (step 3 of the AHP) were finally determined: 0.342, 0.244, 0.097, 0.204, and 0.114, respectively. Consequently, the preferred supplier was the first one, followed by the second (in the case of a DS policy adoption).

The third proposed method for supplier selection is based on formulating and solving a suitable optimization model. A large variety of such models exist. In the following, one of the simplest and most widely used models in the case of a single commodity will be illustrated, corresponding to a multi-objective LP model.

Let m be the number of potential suppliers; n is the number of identified selection criteria, each of which has a weight $w_j, j = 1, \dots, n$; $s_{ij}, i = 1, \dots, m, j = 1, \dots, n$, is the evaluation of supplier i with respect to criterion j; $r_i = \sum_{j=1}^{n} w_j s_{ij}, i = 1, \dots, m$, is the total score of supplier i; d is the quantity required from the suppliers (total demand) during the planning horizon; $c_i, i = 1, \dots, m$, is the unit purchase price from supplier i and $q_i, i = 1, \dots, m$, is the capacity of supplier i, that is, the maximum quantity that the supplier can provide in the planning horizon.

The decision variables are $x_i, i = 1, \dots, m$, representing the quantity of commodity purchased from supplier i in the planning horizon. The mathematical model is as follows:

$$\text{Minimize} \sum_{i=1}^{m} c_i x_i \tag{4.1}$$

$$\text{Maximize} \sum_{i=1}^{m} v_i x_i \tag{4.2}$$

subject to

$$\sum_{i=1}^{m} x_i = d \tag{4.3}$$

$$0 \leq x_i \leq q_i, i = 1, \dots, m, \tag{4.4}$$

where

$$v_i = \frac{r_i}{\max\limits_{k=1,\dots,m}\{r_k\}}, i = 1, \dots, m. \qquad (4.5)$$

The first objective (4.1) amounts to minimizing the total purchase cost. The second objective (4.2) is to maximize the overall score of the selected suppliers (due to constraint (4.3) and the values of v_i, $i = 1, \dots, m$, determined by equation (4.5), the objective function (4.2) cannot exceed d). Constraint (4.3) means that the whole demand must be satisfied. Constraints (4.4) impose lower and upper bounds on the quantity provided by each supplier.

In order to determine a Pareto optimal solution of problem (4.1)–(4.4), the ε-*constraint method* will be used: one of the two objective functions is transformed into a constraint by bounding its value above (in case of minimization) or below (in case of maximization).

[Ilax.xlsx] Ilax is a leading Japanese company specializing in microprocessors and other components for the communications and electronics industry. For the procurement of quartz sand, an important raw material in this field, five suppliers were evaluated using a weighted scoring method on the basis of 18 different criteria. The results of this evaluation are shown in the second column of Table 4.4. The demand for quartz sand for the next month is 6500 tonnes.

During the negotiation phase, each supplier has indicated the maximum quantity of quartz sand available (in tonnes) and the unit selling price offered (in $). These values are shown in the third and fourth columns of Table 4.4, respectively.

Table 4.4 Score of the Ilax suppliers, capacity and unit selling price of quartz sand.

Supplier	Score	Capacity [tonnes]	Unit selling price [$]
1	8.23	3500	792
2	8.01	4000	767
3	7.57	4000	758
4	8.18	4000	780
5	8.54	2500	803

The corresponding multi-objective supplier selection model (4.1)–(4.4) is the following:

Minimize $792x_1 + 767x_2 + 758x_3 + 780x_4 + 803x_5$

Maximize $(8.23/8.54)x_1 + (8.01/8.54)x_2 + (7.57/8.54)x_3$

$\qquad + (8.18/8.54)x_4 + (8.54/8.54)x_5$ $\qquad (4.6)$

subject to

$x_1 + x_2 + x_3 + x_4 + x_5 = 6500$

$$0 \leq x_1 \leq 3500$$

$$0 \leq x_2 \leq 4000$$

$$0 \leq x_3 \leq 4000$$

$$0 \leq x_4 \leq 4000$$

$$0 \leq x_5 \leq 2500.$$

Ilax considers cost minimization to be more important than the maximization of the supplier score. For this reason, it was decided to transform the objective function (4.6) into the following additional constraint:

$$0.9637x_1 + 0.9379x_2 + 0.8864x_3 + 0.9578x_4 + 1.0000x_5 \geq \epsilon,$$

with $\epsilon = 6\,500\alpha$, where $\alpha \in [0, 1]$ is set to 0.95. The following Pareto optimal solution, determined by using the Excel Solver tool under the Data tab, is obtained:

$$\bar{x}_1 = 0; \bar{x}_2 = 4000; \bar{x}_3 = 0; \bar{x}_4 = 10\,925/6; \bar{x}_5 = 4075/6,$$

with a provision cost of about $ 5 033 621.

Model (4.1)–(4.4) deals with a single commodity to be supplied from the suppliers (e.g., corn) in one "shot" (the quantity ordered is delivered once) and need not be an integer (e.g., 345.76 kg of corn). The model can be extended to consider additional features:

1. integer quantities to be supplied (e.g., if the unit load is palletized and not expressed in kg, 450 palletized unit loads could be a feasible solution, whereas 450.56 palletized unit loads is not correct);
2. multi-commodity (e.g., corn and, simultaneously, other cereals);
3. the quantity ordered can be shipped by the suppliers in different time periods (e.g., 100 palletized unit loads of a product are ordered with a deadline of two weeks: the first 30 arrive at the end of the first week and the remaining 70 at the end of the second week).

These cases are proposed as exercises.

4.5 Supplier Relationship Management Software

When an organization uses multiple suppliers, an SRM software can be helpful in managing the supplier relationships and in automating and systematizing the tasks of purchasing products and services, reducing the errors and thereby increasing the overall efficiency of the logistics system. Specifically, the SRM software contributes to the reduction of supply costs and lead time and increase security and flexibility, supporting the automation of different processes:

- *Supplier selection*. Submitting questionnaires for qualifying new suppliers, collecting and cataloguing strategic information, managing selection and performance evaluation considering different criteria.

- *Operational procurement.* Managing supplying planning, order processing, payment systems, transactions, saving historical data. All these activities are usually conducted using consolidated standard tools like EDI, CRP, and CPFR (see the Barilla case described in Problem 1.3).
- *Relationship management.* Managing contracts and agreements, facilitating communication between parties.
- *Performance monitoring.* Using analytical tools for measuring performances and improving the overall process.

Such software often integrates modules for *e-procurement* (online purchase of products and services), including e-sourcing systems and the possibility of saving the most used catalogues (*e-catalogue*), as well as managing auctions and dynamic negotiation.

From the logistics perspective, an SRM software is able to interface with different modules of the ERP (see Section 1.12 and, in particular, Figure 1.28), supporting the management of the inbound materials. In detail, the SRM software receives information about order quantities from the material resource planning module, and sends information about inbound products to the WMS (see Section 5.9) to easily manage the unloading bays. In large companies, SRM software allows support of logistics alliances (see Section 1.1), LSPs (see Section 1.4), and integrated logistics paradigms.

Electrolux, a global manufacturer of household appliances, uses a strategy of strong supplier integration for materials management. Over the past two decades, the company has developed solutions for the digitalization of the procurement, introducing a Web-based SRM platform. A specific module was designed for implementing a *vendor-managed inventory* (VMI) solution, as the main output of a joint project named *Replenishment*. The project involves Electrolux providing monthly or weekly sale levels, production plans, and forecasting to its main suppliers. From their side, suppliers automatically replenish the company's warehouses, considering the minimum stock level previously agreed. This system contributes to reducing errors and stock levels, building a win–win strategy for all parties, decreasing operational logistics costs, and improving efficiency.

4.6 Case Study: the System for the Selection of Suppliers at Baxter

Baxter Healthcare Corporation is a worldwide company operating in the healthcare sector in over 110 countries with over 48 000 employees. The company was founded in 1931 as a manufacturer of intravenous solutions. Baxter provides, through subsidiaries, products and services for the care of patients who are in a critical condition: those affected by hemophilia, immune deficiencies, infectious diseases, kidney disease, and trauma. In particular, BTT (Baxter Transfusion Therapy), a leader in transfusion medicine for more than 40 years, is a manufacturing company and provider of services for the collection, storage, and distribution of blood products. In 2007, BTT's manufacturing facility in San German, Puerto Rico, decided to request quality certification according to the US Food and Drug Administration (FDA) policy.

In particular, the FDA approval requires that each certified company adopts specific rules for purchasing. These rules require that: the company should use written procedures which specify the requirements that suppliers and consultants must meet; the company must evaluate and select potential suppliers and consultants on the basis of their ability to meet specified requirements; the necessary control of the product or service to be delivered should be properly defined and based on analytical data evaluation processes; there must be adequate systems in the company for the registration of suppliers and consultants. To meet these standards, BTT decided to adopt a structured system for the selection of suppliers. The system, briefly described in this section, was based on an SS policy for each product.

Every trimester, BTT prepares an internal report on the performance of each supplier. The report is sent to the vendors in order to let them review it and modify their actions if necessary. For each delivery, the incoming goods are subject to quality controls which can impose corrective actions in case of non-compliance, and preventive actions to reduce the risks of non-compliance.

To each qualified supplier $i = 1, \ldots, m$, is assigned a quality index, denoted as the *supplier quality index* (*SQI*), which is updated monthly. For each type of product to be purchased, the company maintains a list of suppliers, sorted by non-increasing values of *SQI*. The definition of the quality index is made using the weighted scoring method, based on three criteria:

1. quality (Q), with a weight equal to 50%;
2. punctuality (P), weight 40%;
3. company–supplier relationship (R), weight 10%.

The *SQI* calculation rule is

$$SQI_i = 0.5Q_i + 0.4P_i + 0.1R_i, i = 1, \ldots, m.$$

To be selected a supplier must maintain its *SQI* value above nine. For each scheduled delivery, the BTT purchase manager selects from the list the first supplier who is able to accept the terms (quantity and timing) of the provision itself.

The quality of supplier $i = 1, \ldots, m$, is established for each supply based on the verification of compliance with respect to the requirements of the goods supplied and on the supplier's capability to give prompt and effective responses to the received requests for corrective and preventive actions. The quality indicator is therefore calculated by means of three parameters: responsiveness to the verification of incoming goods (QA); responsiveness to comments arising from the quality control procedure (QB); responsiveness to requests for corrective and preventive actions (QC).

The procedure for monitoring incoming goods is described in the following. One out of every five lots is tested if there were no problems with the last 10 consecutive deliveries received; otherwise, all the lots are tested and inspected. For each lot, 40% of the components are tested through a random check of the component's specifications. If the test ends successfully, the unit is certified as "component meeting the requirements". The remaining 60% of components are directly certified or tested. The direct certification is granted if there was no problem with the vendor during the entire previous year. The chemical raw materials, however, cannot be directly certified. A lot is certified if no problem is detected, that is, if all its components meet the requirements. For this reason, parameter QA for supplier $i = 1, \ldots, m$, corresponds to the ratio between the number of accepted lots and the number of lots inspected, multiplied by 10 (this ensures that $QA_i \in [0, 10], i = 1, \ldots, m$).

The calculation of QB is based on the number of days g necessary to receive from the supplier a formal response to comments arising from the quality control procedure. Responses must be received by BTT within at most 10 days; otherwise suppliers are charged a penalty of 5% on the cost of components supplied. For these reasons, QB is calculated as

$$QB_i = \max\{0,\ 11 - g\}, i = 1, \ldots, m. \tag{4.7}$$

Formula (4.7) ensures that $QB_i \in [0, 10], i = 1, \ldots, m$. The same rule is used for the calculation of QC, since BTT gives a maximum time of 10 days to each vendor to respond to requests for corrective and preventive actions on supplies.

The quality indicator for each supplier is obtained as follows:

$$Q_i = 0.6QA_i + 0.2QB_i + 0.2QC_i, i = 1, \ldots, m.$$

The indicator of punctuality P for supplier $i = 1, \ldots, m$, corresponds to the ratio between the number of deliveries made on time and the total number of deliveries made in the month, multiplied by 10. This ensures that $P_i \in [0, 10], i = 1, \ldots, m$. A delivery is made on time if it takes place at most one day after the planned date.

The indicator on the company–supplier relationship R is calculated using five parameters, to each of which the BTT purchase manager assigns a score (between a minimum and a maximum value) representing the level of effectiveness and efficiency of communications between the company and the supplier. The parameters are

1. proactivity (RA), which measures the ability to initiate communication about potential non-compliances and on issues that may affect the quality of a delivery; $RA_i \in [0, 10], i = 1, \ldots, m$;
2. reactivity (RB), which measures the ability to quickly, effectively, and efficiently change delivery procedures and possibly also the goods delivered following specific requests coming from the company; $RB_i \in [0, 15], i = 1, \ldots, m$;
3. the ability to organize emergency and extraordinary events (RC), which measures the ability to accommodate requests of extraordinary quality visits of the plants, to organize special and extraordinary deliveries and so on; $RC_i \in [0, 5], i = 1, \ldots, m$;
4. accessibility (RD), which measures the ability to respond promptly, efficiently, and courteously to company inquiries; $RD_i \in [0, 10], i = 1, \ldots, m$;
5. flexibility (RE), which measures the supplier's ability to adopt adequate methods and supply contents to specific needs declared by the company; $RE_i \in [0, 10], i = 1, \ldots, m$.

Therefore,

$$R_i = \frac{RA_i + RB_i + RC_i + RD_i + RE_i}{5}, i = 1, \ldots, m.$$

The described BTT system for the selection of suppliers obtained the FDA's approval for the San German facility.

4.7 Case Study: the Supplier Selection at Onokar

Onokar is a Turkish company producing frequency inverters, decentralized control units, and electric motors. The company sources a lot of materials from Germany and Turkey. Initially, the supplier selection system was based only on price and on the buyer's personal preference versus the suppliers. Recently, the company recognized several inefficiencies within this process, in particular for the procurement of bearings, one of the most important components because of frequent orders and high purchase costs. The inefficiencies were due to the impossibility of exploiting quantity discounts, because of the high number of products and suppliers. Furthermore, it was decided to consider other selection criteria in addition to price: quality, after-sales service, and delivery performance.

In order to design a more structured system, the management decided to adopt an optimization model for supplier selection. In the model it was assumed that each supplier is characterized by a capacity constraint and offers price discounts based on the quantity purchased. Onokar decided to select a given number of potential suppliers, each of which was evaluated on the basis of the aforementioned criteria for each product. Furthermore, the company imposed some upper bounds on lead time and defect rate. The optimization model had the scope to maximize the total score of the procurement, selecting the suppliers associated with the highest scores and defining the quantities to be ordered. The MIP model formulated by the Onokar is described in the following.

Let I be the set of suppliers, J the set of products, and C the set of criteria. Note that C does not include the price criterion (that is expressed considering the discount quantities). Let $K^i, i \in I$, be the set of pricing levels according to the different quantity discounts associated with supplier i. The following input parameters are defined: q_{ij}, $i \in I, j \in J$, is the capacity of supplier i for product j; $d_j, j \in J$, is the demand of product j; $t_{ij}, i \in I, j \in J$, is the average delivery lateness of supplier i when providing product j (expressed as percentage deviation from a nominal delivery time); $t_j^{\max}, j \in J$, is the maximum delivery lateness allowed for product j; $f_{ij}, i \in I, j \in J$, is the defect rate of supplier i when providing product j; $f_j^{\max}, j \in J$, is the maximum defect rate allowed for product j. Moreover, $s_{ijc}, i \in I, j \in J, c \in C$, is the score associated with each supplier i, product j, and criterion c; $w_{jc}, j \in J, c \in C$, is the weight for product j and criterion c; $s_{ijk}, i \in I, j \in J, k \in K^i$, is the score of each supplier i in the provision of product j when a discount price level k is applied; $v_j, j \in J$, is the weight of price criterion for product j. Note that the scores and the weights are previously computed through the AHP. Finally, m is the maximum number of suppliers to be selected; b_{ijk}, $i \in I, j \in J, k \in K^i$, is the upper bound value of the quantity discount intervals, defined for supplier i, product j and pricing level k (note that the initial level is $b_{ij0} = 0$).

The decision variables are defined as follows: $x_{ij}, i \in I, j \in J$, is the quantity of product j provided by supplier i; $l_{ijk}, i \in I, j \in J, k \in K^i$, is the quantity of product j provided by supplier i at the price level k; $y_i, i \in I$, is a binary decision variable equal to 1 if supplier i is selected, 0 otherwise; $z_{ijk}, i \in I, j \in J, k \in K^i$, is a binary decision variable equal to 1 if pricing level k for product j and supplier i is selected, 0 otherwise.

The procurement problem is formulated as follows:

$$\text{Maximize} \sum_{i \in I} \sum_{j \in J} \sum_{c \in C} w_{jc} s_{ijc} x_{ij} + \sum_{i \in I} \sum_{j \in J} \sum_{k \in K^i} v_j s_{ijk} l_{ijk} \tag{4.8}$$

subject to

$$\sum_{k \in K^i} l_{ijk} = x_{ij}, i \in I, j \in J \tag{4.9}$$

$$\sum_{i \in I} x_{ij} = d_j, j \in J \tag{4.10}$$

$$x_{ij} \le q_{ij}, i \in I, j \in J \tag{4.11}$$

$$\sum_{i \in I} t_{ij} x_{ij} \le t_j^{\max} d_j, j \in J \tag{4.12}$$

$$\sum_{i \in I} f_{ij} x_{ij} \le f_j^{\max} d_j, j \in J \tag{4.13}$$

$$\sum_{j \in J} x_{ij} \le M y_i, i \in I \tag{4.14}$$

$$\sum_{j \in J} \sum_{k \in K^i} l_{ijk} \le M y_i, i \in I \tag{4.15}$$

$$l_{ijk} \le M z_{ijk}, i \in I, j \in j, k \in K^i \tag{4.16}$$

$$\sum_{k \in K^i} z_{ijk} \le 1, i \in I, j \in J \tag{4.17}$$

$$\sum_{i \in I} y_i \le m \tag{4.18}$$

$$b_{ijk-1} + M(z_{ijk} - 1) \le l_{ijk}, i \in I, j \in J, k \in K^i \tag{4.19}$$

$$l_{ijk} \le b_{ijk} + M(1 - z_{ijk}), i \in I, j \in J, k \in K^i \tag{4.20}$$

$$x_{ij} \ge 0, i \in I, j \in j \tag{4.21}$$

$$l_{ijk} \ge 0, i \in I, j \in J, k \in K^i \tag{4.22}$$

$$y_i \in \{0, 1\}, i \in I \tag{4.23}$$

$$z_{ijk} \in \{0, 1\}, i \in I, j \in J, k \in K^i. \tag{4.24}$$

Objective function (4.8) maximizes the total weighted score of the procurement. Constraints (4.9) impose that, for each supplier and each product, the sum of the product quantities purchased at different price levels is equal to the total quantity purchased. Constraints (4.10) impose the satisfaction of the demand for each product. Constraints (4.12) and (4.13) mean that the maximum delivery lateness and defect rate are satisfied for each product, respectively. Constraints (4.14)–(4.16) state the possibility of purchasing a quantity from each supplier only if selected (M is an arbitrarily large positive constant greater than or equal to the total demand for all products). Constraints (4.17) impose the selection of at most one pricing level for each product provided by each supplier. Constraint (4.18) defines the maximum number of suppliers to be selected. Constraints (4.21)–(4.22) define the non-negative conditions of some decision variables, whereas constraints (4.23)–(4.24) require that the remaining decision variables are binary.

The company adopted the model for the procurement of five types of bearings from six different suppliers, with respect to the four criteria already described. The selection of a maximum number of three suppliers was required. The maximum delivery lateness and defect rate were 5% and 2%, respectively, for each bearing type. Among the six suppliers, the model suggested the best three to be selected.

At the first application, in the subsequent trimester, the company achieved a reduction in delivery delays of 15%, and a saving on the total procurement costs of around 20% by making the best use of quantity discounts. The company definitively introduced the model to support its procurement activities, since it has empirically proven to be adequate to take into account different product categories and supplier features.

4.8 Questions and Problems

4.1 Your company needs electric energy for a warehouse. Consult the Web pages of at least three potential providers and evaluate them in terms of their average price during the morning shift, payment conditions, and reputation. Which other criteria could be taken into account? Adopt a weighted scoring method to select the provider.

4.2 Perform a Web search to find an example of a long-term partnership agreement between a company and its suppliers, in order to manage the procurement of strategic items.

4.3 A company has identified the following criteria for selecting its suppliers: product cost (PC), transportation cost (TC), performance awards (PA), product quality (PQ), supplier attitude (SA), supplier reliability (SR), supplier experience (SE), and lead times (LT). Apply the AHP to determine the weights of each criterion on the basis of the pairwise comparative values reported in Table 4.5.

4.4 [Alphen.xlsx] Alphen is an Austrian company producing yoghurt. In order to produce a new product for coeliacs, the company has to select a supplier of goat's milk (which does not contain gluten). The company has decided to consider two selection criteria: the price and the reliability of the potential supplier.

Table 4.5 Values associated with a pairwise comparison of the selection criteria indicated in Problem 4.3.

	PC	TC	PA	PQ	SA	SR	SE	LT
PC	1	5	1/5	1/8	1	1/7	1/5	1/3
TC	1/5	1	1/7	1/8	1/5	1/6	1/9	1/9
PA	5	7	1	1/5	1	1/3	1/4	1/6
PQ	8	8	5	1	1	1	1/2	1
SA	1	5	1	1	1	1/2	1/3	1/6
SR	7	6	3	1	2	1	1	1/3
SE	5	9	4	2	3	1	1	1
LT	3	9	6	1	6	3	1	1

Table 4.6 Potential suppliers and absolute judgements on the two criteria adopted by Alphen.

Supplier	Price	Reliability
1	4.0	1.8
2	5.0	1.5
3	5.5	1.2
4	3.8	1.9
5	5.2	1.6
6	4.8	1.7

After a brief market analysis, the candidates have been identified and ratings for each of the two criteria have been computed (see Table 4.6). The supplier selection has been conducted by using two different methods, which are variants of the weighted scoring method: the *absolute* weighted scoring method (AWSM) and the *relative* weighted scoring method (RWSM). In AWSM, the score of a supplier is given by considering the judgement value expressed on each criterion independently from the values given for the same judgement to the other suppliers. In RWSM, the judgement value on a criterion depends on the worst value assigned the other suppliers for the same criterion. The company has decided to give a 40% ($\alpha = 0.4$) weight to the price criterion and a 60% (i.e., $1-\alpha$) weight to the reliability. Determine the supplier to be selected by using AWSM and RWSM. Discuss how the solution changes for different values of parameter α.

4.5 [ValueEquation.xlsx] To select suppliers, some companies adopt the so-called *value equation*, generally expressed as the ratio between supplier performance and price offered. Typically, the performance represents all non-price factors (i.e., quality of service, technical competence, experience, and willingness to agree to the contractual terms proposed) and its value is selected on a scale from 1 to 100 (the higher the number, the better the performance). The price represents the score given to the supplier pricing proposal, on a scale of 1 to 100 (the lower the price, the higher the score). On the basis of the list of potential suppliers (and corresponding performance and price) given in Table 4.7, select the best supplier by using the value equation. What can you conclude about this method for selecting suppliers?

4.6 [Suntech.xlsx] In the Suntech Solar problem, assume that you know the cumulative evaluation, given by different experts, of the five suppliers for each criterion indicated in Figure 4.2. The scores are reported in Table 4.8. Apply the WEIGTHED_SCORING procedure to select the best supplier, considering the weight of each criterion as that already computed by the AHP.

4.7 [Skyfly.xlsx] Skyfly is a helicopter manufacturer that has to evaluate three suppliers ($S1, S2, S3$) in order to choose the best solution for the procurement of custom-made radars. To this end, the company contacted two experts who, together with the purchase manager, listed the appropriate selection criteria and

Table 4.7 Performance and prices of the potential suppliers for the value equation method illustrated in Problem 4.5.

Supplier	Performance	Price
A	80	90
B	50	50
C	70	30
D	90	60
E	65	40
F	45	20
G	25	30
H	65	80
I	70	20

Table 4.8 Evaluation grid of five suppliers of photovoltaic cells for the Suntech Solar problem.

	Score				
Criterion	Supplier 1	Supplier 2	Supplier 3	Supplier 4	Supplier 5
Product conformity	8	5	4	6	4
Post-sale assistance	5	6	4	9	7
Technical capabilities	7	10	5	7	5
Product price	8	8	6	7	6
Financial information	7	5	5	6	9
Performance awards	8	5	5	7	8
Green packaging	9	6	6	8	6
Environmental certificates	5	9	7	7	4
Recycling	8	5	7	6	5
Resource consumption	6	9	8	6	5
Geographic position	7	6	7	5	6

their weights, and expressed their evaluation of each supplier for each criterion (see Table 4.9). Determine which is the best supplier to select to implement an SS policy, using the WEIGHTED_SCORING procedure. Note that the manager's opinion weighs 30% of the total evaluation, while the same importance has to be given to each of the experts' opinions.

4.8 Assume there are n commodities and let m be the number of potential suppliers; let $p_{ij}, i = 1, \dots, m, j = 1, \dots, n$, be the price required by supplier i for commodity j and $c_i, i = 1, \dots, m$, the fixed cost which is applied only if the supplier i is

Table 4.9 Evaluation criteria, weights and scores of three Skyfly suppliers of custom-made radars.

Criterion	Weight	Manager S1	S2	S3	Expert 1 S1	S2	S3	Expert 2 S1	S2	S3
Price	0.15	6	7	9	8	8	5	7	8	4
Quality	0.20	5	8	7	7	6	7	7	8	5
Technology	0.25	7	8	7	5	8	7	6	7	5
Flexibility	0.10	6	7	7	5	9	4	4	8	8
Guarantee terms	0.18	7	9	9	4	10	5	5	8	10
Recycling	0.12	6	7	6	5	7	7	6	7	5

selected (for the provision of one commodity at least). Define a mathematical model for the supplier selection problem.

4.9 [Ilax.xlsx] Perform a sensitivity analysis on parameter $\alpha \in [0, 1]$ in the Ilax problem.

4.10 Modify the supplier selection model (4.1)–(4.4) for the multi-commodity case.

4.11 [Ilax.xlsx] In the Ilax problem, assume that the supply of quartz sand for the next month should be planned on a weekly basis. Furthermore, assume that the capacity and the unit selling cost vary on a weekly basis, according to the values reported in Table 4.10. Formulate the multi-period version of the Ilax supplier selection problem, and determine an optimal solution by using the ϵ-constraint method (with the same value of ϵ as the one used in the original Ilax problem). How does the new solution differ from the previous one?

4.12 [Visioncare.xlsx] Visioncare is a company that produces semi-finished lenses for prescription glasses. The purchase of polycarbonate, a durable resin-based material, takes place once every three months. The portfolio of possible suppliers consists of six companies that have already been evaluated on the basis of various criteria using the AHP. Since polycarbonate is a bottleneck item, the company has also defined a supply risk parameter which takes into account the

Table 4.10 Weekly capacity and unit selling price for the Ilax suppliers.

Supplier	Week 1 Capacity [tonnes]	Price [$]	Week 2 Capacity [tonnes]	Price [$]	Week 3 Capacity [tonnes]	Price [$]	Week 4 Capacity [tonnes]	Price [$]
1	830	791	860	793	910	789	900	788
2	800	770	1150	765	1150	765	900	769
3	750	765	680	772	950	756	1620	750
4	650	800	795	798	1110	775	1445	765
5	600	799	450	806	700	805	750	803

Table 4.11 Score, supply risk, capacity, and unit selling price of the Visioncare suppliers.

Supplier	Score	Supply risk	Capacity [kg]	Unit selling price [€/kg]
1	8.27	6.00	2500	45
2	8.12	4.50	2000	50
3	7.89	3.20	1000	47
4	8.25	6.50	2500	53
5	8.59	3.70	1500	51
6	7.78	5.00	2000	57

lead times of the purchase history and the reliability of the supplier. The quantity required for the next three months is 4500 kg. The scores, the supply risk parameters, the availability quantities, and the unit selling prices for each supplier are reported in Table 4.11. Formulate and solve a multi-objective supplier selection problem considering simultaneously the minimization of the total cost, the maximization of the total score, and the minimization of the procurement risk.

4.13 An optimization model, suitable for the selection of suppliers of a single commodity, can be defined if the following data are available. Let T be the number of time periods in the planning horizon considered for the provision of a commodity; $t = 1, \ldots, T$, indicates the time period in which a provision takes place. Let m be the number of potential suppliers; r_i, $i = 1, \ldots, m$, is the score of the supplier i; d is the total quantity required from the suppliers (total demand) to be supplied during the planning horizon; o_{it}, $i = 1, \ldots, m, t = 1, \ldots, T$, is the minimum quantity to be possibly ordered from the supplier i at time period t; O_{it}, $i = 1, \ldots, m, t = 1, \ldots, T$, is the maximum quantity that can be possibly ordered from the supplier i at time period t; $c_{it}, i = 1, \ldots, m, t = 1, \ldots, T$, is the unit purchase cost of supplier i at time period t; f_i is the fixed order cost for each supplier; n is the maximum number of suppliers to be activated; b is the available budget for purchasing the quantity d required during the planning horizon, and α ($0 \leq \alpha \leq 1$) is a tolerance measure, corresponding to the percentage of the ordered quantity that can be delivered late, after the end of the planning horizon, and $\alpha_i, i = 1, \ldots, m$, is the average percentage of ordered quantity delivered late by supplier i. Formulate the supplier selection model. (Hint: make use of the decision variables: $x_{it}, i = 1, \ldots, m, t = 1, \ldots, T$, representing the quantity of commodity replenished by supplier i at time period t and of the following binary decision variables: $y_i, i = 1, \ldots, m$, equal to 1 if supplier i is selected, and 0 otherwise, and $z_{it}, i = 1, \ldots, m, t = 1, \ldots, T$, equal to 1 if supplier i is chosen at time period t, and 0 otherwise.)

4.14 [SaintGold.xlsx] Use the supplier selection model of Problem 4.13 to solve the following problem. Saint-Gold is a British company specializing in the production of ceiling fans. For the production of the Airmax model, the company needs a specific numerical controller. There are four potential suppliers whose scores, determined using the weighted scoring method, are 6.97, 6.75, 6.31, and 6.92, respectively. During the negotiation phase, each supplier has imposed to

Table 4.12 Minimum and maximum daily supply quantities, percentage of late deliveries, and fixed order costs of the four suppliers in the Saint-Gold problem.

Supplier	Minimum supply [items]	Maximum supply [items]	Late deliveries [%]	Fixed order cost [£]
1	200	3000	3.50	25
2	300	3600	2.80	35
3	300	3000	4.30	40
4	150	4800	2.50	30

Table 4.13 Unit prices (in £) of the numerical controller offered by the four suppliers in the Saint-Gold problem.

	Day					
Supplier	1	2	3	4	5	6
1	0.23	0.22	0.23	0.22	0.19	0.24
2	0.21	0.24	0.22	0.18	0.21	0.23
3	0.19	0.16	0.20	0.17	0.18	0.19
4	0.18	0.14	0.17	0.19	0.20	0.24

the company a daily minimum and maximum quantity (number of items) to be ordered, as well as the fixed order cost. These values are shown in the second and third columns of Table 4.12. The planning horizon for the supply of 5000 controllers is equal to a six-day working week. Two suppliers have to be selected. The suppliers have presented an offer to the company which indicates the unit price (inclusive of the transportation cost) shown in Table 4.13. Analysing the historical data on the suppliers' efficiency, the manager of Saint-Gold has computed the average rates of the delivery late for each potential supplier (see the fourth column of Table 4.12). The management of Saint-Gold does not tolerate late deliveries for more than 3.5% of products during the planning period. The total budget available for the weekly purchase of the cards is £ 1500.

4.15 Learn more about the key features of a well-known SRM software integrated in an ERP system (e.g., SAP SRM, integrated with the materials management (MM) module in S/4HANA, see Section 1.12) to achieve the end-to-end procurement business process.

5

Managing a Warehouse

5.1 Introduction

Inventories are stockpiles of goods waiting to be manufactured, transported, cleared by customs, or sold. They include:

- components and semi-finished products waiting to be manufactured or assembled in a plant;
- merchandise (raw material, components, finished products) transported through the supply chain (*in-transit inventory*);
- finished products stocked in distribution warehouses and retail points prior to being sold;
- finished products stored by end users (consumers or industrial users) to satisfy their future needs.

Warehousing refers collectively to the activities involving storage and handling of inventories in a systematic and orderly manner in a warehouse. Warehouses are facilities providing protection to goods against atmospheric agents, spoilage, robberies, etc. They may also perform additional functions such as consolidation, deconsolidation, and sorting (see Section 5.1.1). The merchandise may belong to a unique company or to a variety of firms (as it happens in DCs and in EFCs). See Section 5.2.1 for more details.

Warehousing also takes place in port, air cargo, rail, and road freight terminals (see Section 6.3) as well as in intermodal logistics centres (i.e., *freight villages*), where goods (raw materials, components, and finished products) may be held for a few days before being transferred from one mode of transportation to another. The storage capacity of such facilities may be large: e.g., the port of Gioia Tauro, Italy, one of the largest container terminals on the Mediterranean, has a storage zone of $1\,500\,000$ m^2. Freight forwarders use part of this capacity as warehouses to manage their freight.

5.1.1 Warehouse Operations

The fundamental role played by a warehouse (see Figure 5.1) is to receive shipments from an external supplier or from a production line, store them for subsequent retrieval (inbound operations); then, in response to customer orders, recover and ship products to customers (outbound operations).

Introduction to Logistics Systems Management: With Microsoft® Excel® and Python® Examples, Third Edition. Gianpaolo Ghiani, Gilbert Laporte, and Roberto Musmanno.
© 2022 John Wiley & Sons Ltd. Published 2022 by John Wiley & Sons Ltd.

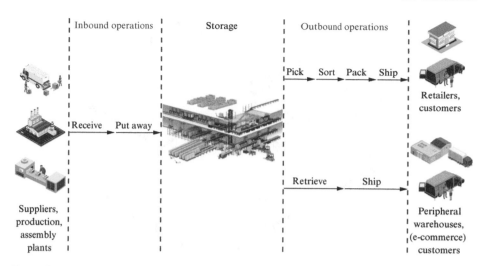

Figure 5.1 Warehouse inbound and outbound operations.

In addition, some warehouses perform deconsolidation and consolidation tasks. *Deconsolidation* means that the merchandise arrives packaged on a larger scale (e.g., as a palletized unit load, see Section 5.4.4) and it is broken down into several smaller unit loads (e.g., cases, see Section 5.4.3). Deconsolidation can be implemented when a new load arrives at the warehouse or at picking time.

Consolidation is the reverse process of deconsolidation. Multiple unit loads retrieved from the storage zone are packaged together before shipping to form unit loads of a higher dimension (e.g., palletized unit loads).

The main warehouse operations are detailed in the following.

- *Receive.* Freight is unloaded from trucks (or rail cars, etc.) at the receiving docks or rail sidings. Quantities are verified and quality checks are performed (in general, randomly). Then, a label is attached to each unit load (see Section 5.7).
- *Put away.* Unit loads are moved to a storage location. If the unit loads for internal use (e.g., cases) differ from the incoming unit loads (e.g., palletized unit loads), then the incoming unit loads are disassembled at this stage. Alternatively, unit loads may be deconsolidated at picking time (see below).
- *Storage.* Storage may last for years (e.g., in humanitarian logistics, see Section 1.5.2), months (e.g., for agricultural seasonal products), weeks (for fast-moving products) or a few hours (in crossdocking, see Section 5.2).
- *Retrieve/pick.* An order lists the products and quantities requested by the subsequent nodes of the logistics system (e.g., a retailer or an assembly station in the case of a production warehouse). The process of retrieving an order is called *order picking*. For small items, a single picking operation may collect items of more than one order (*batch picking*). There are fundamentally two types of order-picking methods: in *picker-to-parts* systems, the order picker either walks or rides a vehicle to the picking location; in *parts-to-picker* systems, material handling equipment brings the requested unit load to the order picker. Picker-to-parts processes can be made easier by a number of technological advances, including *pick-to-light* systems, which indicate to picking operators which unit loads to pick by shining a light on them, and

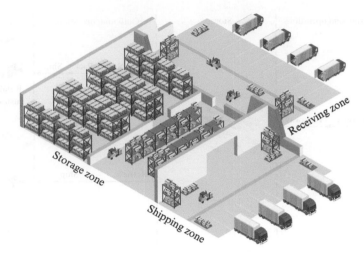

Figure 5.2 Main functional zones of a warehouse.

voice picking systems that let operators know which unit loads to pick via headsets. See Section 5.6 for more details. Conversely, the retrieval process concerns the entire unit loads that are picked from the storage zone and shipped directly as they are.

- *Sort*. When an order contains multiple items, they must be accumulated and sorted before being moved to the shipping zone or to the production floor. These two operations may be carried out during or after the order-picking process. To this end, specialized automatic *sortation systems* may be used.
- *Pack*. Once an order has been taken to the shipping zone, and separated from other orders, a check for accuracy is performed and protective packing (such as polystyrene, air-filled bags, shredded paper, etc.) may be added.
- *Ship*. Personnel may perform quality control and finally outgoing merchandise is loaded onto a vehicle.

5.1.2 Warehouse Functional Zones

A warehouse can be generally divided into three main functional zones (see Figure 5.2): a receiving zone, a storage zone and a shipping zone.

Receiving zone

In the *receiving zone* incoming unit loads are received and temporarily held. Empty pallets and containers can also be held. Freight unloading operations can be carried out outside or inside the receiving zone. In the latter case, *loading and unloading bays* (also called *docks* or *gates*) are integrated in the receiving zone.

Most bays in a warehouse are built 1.2 m above the ground to ease the loading and unloading of commercial vehicles, whose loading compartment is 0.5 m above the ground for most vans and 1.4 m for most trucks. The doors of the bays are generally 3.0 m high and 2.7 m wide. The distance between two adjacent doors is often 1.5 m. The doors are often equipped with insulating panels, a porthole for external control or more portholes for the passage of light, and are light-coloured to avoid heat accumulation.

In the receiving zone it is generally possible to identify a *staging zone*, in which the following activities are carried out:

- quantitative and qualitative controls of the incoming goods. The control level depends on the type of incoming goods (e.g., pharmaceutical products are subject to severe checks according to specific guidelines satisfying local regulatory requirements). In the event that some goods do not meet the required standards, they are temporarily stored in a *non-conforming material subzone* and then picked up later to be returned to the sender;
- deconsolidation and repackaging of the incoming goods in case they are of different size with respect to the unit loads handled by the storage system.

Storage zone

The *storage zone* is the place where the items are stored using appropriate equipment (see Section 5.5) depending on the type, size, and demand for the products. The storage space can be exploited in different ways, either horizontally and vertically, but always considering the characteristics of the stored goods. In some warehouses, it is subdivided into a *static storage zone* (also referred to as *reserve storage zone*) and a *dynamic storage zone* (also called *picking zone* or *forward picking zone*). The former is where the majority of items are stored, typically on palletized unit loads. The latter is an easy-to-access area from which items are picked for order fulfilment. Once the picking zone runs out of a product, it is replenished from the reserve storage zone. The existence of a picking zone means that, starting from incoming full unit loads in the warehouse, mixed unit loads can be formed for shipping.

Shipping zone

The *shipping zone* is generally divided into two subzones. The first one is a staging zone in which the picked unit loads are consolidated and packed. Control of the goods is also carried out in this zone, to avoid errors in the fulfilled orders. In the second one goods are placed in such a way to be loaded onto the vehicles, and transportation documents, conveying information about the cargo being shipped, are prepared.

Other zones

Other warehouse zones are

- a maintenance zone, for periodic inspection and repair of the material handling equipment used in the warehouse; material handling equipment includes either the storage systems and the systems used for moving the loads inside the warehouse (*internal transportation equipment*);
- an administrative zone that hosts the warehouse management offices, including customer service;
- a service zone, including changing rooms and an infirmary.

The above mentioned zones characterize most warehouses. However, they may not all be present in every warehouse (see Section 5.2). Some types of warehouses may include zones where other specialized activities (e.g., customs or health controls) are carried out.

5.1.3 Advantages of Warehousing

While holding stocks may be a necessary evil (see Section 1.6.1) in some circumstances (this is the case of in-transit inventory), there are several reasons why warehousing can be beneficial:

- *Reducing overall logistics cost.* Freight transportation is characterized by economies of scale because of high fixed costs (see Chapter 6). As a result, rather than frequently delivering small orders to a multitude of customers over long distances, a company may find it more convenient to ship large amount of merchandise to a warehouse, and satisfy customer demand from the warehouse (see Sections 3.2–3.3 for more details).
- *Improving service level.* Having a stock of finished goods in a warehouse close to a market yields shorter lead times and improves the overall service level (see Section 1.8.3).
- *Coping with randomness in customer demand and lead times.* Inventories of finished goods help satisfy customers in cases of unexpected peaks of demand or if delivery delays (due, e.g., to unfavourable weather or traffic conditions) occur.
- *Making seasonal products available all over the year.* Most crops are harvested during certain seasons, then stored in warehouses and sold throughout the year.
- *Smoothing the manufacturing of products with seasonal demand.* Some products (e.g., umbrellas) are demanded seasonally; in order to smooth out the ups and downs of production flows, these products are manufactured throughout the year and, in low-demand periods, unsold products are stocked in anticipation of demand peaks.
- *Feeding production lines.* To avoid downtime in manufacturing, sufficient stock of raw material and components needs to be kept in plants' warehouses.
- *Speculating on price patterns.* Products whose price varies greatly during the year (e.g., crude coffee) can be purchased when prices are low, then stored and finally sold when prices go up.
- *Overcoming inefficiencies in managing the logistics system.* Inventories may be used to overcome inefficiencies in managing the logistics system (e.g., a distribution company may hold a stock because it is unable to coordinate supply and demand).

5.2 Types of Warehouses

Warehouses can be categorized with respect to several criteria.

5.2.1 Classification with Respect to the Position in the Logistics System

With respect to the position in the logistics system, warehouses can be defined as follows.

Production warehouses

They store raw materials, semi-finished products and finished products.

Spare parts warehouses

They are used to stock consumables, tools and spare parts necessary for keeping operations running without significant disruption in manufacturing plants. They are also used in the service sector by rail companies, public transportation firms, military corps, as well as by companies providing MRO to airlines.

Distribution warehouses

They are used to store finished products near the end of a logistics system. In some cases, a unique warehouse serves the needs of production and distribution. Distribution warehouses, in turn, are classified as follows.

- *Central warehouses.* These are major warehouses which supply (almost exclusively) peripheral warehouses.
- *Peripheral warehouses.* These are smaller warehouses served by a central warehouse and supplying retailers (or customers). They are sometimes called *regional* or *local warehouses.*
- *Distribution centres.* A distribution centre houses goods from *multiple* manufacturers for a short period of time (typically just a few days) before they are sent to retailers. In a two-tier distribution system, there are CDCs and peripheral DCs, often called RDCs.
- *E-fulfilment centres.* An EFC is a large distribution centre operated by an e-commerce player, where products from a variety of e-commerce partners are stocked in anticipation of customer orders. EFCs eliminate the need for e-commerce manufacturers to maintain their own warehouse.

Amazon.com (see Section 1.6.5) operates over 50 automatic fulfilment centres worldwide. One of them, with an area of 80 000 m², is located in North Haven, USA, close to a parcel sortation centre in Wallingford, a last-mile Amazon logistics delivery station in Bristol and an Amazon air hub in Hartford.

The fulfilment centre in North Haven has 62 loading and unloading bays, stores over one million SKUs (see the definition in Section 5.4.2) with a throughput (see Section 5.8) over one million orders per day. The centre is operated in two 10-hour shifts per day, while four hours per day are devoted to maintenance. The centre employs 2500 full-time operators, each working four shifts per week.

The internal transportation system includes a robotic parts-to-picker system, over 10 miles of conveyors and sortation equipment, an automatic cubing, weighing and labelling equipment, as well as a trailer-loading system.

- *Crossdocking terminals.* A crossdocking terminal (or, simply, a crossdock) is a facility, increasingly common in retail distribution systems, where many transportation lines (typically, each from a single supplier) converge and the loads on inbound vehicles are moved directly onto outbound vehicles (typically, each supplying one or more sales points). This requires strong coordination of incoming and outgoing flows.

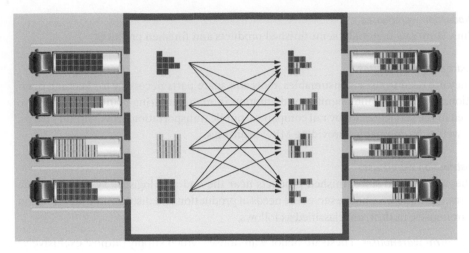

Figure 5.3   A crossdock.

Sometimes, incoming goods may stop for a few hours before they are sent to their destinations. In any case, goods are not accumulated in a storage zone (see Figure 5.3). *Transit point* is often used as a synonym of crossdock, even though, in this facility inbound and outbound vehicles are trucks (not vans).

Luís Simões is a major Portuguese logistics provider serving companies in Portugal and Spain. Its crossdocking platform located in Azambuja, at the outskirts of Lisbon, fulfils the needs of around 50 customers. The facility has a U-shaped layout (see Section 5.10.2) with 18 gates on a single side. In the crossdock, which has a total area of 20 000 m², materials are palletized and handled by forklift trucks (see Sections 5.4.4 and 5.6).

5.2.2   Classification with Respect to Ownership

With respect to ownership, there are four main types of warehouses.

Organization-owned warehouses
They require a capital investment in storage space, as well as in material handling equipment and personnel. This is usually the least-expensive solution in the long run in case of a sustained demand over a long period. Moreover, they are preferable when a high degree of control is required to ensure excellence in service level, or when specialized personnel and equipment are needed. Finally, they can be employed as a depot for the company's vehicles or as a base for a sales office.

Public warehouses
They are operated by firms providing warehousing services to other companies on a short-term basis. They are frequently found in locations such as ports, airports, and intermodal logistics centres. As a rule, public warehouses have standard equipment capable of handling and storing specific types of merchandise (e.g., bulk materials,

temperature-controlled goods, etc.). Costs are directly proportional to the area occupied in the warehouse, to the utilization time and to the services used. Hence, a public warehouse is the option of choice for companies with short-term warehousing needs (e.g., for accommodating seasonal inventories).

Leased warehouse space

This is an intermediate choice between the short-term space rental in a public warehouse and the long-term commitment of an organization-owned warehouse. It is sometimes the preferred solution whenever a reorganization of the distribution system is foreseen in the medium term.

Government warehouses

These warehouses are owned, managed, and controlled by central or local authorities. This is, e.g., the case of the Central Warehousing Corporation, a public warehouse operator established by the Government of India to support the agricultural sector. Government warehouses include *customs depots*, where goods imported from abroad are stored while they await the completion of regular customs operations. They also include *quarantine warehouses*, where products, whose customs duties are still unpaid, are stored.

5.2.3   Classification with Respect to Climate-control

Climate-controlled warehouses are most often used to store perishable products. They can be further classified as follows.

Ambient warehouses

They store goods in the range from 6 °C to 25 °C, sometimes in a humidity-controlled and well-ventilated environment.

Refrigerated warehouses

These warehouses allow the storage of temperature-sensitive products, including pharmaceuticals and food, between 0 °C and 5 °C.

Frozen warehouses

These warehouses allow the storage of frozen products, controlled by systems and procedures to guarantee that the temperature is kept between −50 °C to −1 °C at all times.

5.2.4   Classification with Respect to the Level of Automation

Automation is defined as a wide range of technologies aimed at substituting human labour. Warehousing includes several tasks that are process-oriented, repetitive, and prone to human errors. Hence, warehouses are a good fit for automation, especially if labour cost is high (see Sections 5.3 and 5.10). In the following, a non-exhaustive list of automation techniques is provided.

Product identification systems

Warehouse operations rely on accurate documentation to keep track of inventory levels, as well as to locate goods in the storage zone when a replenishment or a pickup is required. The use of product identification systems (see Section 5.7 for more details) allows automated or semi-automated data collection, and can save many working hours in the warehouse documentation processes.

Material handling

Material handling can range from manual to semi-automatic, to fully automatic. In a *manual system*, filling, lifting, carrying, and emptying are performed by human operators. In particular, goods are moved in and out of the storage zone by operators on foot. In a *semi-automatic system*, equipment is used to reduce the human effort. Often, goods are moved by operators riding a vehicle among storage locations. A wide variety of vehicles can be used to this purpose. Such material handling systems are labour-intensive: labour cost constitutes, on average, 50% of traditional warehousing cost. Alternatively, goods are deposited to, and retrieved from, a storage zone by a computer-controlled mechanism. An operator occupies a fixed position (*bay*) at the front of the zone. Upon request, the mechanism automatically inserts the unit load handed by the operator, or takes the requested unit load to the position of the picker (according to the parts-to-picker paradigm). See Section 5.6 for a review. Automation is also used in sortation, packaging (wrapping machines) and measuring (weighing platforms, etc.). With fully *automatic systems*, there is no need for human intervention and the entire warehouse is controlled and monitored through the use of software. These solutions are currently too expensive, or not technologically feasible, in most sectors.

By replacing a worker with an automated system, a company makes an initial capital investment instead of paying the salary of the unit of personnel for a number of years. On the flip side, automation is not very flexible: in fact, it can be very costly and time-consuming to adapt an automatic system to perform a task different from those for which it was designed originally. As of 2016, more than 10% of US warehouses were already using automated warehousing equipment. Worldwide sales of warehouse automation technology reached $ 2.4 billion in 2021.

5.3 Warehousing Costs

Warehousing consumes space, equipment and labour time, each of which is an expense. According to a 2018 survey by the consulting firm McKinsey, companies spend globally around € 300 billion a year on warehousing, managed either in-house or through 3PL providers.

As observed in Section 5.2.2, in public and government warehouses costs depend on the occupied area (or the number of unit loads stored), the utilization time and the services used (e.g., a *pick-and-pack* service). Public warehouse fees can be a combination of storage fees and inbound and outbound transaction fees. In 2017, the average annual fee for the storage of one palletized unit load was around $ 13 in the USA. Discounted storage solutions were commonly offered and the average discount was approximately 14.17% at 250 palletized unit loads. The average *pick-and-pack* fee for a B2C order with a single item was $ 2.64, while the average fee for a B2B order was $ 3.74.

In organization-owned warehouses, the main elements of cost can be classified into two broad categories: *investment costs* and *recurring costs*.

Warehousing investment costs

Setting up an organization-owned warehouse may need an initial *investment* corresponding to the financial resources to cover the following expenses:

- *Building cost.* This is given by the total needed building area multiplied by the cost per m². The former depends on the required warehouse capacity and the height of the storage zone. The latter is related to the country and the area. It also includes permit fees, architectural design, and engineering costs. As an order of magnitude, in Italy, a building of 500 m² of surface and 5 m high may cost about € 180–300 per m² plus the land cost.
- *Cost of the material handling systems.* This is incurred to acquire and install the material handling systems. As of 2021, the basic cost of a steel racking system (see Section 5.5.2) is around \$ 15–25 per palletized-unit-load slot. The range depends on the complexity of the structures and their height. The material handling system is expensive for automated workflows (between \$ 150 and \$ 200 per m²), when compared with manual handling of the loads (less than \$ 10 for m²).
- *Financial cost.* The financial cost includes the borrowing expenses associated with obtaining a business loan, if needed, to build or purchase assets.

Warehousing recurring costs

Recurring cost are the following expenses that an organization experiences at regular intervals to operate its business:

- *Labour cost.* This depends on the average salary of a warehouse operator as well as on the number of workers needed. The average salary depends primarily on the country (see Table 5.1). For instance, as of 2021, the annual labour expense for a warehouse with 100 employees is nearly \$ 2.9 million in the USA. Labour cost is also related to many other factors, including education, certifications, additional skills, and the number of years spent in the profession.
- *Maintenance cost.* This covers the maintenance of the building, which depends on the warehouse surface, and the maintenance of the material handling equipment, which depends on the technology adopted.

Table 5.1 Annual average labour expense for a warehouse worker in four countries (as of 2021).

Country	Average labour cost
USA [\$]	29 263
UK [£]	18 579
Italy [€]	24 000
India [\$]	2 660

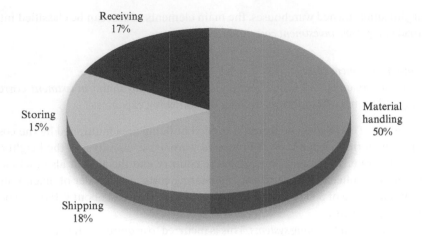

Figure 5.4 Typical breakdown of warehousing cost with respect to the fundamental warehousing activities.

- *Operating costs.* These comprise the cost of electricity (or fuel) to power the material handling system, the heating or refrigerating system, etc.
- *Overhead costs.* These expenses include administrative and general management costs.

Recurring costs may end up exceeding the cost of investment over the lifetime of a warehouse.

From another perspective, costs may be associated with the fundamental warehousing activities (receiving, storing, material handling, and shipping). See Figure 5.4 for a typical cost breakdown with respect to this criterion.

Inventory costs

In addition to warehousing costs, organizations incur inventory costs. They include the *opportunity cost*, represented by the *return on investment* that it would have realized if the financial resources had been better invested. Inventory costs are also related to the *risk* that products in stock become obsolete or damaged, and the lost income associated with a shortage of inventory (*stockout cost*). See Section 5.12 for an in-depth analysis of inventory management.

5.4 Unit Loads

This section first provides a definition of *bulk freight* and *general* (or *packaged*) *freight*. It then introduces the fundamental concepts on which the storage, handling, and transportation of packaged freight is based.

5.4.1 Freight Classification

Broadly speaking, freight can be classified into bulk freight and general freight. Bulk freight is shipped and stored loosely and unpacked in any quantity. It is typically poured as a liquid or solid into merchant ships, railway cars, tanker trucks, tanks, and silos. Bulk cargo can be classified as reported below.

- Liquid bulk freight, including petroleum, liquefied petroleum gas (LPG), liquefied natural gas (LNG), and chemicals.
- Dry bulk freight, comprising two categories: major bulks and minor bulks. Examples of the former, which account for nearly two-thirds of global dry bulk trade, are ores, coal, and grain. Minor bulks include steel products, sugar, cement, and cover the remaining one-third of global dry bulk trade.

Storing bulk freight is relatively simple from a managerial viewpoint and will not be discussed further in this chapter. On the other hand, general freight is transported, handled, and stocked in boxes, cases, barrels, etc., on palletized unit loads or in *intermodal containers*. The remainder of this chapter (except for Section 5.12) is devoted to warehouses specifically designed for such types of freight.

5.4.2 Unit Loads and Stock Keeping Units

Transportation, storage, and handling of packaged freight is based on two fundamental concepts: *unit load* and *stock-keeping unit*.

A unit load is a set of homogeneous (or heterogeneous) items assembled as a single unit, in order to facilitate transportation, storage, and handling. Examples of unit loads are secondary and tertiary packages of both consumer and industrial goods (Section 5.4.3), palletized unit loads (5.4.4), containerized unit loads (Section 5.4.5). Other remarkable examples are the *unit load devices* (ULDs), special pallets or containers used to load freight on wide-body airplanes.

In general, the smaller the unit load, the greater the handling cost. For instance, one palletized unit load of mineral water may contain 76 cases, for a total of 456 bottles. Moving these bottles individually is much more expensive than handling 76 cases which, in turn, is much more expensive than moving a single palletized unit load.

A *stock-keeping unit* (SKU) is a product characterized by a set of attributes (packaging, size, colour, etc.) that distinguishes it from other products or from the same product with different attributes. For instance, a one litre bottle and a six-bottle case of mineral water of a particular brand correspond to distinct SKUs.

A SKU is characterized by a unique code called the SKU *code* or just SKU (see Section 5.7.1) to track inventory level.

5.4.3 Packaging

A *primary package* comes into direct contact with the product. It may constitute a *sales unit* (e.g., a bottle of wine) or part of a sales unit (e.g., a blister for medication, see Figure 5.5(a)).

The main purpose of the primary package is to protect and contain the product. Sometimes it is also used to inform consumers and make a final attempt to convince them to buy the product. The characteristics (shape, colour, size, etc.) of a primary package are often chosen on the basis of marketing considerations regarding: the amount of product customarily purchased by consumers, visual appeal when placed on a retailer's shelf, practicality, ease of storage, recyclability, regulations of the industrial sector, etc.

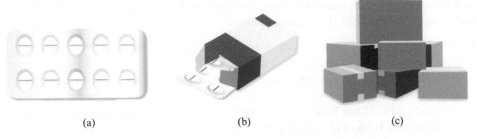

(a)   (b)   (c)

Figure 5.5 An example of (a) primary, (b) secondary and (c) tertiary packages.

A *secondary package*, often made of cardboard printed with well-thought-out branding and design, collates one or more primary packages to create a sales unit (or a unit load) for branding display and logistics purposes. In the latter case, it is often used to speed up shelf restocking in retail points and to display primary packages on shelves, especially in the beverage, food, and cosmetic sectors.

The number of primary packages in a secondary package and its layout depend on the distribution strategy. In this respect, it is worth noting that a secondary package is often used as a shipping container for small shipments (this solution is broadly adopted especially in e-commerce, see Section 1.7.2).

Examples of secondary packages are a carton containing six bottles of wine and a blister pack for medication (see Figure 5.5(b)).

> Marita is a family-run business based in Campi Salentina, Italy. The firm produces extra virgin olive oil which is distributed nationwide and abroad. The oil is bottled in a typical 330 ml glass bottle (primary package). Six bottles are packaged in cardboard boxes (secondary packages) with length 24 cm, width 16 cm, height 18 cm and a total weight of 2.9 kg. These boxes are sold through an e-commerce channel.

A *tertiary package* is used to group more secondary packages in a single unit load (see Figure 5.5(c)) with the aim of easing storage and handling in warehouses, as well as making shipping more efficient. This type of packaging is rarely seen by consumers.

5.4.4 Palletized Unit Loads

Tertiary packages are often arranged, secured, strapped, or fastened onto a portable platform made of wood, corrugated cardboard or plastic (called a *pallet*) so that the entire load is handled as a unit load named *palletized unit load* (or, more simply, *pallet*, thus not distinguishing the unit load from its flat base, see Figure 5.6).

Two-way pallets are only accessible to a forklift from two sides, while four-way pallets can be accessed from all four sides.

Palletized unit loads are the *de facto standard* for wholesale warehouses, DCs and crossdocks, where they are never broken apart. Palletized unit loads also constitute the most popular unit load in road transportation (see Chapter 6). Palletized unit

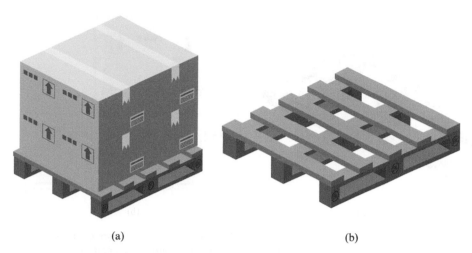

(a)   (b)

Figure 5.6 (a) Palletized unit load; (b) pallet.

loads allow space utilization in warehouses and vehicles to be maximized, as well as minimizing handling time through specialized equipment (see Section 5.6).

There are tens of different pallet sizes around the globe. The International Organization for Standardization (ISO) defines six pallet dimensions:

- 101.6 cm × 121.9 cm, this is the size of the most commonly used (by far) pallet in North America, established by the Grocery Manufacturers Association (GMA);
- 80 cm × 120 cm, this is the size of the most popular pallet in Europe, corresponding to the size of the EUR1-pallet (or Euro-pallet), regulated by the European Pallet Association (EPAL); this pallet is also known as the *EPAL-pallet* and will be referred as such in the following;
- 100 cm × 120 cm, the size of the EUR2-pallet, another size established by EPAL;
- 116.5 cm × 116.5 cm, commonly utilized in Australia;
- 106.7 cm × 106.7 cm, used in North America, Europe, Asia;
- 110 cm × 110 cm, utilized mainly in Asia.

With around 450–500 million units currently in circulation, the EPAL-pallets are exchanged according to the rules of the European Pallet Pool (EPP), a system in which truckers need to bring empty pallets to exchange when they pick up palletized unit loads; otherwise, a pallet-exchange fee is charged. The fee is determined periodically by the EPP. As of 2012, the fee has ranged from a minimum of € 7.07 to a maximum of € 7.98 for a single pallet. Similar mechanisms are implemented by various delivery companies and organizations around the world. For instance, the world's leading logistics company DHL implements the following policy: if a company owes pallets after the 20th day of the month, it will receive an invoice for an equivalent amount.

The EPAL-pallet is made up of nine blocks and three skids on the bottom as a support base. The three skids along the support base run parallel to the 120 cm edge. The top side is composed of five skids, running perpendicularly to the skids of the support base. The EPAL-pallets are normally handled sidewise by the short edge (see Figure 5.7(a)), since, when stored on pallet racks (see Section 5.5.2), the skids of the support base

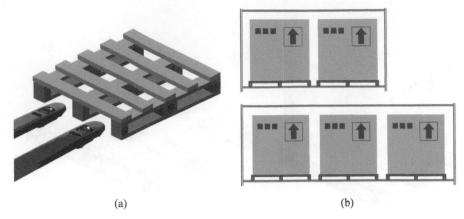

(a)   (b)

Figure 5.7 (a) An EPAL-pallet handled sidewise; (b) storage on a rack of two and three EPAL-palletized unit loads handled sidewise. The minimum length of the beam is, respectively, 182.5 cm and 270.0 cm, considering a minimum clearance between EPAL-pallets or between EPAL-pallet and frame of 7.5 cm.

stay perpendicular to the support beams (see Figure 5.7(b)). Only occasionally, they are handled by their wider edge. The EPAL-palletized unit loads may weigh up to 1500 kg when the load is well balanced or 1000 kg otherwise.

The *pallet loading problem* amounts to finding optimal layouts for loading rectangular packages on a pallet. The objective pursued is the maximization of the pallet surface utilization. Constraints may be related to load stability, total volume and weight. More details can be found in Section 5.15. In general, the size of the packages mounted on a pallet is chosen in such a way as to fit the pallet size. For example, EPAL-pallets are often used in combination with packages of 60 cm × 40 cm (or submultiples), so that the available surface can be completely filled. Such packages include the *euroboxes*, stackable plastic containers that are particularly useful for storing small products, such as components or spare parts in industry.

The use of EPAL-pallets has also made it possible to standardize truck compartments and intermodal containers (see Section 5.4.5). For example, a truck with a 13.2 m × 2.5 m × 2.7 m trailer compartment allows 33 EPAL-pallets to be loaded, arranged as shown in Figure 5.8.

The six-bottle packages produced by Marita are supplied to wholesalers as stretch-wrapped palletized unit loads, based on wooden EPAL-pallets, consisting of 35 secondary packages each, arranged in seven layers of five boxes covering the entire surface of the EPAL-pallet. The gross height of the EPAL-palletized unit load is $(7 \times 18\ cm + 14.4\ cm) = 140.4\ cm$, while the gross weight is $(35 \times 2.9\ kg + 25\ kg) = 126.5\ kg$ (the height and weight of the wooden EPAL-pallets are 14.4 cm and 25 kg, respectively). Note that the net weight (101.5 kg) is significantly less than the safe load of an EPAL-pallet.

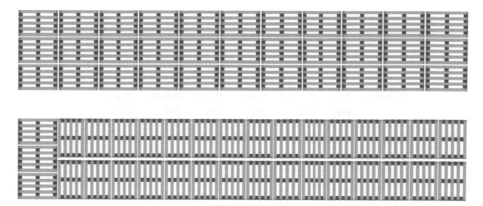

Figure 5.8 Two ways of loading 33 EPAL-pallets in a truck with a 13.2 m × 2.5 m × 2.7 m trailer compartment.

Figure 5.9 An intermodal container.

5.4.5 Containerized Unit Loads

A *containerized unit load* is composed of merchandise, possibly arranged in packages of different types and sizes, loaded into an *intermodal container*, or more simply a *container* (see Figure 5.9). A container is a large metal box (made from steel, aluminum, etc.) designed and built to be used across different modes of transportation, from ship to rail to truck without unloading and reloading their cargo (*intermodal freight transportation*, see Chapter 6).

Intermodal containers come in many types and a number of standardized sizes. General purpose containers are closed on all four lateral sides and can have doors at one or both ends. The most common are the 20 ft and 40 ft ISO standard containers. Their main features are summarized in Table 5.2.

It is worth noting that a 20 ft container can hold 10 GMA-pallets, or 11 EPAL-pallets, or 10 EUR2-pallets in one tier, while a 40 ft container can accommodate 20 GMA-pallets, or 24 EPAL-pallets, or 21 EUR2-pallets in one tier (see Figure 5.10).

Table 5.2 Main features of 20 ft and 40 ft ISO standard containers.

Container	Tare [kg]	Payload [kg]	External dimensions			Capacity [m³]
			Length [m]	Width [m]	Heigth [m]	
20 ft ISO	2300	25 400	6.10	2.44	2.59	32.6
40 ft ISO	3750	26 300	12.19	2.44	2.59	67.7

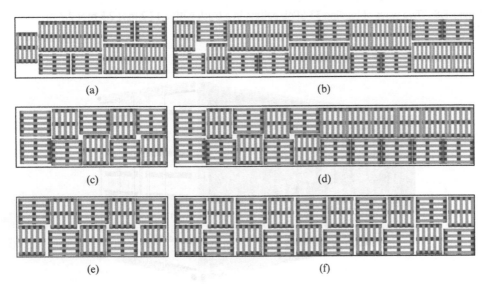

(a) (b)

(c) (d)

(e) (f)

Figure 5.10 Loading of one tier of a 20 ft ISO standard container with: (a) 11 EPAL-pallets; (c) 10 EUR2-pallets; (e) 10 GMA-pallets. In the case of a 40 ft ISO standard container, loading of one tier with: (b) 24 EPAL-pallets; (d) 21 EUR2-pallets; (f) 20 GMA-pallets.

Other common lengths of the containers are 48 ft (14.6 m) and 53 ft (16.2 m), while other common heights are 2.9 m (*hi-cube* containers) and 1.3 m (*half-height* containers).

The capacity of a 20 ft ISO standard container, corresponding to 1 TEU, is used as an inexact unit to describe the cargo capacity of container ships, cargo airplanes, and container terminals. According to this definition, the capacities of 40 ft, 48 ft, and 53 ft containers are 2, 2.4, and 2.65 TEU, respectively. The capacity of a 40 ft ISO standard container is also equal to one *forty-foot equivalent unit* (FEU for short), corresponding, clearly, to 2 TEU.

Apart from their dimensions, general purpose containers exist in many variations: refrigerated containers (for perishable goods), ventilated containers, temperature-controlled containers, tank containers (for bulk liquids), open-top, and open-side containers (for easy loading of heavy machinery), etc.

5.5 Storage Systems

Storage systems are used for maintaining stocks of unit loads over a period of time. The main storage systems are briefly illustrated below.

5.5.1 Block Stacking

In block stacking, unit loads are stacked directly on the floor, in blocks separated by aisles whose width depends on the type of internal transportation equipment used. On the plus side, warehouse costs are minimal because no storage equipment is needed. On the flip side, there is a lack of convenient access since only stock on the top of the stack can be retrieved immediately. Moreover, unit loads at the bottom must be able to hold the weight of unit loads above them (which is often the case of containers, see Figure 5.11). Block stacking is common when groups of unit loads have to be moved together, as it happens, e.g., in port yards and for seasonal products.

5.5.2 Pallet Racks

Racks are the most popular storage equipment. They are steel structures comprised of upright frames interconnected by horizontal beams. Between each upright frame, beams support individual unit loads in *storage bays* (or, simply, *bays*). Racks allow the storage of unit loads in horizontal rows with multiple tiers, as shown in Figure 5.12.

The type of internal transportation equipment restrains the width of the aisles as well as the height of the racks, and then determines the trade-off between storage density (see Section 5.8) and accessibility of the storage locations. In particular, the minimum aisle clearance and the maximum height of racks are 1.5 m and 40 m if material handling is performed by stacker cranes (see Figure 5.13 and Section 5.6.3); on the other hand, if the internal transportation is performed by forklift trucks (see Section 5.6) the maximum height of racks is much lower (7–8 m) and the width of the aisles must be

Figure 5.11 Block stacking of containers.

Figure 5.12   Pallet racks. Reproduced by permission of Mecalux.

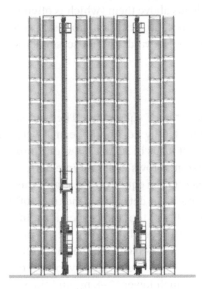

Figure 5.13   Racks with narrow aisles served by stacker cranes. Reproduced by permission of Mecalux.

greater (e.g., 2.5 m) to allow vehicles to manoeuvre (see Figure 5.14). The most common rack types are reviewed in the following.

Selective racks

Selective (or single-deep) racks allow each unit load to be directly accessed without moving other unit loads. Racks have a single-entry if they are fitted to the walls, or on either sides if they are in the middle (see Figure 5.15(a)).

Double-deep racks

Double-deep racks (see Figure 5.15(b)) allow one unit load to be stored behind another on each side of an aisle, thus increasing storage density. Direct access is only available to the first unit load, hence reducing the accessibility of the storage locations. This system requires internal transportation equipment with double-depth telescopic forks. This solution is recommended for storing multiple unit loads per SKU.

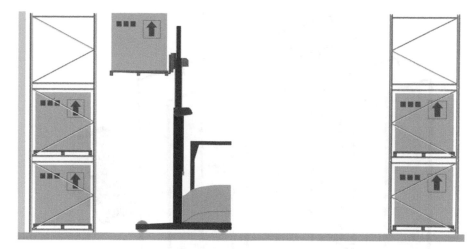

Figure 5.14 Racks with large aisles served by forklift trucks. Reproduced by permission of Mecalux.

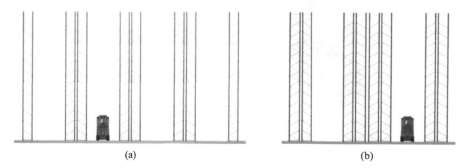

(a)   (b)

Figure 5.15 (a) Selective racks; (b) double-deep racks. Reproduced by permission of Mecalux.

Drive-through racks

In drive-through racks, loads are supported by rails attached to the upright beams. Forklift trucks are driven between the uprights beams (see Figure 5.16). Racks are open at both ends, allowing access from both sides. Hence, the storage zone can also be operated on a FIFO basis, which is common for perishables like food (see Figure 5.17(a)). Drive-through racks have a higher storage density than selective and double-deep racks. Products with the same SKU are often stored in the same lane.

Drive-in racks

Drive-in racks are the same as drive-through racks, except that they are closed at one end, allowing entry and exit only from one endpoint. As a result, the storage zone is operated on a *last-in, first-out* (LIFO) basis, a policy used for products to be held in storage for a long period of time (see Figure 5.17(b)).

Cantilever racks

In cantilever racks long and bulky loads (such as pipes, bar stock, lumber, etc.) are supported by cantilever arms (see Figure 5.18) at different heights. These racks are often found outdoors, in which case they are equipped with a gable roof for protection against bad weather.

Figure 5.16 Drive-through racks. Reproduced by permission of Mecalux.

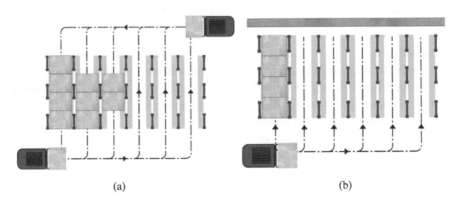

(a)   (b)

Figure 5.17 (a) Replenishment and retrieving operations for (a) drive-through racks (replenishment order: A, B, C, D; retrieving order: A, B, C, D) or (b) drive-in racks (replenishment order: A, B, C, D; retrieving order: D, C, B, A). Reproduced by permission of Mecalux.

Live pallet racks

Live pallet racks use slightly inclined roller beds on each tier to allow palletized unit loads to be moved more easily with the aid of gravity (see Figure 5.19). The palletized

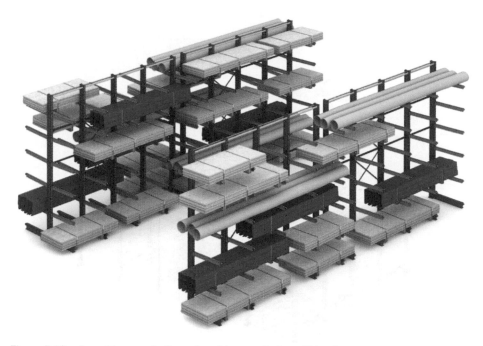

Figure 5.18   A cantilever rack. Reproduced by permission of Mecalux.

unit loads are inserted at one end and picked up at the other end in accordance with a FIFO policy. When a palletized unit load is taken out, the others move forward by one position, so there is always a palletized unit load at the front. Live pallet racks are particularly useful for storing perishable goods.

Push-back racks
Push-back racks use a single loading and unloading point for each counter-sloping storage channel at different heights. Palletized load units are stored by accumulation, pushing back the ones already stored, using sliding rails. Unlike live pallet racks, only one access aisle is required. The push-back racks allow a LIFO policy to be adopted for the storage zone.

Sliding racks
In sliding (or mobile) racks, a single mobile aisle is used to access several rows of racks. The location of the aisle is selected by sliding the racks along guide rails in the floor (see Figure 5.20). This is a relatively expensive storage system suitable for slow moving items. This is often used in refrigerated or frozen warehouses.

5.5.3   Shelves

Warehouse shelves are stretchers used to store lightweight inventory (a few hundred kilograms per shelf). Because they are not compatible with forklift trucks, they are generally manually replenished and picked. Apart from the traditional static version, shelves exist in a number of variants. Two of them, the live shelves (see Figure 5.21(a))

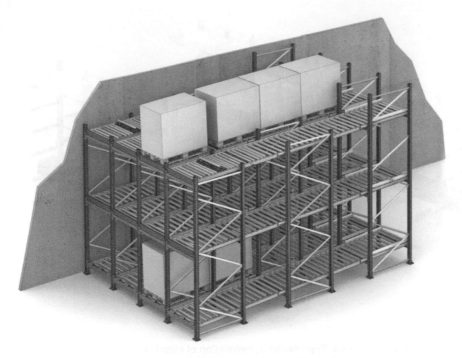

Figure 5.19 Live pallet racks. Reproduced by permission of Mecalux.

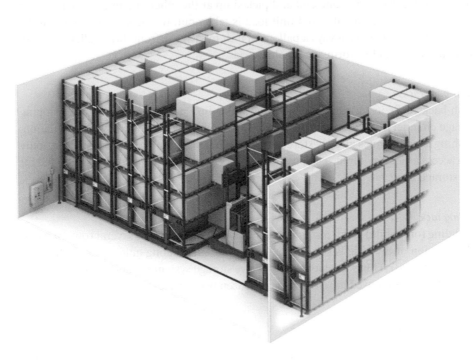

Figure 5.20 Sliding racks. Reproduced by permission of Mecalux.

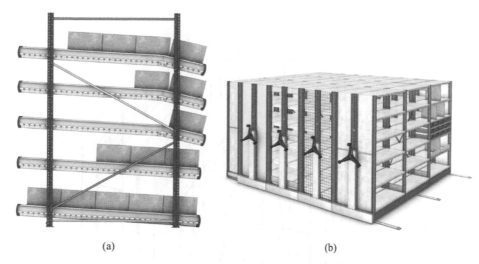

(a)   (b)

Figure 5.21   (a) Live shelves; (b) mobile shelves. Reproduced by permission of Mecalux.

and the mobile shelves (see Figure 5.21(b)) are based on the same principle as the live racks and the sliding racks, respectively.

Multi-level shelving

In multi-level shelving, additional platforms and stairs allow pickers to access the top positions (up to 20 m) of some warehouse shelves (see Figure 5.22). This is a relatively economical solution for increasing storage capacity of shelving systems without the need of extra floor surface.

5.5.4   Cabinet and Carousel Systems

These systems are used to store small parts and pieces of equipment (in particular, small spare parts) in drawers, trays, baskets, or bins.

Cabinets

Cabinets come in a number of variants, including bin storage cabinets, cabinets with vertical drawers, etc.

Storage carousels

Carousels are motorized systems made up of a set of vertically or horizontally revolving trays. A computerized system allows and item to be selected and picked and then commands the carousel to present an operator the right tray (see Figure 5.23). Carousels are an example of parts-to-picker systems in which items move to the operator. In this respect, they can be viewed as a combination of a storage system and internal transportation equipment. Carousels are characterized by reduced retrieval times. On the minus side, they have to be manually replenished (which is time consuming) and may be quite expensive.

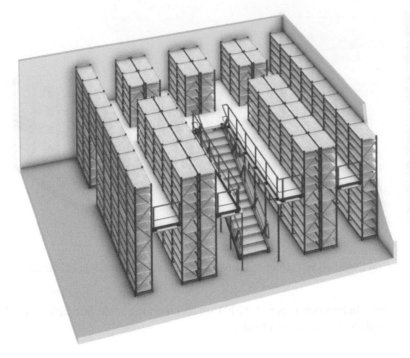

Figure 5.22 Multi-level shelving. Reproduced by permission of Mecalux.

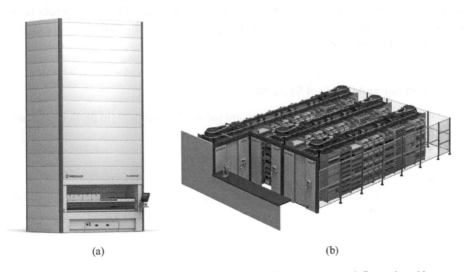

(a)   (b)

Figure 5.23 (a) Vertical storage carousel; (b) horizontal storage carousel. Reproduced by permission of Mecalux.

5.6 Internal Transportation Systems

The internal transportation system is an integral part of any production plant and warehouse. It allows short-distance movements of merchandise within a building or between adjacent buildings (e.g., between workplaces, between a production line and a storage zone, between a storage zone and a loading/unloading bay, etc.).

Internal transportation systems can be manual, semi-automated (if an operator carries out tasks such as driving, loading, and unloading), or fully automated.

5.6.1 Manual Handling and Non-autonomous Vehicles

Manual handling and non-autonomous vehicles are both based on the picker-to-parts principle. In *manual handling*, operators walk along the warehouse to replenish or retrieve inventories (usually lightweight items stored on ground or eye-level tiers).

Carts

In traditional picking, operators push a cart which may be left at the entrance of the aisles in the storage zone. Collected items are placed in the cart which is eventually carried to a staging zone. Carts, with one or more tiers, are often used to carry small items (e.g., fashion items), up to 200 kg, over few tens of metres.

Hand trucks

The *hand trucks*, usually equipped with two or four wheels, are used for easily handling unit loads within the storage zone (see Figure 5.24). The distance covered is of the order of a few tens of metres, while the transportable weight is approximately 200 kg.

Pallet trucks

Pallet trucks are vehicles equipped with forks used for moving one or more pallets at a time. They can be divided into *hand-pallet trucks* and *electric-pallet trucks*.

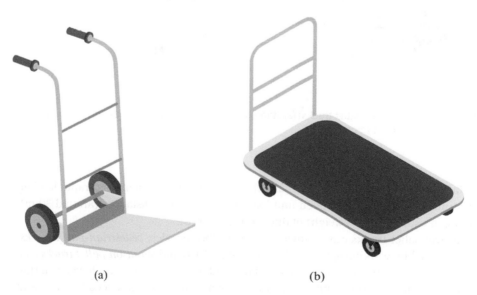

(a)   (b)

Figure 5.24 Hand trucks with (a) two or (b) four wheels.

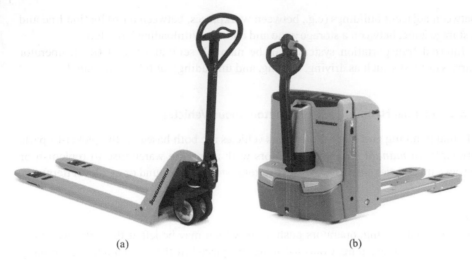

(a)   (b)

Figure 5.25   (a) Hand-pallet truck; (b) pedestrian-pallet truck. Reproduced by permission of Jungheinrich AG.

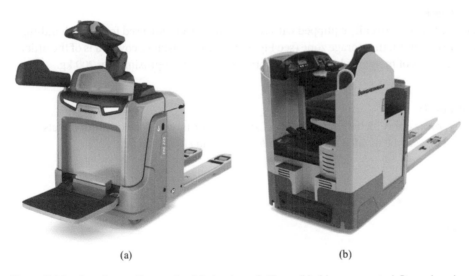

(a)   (b)

Figure 5.26   Stand-on pallet trucks: (a) stand on platform; (b) sideways-seated. Reproduced by permission of Jungheinrich AG.

Hand-pallet trucks (see Figure 5.25(a)) are suitable for frequent and rapid loading and unloading of light palletized unit loads over very-short distances (around 30 m). They can usually carry a weight of three tonnes at most.

Electric-pallet trucks can be divided in two subcategories: *pedestrian-pallet trucks*, manoeuvred by walking operators (see Figure 5.25(b)) and *stand-on pallet trucks* (see Figure 5.26), where operators are allowed to stand on a platform or a sideway seat (the former option being more appropriate in cases where the operators need to get on and off the trucks very often).

(a)   (b)

Figure 5.27   (a) Forklift truck; (b) reach truck. Reproduced by permission of Jungheinrich AG.

Pedestrian-pallet trucks are suitable for short distances (around 50 m) and loads up to three tonnes, whereas stand-on pallet trucks can be efficiently used on medium distances (around 100 m) and are able to handle loads up to four tonnes.

Forklift trucks

Forklift trucks are powered industrial trucks equipped with forks, with a counterweight in the rear of the vehicles (see Figure 5.27(a)) in order to balance the load that will be sustained by the forks. For this reason, a forklift truck is particularly suitable for lifting heavy loads (up to four tonnes) to medium heights (around 8 m).

Forklift trucks are used to load and unload vehicles from the ground and from a loading and unloading bay; they can move on different kinds of floor surface also at considerable slopes, and they can operate on different types of shelving. Moreover, they are typically electrically powered, especially in cases where they have small and compact dimensions, suitable for handling in a storage zone with narrow aisles, on flat floors, and indoors. Others are powered by LPG or diesel, and are more suitable to work outdoors on uneven surfaces. Depending on their characteristics, they can cover distances of a few hundred metres.

Reach trucks (see Figure 5.27(b)) are equipped with a lifting mast that can slide vertically. Their rear wheels, located below the operator, help to distribute the load, thus minimizing the necessity of compensation through counterweights. This allows operators to move in narrow aisles, to reach higher racking (up to 13 m) with a load capacity of three tonnes.

Order pickers (see Figure 5.28) are specifically designed for picking activities in a storage zone. They are often used to pick up items whose sizes and weights are small enough to be handled manually, and are capable of transporting several items in a single trip in the storage zone, since they are loaded in one or more transported containers (pallet, plastic box, roll container, etc.) up to a maximum weight of 1.5 tonne.

(a)   (b)

Figure 5.28   (a) Low-tier order picker, suitable for picking from the ground tier or from shelves with items stored at eye level; (b) medium-tier order picker, in which the operator can be lifted up to 12 m. Reproduced by permission of Jungheinrich AG.

5.6.2 Automated Guided Vehicles

AGVs are battery-powered, computer-controlled electric driverless vehicles used in warehouses, manufacturing plants, and cargo transportation terminals to move loads (see Figure 5.29).

An AGV-based internal transportation system is made up of a fleet of vehicles, a wireless communication subsystem, planning and control software, a user interface for monitoring and supervision, as well as battery chargers. AGV guidance technologies include, depending on the environment (e.g., indoors versus outdoors), lasers and floor-surface mounted magnetic tapes. The planning and control software relies on data (position, speed, status, etc.) coming from a variety of sensors. In particular, most AGVs include an obstacle detection sensing system for safety reasons.

AGVs can be classified as follows:

- *forklift AGVs* are designed to lift palletized unit loads;
- *piggyback AGVs* can carry palletized unit loads, boxes or even intermodal containers; unlike forklift AGVs, they cannot lift the loads directly but require additional equipment (e.g., lateral loading/unloading mechanisms) at the pick-up and drop-off stations;
- *towing AGVs* (also called *tugger AGVs*) tow a number of trailers (e.g., carts) behind them;
- *underride AGVs* can pass under carts and light shelves, lift them slightly and transport them to destination; compared to other AGV types, they require less space to manoeuvre (see Figure 5.30);
- *assembly line AGVs* carry semi-finished products on a path where components and parts are added until a finished product is eventually obtained;

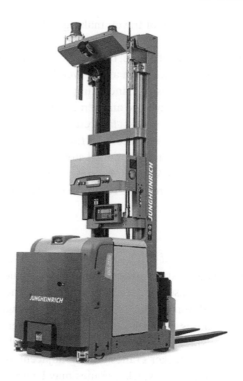

Figure 5.29   An AGV. Reproduced by permission of Jungheinrich AG.

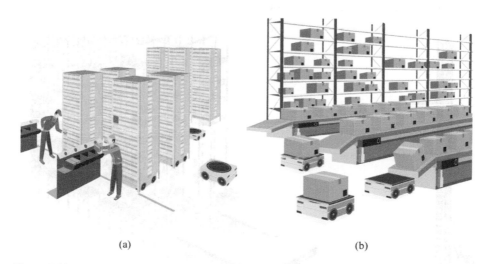

(a)   (b)

Figure 5.30   A fleet of underride AGVs: (a) moving shelves from a storage zone to packaging stations where operators pick specific SKUs out of them; (b) moving parcels out of a sortation system.

- *heavy load AGVs* are special vehicles that can handle heavy weight loads (e.g., rolls of paper or steel coils).

AGVs may have a speed around 1.5 m/s, an acceleration of the order of 0.5 m/s^2 and up to four driving wheels. Some of them can even crab, i.e., move sideways at angles of up to 90° from their main direction of travel, making optimal use of manoeuvring space.

Most AGVs carry a single unit load at a time, but there exist vehicles able to transport two or four palletized unit loads at once, with a total weight of up to four tonnes.

AGVs are a good fit for several industries, including those operating in fast-moving consumer goods markets. An AGV may have an average lifetime of up to 45 000 hours of operation, depending on use. This ensures a fast return on investment. Other advantages of AGVs include their ability to interface with the roller conveyors of palletizers (see Section 5.6.4), stretch wrappers, and automated storage and retrieval systems (see Section 5.6.3). On the minus side, AGVs often need a system built specifically for the customers' needs which requires tailored fine-tuning to optimize warehouse flows.

5.6.3 Stacker Cranes

Stacker cranes, or *storage and retrieval* (S/R) *machines*, are made up of a lifting device (called an elevator), travelling in narrow aisles equipped with guide rails, and a handling device constituted by a telescopic fork. They represent the building block of *automated storage and retrieval systems* (AS/RSs) in high-bay storage zones (see Figure 5.31). Typically, a stacker crane is dedicated to an aisle. However, in systems with a relatively low throughput, a stacker crane can serve several aisles through the use of a mechanism that transports stacker cranes among adjacent aisles. In contrast, if throughput is expected to be high, two stacker cranes may be used for each aisle: one stacker crane to input unit loads from one side of the aisle and a second crane to retrieve unit loads from the other side of the aisle.

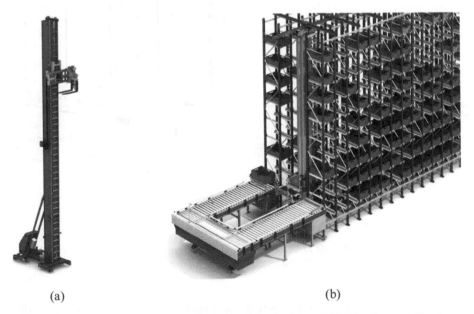

(a) (b)

Figure 5.31 (a) A stacker crane for palletized unit loads; (b) an AS/RS for drawer cabinets. Reproduced by permission of Mecalux.

The space utilization can be improved by storing two (or more) unit loads in depth, at the same front storage location (see double-deep racks in Section 5.5.2). This concept is taken to the extreme in *deep-lane AS/RSs* where up to 10 items are stored per single front storage location, one unit load behind the next.

Stacker cranes can reach heights of 45 m (the most common heights being between 15 m and 25 m) and handle loads up to seven tonnes. Horizontal travel speed is typically up to 3 m/s, while vertical speeds is up to 0.75 m/s.

Order picking can be implemented very efficiently by AS/RS: by applying the parts-to-picker principle, unit loads are automatically carried to an operator at an endpoint of an aisle. AS/RSs are suitable whenever land and labour are expensive and a high warehouse building can be constructed.

There are basically three different types of AS/RSs:

- *palletized-unit-load AS/RSs* handle palletized freight; a single stacker crane performs typically between 20 to 40 storage and retrieval cycles per hour;
- *mini-load container AS/RSs* store items in plastic containers or cartons;
- *mini-load tray AS/RSs* store plastic containers, cartons or items on captive trays; the advantage of these systems is the ability to store unit loads of different shapes and sizes; mini-load and mini-load tray systems are typically installed in buildings with height up to 15 m and with aisle lengths up to 60 m; typically, they can complete from 40 to 80 combined storage and retrieval cycles per hour.

5.6.4 Conveyors

Conveyors are pieces of internal transportation equipment which remain fixed while the loads move, creating a flow along pre-established routes. They come in a number of shapes and configurations including flat belt, roller, slat, chain, and trolley conveyors, to name a few.

Conveyors can move loads of any shape, size, and weight. They constitute one of the cheapest ways to move loads over long distances in a facility. Conveyors can be installed almost anywhere and can be interfaced easily with other internal transportation systems. On the minus side, they require substantial space. Moreover, they are fixed, meaning that a reconfiguration of the warehouse to accommodate increased demand can be difficult and costly to implement.

Conveyors can also be used for sortation within complex circuits to link different origin and destination points (see Figure 5.32). Such circuits are often made up of different sophisticated components including deflectors, push diverters, sliding shoe sorters, pop-up rollers, etc. Circuits are often equipped with intelligent tracking systems and checkpoint technologies (for monitoring unit load weight, dimensions, and condition), connected with a WMS (see Section 5.9).

Another particular type of conveyor, known as an *overhead conveyor*, is based on a track suspended at a certain height from the floor, inside which runs a chain that allows the movement. The track is made up of straight and curved sections to form a closed loop. The unit loads are hooked to the chain by means of linkages. A towing system allows the chain to slide inside the track and, consequently, the handling of the hooked unit loads. This internal transportation equipment is used to move odd-shaped parts, difficult to transport on a traditional belt system.

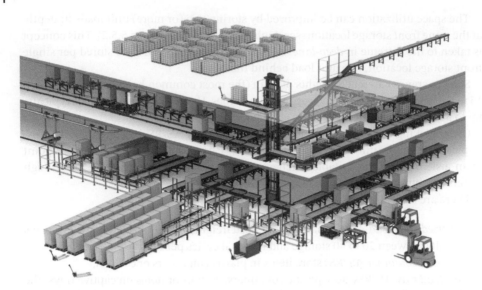

Figure 5.32 A circuit of fixed roller conveyors. Reproduced by permission of Mecalux.

5.7 Product Identification Systems

In every logistics system and, in particular, in every warehouse, it is essential to identify products quickly, precisely, and efficiently in order to update inventory levels, enable traceability, and optimize operations. In the following, the main product identification systems available on the market today will be briefly reviewed.

5.7.1 SKU Codes

A *SKU code* (or simply SKU) consists of a sequence of letters and numbers that identify a particular stock keeping unit. These codes are not standardized: when a company receives a shipment from a supplier, it may maintain the vendor's SKU code or create its own. All the items corresponding to a given SKU code have the same features (functionalities, style, size, colour, and usually, location in the warehouse or retail point) and are interchangeable.

The purpose of the SKU code is to help companies to control the inventory of every item more accurately and quickly.

BNBC is a Bolivian public company specializing in the distribution of beverages. The 150 ml WaterSure mineral water bottle is one of the most popular products. In the warehouse located in Tarija, the SKU code 1WSU-D215-24 corresponds to a cardboard package of 24 transparent PET bottles. The code can be interpreted as follows: 1 identifies the beverage (drinking water), WSU is an abbreviation for the brand name, D defines the amount of product (150 ml) in a bottle, 2 is for the packaging type (cardboard), 1 for the material (PET), and 5 for the colour (transparent); finally, the last two digits correspond to the number of primary packages (24 bottles).

SKU codes include the *Amazon standard identification number* (ASIN), a 10-character alphanumeric unique identifier given by Amazon.com and its partners to any product sold on the Amazon e-marketplace.

5.7.2 Global Trade Item Numbers

While SKU codes are organization-dependent, the *global trade item number* (GTIN) is a standard, global identifier developed by the non-profit organization GS1 that develops and maintains a system of international standards to share information among businesses all over the world.

Three types of GTIN codes are GTIN-12 (*universal product code*, UPC), GTIN-13 (*European article number*, EAN) and the *international standard book number* (ISBN) which serve different purposes. In particular, UPCs appear on almost every retail product in North America. They encode exactly 12 numerical digits and consist of a company prefix (six to nine digits long), a product number and a check digit. In EAN barcodes, which are popular in Europe, there is one more digit. Both UPC and EAN barcodes can be generated as follows: first, a unique company prefix must be obtained from GS1; then the organization can assign numbers to its own products making the length of the code equal to 11 digits for UPC or 12 digits for EAN; then, a final check digit is generated by using a check digit calculator to cope with any typing error in the code.

Another popular GS1 identifier is the *serial shipping container code* (SSCC) used to identify a logistic unit, i.e., any combination of load units put together (in a case, palletized unit load, containerized unit load, etc.). The SSCC enables a shipping unit to be tracked individually.

The GTINs can be encoded in a *barcode*, a *quick response* (QR) code, or an RFID *tag* or *smart label*, illustrated below.

5.7.3 Barcodes

Barcodes are identifiers consisting of an alternating sequence of vertical bars of different widths and spaces that can be read by special scanners (*barcode readers*, see below). Figure 5.33(a) depicts the SKU code, the UPC number and the UPC barcode of a secondary package.

Barcode readers are optical scanners used to read and decode the barcodes, and send the barcode's content to a computer. They use a light source that illuminates the surface of the barcode enabling a sensor to record the variations of the reflected ray, translating the optical impulses into electrical signals, sent to the decoder for further decoding processing.

The barcode readers used in logistics applications can be generally differentiated into two different categories: *laser scanners* or *LED scanners*. Laser scanners

Figure 5.33 (a) SKU code, UPC number and UPC barcode of a secondary package; (b) QR code.

Figure 5.34 (a) Laser barcode scanner; (b) omnidirectional barcode scanner.

(see Figure 5.34(a)) use a laser beam as the light source and they employ a reciprocating mirror or a rotating prism to scan the laser beam back and forth across the barcode. LED scanners use a row of hundreds of micro-light sensors, each of which is measuring the intensity of the front light. A voltage pattern coincident to the barcode pattern is generated by measuring the voltages of the micro-light sensors. The main technological difference between laser and LED scanners is the type of light used in the reading process: in laser scanners the light is generated at a specific frequency by the scanner itself, whereas in LED scanners it is the ambient light to illuminate the barcode.

Laser scanning is the technology employed in *omnidirectional barcode scanners* (see Figure 5.34(b)). They use a rotating polygonal mirror and several fixed mirrors to generate beams in different directions allowing barcodes to be read even on moving packages and at different angles. This allows greater accuracy and speed of scanning and a reduced sensitivity to the characteristics of the reading surface. For these reasons, omnidirectional scanners are more suitable for reading poorly printed or partially dam-aged barcodes. This type of scanner is one of the components of sorting conveyor belts used in warehouses and of checkout counters at retail points.

5.7.4 QR Codes

More recently, the QR code, a two-dimensional version of the barcode, made up of black and white pixel patterns, has become popular (see Figure 5.33(b)). Its main advantage relies on the fact that the QR codes can be read by hand-held devices equipped with internal cameras (e.g., smartphones).

5.7.5 Logistic Labels

A *logistic label* is one of the most common solutions to mark a package. It is a self-adhesive label composed of three parts (see Figure 5.35): the upper part contains information such as the name and address of the sender and the recipient; the central part contains the SSCC and additional information in human-readable form; the lower part contains the barcode representation of the data provided in the central part.

5.7.6 Radio-frequency Identification

RFID is a technology by which digital data encoded in RFID tags and smart labels are captured by a reader via radio waves. An RFID tag consists of an integrated circuit and an antenna (see Figure 5.36(a)). A wrap (generally made of plastic in logistics applica-tions) holds the components together and protects against damage caused by humidity, dust, etc.

An RFID tag can be active or passive. A passive tag is inexpensive (around $ 0.10) and requires no battery. A passive tag uses the electromagnetic energy emitted by a

Figure 5.35   An example of a logistic label.

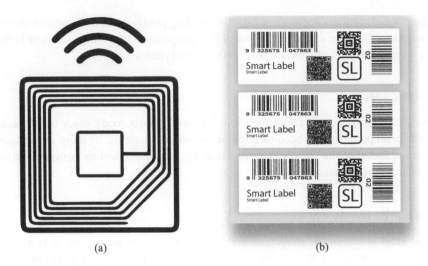

(a) (b)

Figure 5.36   (a) RFID tag; (b) Smart labels.

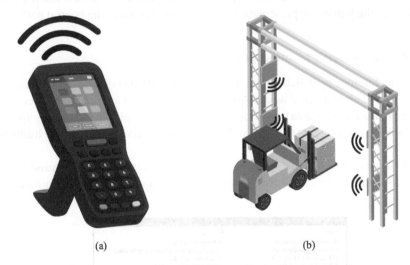

(a) (b)

Figure 5.37   (a) RFID hand-held reader; (b) RFID portal.

RFID reader to reflect back a signal to the reader with a unique ID serial number. The maximum read distance is usually under 1 m. An active tag is equipped with an internal battery and periodically transmits data (not only its ID serial number). It may cost up to $ 15–20 and have a maximum read distance of up to 100 m. RFID tags may be read-only, read-write, or write-once-read-many (WORM).

Since tags have factory-assigned serial numbers, an RFID reader can discriminate between several tags within the reading range and can read them simultaneously.

A smart label is an adhesive label incorporating both an RFID tag (usually a passive tag) and a barcode (see Figure 5.36(b)).

RFID readers can be mobile (see Figure 5.37(a)), hand-held or installed on vehicles, integrated with an antenna. Fixed RFID readers are often integrated in warehouse gates

Table 5.3 Comparison of barcodes and RFID tags.

Barcode	RFID tag
Read-only access mode.	Read-write access mode.
The barcode must be directly visible to the scanner.	The reader and tag do not need visual contact.
The reading of the barcodes is sequential: therefore it is possible to identify one product at a time.	One reader is able to communicate with hundreds of tags almost instantaneously.
The maximum reading distance is a few tens of centimetres.	The maximum reading distance is in the order of 1 m with passive tags and hundreds of metres with active tags.
The maximum quantity of information storable is 100 bytes.	Passive tags store 128 to 8 Kbytes of information; active tags can reach 32 Kbytes and more.
Scanners are extremely sensitive to lights, scratches and stain.	Readers are totally insensitive to dirt and light.
The reading phase requires predefined reading angles; the operation must be carried out at slow speed.	Tags can have any orientation and read-write operations can take while objects are moving.
There are tens of different standards used to code identification information.	Each tag has a unique code at world level. The uniqueness is guaranteed by the manufacturers of the chip.
No precaution is taken to secure information.	Information security access is guaranteed by cryptographic systems.
Duplication is extremely simple.	Duplication is practically impossible.
The cost is virtually zero.	The cost is still prohibitive for some applications.

or portals (see Figure 5.37(b)). Table 5.3 summarizes the main differences between barcodes and RFID tags.

5.8 Warehouse Performance Measures

This section describes a number of indicators measuring the performance of a warehouse. For a more general discussion on KPIs and performance measures in logistics, see Section 1.8.

Warehouse performance measures are used primarily to monitor the functioning of existing warehouses. They are also employed, as requirements, in the design process of new warehouses (see Section 5.10). In the following, such indicators will be described for a homogeneous storage zone where only unit loads of a unique type (generally palletized unit loads) are stored. The extension to the case of multiple-type unit loads is left to the reader.

To evaluate the exploitation of the storage space, the following *storage density* measures are used:

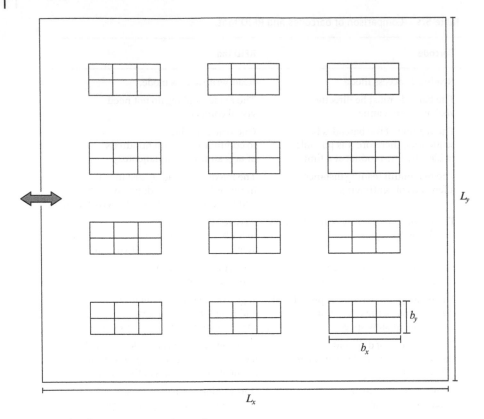

Figure 5.38 The Conkret warehouse layout.

- *surface utilization rate*, defined as the number of unit loads that can be stored per unit of storage surface;
- *volume utilization rate*, defined as the number of unit loads that can be stored per unit of storage volume.

The capability of the warehouse to handle loads is assessed by the following indicators:

- *capacity*, defined as the maximum number of storable unit loads;
- *throughput*, defined as the maximum number of unit loads that can be moved in a time unit (hour, day, week, etc.), either during storing or retrieving operations.

Conkret is an Austrian company that operates in the sector of building materials. Its main warehouse has a storage zone specifically dedicated to stocking EPAL-palletized unit loads of cement bags (see Figure 5.38). Storage is based on a stacking system composed of 12 blocks of 6 EPAL-palletized-unit-load stacks each. Each block has a length $b_x = 3 \times 1.2 \text{ m} + 2 \times 0.075 \text{ m} = 3.75 \text{ m}$, computed considering the 1.2 m length of each EPAL-palletized unit load and a clearance of 0.075 m between stacks in each block. Similarly, the width of each block is equal to $b_y = 2 \times 0.8 \text{ m} + 0.075 \text{ m} = 1.675 \text{ m}$. Along the vertical ($z$) direction, each stack houses $n_z = 5$ EPAL-palletized unit loads, each of which has a gross height of 1.25 m. The width w of the aisles separating the blocks in both directions is equal to 2.5 m. Consequently, the storage zone length can be computed as $L_x = 3b_x + 4w = 21.25$ m, while the width results to be $L_y = 4b_y + 5w = 19.20$ m. The capacity of the storage zone is equal to $12 \times 6 \times 5 = 360$ EPAL-palletized unit loads.

The storage zone is equipped with three forklift trucks. In order to compute the (hourly) throughput, it is necessary to first determine the average completion times (in seconds) of the storage and retrieval cycles, indicated as $T^{(s)}$ and $T^{(r)}$, respectively. In detail, $T^{(r)}$ is given by the sum of the durations of the following activities: empty forklift truck movement; ascent of the forks; EPAL-palletized unit load forking; descent of the forks; loaded forklift truck movement; deforking (for the sake of simplicity, any waiting time is considered negligible). $T^{(s)}$ can be determined with similar considerations. Both $T^{(r)}$ and $T^{(s)}$ depend on the average space covered by the forklift truck and its forks: $d = L_x + L_y/2 = 30.85$ m is the average distance covered by a vehicle in a round trip under the hypothesis that all the storage locations have the same probability of being accessed; $h = 2 \times 1.25$ m $= 2.50$ m is the average vertical movement of the forks which corresponds to the gross height of two EPAL-palletized unit loads (which is equivalent to picking or storing the third palletized unit load in a stack).

Considering parameters $v^{(u)}$, $v^{(l)}$, $v^{(au)}$, $v^{(al)}$, $v^{(dl)}$ and $t^{(f)}$ reported in Table 5.4, $T^{(r)}$ and $T^{(s)}$ are computed as

$$T^{(r)} = \frac{d/2}{v^{(u)}} + \frac{h}{v^{(au)}} + t^{(f)} + \frac{h}{v^{(dl)}} + \frac{d/2}{v^{(l)}} = 79.55 \text{ s},$$

$$T^{(s)} = \frac{d/2}{v^{(l)}} + \frac{h}{v^{(al)}} + t^{(f)} + \frac{h}{v^{(du)}} + \frac{d/2}{v^{(u)}} = 82.88 \text{ s}.$$

Table 5.4 Parameters used for computing the EPAL-palletized-unit-load picking and storing cycle times.

Parameter	Description	Value
$v^{(u)}$	Travel speed of an empty forklift truck	3.50 m/s
$v^{(l)}$	Travel speed of a loaded forklift truck	3.00 m/s
$v^{(au)}$	Unloaded fork ascending speed	0.50 m/s
$v^{(al)}$	Loaded fork ascending speed	0.30 m/s
$v^{(du)}$	Unloaded fork descending speed	0.50 m/s
$v^{(dl)}$	Loaded fork descending speed	0.50 m/s
$t^{(f)}$	Pallet forking and deforking time	60 s

It is worth noting that $t^{(f)}$ represents the time needed for forking of an EPAL-palletized unit load at the ground and deforking it on the stack in a storage operation (or vice versa in a retrieval operation). The hourly throughput of a single forklift truck can be computed as $3600/\overline{T} = 44.33$ EPAL-palletized unit loads per hour, where 3600 is the number of seconds in one hour and

$\overline{T} = (T^{(r)} + T^{(s)})/2 = 81.215$ s is the average of $T^{(r)}$ and $T^{(s)}$. Considering that three forklift trucks are used, the hourly throughput of the storage zone is equal to $3 \times 44.33 = 133$ EPAL-palletized unit loads per hour.

Other relevant warehouse performance indicators are the selectivity and access indices.

- The *selectivity index* is defined, for each storage location, as the ratio between the number of "useful" material handling operations and the total number of material handling operations required to access the unit load stored in that location. Hence, the selectivity index is between zero and one.

  In a selective racking system the selectivity index is one for each storage location, since only a single movement is required to pick a unit load. For a double-deep racking system, the storage location at the front in the working aisle has a selectivity index equal to one, while the storage location at the back has a selectivity index equal to 0.5 (in order to pick up the unit load at the back, it is necessary to pick up the unit load at the front and then the one at the back, resulting in a total of two movements, only one of which is useful). For pallet stacks, each palletized unit load location placed in the first row on the ground floor will have a selectivity index of $1/r$, where r is the number of pallets in the stacks (it will be necessary to first move the $r-1$ overlapping palletized unit loads to make the location available for the useful palletized unit load picking).

  A large selectivity index implies fewer picking operations (and hence lower material handling costs). On the other hand, a high selectivity corresponds to a low storage density.

- The *access index* is defined as the ratio of the throughput (in a given time period) and the capacity of the storage zone. It indicates the average frequency with which material handling operations are carried out per storage location. Storage zones with high access indices correspond, therefore, to highly dynamic warehouses, where the unit loads are frequently handled; vice versa, storage zones with a low access index correspond to more static warehouses, with less frequent material handling operations.

For its warehouse located in Tokyo, Isetan adopts a double-deep racking system, with two single-entry racks on either side fitted to the walls and five double-deep racks down the middle. The number of storage locations for each bay is three. Each rack is composed of 12 bays. The number of shelves is 10. This means that the number of storage locations for each rack is $(3 \times 12 \times 11) = 396$ (including the storage locations on the ground floor of the rack). Since the number of racks is $(2 + 4 \times 5) = 22$, the total storage locations (capacity) is $(396 \times 22) = 8712$. Only half of the storage locations of the 20 double-deep racks are immediately accessible, i.e., $(198 \times 20) = 3960$. Therefore, considering also the two single-entry racks, the selectivity index of the storage zone is $(3960 + 2 \times 396)/8712 = 0.55$. The weekly throughput is 43 000, resulting in an access index equal to $43\,000/8712 = 4.94$ in a week.

Among several issues also concerning the financial capacity to meet the economic investments required for the warehouse management, the selectivity index and the access index can be used to suggest the most appropriate material handling system.

If high selectivity is requested and the access index is also very high, a solution based on racking systems served by stacker cranes can be taken into consideration; in this case, racking allows an almost unit selectivity, whereas the use of stacker cranes enables the operator to deal with the high number of access requests to the storage locations.

On the other hand, when the access index is lower, but high selectivity is also desirable, more economical systems can be used. Therefore, systems based on racking are used to maintain a high selectivity, but are served by forklift trucks. Selectivity remains almost unit, since all the unit loads can be directly picked. A limited number of material handling operations can also be achieved with manual systems, which are less high-performing but more economical compared with automated systems. When the selectivity requested can be lower, but a high value of the access index is desired, systems such drive-through, drive-in, live pallet, push-back racks can be adopted, according to the policy (FIFO or LIFO or both) preferred in the material handling operations.

Finally, when both the selectivity and the access are low, a suitable solution is the adoption of stacks or, in the case of limited space available, of sliding racks.

Another fundamental performance measure is the inventory turnover index:

- the *inventory turnover index* is defined as the ratio of the value of the goods incoming/outgoing into/from the warehouse in a specific time period and the value of the average inventory level in the warehouse. The inventory turnover index is therefore an adimensional number which represents how many times inventory rotates in the considered time period.

> Inseko is a Ukrainian company producing building materials. In the last two years the Kiev warehouse has handled goods for a total value of € 28 980 000 and € 25 490 000, respectively. The average inventory level in the same period was equal to € 4 317 820 and € 4 174 530. Hence, the inventory turnover index during the two years was equal to 6.7 and 6.1, respectively.

Typical values of the inventory turnover index are included between 5 and 10 on an annual basis, however, a good warehouse performance could be also achieved with values of the inventory turnover index outside of the indicated interval, depending on the type of products stocked in the warehouse (see, as an example, Problem 5.3). The calculation of the inventory turnover index can also be done for a single product or for a category of products in the warehouse. In these cases, the inventory turnover index is no longer a measure of the efficiency of the whole warehouse, but remains useful to assess the adequacy of the stock management policies (see Section 5.12) adopted for a single product or a category of products.

The calculation of the inventory turnover index can be achieved in two ways since, besides the value of the products, the quantities can also be considered.

[Fri-X5-A.xlsx] Fri-X5-A is an additive in liquid form for concrete. The product, in iron drums of 200 kg, was among the goods present in the Inseko warehouse last year. Table 5.5 shows the information relative to the stocks of this product in the warehouse during the last year (composed of 10 time periods). Denoted by n_i and q_i, $i = 1, \ldots, 10$, respectively, the number of days and the inventory level of the Fri-X5-A additive for each time period i, the average inventory level of the product can be determined as

$$\bar{q} = \frac{\sum\limits_{i=1}^{10} n_i q_i}{\sum\limits_{i=1}^{10} n_i} = 1577.26.$$

The sales recorded during the last year were equal to 8596 drums. Consequently, the year inventory turnover index of the Fri-X5-A additive, calculated on the quantitative product basis, is equal to $8596/1577.26 = 5.45$.

Table 5.5 Inventory levels of the Fri-X5-A additive in the 10 periods of last year for the Inseko problem.

Period	Time interval	Days	Inventory [drums]
1	01/01 – 20/01	20	1000
2	21/01 – 07/03	46	2200
3	08/03 – 08/04	32	1400
4	09/04 – 05/07	88	800
5	06/07 – 10/07	5	600
6	11/07 – 01/09	53	2100
7	02/09 – 20/10	49	1600
8	21/10 – 22/10	2	600
9	23/10 – 29/12	68	2100
10	30/12 – 31/12	2	1300

In a warehouse characterized by the presence of different storage systems, the inventory turnover indices of the products can be used to establish the most appropriate storage system for each product. Combining the *ABC* classification (see Section 1.11.2) and the inventory turnover index, each product can be classified as of high, medium, or low turnover. The products with high turnover are typically assigned to selective racking systems, possibly served by stacker cranes, characterized by high accessibility and high selectivity; the products with medium turnover can be stocked in double-deep racks, whereas the products with low turnover can be stored in mobile racks or even in stacks.

5.9   Warehouse Management Systems

A WMS is a software system that helps manage, control, optimize, and execute the operations within a warehouse, in order to improve the accuracy and the efficiency of the entire system on a daily basis. To this end, it relies on the following information:

- the main features of each SKU (size, weight, expiration date, packaging, identification codes, etc.);
- the inventory level of each SKU and its locations in the storage zone;
- a description of the warehouse layout, including travel times between pairs of relevant points (loading and unloading bays, storage locations, etc.), the capacity of each storage location, storage restrictions related to hazardous materials, etc.;
- a representation of the internal transportation system (e.g., average number of SKUs per hour picked by an operator);
- arrival times of the inbound materials (from either production lines or external vendors);
- latest ship dates (e.g., the latest point in time that an outgoing vehicle has to be loaded, or a component at stock is provided to a production machine).

On the basis of this information, a WMS implements a number of *business analytics* methodologies:

- the estimation of the resources (operators, vehicles, etc.) needed to manage the inbound and outbound flows in each shift;
- the determination of the sequence in which incoming and outgoing vehicles have to be unloaded and loaded;
- the grouping of orders to be collected in batches (*bach picking*);
- the sequencing of order retrieval tasks;
- the assignment to each load of the most appropriate location within the storage zone;
- dynamic storage;
- the relocation of goods overnight by frequency of purchase.

WMS can incorporate different technologies such as RFID tags that can be used to identify individual items and to track them throughout the entire warehouse; voice picking systems providing operators easy-to-understand voice prompts to instruct them in picking tasks, etc.

Warehouse control systems

In most warehouses operations are run with a combination of automated and manual processes. Each automated piece of equipment (a carousel, an AGV, a stacker crane, a conveyor, etc.) is controlled by a *programmable logic controller* (PLC), a microprocessor-based device (a sort of industrial computer with no screen and keyboard) specifically designed to control complex machines with both sensors and actuators in real time. Sensors (e.g., optical sensors, magnetic sensors, laser scanners, cameras) capture perceptions of the environment and may be used for navigation (this is the case of AGVs), to detect and identify products and to ensure safety. *Actuators* (which include electric motors and valves) are responsible for moving a vehicle or a load, grabbing an object, lifting a part, etc.

A PLC allows to command an electromechanical piece of equipment by providing suitable instructions. For instance, an AGV standing in a given position A may receive the instruction "move along curve $\overarc{AB}$, up to B"; then, a PLC on board the AGV will be in charge to accurately regulate, instant by instant, the voltage/current going into the AGV motor, on the basis of the curvature of $\overarc{AB}$.

A WMS (which is focused on the business logics) and the PLCs of the automated devices (which may belong to different vendors) are integrated by an intermediate software layer called a *warehouse control system* (WCS). A WCS supervises real-time activities. For instance, in an AGV-based warehouse, the WMS may order to transport some loads from given storage locations to specified loading doors while the WCS will be in charge of defining precise *conflict-free* routes and schedules for the AGVs in order to optimize some performance measure (e.g., minimize the overall lateness with respect to some due dates).

A conflict-free solution regulates track contention (i.e., the usage of floor space) and prevents deadlocks (a state in which no AGV can progress because it is waiting for other AGVs to move). The problem is real time in nature for a number of reasons: travel times are not known exactly because of dirt on the warehouse floor or because of moving obstacles (e.g., operators) along the AGV routes, loads are released by machines at random instants, it may take longer than expected to palletize a load, etc. Any time a new update on uncertain information arrives, the WCS modifies its current plan and changes accordingly the commands for the PLCs of the various pieces of automated equipment. Other real-time decisions made by the WCS include the real-time selection of the home positions and dwell points where idle AGVs and stacker cranes should be relocated in anticipation of future requests.

In complex warehouses, there may be pieces of automated equipment from the different vendors. In such a case, an additional software layer, called *material flow control* (MFC), connects the devices from the same supplier (e.g., a number of AGVs). A *system integrator* uses the MFC applications to bring together the subsystems from different vendors and ensuring that they work together as expected. Figure 5.39 shows the architecture of a WMS.

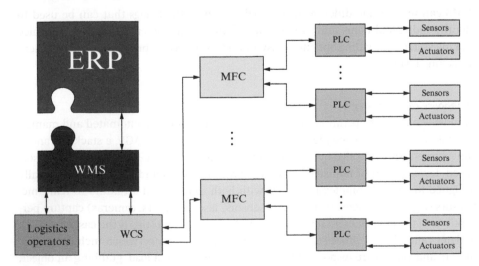

Figure 5.39 Architecture of a WMS. Sensors and actuators are embedded in automated equipments and controlled by PLCs.

SAP extended warehouse management (EWM) is an embedded application in S/4HANA (see Section 1.12) used to control the inbound and outbound warehouse processes. The key functions of EWM are:

- layout definition, to determine the size and location of the storage zones;
- workload computation, automatic resources assignment, and prioritization for the unit loads;
- resource management, integrated with RFID as a support of the mobile devices used in the warehouse processes;
- yard management, to monitor and manage the incoming and outgoing vehicles approaching the loading and unloading bays of the warehouse;
- wave management, consisting of grouping the outgoing unit loads in waves to be processed and picked at the same time;
- replenishment, for the management of the picking zone;
- environmental health and safety, to provide support to properly handle hazardous materials in the warehouse;
- labour management, used to monitor the performance of warehouse employees;
- inventory control.

5.10   Warehouse Design

Warehouse design is a fundamental logistics strategic decision. It amounts to

- selecting the internal transportation technology (and, in particular, its level of automation);
- choosing a warehouse layout;
- sizing the receiving, staging, storage, and shipping zones;
- sizing the internal transport system (including the warehouse staff).

The decision is based on the following data:

- the type of warehouse (crossdock, production warehouse, distribution warehouse, etc.);
- labour cost;
- interest rates;
- the number of SKUs to be kept in stock;
- the inventory level (number of load units to be stored) for each SKU and, consequently, the overall warehouse storage capacity;
- the required throughput (number of loads to be handled in an hour);
- the dimensions (including allowances) and weight of the load units to be stored;
- the expected warehouse utilization.

Moreover, the design process must consider whether the warehouse will be located in an existing building or in a new site. In the former case, the dimensions of the existing building constrain the decision while in the latter case one must consider, in addition to the labour cost and discount rate (already mentioned), the maximum investment cost that the organization is willing to pay.

There is no comprehensive general framework for designing a warehouse. The various decisions are interlinked and there is no one-fits-all sequence in which they should be made. As a rule, several alternative designs have to be developed and assessed before a final decision is made. Analytical methods, like the ones reported in the remainder of this section, are usually proposed to create a short-list of alternative designs which are then evaluated with detailed *discrete event* simulation models (see Section 1.10.3) taking into account tactical and operational issues (see Sections 5.11–5.15).

5.10.1 Internal Transportation Technology Selection

Internal transportation technology selection is often the first decision to be made in the design process. As observed in Section 5.2.4, the adoption of automated equipment is expected to become more and more popular in the coming years as a substitute for human labour in repetitive warehousing operations. However, automation is not always a good fit. In particular, it is not suitable if freight is oddly shaped (as often happens in the LTL load industry, see Chapter 6), or if demand is expected to be unstable in the medium-long run, or if item picking is required.

Automated equipment usually requires a large initial investment. Its economic convenience depends on the labour cost and the available interest rate on loans. The *net present value* (*NPV*) formula can be used to determine whether it is appropriate to invest in a piece of equipment (with a useful life of n years) replacing a warehouse operator. The *NPV* formula assumes that 1 \$ (or €, etc.) today is worth $(1 + r)$ at the end of the year, $(1 + r)^2$ at the end of two years, and so on, where r is an annual interest rate. Conversely, 1 \$ after t years would be worth $1/(1 + r)^t$ in the present (see also Section 2.5.5). Hence, assuming that the equipment has no salvage value, an organization would be willing to pay for it up to

$$NPV = \sum_{t=1}^{n} \frac{w_t}{(1 + r)^t},$$ (5.1)

where w_t is the expected wage of a warehouse operator in t years. The *NPV* is the value *in the present* of the total labour cost (over a period of n years) for a warehouse operator. Hence, the introduction of an automated piece of equipment replacing one warehouse operator is economically convenient only if it costs less than *NPV*. Of course, formula (5.1) can be easily adapted to the case when multiple operators can be replaced by introducing automated equipment.

[Merryl.xlsx] Merryl manufactures and distributes beer in over 80 countries. When designing its warehouses in the USA, the global supply chain manager assumed that $n = 7$, $r = 0.05$ and $w_t = $ \$ 29 263, $t = 1, \dots, n$ (see Table 5.1), which gave rise to a *NPV* equal to \$ 169 327. On the basis of this result, they decided to construct highly automated warehouses, based on an AGV technology.

On the other hand, the *NPV* in India (where $w_t = \$ 2660$, $t = 1, \ldots, n$, see again Table 5.1), turned out to be much less (*NPV* = \$ 15 392). Hence, automation was not worth the cost and a manual material handling system was maintained in every warehouse in that country.

AS/RS versus AGV-based internal transportation

The two fundamental automatic internal transportation technologies for palletized goods are AGVs and AS/RSs, being carousels suitable only for small parts.

AGVs are often used if the warehouse height is under 10 m (although particular AGVs can lift and fetch a load at 12–17 m). Another peculiar aspect of AGVs is that they need relatively large aisles (around 3.5 m), although some vehicles can operate in particularly narrow aisles (only 1.5 m wide).

AS/RSs are part of high-density automated warehouses and are best suited when land cost is high or a very tall building (up to 45 m) is available or can be built. Stacker cranes move along a rail, which limits their mobility. Therefore, they are often paired with conveyors or other automatic internal transportation systems (including AGVs) to deliver the loads to their final destinations within a facility.

5.10.2 Layout Design

Layout design is mainly influenced by the type of warehouse and by the selected internal transportation technology.

Layout of a crossdock

Crossdocks have no storage zone except for small buffers where merchandise wait for a few hours before being transshipped onto an outgoing vehicle.

Crossdocks have two kinds of doors: receiving (or *strip*) doors and shipping (or *stack*) doors. In both cases, a door is assigned a share of floor space. The number of shipping doors is customarily set equal to the number of destinations that have to be served, although destinations requiring a very intensive flow of goods may be assigned more than one door in order to accommodate multiple outgoing vehicles at the same time.

As far as the choice of the crossdock shape is concerned, it is worth observing that crossdocks may have hundreds of doors so that the distances travelled by internal transportation equipment (i.e., forklift trucks) in a trip may exceed several hundred metres. It is then crucial to choose a shape that minimizes the average travelled distance (defined as the total distance between inbound and outbound vehicles weighted by the corresponding intensity of freight flow) which can be assumed as a good approximation of crossdock workload.

The I, L, T, H, and X are the most common shapes (see Figure 5.40). In order to compare the different shapes, two performance measures, named the *diameter* and the *centrality* of a crossdock, can be used. The diameter of a crossdock is the largest distance between any pair of doors. The centrality is the rate at which the diameter grows as the number of doors increases. For an I-shaped terminal the centrality is two. In fact, if two doors are added at each end, the diameter increases by two doors so that the centrality is $4/2 = 2$. The centrality of the most common shapes are reported in Table 5.6. Larger

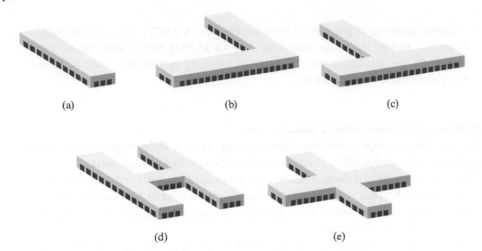

Figure 5.40 Common crossdock layouts: (a) I-shape; (b) L-shape; (c) T-shape; (d) H-shape; (e) X-shape.

Table 5.6 Features of the most common crossdock shapes.

Shape	Centrality	Number of corners (inside - outside)
I	4/2 = 2	0 - 4
L	4/2 = 2	1 - 5
T	6/2 = 3	2 - 6
H	8/2 = 4	4 - 8
X	8/2 = 4	4 - 8

values of centrality are better for two reasons: the workload is smaller and the forklift traffic congestion is lighter. For example, for an I-shape, forklift traffic may be very heavy in the middle of the crossdock (in fact, it varies with the square of the number of doors) while for an H-shape it is much lighter.

On the other hand, shapes other than the I-shape show a deterioration in efficiency for the crossdocks having the same area, due to the larger number of *corners*. Outer and inner corners (see Figure 5.41) reduce the potential number of doors along the perimeter. Each door needs an adequate amount of surface area, otherwise interference occurs between adjacent doors. Doors in the outer corners have less space available, as shown in Figure 5.41, and therefore there should be fewer of them. Doors in the inner corners are unusable because the vehicle manoeuvring areas would overlap. As a result, by increasing the corners to maintain the same number of doors, the surface area, as well as the diameter, of the crossdock would have to increase thus losing efficiency.

As a guideline, the I-shape is best for small to mid-sized crossdocks. The T-shape is best for crossdocks with about 150 to 250 doors, while for larger crossdocks the H or X-shape should be preferred.

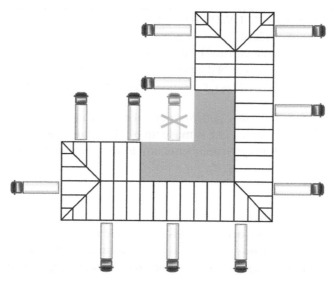

Figure 5.41 An L-shaped crossdock. For each door, the crossdock surface area assigned for freight handling is represented. In the outer corners, the surface area assigned to each door is smaller. The doors cannot be opened in the inner corner at the shadowed area.

An alternative and more rigorous way of selecting the crossdock layout is defined by the following CROSSDOCK_LAYOUT procedure, which can be regarded as a kind of Monte Carlo simulation (see Section 1.10.3).

Layout of a warehouse

The main aspect to be considered when designing a warehouse layout is the degree of complementarity between pairs of activities. The greater the extent of the material flows between two activities, the greater their degree of complementarity, and therefore the greater the necessity to carry them out in zones that are close together. The layout of a warehouse can generally be designed in three different ways.

The first layout (see Figure 5.42), is called *flow-through* (or I-shaped) and arranges the receiving zone and the shipping zone at opposite ends of the warehouse. This kind of layout presumes that the majority of the unit loads traversing the warehouse will need the same operations and should therefore be processed in the same sequence: the unit loads traverse the warehouse from one part (receiving zone) to the other (shipping zone), passing through the storage zone. The flow-through layout is suitable for long, narrow spaces through which a high number of unit loads transits.

The second layout (see Figure 5.43) is called U-flow. It has a U shape, which means that its receiving and shipping zones are located on the same side of the warehouse. The U-flow layout adds flexibility to warehouse operations: the same loading and unloading bays can be used for receiving and shipping operations, according to the needs of the moment. This layout is particularly suitable for low material flows and enables the expansion of the warehouse along three sides (those sides without loading and unloading bays).

In an L-shaped warehouse layout (see Figure 5.44) the receiving and shipping zones are on contiguous sides of the warehouse. This layout is especially useful in situations characterized by low material flows and in warehouses located in square buildings. The

---

1: **procedure** CROSSDOCK_LAYOUT(L, S, l^*)

2:     # L is the set of candidate crossdock layouts;

3:     # l^* is the best layout returned by the procedure;

4:     # $m_l, l \in L$, is the number of doors of the crossdock layout l;

5:     # $d_{ij}^l, i, j = 1, \ldots, m_l, l \in L$, is the distance between doors i and j in the crossdock layout l;

6:     # S is the set of *freight flow patterns* identified as the most likely to occur;

7:     # A freight flow pattern $s \in S$ is made up of a list of n_s vehicles and the corresponding freight flows f_{hk}^s (number of unit loads to be transferred from vehicle h to vehicle k), $h, k = 1, \ldots, n_s$;

8:     **for** $l \in L$ **do**

9:         **for** $s \in S$ **do**

10:             **if** $m_l \geq n_s$ **then**

11:                 # The crossdock layout l is compatible with the flow pattern s;

12:                 Solve a crossdock door assignment problem (see Section 5.13), by considering $d_{ij}^l, i, j = 1, \ldots, m_l$, and $f_{hk}^s, h, k = 1, \ldots, n_s$, determining z_{ls}^*;

13:                 # z_{ls}^* is the optimal crossdock workload, defined as the total distance covered by the internal transportation equipment for moving freight flows $f_{hk}^s, h, k = 1, \ldots, n_s$, from incoming to outgoing vehicles in layout $l, l \in L$;

14:             **else**

15:                 # The number m_l of doors is not compatible with the number n_s of required vehicles;

16:                 $z_{ls}^* = \infty$;

17:             **end if**

18:         **end for**

19:         # Determine the mean workload associated with crossdock layout l;

$$z_l^* = \frac{1}{l} \sum_{s \in S} z_{ls}^*;$$

20:     **end for**

21:     # Determine the best layout;

22:     $l^* = \arg \min_{l \in L} \{z_l^*\}$;

23:     **return** l^*;

24: **end procedure**

---

orientation of the aisles in the storage zone can be longitudinal, that is, with shelving laid out perpendicularly to the shipping zone, or transversal, with the shelving laid out parallel to the shipping zone.

In some rare cases, in particular when the storage system is based on block stacking, independent of the solution adopted for the warehouse layout, the orientation of the aisles in the storage zone can be diagonal, which leads to layouts such as the V-shaped layout, also called the *seagull wing* layout (see Figure 5.45), or the one named the *fishbone* layout (see Figure 5.46).

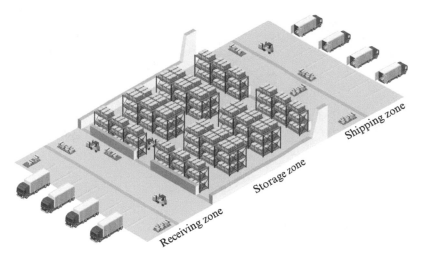

Figure 5.42 Flow-through warehouse layout.

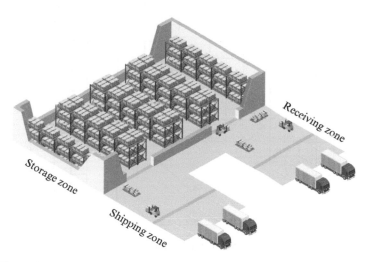

Figure 5.43 U-shaped warehouse layout.

5.10.3 Sizing of the Storage Zone

The dimensions of the storage zone depend on the number of SKUs, as well as on their reorder quantities and safety stocks (see Section 5.12). They are also affected by the material handling equipment which has an impact on the width of the aisles as well as on the height of the warehouse. In the following it is assumed that the warehouse stores unit loads (e.g., palletized unit loads) and that a storage location can hold a single unit load. Should these hypotheses not be verified, the following formulae may be adapted straightforwardly.

The size of the storage zone also depends on the storage policy. In a *dedicated-storage policy*, each SKU is assigned to a pre-established set of storage locations. This approach is easy to implement but results in an underutilization of the storage space. In fact, the space required corresponds to the sum of the maximum inventory levels over time of

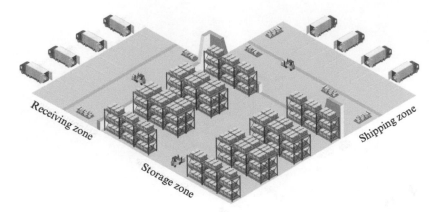

Figure 5.44 L-shaped warehouse layout.

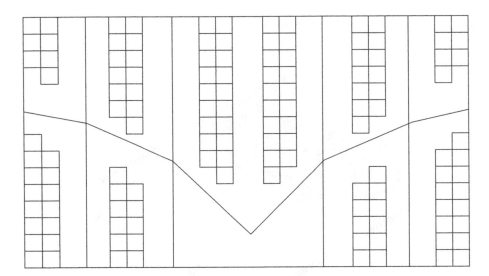

Figure 5.45 V-shaped layout of aisles in a storage zone.

the various SKUs. Let n be the number of SKUs and let $I_j(t)$, $j = 1, \ldots, n$, be the inventory level (in terms of unit loads) of SKU j at time t. The capacity m_d in a dedicated-storage policy must be

$$m_d = \sum_{j=1}^{n} \max_t I_j(t). \tag{5.2}$$

In a *random-storage policy*, the allocation of the unit loads is decided dynamically on the basis of the current occupation of the storage locations and on future arrival and request forecasts. Therefore, the positions assigned to a SKU are variable in time. In this case, the required capacity m_r is

$$m_r = \max_t \sum_{j=1}^{n} I_j(t) \leq m_d. \tag{5.3}$$

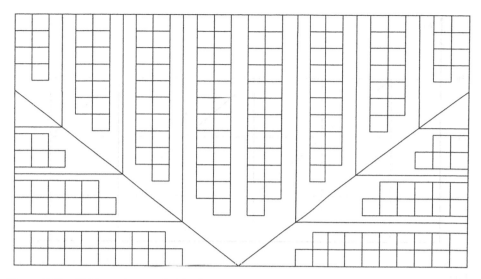

Figure 5.46   Fishbone layout of aisles in a storage zone.

The random-storage policy allows a higher utilization of the storage space to be achieved, but requires that each unit load is automatically identified through a barcode (or another identification technique), and a database of the current positions of all stocked unit loads is updated at every storage and retrieval.

Potan Up bottles two types of mineral water. In the warehouse located in Hangzhou, China, inventories are managed according to a reorder point policy (see Section 5.12). Order sizes and safety stocks are reported in Table 5.7. Inventory levels as a function of time are illustrated in Figures 5.47 and 5.48. The company is currently using a dedicated-storage policy. Therefore, the capacity of the storage zone is, according to formula (5.2),

$$m_d = 600 + 360 = 960.$$

Table 5.7   Order sizes and safety stocks in the Potan Up problem.

SKU	Order size [palletized unit loads]	Safety stock [palletized unit loads]
Natural water	500	100
Sparkling water	300	60

The firm is now considering the opportunity of using a random-storage policy, in which case, according to formula (5.3), the capacity of the storage would be

$$m_r = 600 + 210 = 810.$$

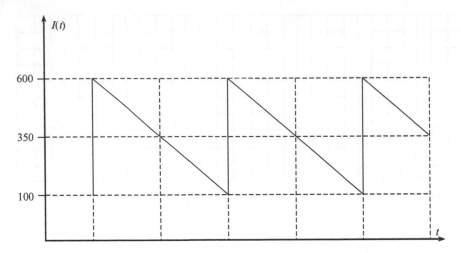

Figure 5.47 Inventory level of natural mineral water (in palletized unit loads) in the Potan Up problem.

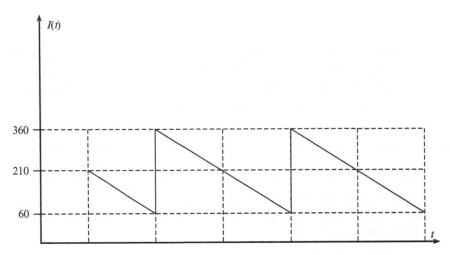

Figure 5.48 Inventory level of sparkling mineral water (in palletized unit loads) in the Potan Up problem.

In a *class-based storage policy*, SKUs are divided into a number of categories according to their demand, and each category is associated with a set of zones where SKUs are stored according to a random-storage policy. The class-based storage policy reduces to the dedicated-storage policy if the number of categories is equal to the number of SKUs, and to the random-storage policy if there is a single category.

Sisa adopted a class-based storage policy (with six SKU categories) for its DC in Gioia Tauro, Italy. The description, number of palletized unit loads per category and storage occupation are shown in Table 5.8.

Table 5.8 SKU categories in the Sisa DC.

Category	Description	Palletized unit loads	Occupation of storage zone [%]
1	Food	1105	17
2	Beverage	2782	42
3	Appliances	247	4
4	Domestic cleansing products	504	8
5	Personal hygiene products	699	11
6	Others	1163	18
	Total	6500	100

Sizing the storage zone for a flow-through layout

Once the number of required storage locations has been computed, the optimal dimensions of the storage zone can be determined. As already observed, the solution to this problem depends on the selected warehouse layout as well as on the material handling equipment.

In the following the problem will be defined for the flow-through layout in Figure 5.49 under the hypotheses that:

- the internal transportation system moves unit loads (e.g., palletized unit loads);
- all the storage locations have the same probability of being accessed;
- a picker (eventually supported by a vehicle of the internal transportation) performs a *single cycle*, i.e., either a put-away or a retrieval operation; in particular, for a put-away operation, a picker transports a unit load from the receiving zone to a storage

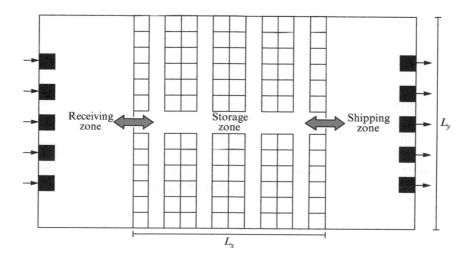

Figure 5.49 Warehouse layout chosen to illustrate the storage zone sizing problem.

location, drops off the load and comes back empty to the receiving zone; similarly, for a retrieval operation, a picker reaches the storage location, picks up the load and then moves to the shipping zone.

If some of these hypotheses do not hold, the following procedure can be modified straightforwardly (see, e.g., Problem 5.10).

Sizing the storage zone consists of determining the length L_x and width L_y of the storage zone, the height of the warehouse being determined by the material handling equipment (and by the height of the building, whenever it already exists). Let m be the required capacity (expressed in terms of unit loads); α_x and α_y the gross occupation (i.e., occupation including allowances) of a storage location along the directions x and y, respectively; w_x and w_y, the width of the side aisles and of the central aisle, respectively, and v the average speed of an operator or a vehicle. The decision variables are n_x, the number of storage locations in the x direction, and n_y, the number of storage locations in the y direction. The number n_z of storage locations in the vertical direction is determined by the material handling equipment or by the height of the building, as previously observed.

The extension L_x of the storage zone along direction x is given by the following relation

$$L_x = \left(\alpha_x + \frac{w_x}{2}\right) n_x$$

where, according to Figure 5.49, n_x is assumed to be an even number. Similarly, L_y is given by

$$L_y = \alpha_y n_y + w_y.$$

Therefore, the average distance covered by a picker for a put-away or a retrieval operation is $L_x + L_y/2$. Hence, the problem of sizing the storage zone can be formulated as follows:

$$\text{Minimize } T_{SC} = \left(\alpha_x + \frac{w_x}{2}\right) \frac{n_x}{v} + \frac{\alpha_y n_y + w_y}{2v} \tag{5.4}$$

subject to

$$n_x n_y n_z \geq m \tag{5.5}$$

$$n_x,\ n_y \geq 0,\ \text{integer} \tag{5.6}$$

$$n_x\ \text{even}, \tag{5.7}$$

where objective function (5.4) is the average travel time T_{SC} of an operator or a vehicle in a *single cycle* (either a storage or retrieval operation), while inequality (5.5) states that the number of storage locations is at least equal to m.

Problem (5.4)–(5.7) can be solved in the following way. The integrality constraints on variables n_x and n_y and constraint (5.7) are relaxed. Hence, inequality (5.5) will be satisfied at the optimum for the relaxed problem as an equality

$$n_x = \frac{m}{n_y n_z}. \tag{5.8}$$

Therefore, n_x can be removed from the relaxed problem, whose formulation becomes

$$\text{Minimize } T_{SC} = \left(\alpha_x + \frac{w_x}{2}\right)\frac{m}{n_y n_z v} + \frac{\alpha_y n_y + w_y}{2v} \tag{5.9}$$

subject to

$$n_y \geq 0.$$

Since the objective function (5.9) is convex, its minimizer n'_y can be found through the following relation:

$$\frac{d}{d n_y}\left(\left(\alpha_x + \frac{w_x}{2}\right)\frac{m}{n_y n_z v} + \frac{\alpha_y n_y + w_y}{2v}\right) = 0.$$

Hence,

$$n'_y = \sqrt{\frac{2m\left(\alpha_x + \frac{w_x}{2}\right)}{\alpha_y n_z}}. \tag{5.10}$$

Finally, n'_x is obtained from formula (5.8):

$$n'_x = \sqrt{\frac{m\alpha_y}{2n_z(\alpha_x + \frac{w_x}{2})}}. \tag{5.11}$$

The optimal values n_x^* and n_y^* of decision variables n_x and n_y are obtained by suitably rounding n'_x and n'_y to integer quantities, so as to guarantee the satisfaction of inequality (5.5) and that n_x^* be even.

Wagner Bros plans to build a new warehouse near Sidney, Australia, in order to supply its sales points in New South Wales. On the basis of a preliminary analysis of the problem, it has been decided that the facility should accommodate at least 780 palletized unit loads stored onto selective racks and transported by means of reach trucks. Each rack has four shelves ($n_z = 4$), each of which can store a single palletized unit load occupying a $1.05 \times 1.05 \text{ m}^2$ gross area. The racks are arranged as in Figure 5.49, where side aisles are 3.5 m wide, while the central aisle is 4 m wide. The average speed of a reach truck is 5 km/h (i.e., 1.39 m/s). Using formulae (5.10) and (5.11), the following values are obtained:

$$n'_x = \sqrt{\frac{780 \times 1.05}{2 \times 4 \times \left(1.05 + \frac{3.5}{2}\right)}} = 6.05;$$

$$n'_y = \sqrt{\frac{2 \times 780 \times \left(1.05 + \frac{3.5}{2}\right)}{1.05 \times 4}} = 32.25.$$

By rounding these values appropriately results in $n_x^* = 6$ and $n_y^* = 33$; the capacity turns out to be $6 \times 33 \times 4 = 792$, while $L_x = [1.05 + (3.5/2)] \times 6 = 16.80 \text{ m}$ and $L_y = 1.05 \times 33 + 4 = 38.65 \text{ m}$. The optimal single cycle time is

$$T_{SC}^* = \left(1.05 + \frac{3.5}{2}\right)\frac{792}{33 \times 4 \times 1.39} + \frac{1.05 \times 33 + 4}{2 \times 1.39} = 26.01 \text{ s}.$$

The surface utilization rate (given by the ratio between the number of storage locations and the surface of the storage zone, see Section 5.8) corresponds to $792/(16.80 \times 38.65) = 1.22$ palletized unit loads m^{-2}. Alternatively, the index mentioned above can be expressed as the surface percentage occupied by the storage locations, computed as $[6 \times 33 \times 1.05 \times 1.05/(16.80 \times 38.65)] \times 100 = 33.62\%$.

5.10.4 Sizing of the Receiving and Shipping Zones

In order to size the receiving zone, the first step is to evaluate the number p of unloading bays. Assuming that a team of operators is in charge of disembarking the merchandise (or an automatic unloading system is used), while another team of operators (or a fleet of AGVs) transports the unit loads to the storage zone, p can be computed as

$$p = \left\lceil \frac{d\,\overline{T}}{3600\,h\,\eta} \right\rceil, \tag{5.12}$$

where d represents the average number of unit loads entering the warehouse in a given time interval (e.g., a week), $\overline{T}$ is the average time (in seconds) for unloading a unit load from a vehicle and h is the number of working hours available for unloading operations in the considered time interval. Note that $\overline{T}$ may also include the time required to free the receiving zone from the unloaded package; $\eta \in [0, 1]$ is an efficiency measure taking into account the time a vehicle needs to approach/detach from the bays, idle time, queueing, traffic congestion (also in the yard), etc.

Alternatively, p may be computed by considering the worst case scenario, i.e., the time interval when the maximum number of vehicles have to be unloaded. However, this may lead to an oversizing of the receiving zone with a consequent significant cost increase.

At the Sisa DC in Gioia Tauro (Italy), the average number of incoming trucks is 34 from Monday to Saturday, with an average number of palletized unit loads per vehicle equal to 20. As a result, $34 \times 20 \times 6 = 4080$ palletized unit loads are expected each week. The average time to disembark a pallet is 192 s. Incoming vehicles can be unloaded from 6:00 to 12:00. Hence, the weekly average time available for the vehicle-unloading operations is 36 hours, corresponding to 129 600 s. Therefore, considering an efficiency factor $\eta = 0.8$, the number of unloading bays is computed as

$$p = \left\lceil \frac{4080 \times 192}{129\,600 \times 0.8} \right\rceil = 8.$$

Once p has been computed, an estimate of the size of the receiving zone can be obtained by considering that, as a rule of thumb, a space of about 50 m^2 is needed for loading/unloading a vehicle with a capacity of up to 50 palletized unit loads. In particular, the width of a loading/unloading bay is around 2.7 m and the distance between two adjacent bays is about 1.5 m (see Section 5.1.2). Based on these data, the dimensions L_x and L_y of the receiving zone can be computed as

$$L_y = 2.7p + 1.5(p + 1), \tag{5.13}$$

$$L_x = \frac{50p}{L_y}. \tag{5.14}$$

The receiving zone of the Sisa DC at Gioia Tauro measures about $50 \times p = 400$ m². The L_y dimension, on the basis of (5.13), is 35.10 m, whereas L_x, according to (5.14), is about 11.40 m.

The shipping zone can be sized by using a procedure similar to the one illustrated for the receiving zone. In particular, the number of loading bays can be determined by using formula (5.12), where the parameters refer to the outbound operations.

Finally, the size of the staging zone (if it is present) depends on the system used to sort, consolidate and package the items to be shipped.

There are seven loading bays in the shipping zone of the Sisa DC in Gioia Tauro. These bays can be sometimes used for receiving goods during the peak season. The size L_y of the shipping zone is obtained by using formula (5.13) and results in 30.90 m. Since the warehouse is U-shaped, L_x coincides with that of the receiving zone and results in about 11.40 m. Consequently, the shipping zone measures approximately 352.26 m², i.e., 50.32 m² for each loading bay.

The above dimensions of the receiving and shipping zones do not take into account the surface of any staging subzones, which are dependent on the equipment used to sort, deconsolidate, consolidate, and pack the unit loads.

5.10.5 Sizing of an AS/RS

The main decisions when sizing an AS/RS (see Figure 5.50) are the determination of the number of aisles n_a, the number of stacker cranes n_s as well as the length L_y and the height L_z of the racks. The objective is to minimize the number of stacker cranes (which contribute the most to the system cost) subject to capacity and throughput constraints.

Let m be the required capacity (expressed in terms of unit loads); α_x, α_y and α_z the gross occupation of a storage location along directions x, y, and z, respectively; w_x the width of each aisle along the x direction. Let n_x, n_y, and n_z be the number of unit loads along the x, y, and z directions, respectively. The following relations hold:

$$L_x = (2\alpha_x + w_x)n_a;$$

$$L_y = \alpha_y n_y; \tag{5.15}$$

$$L_z = \alpha_z n_z; \tag{5.16}$$

$$n_x = 2n_a. \tag{5.17}$$

In the following, it is assumed for the sake of simplicity that

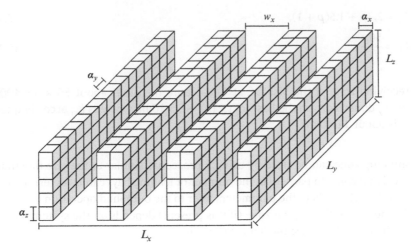

Figure 5.50 Storage zone of an AS/RS.

- each storage location can accommodate a single unit load;
- L_y and L_z have been given tentative values, satisfying technological and space occupation constraints.

The capacity constraint (see relation (5.5)) imposes that

$$n_x n_y n_z \geq m. \tag{5.18}$$

Based on relations (5.15)–(5.17), inequality (5.18) can be written as:

$$n_a \geq \frac{m/2}{(L_y \, L_z)/(\alpha_y \, \alpha_z)},$$

from which the number n_a of aisles can be estimated as

$$n_a = \left\lceil \frac{m/2}{(L_y \, L_z)/(\alpha_y \, \alpha_z)} \right\rceil. \tag{5.19}$$

Throughput of a stacker crane

The expected throughput of a stacker crane depends on the building height L_z and on the length L_y of the racks, as well as on the horizontal and vertical speeds (v_y and v_z, respectively) which are assumed to be constant.

Since the stacker crane can move simultaneously and independently along the horizontal axis y and along the vertical axis z, the travel time of the S/R machine between two locations of the rack follows the Chebyshev metric and is given by

$$\max\left\{ \frac{\Delta y}{v_y}, \frac{\Delta z}{v_z} \right\},$$

where Δy and Δz are the distances between the two locations in the horizontal and vertical directions, respectively.

Al Fokan distributes soft drinks in the Persian Gulf area. Its warehouse located in Bahrain includes a storage zone served by an AS/RS with aisles 30 m long and 18 m high. The horizontal and vertical speeds of a stacker crane are 220 cm/s and 40 cm/s, respectively. Hence, the maximum completion time of a single storage or retrieval operation is

$$\max\left\{\frac{30}{2.20},\frac{18}{0.40}\right\}=45 \text{ s}.$$

A stacker crane can perform

- a *single cycle*, in which either a storage or a retrieval operation is performed (see Figure 5.51(a));
- a *dual cycle*, in which pairs of storage and retrieval operations are combined in an attempt to reduce the overall travel time (see Figure 5.51(b)).

The single and the dual command travel times are denoted as T_{SC} and T_{DC}, respectively. Since storage/retrieval requests are not known in advance, both T_{SC} and T_{DC} are random variables. Under the hypothesis that

- the m storage locations are randomly assigned to unit loads,
- there is a unique input–output (I/O) station at coordinates $(0,0)$ of the Cartesian plane yz, denoted as 0,

the expected values of T_{SC} and T_{DC} are:

$$E[T_{SC}]=\frac{1}{m}\sum_{i=1}^{m}2t_{0i}; \tag{5.20}$$

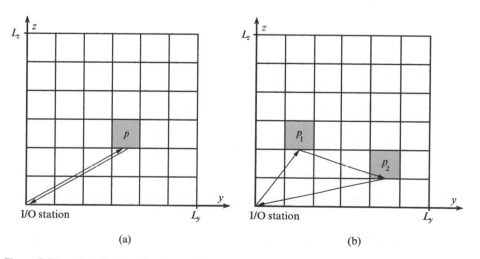

(a)   (b)

Figure 5.51 (a) A single cycle of a stacker crane (p is either a storage or a retrieval location); (b) a dual cycle of a stacker crane (p_1 is a storage location and p_2 the subsequent retrieval location).

$$E[T_{DC}] = \frac{1}{m(m-1)} \sum_{i=1}^{m-1} \sum_{j=i+1}^{m} 2(t_{0i} + t_{ij} + t_{j0}), \qquad (5.21)$$

where t_{ij} $(= t_{ji})$, $i, j = 0, \ldots, m$, denotes the travel time of the stacker crane between storage locations i and j, when $i, j \neq 0$, or between a storage location and the I/O station, when $i = 0$ or $j = 0$.

Instead of formulae (5.20) and (5.21), the closed-form approximation by A. Y. Bozer and J. A. White can be used in which it is assumed that a stacker crane can store and retrieve a unit load in any point of coordinates (y, z) of the Cartesian plane yz (see Figure 5.51).

Let T_y be the horizontal travel time required to go to the end of the farthest column of the rack from the I/O station and let T_z be the vertical travel time required to go to the upper end of the rack. These travel times are expressed by the following relations: $T_y = L_y/v_y$ and $T_z = L_z/v_z$. Since the stacker crane travels simultaneously and independently in the horizontal and vertical directions, $T = \max\{T_y, T_z\}$ is the time required to reach the farthest storage location from the I/O station. Let $b = \min(T_y/T, T_z/T)$. In the following, without loss of generality, it is assumed that $T = T_y$ and that, consequently, $b = T_z/T$.

The expected single command travel time can be computed as follows. The stacker crane will take a time equal to $t_{yz} = \max\{t_y, t_z\}$ to reach (y, z), where $t_y = y/v_y$ and $t_z = z/v_z$. Under the hypothesis that each storage location is accessible with the same probability, the coordinates are independent random variables. Hence,

$$F_{t_{yz}}(t) = Pr(t_{yz} \leq t) = Pr(t_y \leq t, t_z \leq t) = Pr(t_y \leq t)Pr(t_z \leq t).$$

Moreover, the coordinates y and z are uniformly distributed:

$$Pr(t_y \leq t) = t/T, \ 0 \leq t \leq T;$$

$$Pr(t_z \leq t) = \begin{cases} t/bT, & \text{if } 0 \leq t \leq bT \\ 1, & \text{if } bT < t \leq T. \end{cases}$$

As a result, the probability density function of t_{yz} is

$$f_{t_{yz}}(t) = \begin{cases} 2t/bT^2, & \text{if } 0 \leq t \leq bT \\ 1/T, & \text{if } bT < t \leq T. \end{cases}$$

As a consequence, the expected single command travel time is

$$\begin{aligned} E[T_{SC}] &= 2\int_0^T t f_{t_{yz}}(t)\, dy = \frac{4}{bT^2} \int_0^{bT} t^2\, dy + \frac{2}{T} \int_{bT}^T t\, dy \\ &= (b^2/3 + 1)T. \qquad (5.22) \end{aligned}$$

Similarly, the expected dual command travel time can be computed as (see Problem 5.13):

$$E[T_{DC}] = (4/3 + b^2/2 - b^3/30)T. \qquad (5.23)$$

Once computed the expected single or dual command travel time, the throughput R_s of a stacker crane can be determined as follows. Assume that $E[T_{SC}]$ or $E[T_{DC}]$ are computed in seconds and R_s is expressed in unit loads per hour. Hence,

$$R_s = \frac{3600}{E[T_{SC}] + T^{(s/r)}} \tag{5.24}$$

or

$$R_s = 2\left(\frac{3600}{E[T_{DC}] + 2T^{(s/r)}}\right), \tag{5.25}$$

depending on whether the cycle is single or dual. $T^{(s/r)}$ is the time, usually deterministic, spent by the stacker crane to extract a load from the I/O station, plus the time to insert the load in a storage slot during a put-away operation (or vice versa in the case of a retrieval operation). The multiplicative constant 2 in formula (5.25) is due to the fact that in a dual cycle two unit loads are handled at a time, one to be stored and another one to be retrieved.

[Wert.xlsx] Wert is a Dutch company specializing in the development of automated logistic solutions for warehouses. Last February, Wert acted as consultant for a primary beverage company with the aim of designing an AS/RS. Among the alternative designs considered by the company, there was a stacker crane characterized by the following data: $T^{(s/r)} = 20.00$ s, $v_y = 3.00$ m/s and $v_z = 0.75$ m/s, to serve selective pallet racks of $L_y = 180$ m long and $L_z = 36$ m high.

Wert computed the following parameters: $T_y = L_y/v_y = 60.00$ s, $T_z = L_z/v_z = 48.00$ s, $T = \max\{T_y, T_z\} = 60.00$ s and $b = \min\{T_y/T, T_z/T\} = 0.80$. If the stacker crane performs single cycles, the expected travel time was estimated, by means of formula (5.22), as $[(0.80)^2/3 + 1] \times 60.00 = 72.80$ s. Hence, the throughput of the considered stacker crane, given by formula (5.24), is $R_s = \lceil 3600/(72.80 + 20.00) \rceil = 39$ palletized unit loads per hour.

Number of stacker crane
If the whole storage zone is required to have a throughput R, the number of stacker cranes can be easily estimated as

$$n_s = \left\lceil \frac{R}{R_s} \right\rceil. \tag{5.26}$$

Formulae (5.19) and (5.26) allow the storage zone served by an AS/RS to be designed.

[Wert.xlsx] In the Wert problem, the storage zone of the primary beverage company should accommodate 40 000 palletized unit loads, each of them occupying a space in the racks of $\alpha_y = 1.28$ m wide and $\alpha_z = 2.00$ m high. Formula (5.19) implies that

$$n_a = \left\lceil \frac{40\,000/2}{(180 \times 36)/(1.28 \times 2.00)} \right\rceil = 8.$$

If the required throughput of the storage zone is 230 palletized unit loads per hour, the number n_s of required stacker cranes, determined by using formula (5.26), is

$$n_s = \left\lceil \frac{230}{39} \right\rceil = 6.$$

Since $n_a > n_s$, the company has to decide whether each of the eight aisles has to be assigned a dedicated stacker crane (in which case a relevant increase in costs is expected) or the six stacker cranes have to be shared among the eight aisles (in which case a degradation in throughput is expected).

5.10.6 Sizing a Vehicle-based Internal Transportation System

In warehouses where internal transportation is performed by vehicles (such as pallet trucks and AGVs), a fundamental design decision is to determine the size of the fleet. An accurate evaluation of this parameter must consider a number of aspects, including:

- the items to picked (e.g., cases) which may be different from the unit loads stored when a replenishment arrives (e.g., palletized unit loads);
- the capacity of the vehicles (which may vary from tens of small items to a single or multiple palletized unit loads, up to a containerized unit load);
- the compatibility between vehicles (especially AGVs) and individual pieces of automated equipment;
- congestion, which may be severe in some parts of the warehouse (e.g., close to palletizers or next to aisles where high-demand SKUs are stored);
- allocation policies (which may determine the intensity of vehicle congestion);
- priorities and time windows associated with storage and retrieval operations;
- dispatching, routing, and scheduling policies.

When all these aspects have to be taken care of, discrete event simulation is the tool of choice. Here, an estimate of the number n_v of vehicles needed to support a given throughput is provided by a simple closed-form expression under the hypotheses that the vehicles have unit capacity and perform single cycles, and no priorities and time windows are considered. The formula is the following:

$$n_v = \left\lceil \frac{R \, \overline{T}}{3600 \, h \, \eta} \right\rceil,$$

where R is the required throughput, expressed in unit loads to be moved (either stored or retrieved) in a day, $\overline{T}$ is the average duration of a single cycle (in seconds), h is the number of working hours per day and $\eta \in [0, 1]$ is an efficiency measure modelling vehicle unavailability, idle time, queueing, congestion, etc.

Foodway, a company operating in the distribution of groceries, is building a new warehouse, whose storage zone is equipped with selective racks and forklift trucks. During the design phase, Foodway estimates a daily throughput R of 400 palletized unit loads. In order to quantify the number of forklift trucks to use, the logistics manager has to evaluate different parameters affecting throughput. They estimate that the average duration of a single cycle is 128 s. The number of working hours per day is equal to $h = 7$ and the truck efficiency level is equal to $\eta = 0.75$. The number of forklift trucks is then computed as follows:

$$n_v = \left\lceil \frac{400 \times 128}{3600 \times 7 \times 0.75} \right\rceil = 3.$$

5.11 Storage Space Allocation

The allocation of SKUs to storage locations is a tactical decision problem. It is based on the principle that fast-moving goods must be placed closer to the I/O points of the storage zone in order to minimize the average handling time. In the following, the problem is modelled and solved under the hypothesis that each storage location can accommodate a single unit load. It is also assumed that a dedicated-storage policy is used. The extension to the general case is straightforward and left to the reader.

Formally, the storage space allocation problem amounts to assigning each unit load to a different storage location of the m_d available. Let n be the number of SKUs; m_j, $j = 1, \dots, n$, the number of unit loads of SKU j to be stored; n_p the number of I/O points of the storage zone; o_{jh}, $j = 1, \dots, n$, $h = 1, \dots, n_p$, the average number of handling operations on SKU j through I/O point h per time period; t_{hk}, $h = 1, \dots, n_p$, $k = 1, \dots, m_d$, the travel time from I/O point h to storage location k.

Under the hypothesis that all the storage locations assigned to a SKU have an identical utilization rate, the cost c_{jk}, $j = 1, \dots, n$, $k = 1, \dots, m_d$, of assigning the SKU j to the storage location k is equal to

$$c_{jk} = \sum_{h=1}^{n_p} \frac{o_{jh}}{m_j} t_{hk}, \tag{5.27}$$

where o_{jh}/m_j represents the average number of handling operations per storage location and time period on SKU j through I/O point h. Consequently, $(o_{jh}/m_j)t_{hk}$ is the total average travel time between the I/O point h and the storage location k to perform the handling operations on SKU j, if SKU j is assigned to storage location k.

Let x_{jk}, $j = 1, \dots, n$, $k = 1, \dots, m_d$, be a binary decision variable, equal to 1 if SKU j is assigned to storage location k, 0 otherwise. The problem of seeking the optimal assignment of SKUs to storage locations can be then modelled as follows:

$$\text{Minimize} \sum_{j=1}^{n} \sum_{k=1}^{m_d} c_{jk} x_{jk} \tag{5.28}$$

subject to

$$\sum_{k=1}^{m_d} x_{jk} = m_j, \; j = 1, \dots, n \tag{5.29}$$

$$\sum_{j=1}^{n} x_{jk} \leq 1, \; k = 1, \dots, m_d \tag{5.30}$$

$$x_{jk} \in \{0, 1\}, \; j = 1, \dots, n, \; k = 1, \dots, m_d, \tag{5.31}$$

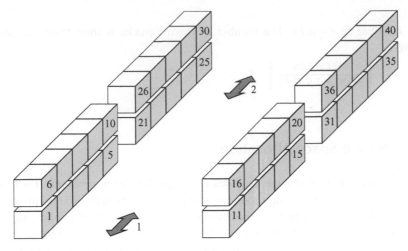

Figure 5.52 Storage zone of the Malabar warehouse.

where constraints (5.29) state that all the unit loads must be allocated, while constraints (5.30) impose that each storage location $k = 1, \ldots, m_d$, can store at most one unit load. It is worth noting that because of the particular structure of constraints (5.29) and (5.30), relations (5.31) can be replaced with the simpler non-negativity conditions

$$x_{jk} \geq 0, \ j = 1, \ldots, n, \ k = 1, \ldots, m_d, \tag{5.32}$$

since it is known a priori that there exists an optimal solution of problem (5.28)–(5.30), (5.32) in which all the variables take 0 or 1 values.

[`Malabar.xlsx`, `Malabar.py`] Malabar is an Indian company whose warehouse located in Ponnani, in the State of Kerala, has a storage zone with two I/O points and 40 storage locations, arranged in four selective racks (see Figure 5.52). The features of the SKUs are reported in Table 5.9, while the distances between the two I/O points and the storage locations are given in Tables 5.10 and 5.11. The optimal assignment of unit loads to the storage locations can be found by solving the optimization model (5.28)–(5.30), (5.32), in which $n = 5$, $m_d = 40$, while $m_j, j = 1, \ldots, 5$, are calculated on the basis of the second column of Table 5.9. The cost coefficients $c_{jk}, j = 1, \ldots, 5, k = 1, \ldots, 40$, reported in the `Malabar.xlsx` file, are calculated using formula (5.27), where it is assumed that travel time t_{hk} between I/O point $h = 1, 2$, and storage location $k = 1, \ldots, 40$, is directly proportional to the corresponding distance. The optimal solution, determined by using the `Excel Solver` under the `Data` tab, is reported in Table 5.12. It is worth noting that two storage locations (locations 26 and 27) are not used since the positions available are 40, while $\sum_{j=1}^{5} m_j = 38$.

Table 5.9 Features of the SKUs in the Malabar problem.

		Number of daily storage and retrieval operations	
SKU	Number of required storage locations	I/O point 1	I/O point 2
1	12	25	18
2	6	16	26
3	8	14	30
4	4	24	22
5	8	22	22

Table 5.10 Distance between storage locations and I/O point 1 in the storage zone of the Malabar warehouse.

Storage location	Distance [m]	Storage location	Distance [m]	Storage location	Distance [m]	Storage location	Distance [m]
1	2	11	2	21	14	31	14
2	4	12	4	22	16	32	16
3	6	13	6	23	18	33	18
4	8	14	8	24	20	34	20
5	10	15	10	25	22	35	22
6	3	16	3	26	15	36	15
7	5	17	5	27	17	37	17
8	7	18	7	28	19	38	19
9	9	19	9	29	21	39	21
10	11	20	11	30	23	40	23

Table 5.11 Distance between storage locations and I/O point 2 in the storage zone of the Malabar warehouse.

Storage location	Distance [m]	Storage location	Distance [m]	Storage location	Distance [m]	Storage location	Distance [m]
1	22	11	22	21	10	31	10
2	20	12	20	22	8	32	8
3	18	13	18	23	6	33	6
4	16	14	16	24	4	34	4
5	14	15	14	25	2	35	2
6	23	16	23	26	11	36	11

(Continued)

Table 5.11   (Continued)

Storage location	Distance [m]	Storage location	Distance [m]	Storage location	Distance [m]	Storage location	Distance [m]
7	21	17	21	27	9	37	9
8	19	18	19	28	7	38	7
9	17	19	17	29	5	39	5
10	15	20	15	30	3	40	3

Table 5.12   Optimal assignment of the SKUs to the storage locations in the Malabar problem.

SKU (j)	Storage location (k)
1	1, 6, 7, 8, 9, 10, 11, 16, 17, 18, 19, 20
2	22, 23, 24, 32, 33, 34
3	25, 28, 29, 30, 35, 38, 39, 40
4	2, 3, 12, 13
5	4, 5, 14, 15, 21, 31, 36, 37

If the storage zone has a single I/O point ($n_p = 1$), an optimal solution of problem (5.28)–(5.31) can be found with a straightforward procedure. In fact, under this hypothesis, cost coefficients c_{jk}, $j = 1, \dots, n$, $k = 1, \dots, m_d$, take the following form:

$$c_{jk} = \frac{o_{j1}}{m_j} t_{1k} = a_j b_k,$$

where $a_j = o_{j1}/m_j$ and $b_k = t_{1k}$ depend only on SKU j and on storage location k, respectively. Then, the optimal assignment of unit loads to storage locations can be determined by using the following SKU_ALLOCATION procedure.

This procedure is based on the observation that the minimization of the scalar product of two vectors $\boldsymbol{\alpha}$ and $\boldsymbol{\beta}$ may be achieved by ordering $\boldsymbol{\alpha}$ by non-increasing values and $\boldsymbol{\beta}$ by non-decreasing values.

[Malabar.xlsx] Assume that the storage zone of the warehouse of Malabar has a single I/O point (corresponding to I/O point 1 in Figure 5.52) and coefficients a_j, $j = 1, \dots, 5$, are those reported in Table 5.13. The corresponding values of α_i, $\sigma_\alpha(i)$, β_i, $\sigma_\beta(i)$, $i = 1, \dots, 38$, would be those described in Table 5.14. Then, the optimal solution would be the one reported in Table 5.15. It is worth noting that no unit loads would be allocated to the storage locations farthest from the I/O point (locations 30 and 40).

1: **procedure** SKU_ALLOCATION $(\mathbf{a}, \mathbf{b}, \mathbf{X}^*, z^*)$

2:      # $\mathbf{X}^*$ is the optimal solution returned by the procedure;

3:      # z^* is the cost corresponding to the optimal solution;

4:      Construct a vector α of $\sum_{j=1}^{n} m_j$ components, in which there are m_j copies of each $a_j, j = 1, \ldots, n$;

5:      Sort α by non-increasing values of its components;

6:      Define $\sigma_\alpha(i)$ in such a way that $\sigma_\alpha(i) = j$ if $\alpha_i = a_j, i = 1, \ldots, \sum_{h=1}^{n} m_h$;

7:      # Let $\mathbf{b}$ be the vector of m_d components corresponding to values b_k, $k = 1, \ldots, m_d$;

8:      Sort $\mathbf{b}$ by non-decreasing values of its components;

9:      Let β be the vector of $\sum_{j=1}^{n} m_j$ components, corresponding to the first $\sum_{j=1}^{n} m_j$ components of $\mathbf{b}$;

10:      Define $\sigma_\beta(i)$ in such a way that $\sigma_\beta(i) = k$ if $\beta_i = b_k, i = 1, \ldots, \sum_{h=1}^{n} m_h$;

11:      # Determine the optimal solution of problem (5.28)–(5.31);

12:

$$x^*_{\sigma_\alpha(i), \sigma_\beta(i)} = 1, \ i = 1, \ldots, \sum_{j=1}^{n} m_j;$$

13:      $x^*_{jk} = 0$, for all the remaining components;

14:      $z^* = \sum_{j=1}^{n} \sum_{k=1}^{m_d} a_j b_k x^*_{jk}$;

15:      **return** $\mathbf{X}^*, z^*$;

16: **end procedure**

Table 5.13 $a_j, j = 1, \ldots, 5$, coefficients in the Malabar problem (case of a single I/O point).

SKU (j)	Number of required storage locations	Daily storage and retrieval operations	Coefficient (a_j)
1	12	43	3.58
2	6	42	7.00
3	8	44	5.50
4	4	46	11.50
5	8	44	5.50

Table 5.14 Values of α_i, $\sigma_\alpha(i)$, β_i, $\sigma_\beta(i)$, for $i = 1, \ldots, 38$, in the Malabar problem (case of a single I/O point).

i	α_i	$\sigma_\alpha(i)$	β_i	$\sigma_\beta(i)$	i	α_i	$\sigma_\alpha(i)$	β_i	$\sigma_\beta(i)$
1	11.50	4	2	1	20	5.50	5	11	20
2	11.50	4	2	11	21	5.50	5	14	21
3	11.50	4	3	6	22	5.50	5	14	31
4	11.50	4	3	16	23	5.50	5	15	26
5	7.00	2	4	02	24	5.50	5	15	36
6	7.00	2	4	12	25	5.50	5	16	22
7	7.00	2	5	7	26	5.50	5	16	32
8	7.00	2	5	17	27	3.58	1	17	27
9	7.00	2	6	3	28	3.58	1	17	37
10	7.00	2	6	13	29	3.58	1	18	23
11	5.50	3	7	8	30	3.58	1	18	33
12	5.50	3	7	18	31	3.58	1	19	28
13	5.50	3	8	4	32	3.58	1	19	38
14	5.50	3	8	14	33	3.58	1	20	24
15	5.50	3	9	9	34	3.58	1	20	34
16	5.50	3	9	19	35	3.58	1	21	29
17	5.50	3	10	5	36	3.58	1	21	39
18	5.50	3	10	15	37	3.58	1	22	25
19	5.50	5	11	10	38	3.58	1	22	35

Table 5.15 Optimal assignment of SKUs to the storage locations in the Malabar problem (case of a single I/O point).

SKU (j)	Storage location (k)
1	23, 24, 25, 27, 28, 29, 33, 34, 35, 37, 38, 39
2	2, 3, 7, 12, 13, 17
3	4, 5, 8, 9, 14, 15, 18, 19
4	1, 6, 11, 16
5	10, 20, 21, 22, 26, 31, 32, 36

5.12 Inventory Management

As explained in Section 5.1, stocks are finished goods, components, and materials held by an organization at different locations in a supply network for later utilization, production, or resale. *Inventory management* aims at sizing the stock levels across a logistics system as well as deciding when, and how much, to order of each SKU (also called product following this section). The objective is to provide a pre-established level of service at minimum cost.

Inventory management is substantially different for demand-independent products (i.e., finished products) and demand-dependent products (components of finished products). In the latter case, a technique known as *material requirements planning* (MRP) is used to schedule purchase and manufacturing activities. The goal is to make raw materials and components available at the right time in order to feed the production process while maintaining the lowest possible inventory level. For an in-depth examination of this topic, the reader is referred to specialized books, whereas the reminder of this section will be focused on demand-independent products.

Inventory management models can be classified as *deterministic* or *stochastic*, depending on whether demand and lead time are known with certainty or uncertainty, respectively. Lead time corresponds to the time interval between the issuing of an order and the delivery of the corresponding goods. Depending on the type of business or industry in which one is involved, it may be used as a synonym of delivery time or, sometimes, of OCT (see Section 1.8.3).

5.12.1 Deterministic models

Deterministic models, besides having several applications (including inventory management in petrochemical companies supplied by pipelines), present general features whose study is fundamental for the development of models under uncertainty.

Deterministic models with a constant demand rate

In the case of a single product with a constant demand rate, the replenishments are periodic and the objective pursued consists in the minimization of the average total cost in each period. Let d be the constant demand rate, T the time lapse between two consecutive orders and q the order size, i.e., the product quantity ordered at each replenishment. This means that

$$q = dT. \tag{5.33}$$

Let c be the unit production cost or purchase price (depending on whether the product is internally produced or bought from an external supplier), supposed to be independent of the order size. Moreover, let k be the fixed reorder (or resupply) cost, h the unit holding (or carrying) cost per unit of time, u the unit shortage (or stockout) cost, independent of the duration of the shortage (e.g., the paperwork cost of managing an urgent order), and v the unit shortage cost per unit of time (e.g., the discount to be offered to the customer for late delivery of the order). The unit holding cost h includes storage cost, opportunity cost (due to the lost interest on capital invested in inventory), taxes on inventory, insurance cost, and other cost components associated with theft, damage, or stocked product obsolescence. Generally, the holding cost is expressed as a percentage of the unit production cost or purchase price,

$$h = pc, \tag{5.34}$$

where p can be relatively high (e.g., 10–20% on an annual basis), especially when the bank interests are high. Moreover, let $I(t)$ be the inventory level at time t, $A(t)$ the shortage level at time t, M the maximum inventory level, m the maximum shortage, l the lead time. The problem is to determine q (or, equivalently, T), and m in such a way that the overall average cost per unit of time is minimum. Two different cases are considered, according to whether the reorder time is instantaneous or not.

Case of non-instantaneous replenishment. Let $T_r \geq 0$ be the *replenishment time*, that is, the time to make a replenishment, and r the *replenishment rate*, that is, the product quantity per unit of time received during T_r. This case also includes manufacturing firms that, in order to fulfill the demand of a product, realize its production internally, rather than contracting with an external supplier. Production takes place at a rate r and therefore, over time equal to T_r, a product quantity q is manufactured equal to:

$$q = rT_r. \tag{5.35}$$

The inventory level $I(t)$ as a function of time t is shown in Figure 5.53. The dashed lines represent the gross inventory during the replenishment phases (their slope is r), i.e., without considering the ordered quantity exiting the warehouse in the meantime. In effect, since the product quantity is picked up at a rate d while a replenishment takes place, the slope of the inventory during T_r is $r - d$. Finally, after a replenishment, the inventory level decreases at a rate d. Let T_1, T_2, T_3, and T_4 be the time the inventory level of the product takes to go from $-m$ to 0, from 0 to M, from M to 0 and from 0 to $-m$, respectively. Note that $T_r = T_1 + T_2$. The maximum inventory level M is given by

$$m + M = (r - d)T_r.$$

Therefore, from formula (5.35),

$$M = (r - d)T_r - m = q(1 - d/r) - m.$$

The total average cost per unit of time $\mu(q, m)$ is

$$\mu(q, m) = \frac{1}{T}(k + cq + h\bar{I}T + um + v\bar{A}T). \tag{5.36}$$

The factor in parentheses on the right-hand side is the *average cost per cycle* (a cycle is the time $T = T_1 + T_2 + T_3 + T_4$ elapsed between two consecutive replenishments), given by the fixed and variable reorder costs of a replenishment (k and cq, respectively), and

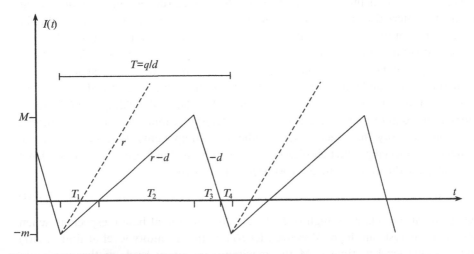

Figure 5.53 Inventory level as a function of time.

the holding cost $h\bar{I}T$, plus the shortage costs (um and $v\bar{A}T$). The holding and shortage costs depend on the average inventory level $\bar{I}$, and on the average shortage level $\bar{A}$, respectively:

$$\bar{I} = \frac{1}{T} \int_0^T I(t)dt = \frac{1}{T}\left(\frac{M(T_2 + T_3)}{2}\right);$$

$$\bar{A} = \frac{1}{T} \int_0^T A(t)dt = \frac{1}{T}\left(\frac{m(T_1 + T_4)}{2}\right).$$

Moreover, since

$$m = (r - d)T_1;$$

$$M = (r - d)T_2;$$

$$M = dT_3;$$

$$m = dT_4,$$

time intervals T_1, T_2, T_3 and T_4 are given by

$$T_1 = \frac{m}{r - d};$$

$$T_2 = \frac{M}{r - d};$$

$$T_3 = \frac{M}{d};$$

$$T_4 = \frac{m}{d}.$$

Consequently,

$$\bar{I} = \frac{M^2}{2q(1 - d/r)} = \frac{\left[q(1 - d/r) - m\right]^2}{2q(1 - d/r)}; \tag{5.37}$$

$$\bar{A} = \frac{m^2}{2q(1 - d/r)}. \tag{5.38}$$

Finally, using formulae (5.33), (5.37), and (5.38), formula (5.36) can be rewritten as

$$\mu(q, m) = kd/q + cd + \frac{h\left[q(1 - d/r) - m\right]^2}{2q(1 - d/r)} + umd/q + \frac{vm^2}{2q(1 - d/r)}. \tag{5.39}$$

If shortages are allowed, the minimum point (q^*, m^*) of the convex function $\mu(q, m)$ can be obtained by solving the equations

$$\frac{\partial\mu(q, m)}{\partial q} = 0;$$

$$\frac{\partial\mu(q, m)}{\partial m} = 0.$$

As a result,

$$q^* = \sqrt{\frac{h + v}{v}}\sqrt{\frac{2kd}{h(1 - d/r)} - \frac{(ud)^2}{h(h + v)}}, \tag{5.40}$$

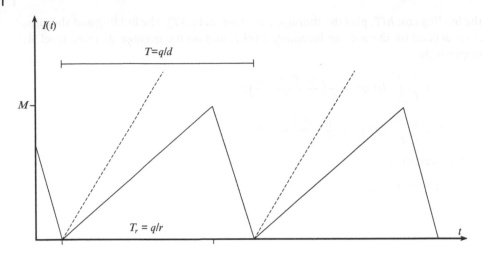

Figure 5.54 Inventory level as a function of time when shortage is not allowed.

and

$$m^* = \frac{(hq^* - ud)(1 - d/r)}{(h + v)}.$$
(5.41)

If shortages are not allowed (see Figure 5.54), formula (5.39) can be simplified since $m = 0$:

$$\mu(q) = kd/q + cd + \frac{hq(1 - d/r)}{2}.$$
(5.42)

Hence, a single equation has to be solved:

$$\frac{d\mu(q)}{dq} = 0,$$

Finally, the optimal order size q^* is (see Figure 5.55):

$$q^* = \sqrt{\frac{2kd}{h(1 - d/r)}}.$$
(5.43)

Golden Food distributes tinned foodstuff in Great Britain. In a warehouse located in Birmingham, the demand rate d for tomato puree is 400 palletized unit loads per month. The value of a palletized unit load, corresponding to the purchase price, is $c = £\,2500$ and the annual unit holding cost can be computed according to equation (5.34), where p is 14.5%. Issuing an order costs £ 30. The replenishment rate r is 40 palletized unit loads per day. Shortages are not allowed. The holding cost corresponds to

$$h = 0.145 \times 2500 = £\,362.50 \text{ (per palletized unit load per year)}$$

$$= £\,30.21 \text{ (per palletized unit load per month)}.$$

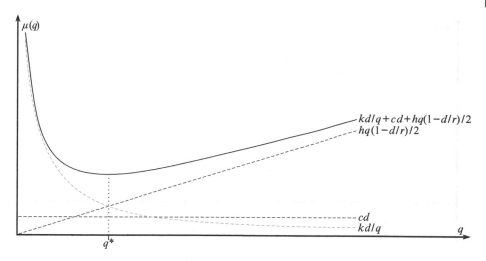

Figure 5.55 Average costs as a function of q.

Therefore, from formula (5.43),

$$q^* = \sqrt{\frac{2 \times 30 \times 400}{30.21\,[1 - 400/(40 \times 20)]}} = 39.86 \approx 40 \text{ palletized unit loads,}$$

where it is supposed that the number of the workdays in a month equals 20 (hence the demand rate d is 20 palletized unit loads per workday). Finally, from formulae (5.33) and (5.35),

$$T^* = 40/400 = 1/10 \text{ month} = 2 \text{ workdays;}$$

$$T_r^* = 40/40 = 1 \text{ workday.}$$

Case of instantaneous replenishment. If replenishment is instantaneous, the optimal inventory policy can be obtained by formulae (5.40), (5.41), and (5.43), taking into account that $r \to \infty$. If shortages are allowed (see Figure 5.56), then

$$q^* = \sqrt{\frac{h + v}{v}} \sqrt{\frac{2kd}{h} - \frac{(ud)^2}{h(h + v)}};$$

$$m^* = \frac{hq^* - ud}{(h + v)}.$$

If shortages are not allowed (see Figure 5.57), formula (5.42) becomes

$$\mu(q) = kd/q + cd + \frac{hq}{2} \tag{5.44}$$

and the optimal order size is given by

$$q^* = \sqrt{\frac{2kd}{h}}. \tag{5.45}$$

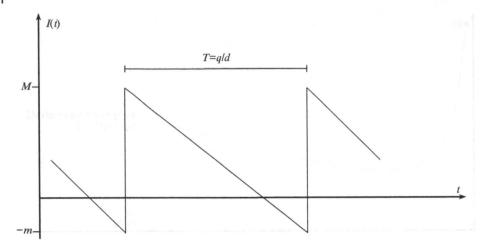

Figure 5.56 Inventory level as a function of time in the instantaneous replenishment case.

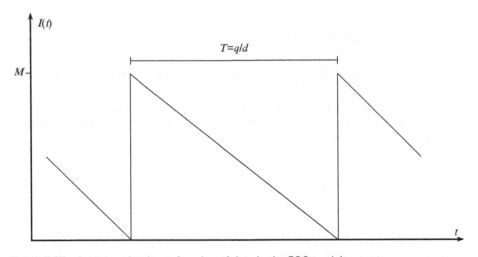

Figure 5.57 Inventory level as a function of time in the EOQ model.

This is the classical *economic order quantity* (EOQ) model introduced by F. W. Harris. The total cost per unit of time of an EOQ policy is

$$\mu(q^*) = \sqrt{2kdh} + cd.$$

Optimal policies with no shortage allowed (in particular, the EOQ policy) satisfy the *zero inventory ordering* (ZIO) property which states that an order is received exactly when the inventory level falls to zero.

Al-Bufeira Motors manufactures spare parts for aircraft engines in Saudi Arabia. Its component YO2PN, produced in a plant located in Jiddah, has a demand of 220 units per year and a production cost of $ 1200. Manufacturing this product requires a time-consuming setup that costs $ 800. The unit annual holding cost is $ 192. Shortages are not allowed. Therefore, from formula (5.45),

$$q^* = \sqrt{\frac{2 \times 800 \times 220}{192}} = 42.82 \approx 43 \text{ units,}$$

and, from formula (5.33),

$$T^* = 42.82/220 = 0.19 \text{ years} = 71.04 \text{ days.}$$

The total cost is given by formula (5.44):

$$kd/q^* + \frac{hq^*}{2} = \frac{800 \times 220}{42.82} + \frac{192 \times 42.82}{2} = \$ 8220.95/\text{year,}$$

plus

$$cd = 220 \times 1200 = \$ 264\,000/\text{year.}$$

Another important aspect regarding the models just presented concerns the lead time l, assumed here to be known with certainty. If it were equal to zero, the reorder could be issued at the moment when the inventory level becomes equal to $-m$, otherwise (i.e., $l > 0$), the replenishment must be anticipated by l units of time so that the shortage level does not exceed m. Starting from l, the *reorder point* s (defined as the value of $I(t)$ in correspondence of which a new replenishment is needed) can be calculated as

$$s = (l - \lfloor l/T \rfloor T)d - m,$$

where $\lfloor l/T \rfloor$ is the number of cycles of delay included in the lead time (see Figure 5.58).

In the Al-Bufeira Motors problem, a setup has to be planned six days in advance. Assuming $T = 67$ days, the reorder point s is

$$s = (6 - \lfloor 6/67 \rfloor \times 67)\frac{220}{365} - 0 = 3.62 \approx 4 \text{ units.}$$

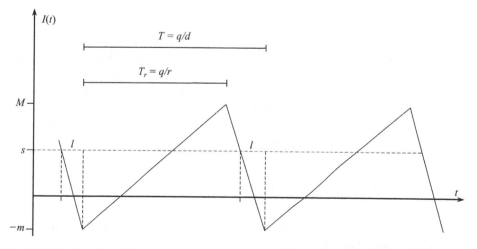

Figure 5.58   Reorder point.

Deterministic models with a time-varying demand rate

If the demand rate is deterministic but time varying, the following procedure can be adopted. Let $1, \dots, T_H$ be a finite and discrete time horizon and let $d_t, t = 1, \dots, T_H$, be the product demand at time period t. The problem is to decide how much to order in each time period in such a way that the sum of fixed and variable reorder costs, plus holding costs is minimized. No shortage is allowed. H. M. Wagner and T. M. Whitin formulated this problem as follows. The decision variables are $q_t, t = 1, \dots, T_H$, that is, the product quantity ordered at the beginning of each time period t, the inventory level $I_t, t = 1, \dots, T_H$, at the end of time period t; in addition let $y_t, t = 1, \dots, T_H$, be a binary decision variable equal to 1 if an order is placed in time period t, 0 otherwise. The problem is then

$$\text{Minimize} \sum_{t=1,\dots,T_H} (ky_t + cq_t + hI_t) \tag{5.46}$$

subject to

$$I_t = I_{t-1} + q_t - d_t, t = 1, \dots, T_H \tag{5.47}$$

$$q_t \le y_t \sum_{j=t,\dots,T_H} d_j, t = 1, \dots, T_H \tag{5.48}$$

$$I_0 = 0 \tag{5.49}$$

$$I_t \ge 0, t = 1, \dots, T_H$$

$$q_t \ge 0, t = 1, \dots, T_H$$

$$y_t \in \{0, 1\}, t = 1, \dots, T_H,$$

where objective function (5.46) is the total cost. Formulae (5.47) are the inventory balance constraints, inequalities (5.48) state that for each time period $t = 1, \dots, T_H$, q_t is zero if y_t is zero, and formula (5.49) specifies the initial inventory.

An optimal solution of the Wagner–Whitin model can be obtained in $O(T_H^2)$ time through a dynamic programming algorithm. This algorithm is based on the theoretical result which states that any optimal policy satisfies the ZIO property, that is,

$$q_t I_{t-1} = 0, t = 1, \dots, T_H.$$

The proof is left to the reader as an exercise (see Problem 5.26). A corollary of the previous proposition is that in an optimal policy, the product quantity ordered at each time period is the total demand of a set of consecutive subsequent time periods.

The algorithm is as follows. Let $G = (V, A)$ be a directed acyclic graph, where $V = \{1, \dots, T_H, T_H + 1\}$ is a vertex set and $A = \{(t, t') : t = 1, \dots, T_H, t' = t + 1, \dots, T_H + 1\}$ is an arc set. With each arc (t, t') is associated the following cost in time period t to satisfy the product demands in time periods $t, t + 1, \dots, t' - 1$:

$$g_{tt'} = k + c \sum_{j=t}^{t'-1} d_j + h \sum_{j=t}^{t'-1} (j - t) d_j.$$

Then a shortest path from vertex 1 to vertex $T_H + 1$ corresponds to a least cost inventory policy (see Problem 5.25).

[SaoVincente.xlsx, SaoVincente.py] Sao Vincente Chemical is a Portuguese company producing lubricants. In the next year its product named Serrado Oil is expected to have a demand of 720, 1410, 830 and 960 palletized unit loads in winter, spring, summer, and autumn, respectively. Manufacturing this product requires a time-consuming setup that costs € 8900. The variable production cost amounts to € 350 per palletized unit load while the initial inventory is zero. The unit holding cost is € 6.50 per season.

Let $t = 1, 2, 3, 4$, represent the winter, spring, summer, and autumn periods, respectively. By solving the Wagner–Within model using the Excel Solver tool, it follows that the optimal policy is to produce at the beginning of winter, spring, and autumn of the next year. In particular, $y_1^* = y_2^* = y_4^* = 1$, $y_3^* = 0$, $q_1^* = 720$, $q_2^* = 2240$, $q_3^* = 0$, $q_4^* = 960$, $I_1^* = 0$, $I_2^* = 830$, $I_3^* = 0$, $I_4^* = 0$. The total holding, production and setup cost amounts to € 1 404 095.

Deterministic models with quantity discounts

In the previous models, it has been assumed that the unit production cost or purchase price is always constant (equal to c). In practice, quantity discounts offered by suppliers, or economies of scale in the manufacturing processes, make the unit production cost or purchase price dependent on the order size q. In the following, the most practical applications of quantity discounts are examined under the EOQ hypothesis: (a) *all-quantity discounts* and (b) *incremental discounts*. In case (a), the production or purchase cost function $f(q)$ is assumed to be piecewise linear (see Figure 5.59):

$$f(q) = c_i q, \quad q_{i-1} \le q < q_i, \quad i = 1, \dots, n,$$

where $q_0 = 0, q_1, \dots, q_n$, are known *discount breaks* $(q_i < q_{i+1}, i = 1, \dots, n-1)$, $f(q_0) = 0$, and $c_i > c_{i+1}$, $i = 1, 2 \dots, n$. Hence, if the order size q is included between discount breaks q_{i-1} and q_i, the unit production or purchase cost is c_i, $i = 1, \dots, n$. It is worth noting that, depending on c_i coefficients, $f(q)$ can be greater than $f(q')$ for $q < q'$

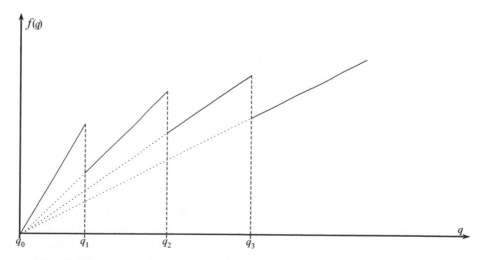

Figure 5.59 Plot of the production or purchase cost function in the case of all-quantity discounts.

(see Figure 5.59). In practice, the *effective* cost function is $\min\{c_i q, c_{i+1} q_i\}, q_{i-1} \leq q <$ $q_i, i = 1, \ldots, n$, i.e., if the minimum cost corresponds to $c_{i+1} q_i$, this means that it is more convenient to product or purchase a greater quantity q_i instead of q.

The total average cost function $\mu(q)$ can be written as

$$\mu(q) = \mu_i(q), \; q_{i-1} \leq q < q_i, \; i = 1, \ldots, n,$$

where, since $h_i = pc_i, i = 1, \ldots, n$,

$$\mu_i(q) = kd/q + c_i d + \frac{h_i q}{2}, \; i = 1, \ldots, n. \tag{5.50}$$

Then, the optimal order size q^* can be obtained through the following EOQ_ALL-QUANTITY_DISCOUNTS procedure.

1: **procedure** EOQ_ALL-QUANTITY_DISCOUNTS($k, d, \mathbf{q}, \mathbf{h}, q^*$)

2:    # Determine the order size q_i' that minimizes $\mu_i(q), i = 1, \ldots, n$, by imposing

$$\frac{d\,\mu_i(q_i)}{d\,q_i} = 0, \; i = 1, \ldots, n;$$

3:    **for** $i = 1, \ldots, n$ **do**

4:

$$q_i' = \sqrt{\frac{2kd}{h_i}}; \tag{5.51}$$

5:    **end for**

6:    $q_0 = 0$;

7:    **for** $i = 1, \ldots, n$ **do**

8:

$$q_i^* = \begin{cases} q_{i-1}, & \text{if } q_i' < q_{i-1} \\ q_i', & \text{if } q_{i-1} \leq q_i' \leq q_i \; ; \\ q_i, & \text{if } q_i' > q_i \end{cases} \tag{5.52}$$

9:    **end for**

10:    # Computation of the optimal solution q^*;

11:

$$i^* = \arg \min_{i=1,\ldots,n} \{\mu(q_i^*)\};$$

12:    $q^* = q_{i*}^*$;

13:    **return** q^*;

14: **end procedure**

Maliban runs more than 200 stationery outlets in Spain. The firm buys its products from a restricted number of suppliers and stores them in a warehouse located near Seville. Maliban expects to sell 3000 boxes of the Prince Arthur pen during the next year. The percentage p with which to calculate the annual holding cost is 30%. Placing an order costs € 50. The supplier offers a box at € 3, if the amount bought is less than 500 boxes. The price is reduced by 1% if 500 to 2000 boxes are ordered. Finally, if more than 2000 boxes are ordered, an additional 0.5% discount is applied. Then, by using formulae (5.51) and (5.52):

$$q_1' = \sqrt{\frac{2 \times 50 \times 3000}{0.30 \times 3}} = 577.35 \text{ boxes;}$$

$$q_2' = \sqrt{\frac{2 \times 50 \times 3000}{0.30 \times 2.97}} = 580.26 \text{ boxes;}$$

$$q_3' = \sqrt{\frac{2 \times 50 \times 3000}{0.30 \times 2.955}} = 581.73 \text{ boxes;}$$

$$q_1^* = 500 \text{ boxes}, q_2^* = 580.26 \text{ boxes}, q_3^* = 2000 \text{ boxes.}$$

By comparing the corresponding annual average costs given by formula (5.50), the optimal order size is $q^* = 580$ boxes ($\approx q_2^*$), corresponding to an annual average cost of € 9427.

In case (b), the production or purchase cost function $f(q)$ is assumed to be dependent on q as follows (see Figure 5.60):

$$f(q) - f(q_{l-1}) + c_l(q - q_{l-1}), \quad q_{l-1} \le q < q_i, \quad i = 1, \dots, n \tag{5.53}$$

where $q_0 = 0, q_1, \dots, q_n$ are known discount breaks ($q_i < q_{i+1}, i = 1, \dots, n-1$), $f(q_0) = 0$, and $c_i > c_{i+1}$, $i = 1, 2 \dots$. Consequently, if the order size q is included between discount breaks q_{i-1} and q_i, the unit production cost or purchase price of the quantity $(q - q_{i-1})$ is c_i, the unit production cost or purchase price of the quantity $(q_{i-1} - q_{i-2})$ is c_{i-1} etc. The average total cost function $\mu(q)$ is

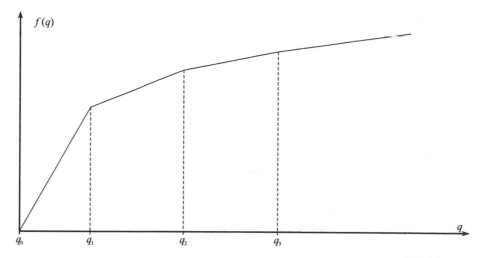

Figure 5.60 Plot of the production or purchase cost function in the case of incremental discounts.

$$\mu(q) = \mu_i(q), \ q_{i-1} \le q < q_i, \ i = 1, \dots, n$$

where, on the basis of formula (5.44),

$$\mu_i(q) = kd/q + f(q)d/q + p\frac{f(q)}{q}q/2, \ i = 1, \dots, n$$

Using formula (5.53), $\mu_i(q), i = 1, \dots, n$, can be rewritten as

$$\mu_i(q) = kd/q + [f(q_{i-1}) + c_i(q - q_{i-1})]d/q$$

$$+ \frac{p}{2}[f(q_{i-1}) + c_i(q - q_{i-1})], \ i = 1, \dots, n \qquad (5.54)$$

The optimal order size q^* can be computed through a procedure, namely EOQ_INCREMENTAL_DISCOUNTS, very similar to that used in the previous case and illustrated below.

1: **procedure** EOQ_INCREMENTAL_DISCOUNTS $(k, d, p, \mathbf{c}, \mathbf{q}, \mathbf{h}, q^*)$

2:   # Determine the order size q_i' that minimizes $\mu_i(q), i = 1, \dots, n$, by imposing

$$\frac{d \mu_i(q_i)}{d q_i} = 0, \ i = 1, \dots, n;$$

3:   $q_0 = 0$;

4:   **for** $i = 1, \dots, n$ **do**

5:

$$q_i' = \sqrt{\frac{2d[k + f(q_{i-1}) - c_i q_{i-1}]}{h_i}} \ ; \qquad (5.55)$$

6:     **if** $q_i' \notin [q_{i-1}, q_i]$ **then**

7:       $\mu_i(q_i') = \infty$;

8:     **end if**

9:   **end for**

10:   **for** $i = 1, \dots, n$ **do**

11:

$$q_i^* = \begin{cases} q_{i-1}, & \text{if } q_i' < q_{i-1} \\ q_i', & \text{if } q_{i-1} \le q_i' \le q_i \ ; \\ q_i, & \text{if } q_i' > q_i \end{cases}$$

12:   **end for**

13:   # Computation of the optimal solution q^*;

14:

$$i^* = \arg\min_{i=1,\dots,n} \{\mu(q_i^*)\};$$

15:   $q^* = q_{i^*}^*$;

16:   **return** q^*;

17: **end procedure**

If Maliban applies an incremental discount policy on the number of boxes of Prince Arthur pens to purchase, then, by using formula (5.55),

$$q_1' = \sqrt{\frac{2 \times 3000 \times 50}{0.30 \times 3}} = 577.35 \text{ boxes};$$

$$q_2' = \sqrt{\frac{2 \times 3000[50 + (3 \times 500) - (2.97 \times 500)]}{0.30 \times 2.97}} = 661.60 \text{ boxes};$$

$$q_3' = \sqrt{\frac{2 \times 3000\{50 + [(3 \times 500) + (2.97 \times 1500)] - (2.955 \times 2000)\}}{0.30 \times 2.955}}$$

$$= 801.86 \text{ boxes}.$$

Consequently, as $q_1' > 500$ and $q_3' < 2000$, the optimal order size is $q^* = 662$ boxes ($\approx q_2'$), corresponding to an annual average cost, given by formula (5.54), equal to € 9501.73.

5.12.2 Stochastic Models

Inventory problems with uncertain demand or lead times have quite a complex mathematical structure. In this section, a restricted number of stochastic models are illustrated. A classical *newsboy problem* is first examined, where a one-shot reorder decision has to be made. Then, (s, S) *policies* are introduced for a variant of the newsboy problem. Finally, the most common inventory policies used by practitioners (namely, the *reorder point*, the *reorder cycle*, the (s, S) and the *two-bin* policies) are reviewed and compared. The first three policies make use of data forecasts, whereas the fourth policy does not require any data estimate.

The newsboy problem

In the newsboy problem, a replenishment decision has to be made at the beginning of a period (before demand for the product is known) for a single product. It deals with the decision made by a newsboy (hence the name of the problem), on the number of newspapers to purchase early in the morning to sell during the entire day, avoiding, on the one hand, having unsold newspapers at the end of the day and, on the other hand, running out of newspapers before the end of the day. The demand d is modelled as a random variable with a continuous cumulative distribution function $F_d(\delta)$. Let c be the unit production cost or purchase price, w and u be the unit selling price and the unit salvage value, respectively. Of course,

$$w > c > u.$$

There is no fixed reorder cost or an initial inventory. In addition, shortage costs are assumed to be negligible. If the company orders a quantity q of the product, the *expected revenue* $\rho(q)$ is

$$\rho(q) = w \int_0^\infty \min(\delta, q) dF_d(\delta) + u \int_0^\infty \max(0, q - \delta) dF_d(\delta) - cq$$

$$= w \left(\int_0^q \delta dF_d(\delta) + q \int_q^\infty dF_d(\delta) \right) + u \int_0^q (q - \delta) dF_d(\delta) - cq.$$

By adding and subtracting $r \int_q^\infty \delta dF_d(\delta)$ to the right-hand side, $\rho(q)$ becomes

$$\rho(q) = wE[d] + r \int_q^\infty (q - \delta) dF_d(\delta) + u \int_0^q (q - \delta) dF_d(\delta) - cq, \tag{5.56}$$

where $E[d]$ is the *expected demand*. It is easy to show that $\rho(q)$ is concave for $q \geq 0$, and $\rho(q) \to -\infty$ for $q \to \infty$. As a result, the maximum expected revenue is achieved when the derivative of $\rho(q)$ with respect to q is zero. Hence, by applying the Leibnitz rule, the optimality condition becomes

$$w(1 - F_d(q)) + uF_d(q) - c = 0,$$

where, by definition, $F_d(q)$ is the probability $Pr(d \leq q)$ that the demand does not exceed q. As a result, the *optimal order quantity S* satisfies the following condition:

$$Pr(d \leq S) = \frac{w - c}{w - u}. \tag{5.57}$$

Emilio Tadini & Sons is a hand-made shirt retailer located in Rome, Italy, close to Piazza di Spagna. This year Mr Tadini faces the problem of ordering a new, brightly coloured shirt made by a Florentine firm. He assumes that the demand is uniformly distributed between 200 and 350 units. The purchase cost is $c = €\,18$ while the selling price is $w = €\,52$ and the salvage value of one unsold shirt is $u = €\,7$. According to formula (5.56), the expected revenue is

$$\rho(q) = 52 \times 275 + 52 \int_{200}^{350} (q - \delta) \frac{1}{350 - 200} d\delta - 18q = 34q,$$

for $0 \leq q \leq 200$;

$$\rho(q) = 52 \times 275 + 52 \int_q^{350} (q - \delta) \frac{1}{350 - 200} d\delta$$

$$+ 7 \int_{200}^q (q - \delta) \frac{1}{350 - 200} d\delta - 18q = -0.15q^2 + 94q - 6000,$$

for $200 < q \leq 350$, and

$$\rho(q) = 52 \times 275 + 7 \int_{200}^{350} (q - \delta) \frac{1}{350 - 200} d\delta - 18q = -11q + 12\,375,$$

for $q > 350$.

According to formula (5.57), $Pr(d \leq S) = (S - 200)/(350 - 200)$ for $200 \leq S \leq 350$. Hence, Mr Tadini should order $S = 313.33 \approx 313$ shirts.

Hence, the corresponding maximum expected revenue is equal to $\rho(313) = €\,8726.65$.

The (s, S) policy for single-period problems

If there is an initial inventory q_0 and a fixed reorder cost k, the optimal replenishment policy can be obtained as follows. If $q_0 \geq S$, no reorder is needed. Otherwise, the best policy is to order a quantity equal to $S - q_0$, provided that the expected revenue associated with this choice is greater than the expected revenue associated with not ordering anything. Hence, two cases can occur:

1. if the expected revenue $\rho(S) - k - cq_0$ associated with reordering is greater than the expected revenue $\rho(q_0) - cq_0$ associated with not reordering, then a quantity $S - q_0$ has to be reordered;
2. otherwise, no order has to be placed.

As a consequence, if $q_0 < S$, the optimal policy consists of ordering a quantity $S - q_0$ if $\rho(q_0) \leq \rho(S) - k$. In other words, if s is the value such that

$$\rho(s) = \rho(S) - k,$$

the optimal policy is to order a quantity $S - q_0$ if the initial inventory level q_0 is less than or equal to s, otherwise not to order. Policies like this are known as (s, S) policies. Parameter s acts as a reorder point, while S is sometimes called the *order-up-to-level*.

> If $q_0 = 50$ and $k = €\ 400$ in the Emilio Tadini & Sons problem, $\rho(s) = \rho(S) - k = €\ 8326.65$ so that $s = 262$. As $q_0 < s$, the optimal policy is to order $S - q_0 = 263$ shirts.

The reorder point policy

In the reorder point policy (or *fixed order quantity policy*), the inventory level is kept under observation in a continuous way. As soon as its net value $I(t)$ (the quantity in stock minus the demand unsatisfied plus the backlog, i.e., the quantity ordered but not yet received) reaches a reorder point s, a constant quantity q is ordered (see Figure 5.61).

The reorder size q is computed through the procedures illustrated in the previous sections, by replacing d with $\overline{d}$. In particular, under the EOQ hypothesis,

$$q = \sqrt{\frac{2k\overline{d}}{h}}. \tag{5.58}$$

The reorder point s is obtained in the following way. Let $\alpha \in [0, 1]$ be the probability of accepting stockout during the lead time l (e.g., $\alpha = 0.05$). This is equivalent to assuming that the demand can only exceed m during the time interval l with probability α. Consequently, the probability of having a non-negative inventory during the lead time l is $(1 - \alpha)$, corresponding to a desired service level (e.g., 95%).

In the following, it is assumed that

- the demand rate d is distributed according to a normal distribution with mean $\overline{d}$ and standard deviation σ_d;
- $\overline{d}$ and σ_d are constant in time;
- the lead time l is deterministic or is distributed according to a normal distribution with mean $\overline{l}$ and standard deviation σ_l;
- the demand rate and the lead time are statistically independent.

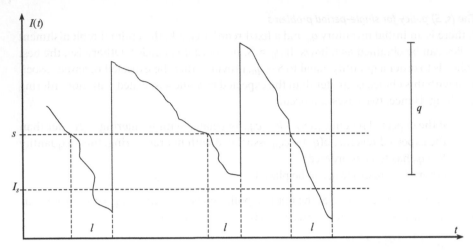

Figure 5.61 Reorder point policy.

The average demand rate $\overline{d}$ can be forecast with one of the methods illustrated in Chapter 2, while the standard deviation σ_d can be estimated as the square root of *MSE*. Analogous procedures can be adopted for the estimation of $\overline{l}$ and σ_l.

Let z_α be the value under which a standard normal random variable falls with probability $(1 - \alpha)$ (for more details, see the Moderan problem in Section 1.10.3). If l is deterministic, then

$$s = \overline{d}l + z_\alpha \sigma_d \sqrt{l},$$ (5.59)

where $\overline{d}l$ and $\sigma_d \sqrt{l}$ are the mean and the standard deviation of the demand in a time interval of duration l, respectively. If l is random, then

$$s = \overline{d}\,\overline{l} + z_\alpha \sqrt{\sigma_d^2 \overline{l}^2 + \sigma_l^2 \overline{d}^2},$$

where $\overline{d}\,\overline{l}$ and $\sqrt{\sigma_d^2 \overline{l}^2 + \sigma_l^2 \overline{d}^2}$ are the mean and the standard deviation of the demand in a time interval of random duration l, respectively.

The reorder point s, minus the average demand in the reorder period, constitutes a *safety stock* I_S. For example, in case l is deterministic, the safety stock is

$$I_S = s - \overline{d}\,l = z_\alpha \sigma_d \sqrt{l}.$$ (5.60)

Papier is a French retail chain. At the outlet located in downtown Lyon, the expected demand for mouse pads is 45 units per month, whereas *MSE* is 23.5. The purchase price of a mouse pad is € 4, and the fixed reorder cost is equal to € 30. The annual unit holding cost is € 0.8, corresponding to € $(0.8/12) = 0.067$ on a monthly basis. Lead time is one month and a service level equal to 95% is required. This means that $\alpha = 0.05$ and $z_{0.05} = 1.645$.

Since $\overline{d} = 45$, from formula (5.58),

$$q^* = \sqrt{\frac{2 \times 30 \times 45}{0.067}} = 201.25 \approx 201 \text{ mouse pads.}$$

Moreover, σ_d can be estimated as

$$\sigma_d = \sqrt{23.5} = 4.85.$$

Since $l = 1$, from formula (5.59), the reorder point s is

$$s = 45 + 1.645 \times 4.85 = 52.97 \approx 53 \text{ mouse pads.}$$

Consequently, the safety stock I_s is equal to

$$I_s = 53 - 45 = 8 \text{ mouse pads.}$$

The reorder cycle policy

In the reorder cycle policy (or *periodic review* policy) the inventory level is kept under observation periodically at time instants t_i ($t_{i+1} = t_i + T$, $T \geq 0$). At time t_i, a product quantity equal to $q_i = S - I(t_i)$ is ordered (see Figure 5.62). Parameter S (referred to as the order-up-to-level) represents the maximum inventory level in case the lead time is negligible.

The review period T can be chosen using procedures analogous to those used for determining q^* in the deterministic models. For instance, under the EOQ hypothesis,

$$T = \sqrt{\frac{2k}{h\overline{d}}}. \tag{5.61}$$

The parameter S is determined in such a way that a stockout will occur with a probability α. Since the *risk interval* is equal to T plus l, S is required to be greater than or equal to the demand in time interval $T + l$, with probability equal to α. If lead time l is deterministic, then

$$S = \overline{d}(T + l) + z_\alpha \sigma_d \sqrt{T + l}, \tag{5.62}$$

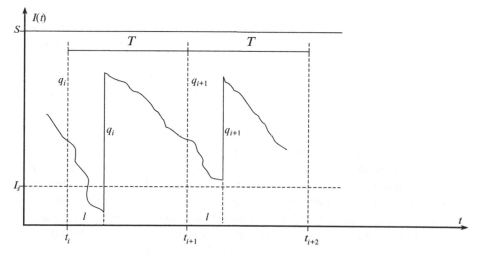

Figure 5.62 Reorder cycle policy.

where $\overline{d}(T + l)$ and $\sigma_d \sqrt{T + l}$, are the mean and the standard deviation of the demand in time interval $T + l$, respectively. If lead time is a random variable, then

$$S = \overline{d}(T + \overline{l}) + z_\alpha \sqrt{\sigma_d^2 (T + \overline{l}) + \sigma_l^2 \overline{d}^2},$$

where $\overline{d}(T + \overline{l})$ and $\sqrt{\sigma_d^2 (T + \overline{l}) + \sigma_l^2 \overline{d}^2}$ are the mean and the standard deviation of the demand in time interval $T + \overline{l}$, respectively.

The difference between S and the average demand in $T + \overline{l}$ makes up a safety stock I_S. For example, if the lead time is deterministic,

$$I_S = z_\alpha \sigma_d \sqrt{T + l}. \tag{5.63}$$

Comparing formula (5.63) with formula (5.60), it can be seen that the reorder cycle inventory policy involves a higher level of safety stock. However, such a policy does not require a continuous monitoring of the inventory level.

In the Papier inventory problem, the parameters of the reorder cycle policy, computed through formulae (5.61) and (5.62), are

$$T = \sqrt{\frac{2 \times 30}{0.067 \times 45}} = 4.47 \text{ months,}$$

$$S = 45 \times (4.47 + 1) + 1.645 \times 4.85 \times \sqrt{4.47 + 1} = 264.90 \approx 265 \text{ mouse pads.}$$

The associated safety stock, given by formula (5.63), is equal to

$$I_S = 1.645 \times 4.85 \times \sqrt{4.47 + 1} = 18.65 \approx 19 \text{ mouse pads.}$$

The (s, S) policy

The (s, S) inventory policy is a natural extension of the (s, S) policy illustrated for the single-period case. At time t_i, a quantity $S - I(t_i)$ is ordered if $I(t_i) < s$ (see Figure 5.63). If s is large enough ($s \rightarrow S$), the (s, S) policy is similar to the reorder cycle policy. On the other hand, if s is small ($s \rightarrow 0$), the (s, S) policy is similar to a reorder point policy with a reorder point equal to s and a reorder quantity $q \cong S$. On the basis of these observations, the (s, S) policy can be seen as a good compromise between the reorder point and the reorder cycle policies. Unfortunately, parameters T, S, and s are difficult to determine analytically. Therefore simulation is often used in practice (see Problem 5.28).

[Browns.xlsx] A supermarket close to Los Alamos, New Mexico, United States, belonging to the Browns supermarkets chain, makes use of an (s, S) policy for managing the inventory of 32 oz tomato juice bottles of a well-known brand. During a workday, if the inventory of this product falls below a certain minimum $s = 300$ bottles, then, at the end of the same day, a replenishment order is issued to restore the inventory of tomato juice to a maximum quantity $S = 600$ bottles. The lead time is one day. The inventory level of tomato juice at the end of November was 420 bottles. Considering the daily sales of this product recorded from 1–7 December (see the second column of Table 5.16), the adoption of the (s, S) policy described above implies the inventory levels reported in the third column of Table 5.16. In particular, at the end of 2 December, the inventory level was $238 < 300$ bottles. Consequently, a replenishment order of $600 - 238 = 362$ bottles was issued. At the end of subsequent day the inventory level was $238 - 91 + 362 = 509$ bottles. Similarly, on 6 December, the inventory level reached the value of 234 bottles. A replenishment order of $600 - 234 = 366$ bottles was placed. Considering the lead time of one day, the inventory level on 7 December was $234 - 85 + 366 = 515$ bottles.

Table 5.16 Daily sales and inventory levels of 32 oz tomato juice bottles in the first week of last December in a Browns supermarket.

		Inventory level	
Day	Sales	(s, S) policy	Two-bin policy
30 Nov	–	420	420
1 Dec	95	325	325
2 Dec	87	238	638
3 Dec	91	509	547
4 Dec	93	416	454
5 Dec	88	328	366
6 Dec	94	234	672
7 Dec	85	515	587

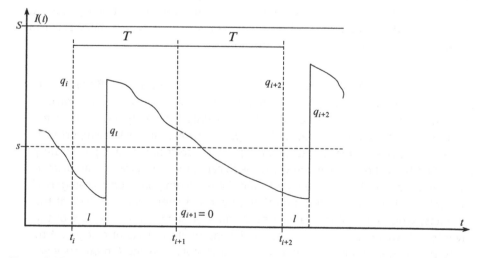

Figure 5.63 (s, S) policy.

The two-bin policy

The two-bin policy can be seen as a variant of the reorder point policy where no demand forecast is needed, and the inventory level does not have to be monitored continuously. The product in stock is supposed to be stored in two identical bins. As soon as one of the two bins becomes empty, an order is issued for an amount equal to the bin capacity.

[Browns.xlsx] If the two-bin policy is adopted for managing the inventory of 32 oz tomato juice bottles, the inventory levels in the first week of last December in the Browns supermarket in Los Alamos are reported in the fourth column of Table 5.16, assuming the capacity of each bin equal to 400 bottles. In particular, it is worth observing that during the week two replenishment orders were placed, at the end of 1 and 5 December respectively, when the inventory level was below 400 bottles (i.e., when a bin became empty). At the end of the subsequent days (i.e., 2 and 6 December), the inventory levels were $325 - 87 + 400 = 638$ bottles and $366 - 94 + 400 = 672$ bottles, respectively.

5.12.3 Selecting an Inventory Policy

It is quite common for a warehouse to contain thousands of products, each of which stocked in several hundreds (or even thousands) units. In such a context, products having a strong impact on the total cost have to be inventory-managed carefully while for less important products it is wise to resort to simple and low-cost inventory policies.

The problem is generally tackled by clustering the products into three categories using the *ABC* classification introduced in Section 1.11.2, on the basis of the average value of the items in stock. In particular, category *A* is made up of products corresponding to a high percentage (e.g., 80%) of the total warehouse value. Category *B* is constituted by a set of products associated with an additional 15% of the warehouse value, while category *C* is formed by the remaining products. The products of categories *A* and *B* should be managed with inventory policies based on forecasts and a frequent monitoring (e.g., category *A* by means of the reorder point policy and category *B* through the reorder cycle policy). Products in category *C* can be managed using the two-bin policy that does not require any forecast.

[WTC.xlsx] The Walloon Transport Consortium (WTC) operates a Belgian public transport service in the Walloon region. Buses are maintained in a facility located in Ans, close to a vehicle depot. The average inventory levels, the average unit values and the total value of the spare parts kept in stock are reported in Table 5.17. It was decided to allocate the products corresponding approximately to the first 80% of the total inventory value to category *A*, the products associated with the following 15% to category *B*, and the remaining products to category *C* (see Table 5.18). It is worth noting that category *A* contains about 30% of the products, while each of the categories *B* and *C* accounts for about 35% of the inventory. It was decided to manage the inventory of products of category *A* by using the reorder point policy, whereas the inventory of *B* and *C* products were managed by using the reorder cycle policy and the (*s*, *S*) policy, respectively.

Table 5.17 Features of the spare parts stocked by WTC.

SKU code	Average stock [units]	Average unit value [€]	Total value [€]
AX24	137	50	6 850
BR24	70	2 000	140 000
BW02	195	250	48 750
CQ23	6	6 000	36 000
CR01	16	500	8 000
FE94	31	100	3 100
LQ01	70	2 500	175 000
MQ12	18	200	3 600
MW20	75	500	37 500
NL01	15	1 000	15 000
PE39	16	3 000	48 000
RP10	20	2 200	44 000
SP00	13	250	3 250
TA12	100	2 500	250 000
TQ23	10	5 000	50 000
WQ12	30	12 000	360 000
WZ34	30	15	450
ZA98	70	250	17 500

Table 5.18 *ABC* classification of the spare parts in the WTC problem.

SKU code	Average stock [units]	Total value [€]	Cumulated average stock [%]	Total cumulated value [%]	Class
WQ12	30	360 000	3.25	28.87	A
TA12	100	250 000	14.10	48.92	A
LQ01	70	175 000	21.69	62.95	A
BR24	70	140 000	29.28	74.18	A
TQ23	10	50 000	30.37	78.19	A
BW02	195	48 750	51.52	82.10	B
PE39	16	48 000	53.25	85.95	B
RP10	20	44 000	55.42	89.47	B
MW20	75	37 500	63.56	92.48	B
CQ23	6	36 000	64.21	95.37	B
ZA98	70	17 500	71.80	96.77	C

(Continued)

Table 5.18 (Continued)

SKU code	Average stock [units]	Total value [€]	Cumulated average stock [%]	Total cumulated value [%]	Class
NL01	15	15 000	73.43	97.98	C
CR01	16	8 000	75.16	98.62	C
AX24	137	6 850	90.02	99.17	C
MQ12	18	3 600	91.97	99.45	C
SP00	13	3 250	93.38	99.72	C
FE94	31	3 100	96.75	99.96	C
WZ34	30	450	100.00	100.00	C
Total	922	1 247 000			

5.12.4 Multiproduct Inventory Models

When several products are kept in stock, their inventory policies are intertwined because of common constraints and joint costs, as discussed now in two separate cases. In the first case, a limit is placed on the total investment in inventory, or on the warehouse space. In the second case, products share joint reorder costs. For the sake of simplicity, both analyses will be performed under the EOQ model hypothesis.

Models with capacity constraints

Let n be the number of products in stock and q_j, $j = 1, \dots, n$, the quantity of product j ordered at each replenishment. The inventory management problem can be formulated as

$$\text{Minimize } \mu(q_1, \dots, q_n) \qquad (5.64)$$

$$\text{subject to}$$

$$g(q_1, \dots, q_n) \leq b \qquad (5.65)$$

$$q_1, \dots, q_n \geq 0, \qquad (5.66)$$

where the objective function (5.64) is the total average cost per unit of time. Under the EOQ hypothesis, the objective function $\mu(q_1, \dots, q_n)$ can be written as

$$\mu(q_1, \dots, q_n) = \sum_{j=1}^{n} \mu_j(q_j),$$

where, on the basis of formula (5.44),

$$\mu_j(q_j) = k_j d_j / q_j + c_j d_j + \frac{h_j q_j}{2}, \quad j = 1, \dots, n,$$

and quantities d_j, c_j, h_j, $j = 1, \dots, n$, are the demand rate, the unit production cost (or purchase price), and the unit holding cost of product j per unit of time, respectively. As is customary, $h_j = p_j c_j$, $j = 1, \dots, n$, i.e., h_j is computed as a percentage p_j of c_j.

Formula (5.65) is a side constraint (referred to as a "capacity constraint") representing both a budget constraint and a warehouse constraint. It can usually be considered as linear:

$$\sum_{j=1}^{n} a_j q_j \leq b, \tag{5.67}$$

where a_j, $j = 1, \ldots, n$, and b are constants. As a result, problem (5.64)–(5.66) can be solved through iterative algorithms for non-linear programming problems. Alternatively, the following simple heuristic, namely EOQ_WITH_CAPACITY_CONSTRAINTS procedure, can be used if the capacity constraint is linear and the percentage values p_j, $j = 1, \ldots, n$, are identical for all the products and equal to p.

[NewFrontier.xlsx] New Frontier distributes knapsacks and suitcases in most US states. Its most successful models are the Preppie knapsack and the Yuppie suitcase. The Preppie knapsack has an annual demand of 150 000 units, a unit purchase price of $ 30 and an annual holding cost equal to 20% of its purchase price. The Yuppie suitcase has an annual demand of 100 000 units, a unit purchase price of $ 45 and an annual holding cost equal to 20% of its purchase price. In both cases, placing an order for each of the two products costs $ 250. The company management requires that the average capital invested in inventories does not exceed $ 75 000. This condition can be expressed by the following constraint:

$$30q_1/2 + 45q_2/2 \leq 75\,000,$$

where it is assumed, as a precaution, that the average inventory level is the sum of the average inventory levels of the two products. The EOQ order sizes, given by formula (5.68),

$$q_1' = \sqrt{\frac{2 \times 250 \times 150\,000}{0.2 \times 30}} = 3535.53 \text{ units};$$

$$q_2' = \sqrt{\frac{2 \times 250 \times 100\,000}{0.2 \times 45}} = 2357.02 \text{ units},$$

do not satisfy the budget constraint. The optimal solution can be found by using the GRG (*generalized reduced gradient*) algorithm available in the Excel Solver tool:

$$q_1^* = 2500 \text{ units};$$

$$q_2^* = 1666.67 \text{ units},$$

whose total cost is $ 9 045 000. Applying the EOQ_WITH_CAPACITY_CONSTRAINTS procedure, by using formula (5.70), the same solution is obtained. In effect:

$$\delta^* = \left[\frac{1}{75\,000} \left(\frac{30}{2} \sqrt{\frac{2 \times 250 \times 150\,000}{30}} + \frac{45}{2} \sqrt{\frac{2 \times 250 \times 100\,000}{45}} \right) \right]^2$$

$$- 0.2 = 0.2,$$

hence:

$$\bar{q}_1 = \sqrt{\frac{2 \times 250 \times 150\,000}{(0.2 + 0.2) \times 30}} = 2500 \text{ units};$$

$$\bar{q}_2 = \sqrt{\frac{2 \times 250 \times 100\,000}{(0.2 + 0.2) \times 45}} = 1666.67 \text{ units}.$$

1: **procedure** EOQ_WITH_CAPACITY_CONSTRAINTS $(\mathbf{k}, \mathbf{d}, p, \mathbf{c}, \mathbf{h}, \mathbf{q})$

2:     # $\mathbf{q}$ is the order size vector of n components, one for each product, returned by the procedure;

3:     # Using relation (5.45), compute the EOQ disjoint order sizes q'_j, $j = 1, \ldots, n$;

4:     **for** $j = 1, \ldots, n$ **do**

$$q'_j = \sqrt{\frac{2 k_j d_j}{p c_j}}; \tag{5.68}$$

5:     **end for**

6:     **if** capacity constraint (5.67) is satisfied **then**

7:         # Return optimal order sizes q'_j, $j = 1, \ldots, n$;

8:         $\mathbf{q} = \mathbf{q}'$;

9:     **else**

10:         Increase the value of p by δ (to be determined);

11:         **for** $j = 1, \ldots, n$ **do**

$$q_j(\delta) = \sqrt{\frac{2 k_j d_j}{(p + \delta) c_j}}; \tag{5.69}$$

12:         **end for**

13:         # Determine a solution δ^* to equation

$$\sum_{j=1}^{n} a_j q_j(\delta^*) = b;$$

14:

$$\delta^* = \left[\frac{1}{b} \sum_{j=1}^{n} \left(a_j \sqrt{\frac{2 k_j d_j}{c_j}} \right) \right]^2 - p; \tag{5.70}$$

15:         Insert δ^* into relation (5.69) to compute order sizes $\bar{q}_j$, $j = 1, \ldots, n$;

16:         # Return order sizes $\bar{q}_j$, $j = 1, \ldots, n$;

17:         $\mathbf{q} = \bar{\mathbf{q}}$;

18:     **end if**

19:     **return q**:

20: **end procedure**

Models with joint reorder costs

For the sake of simplicity, it is assumed in this section that only two products are kept in stock. Let k_1 and k_2 be the fixed costs for reordering the two products at different moments in time, and let k_{1-2} be the fixed cost for ordering both products at the same time ($k_{1-2} < k_1 + k_2$). In addition, let T_1 and T_2 be the time lapses between consecutive replenishments of products 1 and 2, respectively (see Figure 5.64). Then,

$$q_1 = d_1 T_1; \tag{5.71}$$

$$q_2 = d_2 T_2. \tag{5.72}$$

The periodicity of a joint replenishment policy is

$$T = \max\{T_1, T_2\}.$$

In each time period T, the orders issued for the two products are

$$N_1 = T/T_1;$$

$$N_2 = T/T_2.$$

Here, N_1 and N_2 are positive integer numbers, one of them being equal to one (in the situation depicted in Figure 5.64, $N_1 = 3$ and $N_2 = 1$). During each time period T, two products are ordered simultaneously exactly once. Moreover, $N_j - 1$ single orders are placed for each product $j = 1, 2$. Hence, the total average cost per unit of time is

$$\mu(T, N_1, N_2) = \frac{k_{1-2} + (N_1 - 1)k_1 + (N_2 - 1)k_2}{T}$$

$$+ c_1 d_1 + c_2 d_2 + \frac{h_1 d_1 T}{2N_1} + \frac{h_2 d_2 T}{2N_2}. \tag{5.73}$$

By solving equation

$$\frac{\partial \mu(T, N_1, N_2)}{\partial T} = 0,$$

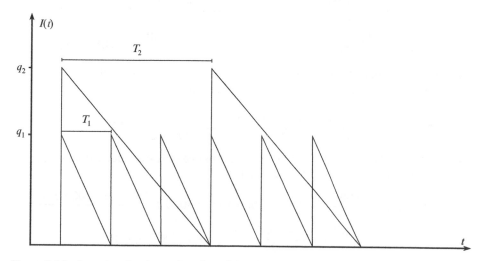

Figure 5.64   Inventory level as a function of time in the case of synchronized orders.

the value $T^*(N_1, N_2)$ that minimizes $\mu(T, N_1, N_2)$ is obtained:

$$T^*(N_1, N_2) = \sqrt{\frac{2N_1 N_2 [k_{1-2} + (N_1 - 1)k_1 + (N_2 - 1)k_2]}{h_1 d_1 N_2 + h_2 d_2 N_1}}, \tag{5.74}$$

as a function of N_1 and N_2.

Shamrock Microelectronics is an Irish company which assembles printed circuit boards (PCBs) for a number of major companies in the appliance sector. The Y23 PCB has an annual demand of 3000 units, a unit purchase price of € 30 and an annual holding cost equal to 20% of its purchase price. The Y24 PCB has an annual request of 5000 units, a purchase price of € 40 and an annual holding cost equal to 25% of its purchase price. The cost of issuing a joint order is € 300 while a reorder of each of the two single products costs € 250. If no joint orders are placed, the order sizes are, according to formula (5.45),

$$q_1^* = \sqrt{\frac{2 \times 250 \times 3000}{0.2 \times 30}} = 500 \, \text{units},$$

$$q_2^* = \sqrt{\frac{2 \times 250 \times 5000}{0.25 \times 40}} = 500 \, \text{units}.$$

From formulae (5.71) and (5.72),

$$T_1^* = 500/3000 = 1/6,$$

$$T_2^* = 500/5000 = 1/10.$$

This means that Shamrock would issue $1/T_1^* = 6$ orders per year of Y23 PCB and $1/T_2^* = 10$ orders per year of the Y24 PCB. Since

$$\mu_1(q_1^*) = \frac{250 \times 3000}{500} + 30 \times 3000 + \frac{0.2 \times 30 \times 500}{2} = € \, 93\,000/\text{year},$$

$$\mu_2(q_2^*) = \frac{250 \times 5000}{500} + 40 \times 5000 + \frac{0.25 \times 40 \times 500}{2} = € \, 205\,000/\text{year},$$

the average annual cost is € 298 000/year. If a joint order is placed and $N_1 = 1$, $N_2 = 2$, the periodicity of joint orders is, according to formula (5.74),

$$T^* = \sqrt{\frac{2 \times 1 \times 2 \times (300 + 250)}{0.2 \times 30 \times 3000 \times 2 + 0.25 \times 40 \times 5000 \times 1}} = 0.16.$$

Shamrock would issue $1/T^* = 6.25$ joint orders per year. The annual average cost, computed through formula (5.73), is equal to

$$\mu(T^*, 1, 2) = \frac{300 + 250}{0.16} + 30 \times 3000 + 40 \times 5000 + \frac{0.2 \times 30 \times 3000 \times 0.16}{2}$$

$$+ \frac{0.25 \times 40 \times 5000 \times 0.16}{2 \times 2} = € \, 296\,877.5/\text{year}.$$

5.13 Crossdock Door Assignment Problem

As illustrated in Sections 5.2 and 5.10.2, crossdocks are facilities where shipments (usually palletized unit loads) are transferred directly from incoming to outgoing vehicles with no (or short) storage in between. Empty pallets generally follow the reverse route.

Material handling costs may be minimized by appropriately assigning incoming vehicles to receiving doors and outgoing vehicles to shipping doors in the attempt to minimize the total distance that the internal transportation equipment must cover (seen as a proxy of the crossdock workload). The crossdock door assignment is an operational problem aimed at just deciding how vehicles should be assigned to doors. Even though in some crossdocks it is preferred to maintain the distinction of the doors for a simpler management of the internal transportation activities (see Problem 5.31), a door can generally act indifferently either as a receiving door or a shipping door.

A crossdock door assignment problem can be formulated as a *quadratic semiassignment problem* in the following way. Let n be the number of incoming and outgoing vehicles that have to be simultaneously assigned to the crossdock doors. For the sake of simplicity, the incoming vehicles are numbered from 1 to n_i ($n_i < n$), whereas the outgoing vehicles are numbered from $n_i + 1$ to n; m is the number of doors such that $m \geq n$. Moreover, let d_{ij} be the distance between the doors i and j, $i, j = 1, \ldots, m$, and let f_{hk} be the number of unit loads to be transferred from incoming vehicle $h = 1, \ldots, n_i$, to outgoing vehicle $k = n_i + 1, \ldots, n$. Let $x_{ih}, i = 1, \ldots, m, h = 1, \ldots, n$, be a binary decision variable equal to 1 if door i is assigned to vehicle h, 0 otherwise. The problem formulation is then the following.

$$\text{Minimize} \sum_{i=1}^{m} \sum_{j=1}^{m} \sum_{h=1}^{n_i} \sum_{k=n_i+1}^{n} d_{ij} f_{hk} x_{ih} x_{jk} \tag{5.75}$$

subject to

$$\sum_{i=1}^{m} x_{ih} = 1, h = 1, \ldots, n \tag{5.76}$$

$$\sum_{h=1}^{n} x_{ih} \leq 1, i = 1, \ldots, m \tag{5.77}$$

$$x_{ih} \in \{0, 1\}, i = 1, \ldots, m, h = 1, \ldots, n. \tag{5.78}$$

Objective function (5.75) represents the total distance that the internal transportation equipment must cover. Constraints (5.76) impose that each vehicle is assigned to a door. Constraints (5.77) impose that each door is assigned to at most one vehicle.

Problem (5.75)–(5.78) is non-linear (because of the objective function) and hard to solve. The formulation can be linearized by introducing additional binary decision variables y_{ihjk}, $i, j = 1, \ldots, n, h = 1, \ldots, n_i, k = n_i + 1, \ldots, n$, such that $y_{ihjk} = x_{ih} x_{jk}$. This means that:

1. if either x_{ih} or x_{jk} (or both) are zero, y_{ihjk} must be zero as well;
2. if x_{ih} and x_{jk} are both one, y_{ihjk} must be one as well.

1: **procedure** ALTERNATING_ASSIGNMENT (*P*)

2:      # *P* is the crossdock door assignment plan;

3:      Sort doors by non-decreasing average distance to all other doors;

4:      Sort incoming vehicles by non-increasing freight flows;

5:      Sort outgoing vehicles by non-increasing freight flows;

6:      **while** There exist vehicles to allocate **do**

7:          Assign *alternatively*, the incoming vehicle and the outgoing vehicle to the door, following the order established for incoming vehicles, outgoing vehicles, and doors;

8:          Update *P*;

9:      **end while**

10:      **return** *P*;

11: **end procedure**

It is easy to show that these conditions are satisfied by the following additional constraints

$$y_{ihjk} \geq x_{ih} + x_{jk} - 1, i, j = 1, \ldots, m, h = 1, \ldots, n_i, k = n_i + 1, \ldots, n,$$

in combination with the fact that the objective function should be minimized.

Hence, the linearized version of problem (5.75)–(5.78) becomes the following MIP problem:

$$\text{Minimize} \sum_{i=1}^{m} \sum_{j=1}^{m} \sum_{h=1}^{n_i} \sum_{k=n_i+1}^{n} d_{ij} f_{hk} y_{ihjk} \tag{5.79}$$

subject to

$$\sum_{i=1}^{m} x_{ih} = 1, h = 1, \ldots, n \tag{5.80}$$

$$\sum_{h=1}^{n} x_{ih} \leq 1, i = 1, \ldots, m \tag{5.81}$$

$$y_{ihjk} \geq x_{ih} + x_{jk} - 1, i, j = 1, \ldots, m, h = 1, \ldots, n_i,$$
$$k = n_i + 1, \ldots, n \tag{5.82}$$

$$x_{ih} \in \{0, 1\}, i = 1, \ldots, m, h = 1, \ldots, n \tag{5.83}$$

$$y_{ihjk} \in \{0, 1\}, i, j = 1, \ldots, m, h = 1, \ldots, n_i, k = n_i + 1, \ldots, n. \tag{5.84}$$

A feasible door assignment can be obtained through a general-purpose *alternating* constructive heuristic, named ALTERNATING_ASSIGNMENT procedure, described as follows.

[Ankommet.xlsx, Ankommet.csv, Ankommet.py] Ankommet is a trucking company operating in Denmark and Germany. Its crossdock located near Hamburg has eight doors (labelled from A to H) which can act as either strip or stack doors. Material handling is performed by forklift AGVs which can move a single palletized unit load at a time, at a speed equal to 1 m/s. The

company used to assign incoming and outgoing trucks to doors by using the
ALTERNATING_ASSIGNMENT procedure. On 12 June, four incoming trucks
(numbered from 1 to 4) and three outgoing trucks (numbered from 5 to 7) were
expected early in the morning. Table 5.19 reports the distance (in m) between
the doors, while Table 5.20 describes the number of palletized unit loads to be
transshipped from each incoming truck to each outgoing truck. The sorted list
of doors (code line 3) is (F, H, E, C, B, D, G, A), the sorted list of incoming trucks
(code line 4) is (1, 4, 2, 3) and the sorted list of outgoing trucks (code line 5) is (5,
6, 7). The corresponding door assignment plan (code lines 6–8) is the following:
$1 \rightarrow F, 5 \rightarrow H, 4 \rightarrow E, 6 \rightarrow C, 2 \rightarrow B, 7 \rightarrow D$ and $3 \rightarrow G$. The unused door is, of
course, A. The overall distance covered by the forklift AGVs is 4778 m, with a
time equal to 4778 s, corresponding to 1.33 hours.

By determining the optimal solution of the problem (5.79)–(5.84), the following crossdock door assignment is obtained (see Ankommet.py file): $4 \rightarrow A, 7 \rightarrow$
$B, 1 \rightarrow C, 3 \rightarrow D, 6 \rightarrow E, 5 \rightarrow F$ and $2 \rightarrow H$. The unused door in this case is G.
The distance covered by forklift AGVs is 4350 m. As a result, the time needed to
handle palletized unit loads is 4350 s, i.e., 1.21 hours.

Table 5.19 Distances (in m) between doors
in the Ankommet crossdock.

Door	A	B	C	D	E	F	G	H
A	0	32	68	97	75	70	75	40
B		0	42	80	53	65	82	47
C			0	45	15	49	79	55
D				0	30	36	65	65
E					0	38	69	53
F						0	31	32
G							0	36
H								0

Table 5.20 Number of palletized unit loads
to be transferred from incoming trucks
to outgoing trucks in the Ankommet problem.

	Outgoing truck		
Incoming truck	5	6	7
1	11	14	8
2	10	6	7
3	8	9	4
4	7	6	15

5.14   Put-away and Order Picking Optimization

As introduced in Section 5.1.1, put away amounts to moving incoming unit loads from the receiving zone to an optimal location in the storage zone and order picking consists in collecting items in a specified quantity from the storage zone and taking them to the shipping zone to fulfill customer orders. Organizing these activities constitutes an operational problem which is to

- allocate put-away and order picking tasks to the resources (operators, vehicles) of the internal transportation system;
- sequence their operations;
- optimize routing and solve conflicts for internal transportation vehicles, such as forklift trucks and AGVs, for which routes are variable.

This problem is intrinsically real time, since put-away and order picking tasks arrive in a dynamic fashion and are often modified (or even cancelled) on the fly.

The objective pursued is the optimizazion of a measure of efficiency (e.g., the maximization of the average number of put-away and order picking tasks performed in an hour) or of a measure of effectiveness (e.g., the minimization of the total lateness accumulated with respect to given due dates) or a combination of the two. By way of example, in a finished product warehouse, which provides a cushion between the manufacturing process and the distribution system, put-away tasks must be performed within strict due dates to avoid that workstation buffers become full and production lines (or palletizers) stop.

When optimizing a pure measure of efficiency, one has to avoid a phenomenon known as *starvation*, which occurs when a low priority task is repeatedly postponed to perform higher priority tasks.

As explained in Section 5.2, put-away and order picking can be done according to two paradigms: parts-to-picker systems versus picker-to-parts and AGV-based systems.

5.14.1   Parts-to-picker Systems

Parts-to-picker systems include carousels and AS/RSs. Unit loads and items are retrieved from the storage zone and taken to an I/O station from which they are sent to the shipping zone. Alternatively, an operator breaks down the unit loads and extracts the quantity requested (e.g., a number of cases from a palletized unit load). Then, the residual unit loads are stored back again.

5.14.2   Picker-to-parts and AGV-based Systems

At any time, the warehouse is characterized by a list of pending operations, either put-away and picking tasks corresponding to customer orders, to be performed by an operator or a vehicle (for the sake of brevity, the term "vehicle" will be used hereafter in this section to identify the internal transportation equipment used).

In its simplest form, the problem can be modelled as a *vehicle routing problem with pickups and deliveries* (VRPPD, see Chapter 6) over a complete directed graph $G = (V, A)$, where the vertex set V is constituted by the set of pickup and delivery points to be visited, as well as by the current positions of the vehicles. With each arc $(i, j) \in A$

is associated the duration of the quickest path from vertex i to vertex j in the warehouse. Vehicles can be identical or, as customary with AGVs, heterogeneous in which case compatibilities between vehicles and tasks must be accounted for.

A vehicle can have a capacity equal to one (or, at most, two or four) as customary with palletized and containerized unit loads, or much greater when picking small packages, such as components or spare parts.

Another relevant aspect is the congestion that may arise among vehicles, especially if they are moving palletized or containerized unit loads. Congestion may be created when two or more vehicles want to store or retrieve at the same location or when a vehicle cannot manoeuvre because of another vehicle or the narrowness of an aisle.

Since the VRPPD is hard to solve and plans need to be generated on the fly, a number of fast heuristics have been proposed over the years. These approaches decompose the whole problem into simpler subproblems (dispatching and routing) for each of which they attempt to seek a solution. It is worth noting that this approach may result in poor plans. For this reason, the availability of more and more powerful computers has prompted the development of more advanced heuristics which guarantee much better solutions (see Chapter 6 for more information).

Dispatching

Dispatching is the assignment of put-away and picking tasks to vehicles. When picking small packages (which is usually performed by human operators), traditional dispatching policies include batch and zone picking.

In *batch-picking* orders with packages in adjacent storage locations are assigned to an operator over a single trip. This policy is best suited for large warehouses filling multiple orders at the same time. Batch pickers usually drive a cart or an order picker where the collected packages are put.

In *zone-picking*, the storage zone is partitioned into a number of zones in each of which picking is performed by a single operator. On the plus side, this policy reduces travel time and does not result in congestion. On the minus side, the risk of errors in the subsequent sorting and consolidation phases is high.

Note that more sophisticated policies can be implemented combining different methods together.

Vehicle Routing

The second subproblem is the determination of a minimum-cost route for each vehicle. As already described, here the "cost" includes both efficiency and effectiveness measures, the former being related to the distance travelled and the latter to the lateness accumulated with respect to store and retrieve due dates.

Vehicle routing is part of a large class of combinatorial optimization problems which will be examined extensively in Chapter 6 in the context of distribution management. In this section, the case of a single picker collecting a (large) number of small packages is considered, resulting in a *road travelling salesman problem* (RTSP). The RTSP is a slight variant of the classical *travelling salesman problem* (TSP) (see Section 6.12.1) and consists of determining a minimum-time tour including a subset of vertices of a graph. The RTSP is hard to solve, but, in the case of warehouses, it is often solvable in polynomial time due to the particular characteristics of the underlying network.

For example, with reference to the storage zone represented in Figure 5.65, the RTSP could be modelled on a graph $G = (A \cup B \cup V, E)$, where $A = \{a_1, \ldots, a_r\}$ and $B = \{b_1, \ldots, b_r\}$ are the sets of vertices representing, respectively, the upper and lower ends of the r side aisles of the storage zone; $V = \{v_0, v_1, \ldots, v_n\}$ is the set of the n storage locations to be visited and of v_0, corresponding to the I/O point of the storage zone; E is the set of edges joining storage locations to be visited, the ends of the side aisles and the I/O point. The graph corresponding to the storage zone of Figure 5.65 is reported in Figure 5.66.

If each aisle has a single entrance, the shortest duration route is obtained by first visiting all required storage locations placed in the upper side aisles and then all required storage locations situated in the lower side aisles (see Figure 5.67). On the other hand, if the side aisles have some interruptions (that is, if there is more than one cross aisle), the problem can be solved to optimality by using a *dynamic programming algorithm*, whose worst-case computational complexity is a linear function of the number of side aisles. However, if there are several cross-aisles, the dynamic programming procedure becomes impractical. Therefore, in what follows, four heuristics are illustrated:

- the s-shape heuristic;
- the largest-gap heuristic;
- the combined heuristic;
- the aisle-by-aisle heuristic.

To describe these heuristics the storage zone is assumed to have a general layout like that of Figure 5.68, consisting of a single I/O point, of a number of blocks numbered from 1 to n (1 is the farthest block from the I/O point, n is the nearest), linked together by main aisles. The storage locations are generally at both sides of a generic main aisle. Between each pair of blocks there is a cross-aisle that can be used to go from one main aisle to another, or from one block to the next. Transversally, there is the rear aisle corresponding with block 1, just as the front aisle is transversal connecting all the main

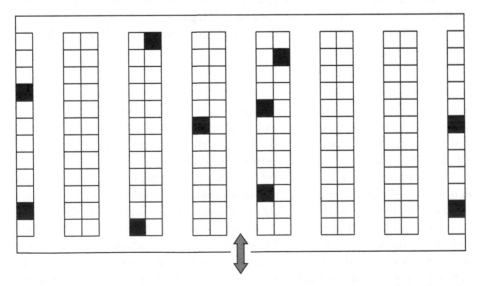

Figure 5.65 Layout of a storage zone. The storage locations to be visited are dark coloured.

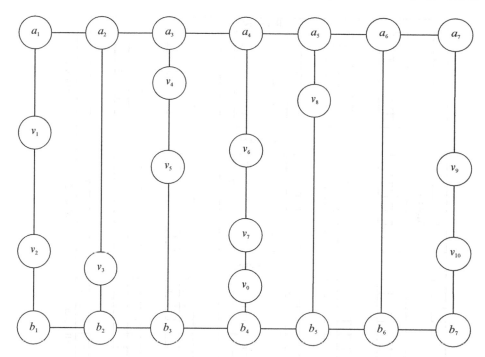

Figure 5.66 Graph $G = (A \cup B \cup V, E)$ associated with the RTSP in the storage zone of Figure 5.65.

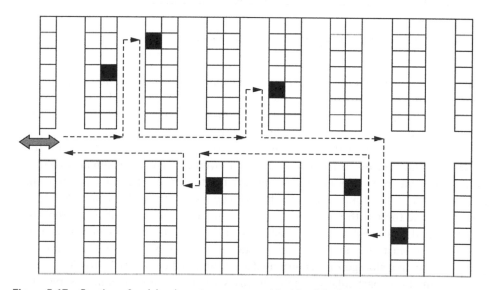

Figure 5.67 Routing of a picker in a storage zone with side aisles having a single entrance.

aisles to block n. It is assumed, moreover, that the pickers can cross the aisles in both directions and can also change directions within them.

S-shape heuristic

The general idea of the S-shape heuristic is that the visit to the storage locations for retrievals take place block by block, starting from block 1 (or from the first block

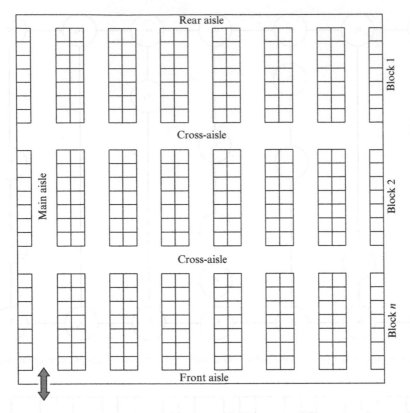

Figure 5.68 Layout of the storage zone divided into blocks.

containing storage locations to visit), gradually approaching block n, the closest to the position of the I/O point. The visit to the storage locations of each block takes place following an S-shaped route (hence the name of the method), beginning from the main aisle most to the left or most to the right containing storage locations to visit.

Every main sub-aisle of the block containing storage locations to visit is entirely traversed without ever inverting the direction of movement. The main sub-aisles where no unit load has been allocated for picking are not traversed. After having picked the last unit load of the picking list in the block to be visited, the picker can, only in this phase, reverse their path in a main sub-aisle. The picker begins their route from the I/O point and proceeds towards the first block to visit, traversing the nearest main aisle containing at least one storage location of the first block to visit. Figure 5.69 shows a picking route in a storage zone determined according to the S-shape heuristic.

The heuristic is particularly suitable in cases where the storage zone is served by many pickers. In fact, the constraint imposed by the heuristic of completely traversing the main sub-aisles to visit eliminates awkward reversing manoeuvres which could hinder other pickers.

Largest-gap heuristic

It is similar to the S-shape heuristic in the principle of visiting the storage locations starting from the farthest blocks from the I/O point, up to block n. The difference is that

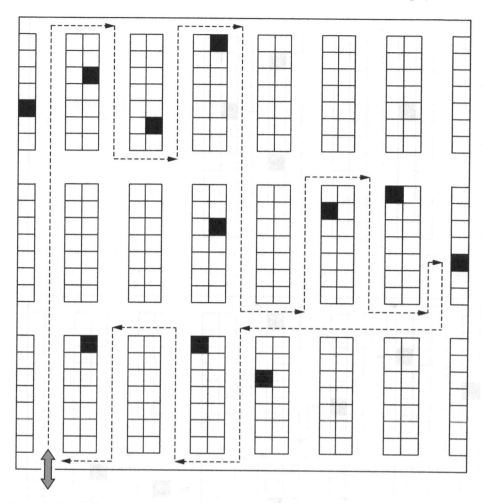

Figure 5.69   Picking route in a storage zone with the S-shape heuristic.

the main sub-aisles of each block are visited up to a certain point (corresponding to the storage location to visit) before returning to the entrance point. To calculate the point of return on the route in each main sub-aisle, it is necessary, for each of them, to determine the largest gap between two adjacent storage locations to visit, or between the side aisle and a storage location to visit (see Figure 5.70). This largest gap corresponds to the portion of the main sub-aisle that will not be visited. The fact that a return can be made to the entrance point of a generic main sub-aisle, means that the same sub-aisle can be accessed a second time, starting from the next cross-aisle. Figure 5.71 depicts a picking route in a storage zone calculated using the largest-gap heuristic.

The possibility of reversing in the main sub-aisle makes the heuristic particularly suitable to cases having a low picking density per aisle, meaning that the largest gap has a greater probability of being long, which yields a reduction in the total distance traversed.

Combined heuristic

Combining the two previous heuristics yields a new procedure that incorporates the advantages of the largest-gap heuristic and of the S-shape heuristic. In the first case,

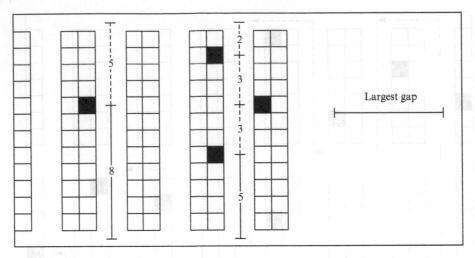

Figure 5.70 Computation of the largest gap in two main sub-aisles of the storage zone.

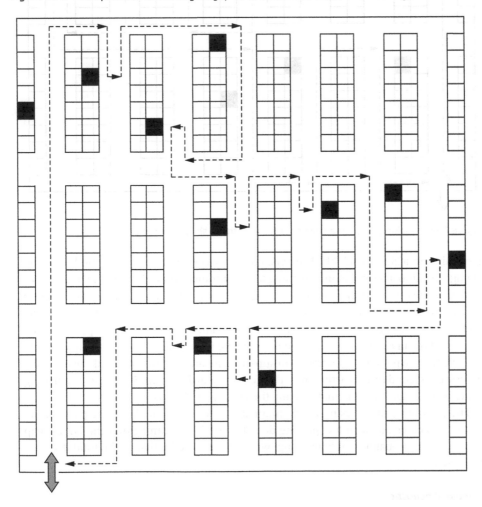

Figure 5.71 Picking route in a storage zone determined by using the largest-gap heuristic.

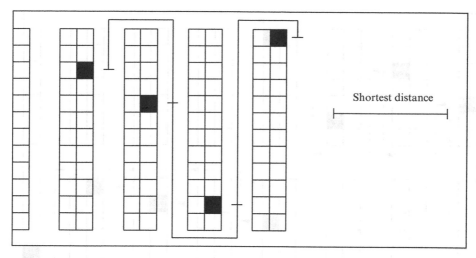

Figure 5.72 Determination of routes between two consecutive storage locations to visit using the combined heuristic.

these derive from the possibility of reversing and returning to the cross-aisle used for entrance. In the second case, they derive from the complete traversal of the main sub-aisles containing storage locations to visit. In the combined heuristic, the main sub-aisle containing storage locations for the picking of loads is visited just once. The objective is to create a minimum gap to traverse between the last picking point of the main sub-aisle currently visited and the first picking point of the next aisle (see Figure 5.72). Figure 5.73 shows a picking route in a storage zone calculated by means of the combined heuristic.

The risk of aisle congestion caused by reversing traffic is less than with the largest-gap heuristic.

Aisle-by-aisle heuristic

This heuristic is based on the general principle that every main aisle is traversed just once. The route begins from the I/O point and proceeds towards the nearest main aisle containing at least one storage location to visit. When all storage locations in this main aisle have been visited, the picker identifies the cross-aisle that allows to reach the next main aisle. In each case, the cross-aisles to go from one main aisle to the next are chosen so as to minimize the distance travelled. Figure 5.74 shows a picking route generated by this heuristic.

5.15 Load Consolidation

As explained in Section 5.1.1, consolidation arises in the shipping zone of a warehouse when small quantities of different SKUs are packed together before being dispatched. Actually, consolidation is a much more general logistics strategy in which a shipper combines multiple small loads directed to a particular geographic area into a single unit load (e.g., a palletized or containerized unit load), or into the hold of a vehicle, that is

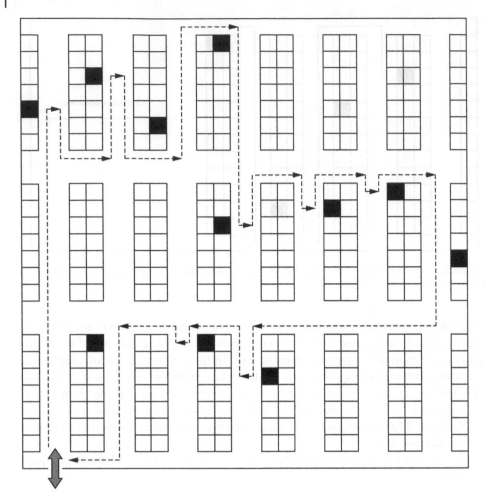

Figure 5.73 Picking route in a storage zone determined by using the combined heuristic.

then hauled to a terminal where the shipments are broken down and sent individually to their final destinations.

All these operational decisions, namely, pallet loading, container loading, truck loading, container ship loading, air cargo loading, require the solution of an optimization problem known as the *bin packing* problem. In the remainder of this section, the loads to be consolidated are indicated as items, whereas the containers into which the items must be packed are called bins.

While some constraints (e.g., geometric and weight constraints) are common to all bin packing problems, some others are peculiar to specific "bins". For instance, in truck loading one has to limit the maximum weight supported by each axle of the truck as well as take care of dynamic stability to avoid cargo displacement when the truck is moving. Similarly, in air cargo loading one has to consider the moment of the load with respect to the centre of gravity of the cargo (see Section 5.17).

In this section, only common constraints are considered with the aim to devising general-purpose algorithms which cannot be applied directly in several real

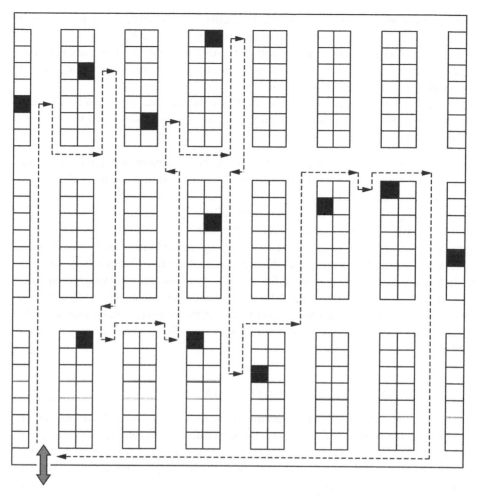

Figure 5.74 Picking route in a storage zone determined by using the aisle-by-aisle heuristic.

applications, especially in the case of problems in three dimensions where constraints like the load stability becomes relevant. However, they can serve, *mutatis mutandis*, as a basis for the design of more sophisticated decision support systems applicable in real contexts.

In any case, even the simplified versions of the bin packing problems considered in the following are mostly hard to solve but must often be solved in a short amount of time. Hence, heuristics, rather than exact algorithms, are frequently used.

Classification

In some bin packing problems, not all the physical characteristics of the items are relevant. For instance, when loading high-density items onto a truck, items can be characterized by their weight only, without any concern for their length, width, and height. As a result, bin packing problems can be classified according to the number of dimensions needed to characterize an item:

- *One-dimensional bin packing problems.* One-dimensional problems often arise when dealing with high density items, in which case weight is binding.
- *Two-dimensional bin packing problems.* Two-dimensional problems usually arise when loading a pallet or a truck with items of the same height or which do not overlap, as well as in air cargo loading.
- *Three-dimensional bin packing problems.* Three-dimensional problems occur when dealing with low density items, in which case volume is binding, and in container ship loading.

In the following it is assumed, that in two-dimensional problems items are rectangles, and that their sides must be parallel or perpendicular to the sides of the bins in which they are loaded. Similarly, in three-dimensional problems, the items are assumed to be parallelepipeds and their surfaces are parallel or perpendicular to the surfaces of the bins in which they are loaded. These assumptions are satisfied in most settings.

Bin packing problems are usually classified as *offline* and *online* problems, depending on whether the items to be loaded are all available or not when packing starts. In the first case, items can be presorted (for example, by non-decreasing weights) in order to improve heuristic performance. A heuristic using such preprocessing is said to be *offline*, otherwise it is called *online*. Clearly, an online heuristic can be used for solving an offline problem, but an offline heuristic cannot be used for solving an online problem.

5.15.1 One-dimensional Bin Packing Problems

The simplest one-dimensional bin packing problem is known as the 1-bin packing (1-BP) problem. It amounts to determining the least number of identical capacitated bins in which a given set of weighted items can be accommodated. Let m be the number of items to be loaded; n the number of available bins (or an upper bound on the number of bins in a feasible solution); $p_i, i = 1, \ldots, m$, the weight of item i; $q \, (\geq p_i, i = 1, \ldots, m)$ the capacity of bin $j = 1, \ldots, n$. This means that an upper bound on the number of bins to be used is m. The problem can be modelled by means of binary decision variables $x_{ij}, i = 1, \ldots, m, j = 1, \ldots, n$, each of which equal to 1 if item i is assigned to bin j or 0 otherwise, and binary decision variables $y_j, j = 1, \ldots, n$, equal to 1 if bin j is used, 0 otherwise. The 1-BP problem can then be modelled as follows

$$\text{Minimize} \sum_{j=1}^{n} y_j \tag{5.85}$$

subject to

$$\sum_{j=1}^{n} x_{ij} = 1, \ i = 1, \ldots, m \tag{5.86}$$

$$\sum_{i=1}^{m} p_i x_{ij} \leq q y_j, \ j = 1, \ldots, n \tag{5.87}$$

$$x_{ij} \in \{0, 1\}, \ i = 1, \ldots, m, \ j = 1, \ldots, n$$

$$y_j \in \{0, 1\}, \ j = 1, \ldots, n.$$

```
1: procedure FF (S, p, q, T)
2:     # S is the list of m items each of them having a weight pᵢ, i ∈ S;
3:     # V is a list of at least m available identical bins, each of capacity q;
4:     # T is the list of bins used.
5:     T = ∅;
6:     while S ≠ ∅ do
7:         Extract item i from the top of S and insert it into the first bin j ∈ T having
           a residual capacity greater than or equal to pᵢ;
8:         if no such bin exists then
9:             Extract a new bin k from V and put it at the bottom of T;
10:            Insert item i into bin k;
11:        end if
12:    end while
13:    return T, and the list of items loaded for each bin j ∈ T;
14: end procedure
```

Objective function (5.85) represents the number of bins used. Constraints (5.86) state that each item is allocated to exactly one bin. Constraints (5.87) guarantee that bin capacities are not exceeded.

A lower bound $\underline{z}(I)$ on the number of bins in any 1-BP feasible solution can be easily obtained as

$$\underline{z}(I) = \left\lceil \sum_{i=1}^{m} p_i/q \right\rceil. \tag{5.88}$$

The lower bound (5.88) can be very poor if the average number of items per bin is low (see Problem 5.38 for an improved lower bound). Such lower bounds can be used in a branch-and-bound algorithm, or to evaluate the performance of heuristics. In the remainder of this section, four heuristics are illustrated. The first two procedures, the *first-fit* (FF) and the *best-fit* (BF), are online heuristics while the others are offline heuristics. The FF procedure is reported in the following.

The BF procedure is the same as the FF except that in code line 7 item i is inserted into the *best compatible* bin $j \in T$, whose residual capacity must be greater than or equal to the value of p_i rounded up.

Both procedures can be implemented so that their computational complexity is equal to $O(m \log m)$.

It is useful to characterize the performance ratios of such heuristics. Recall that the *performance ratio* R^H of a heuristic H is defined as follows

$$R^H = \sup_I \left\{ \frac{z^H(I)}{z^*(I)} \right\}, \tag{5.89}$$

where I is a generic instance of the problem, $z^H(I)$ is the objective function value of the solution provided by heuristic H for instance I and $z^*(I)$ represents the optimal solution value for the same instance. Based on definition (5.89), R^H is such that:

a) $z^H(I)/z^*(I) \leq R^H$, $\forall I$;

b) there is an instance I such that $z^H(I)/z^*(I)$ is arbitrarily close to R^H.

Unfortunately, the performance ratios of the FF and BF procedures are not known, but it has been proved that

$$R^{\mathrm{FF}} \leq 7/4$$

and

$$R^{\mathrm{BF}} \leq 7/4.$$

The FF and BF procedures can be easily transformed into offline heuristics, by preliminary sorting the items by non-increasing weights, yielding the *first-fit decreasing* (FFD) and the *best-fit decreasing* (BFD) procedures. Their complexity is still equal to $O(m \log m)$, while their performance ratios can be proven to be

$$R^{\mathrm{FFD}} = R^{\mathrm{BFD}} = 3/2.$$

Moreover, it can be shown that this is the minimum performance ratio that a polynomial 1-BP heuristic can have.

[AlBahar.xlsx] Al Bahar is an Egyptian trucking company located in Alexandria which must plan the shipment of 17 parcels, whose characteristics are reported in Table 5.21. For these shipments the company can use a single van whose capacity is 600 kg. Applying the BFD procedure, the parcels are sorted by non-increasing weights (see Table 5.22) and the solution reported in Table 5.23 is obtained. The number of trips is six. The lower bound on the number of trips given by formula (5.88) is $\lceil 3132/600 \rceil = 6$. Hence, the solution determined by the BFD procedure is optimal.

Table 5.21 Weight of the parcels in the Al Bahar problem.

Number of parcels	Weight [kg]
4	252
3	228
3	180
3	140
4	120

Table 5.22 Sorted list of parcels in the Al Bahar problem.

Parcel	Weight [kg]	Parcel	Weight [kg]
1	252	10	180
2	252	11	140
3	252	12	140
4	252	13	140

(Continued)

Table 5.22 (Continued)

Parcel	Weight [kg]	Parcel	Weight [kg]
5	228	14	120
6	228	15	120
7	228	16	120
8	180	17	120
9	180		

Table 5.23 Parcel-to-trip allocation in the optimal solution of the Al Bahar problem.

Parcel	Weight [kg]	Trip	Parcel	Weight [kg]	Trip
1	252	1	10	180	5
2	252	1	11	140	3
3	252	2	12	140	5
4	252	2	13	140	5
5	228	3	14	120	5
6	228	3	15	120	6
7	228	4	16	120	6
8	180	4	17	120	6
9	180	4			

5.15.2 Two-dimensional Bin Packing Problems

The simplest two-dimensional bin packing (2-BP) problem amounts to determining the least number of identical rectangular bins in which a given set of rectangular items can be accommodated, with no item rotation allowed. Without loss of generality, it is assumed that each item can be loaded into a bin.

The 2-BP problem can be formulated as a *set partitioning problem*. Let L and W be the length and the width of a bin and let l_i and w_i, be the length and the width of item $i = 1, \dots, m$, respectively. Moreover, let V be the set of all feasible loading plans for a single bin and let a_{ij}, $i = 1, \dots, m$, $j \in V$, be a binary constant equal to 1 if item i belongs to loading plan j, 0 otherwise. Indicating with x_j, $j \in V$, a binary decision variable equal to 1 if loading plan $j \in V$ is chosen (which means that a bin is loaded according to plan j), 0 otherwise, the 2-BP problem can be formulated as follows:

$$\text{Minimize} \sum_{j \in V} x_j \tag{5.90}$$

subject to

$$\sum_{j \in V} a_{ij} x_j = 1, \ i = 1, \dots, m \tag{5.91}$$

$$x_j \in \{0, 1\}, \ j \in V. \tag{5.92}$$

Objective function (5.90) represents the number of bins used, whereas constraints (5.91) ensure that each item $i = 1, \ldots, m$, is loaded in exactly one bin. Model (5.90)–(5.92) is also valid for the 1-BP case, for which the feasibility check of a loading plan j

---

```
1:  procedure FFF (S, l, w, L, W, T)
2:      # S is the list of m items each of them having length l_i and width w_i, i ∈ S;
3:      # V is a list of at least m available identical bins, each of dimensions L and W;
4:      # T is the list of bins used;
5:      T = ∅;
6:      Sort S by non-increasing length of the items;
7:      while S ≠ ∅ do
8:          Extract item i from the top of S;
9:          Insert item i into the leftmost position of the first compatible layer (which
            can accommodate it) of the first compatible bin j ∈ T;
10:         if no such layer exists then
11:             Create a new one in the first compatible bin of T;
12:             Insert item i in the leftmost position of the layer;
13:             if there is no bin of T which can accommodate the layer then
14:                 Extract a new bin k from V and put it at the bottom of T;
15:                 Insert item i into the leftmost position at the bottom of bin k;
16:             end if
17:         end if
18:     end while
19:     return T, and the list of items loaded for each bin j ∈ T;
20: end procedure
```

```
1:  procedure BL (S, l, w, L, W, T)
2:      # S is the list of m items each of them having length l_i and width w_i, i ∈ S;
3:      # V is a list of at least m available identical bins, each of dimensions L and W;
4:      # T is the list of bins used;
5:      T = ∅;
6:      Sort S by non-increasing length of the items;
7:      while S ≠ ∅ do
8:          Extract item i from the top of S;
9:          Insert item i into the leftmost position at the bottom of the first compatible
            bin j ∈ T;
10:         if no such bin exists then
11:             Extract a new bin k from V, and put it at the bottom of T;
12:             Insert item i into the leftmost position at the bottom of bin k;
13:         end if
14:     end while
15:     return T, and the list of items loaded for each bin j ∈ T;
16: end procedure
```

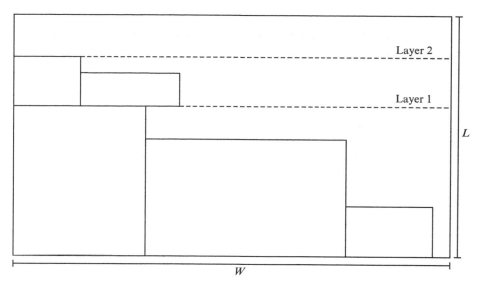

Figure 5.75 Layers of items inside a bin.

simply amounts to verify that $\sum\limits_{i=1}^{m} a_{ij} p_i \leq q$. Despite its simplicity, this set partitioning problem is hard to solve. The number of feasible loading plans to be generated depends on the particular instance and increases exponentially with the number of items. There- fore, the only way to optimally solve problem (5.90)–(5.92) is to dynamically generate the feasible loading plans when necessary. In this case, it is useful to have good lower bounds on the number of bins, one of which is easily obtained by letting

$$ \underline{z}(I) = \left\lceil \sum_{i=1}^{m} l_i w_i / LW \right\rceil . $$

Most heuristics for the 2-BP problem are based on the idea of forming layers of items inside the bins. Each layer has a width W, and a length equal to that of its longest item. All the items of a layer are located on its bottom, which corresponds to the level of the longest item of the previous layer (see Figure 5.75). Here two offline heuristics are illustrated, named *finite first-fit* (FFF) and *finite best-fit* (FBF) procedures. The FFF procedure is reported in the following. The FBF procedure is the same as the FFF except that an item i extracted from the top of S is inserted into the leftmost position of the layer of a bin $j \in T$ whose residual width is greater than or equal to, and closer to, item width.

Such layer heuristics have a low computational complexity. However, they can turn out to be inefficient if the average number of items per bin is relatively small. In such a case, the following *bottom-left* (BL) procedure usually provides better solutions.

[Kumi.xlsx] Kumi is a South Korean company manufacturing customized office furniture in Pusan. Outgoing products for overseas customers are usually loaded into 40 ft ISO containers, whose characteristics are reported in Table 5.2. Once packaged, parcels are 2 m or 1 m high. They are loaded on EPAL-pallets and cannot be rotated at loading time. The list of parcels shipped on 14 May is reported in Table 5.24. Parcels that are 1 m high are coupled in order to form six pairs of $(1 \times 1)\,m^2$ parcels and five pairs of $(0.8 \times 0.5)\,m^2$ parcels. Then, each such pair is considered as a single item. Applying the FBF procedure, the solution reported in Figure 5.76 is obtained.

Table 5.24 Features of the parcels shipped by Kumi on 14 May.

Quantity	Length [m]	Width [m]	Height [m]
6	1.50	1.50	2.00
5	1.20	1.70	2.00
13	1.00	1.00	1.00
11	0.80	0.50	1.00

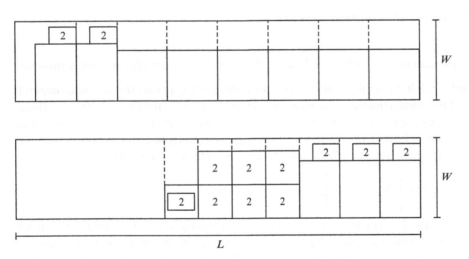

Figure 5.76 Parcels loaded into the two containers shipped by Kumi (2 indicates two overlapping parcels).

5.15.3 Three-dimensional Bin Packing Problems

The simplest three-dimensional bin packing (3-BP) problem amounts to determining the least number of identical parallelepipedic bins in which a given set of parallelepipedic items can be accommodated, with no item rotation allowed. Let L, W, and H be the length, the width and the height of a bin, and let l_i, w_i, and h_i be the length, the width and the height of item $i = 1, \dots, m$, respectively. As for previous bin packing problems, it is assumed that each item can be loaded into a bin. The 3-BP problem can be formulated using the same set partitioning model (5.90)–(5.92) illustrated in the two-dimensional case. A lower bound $\underline{z}(I)$ on the number of bins is

$$\underline{z}(I) = \left\lceil \sum_{i=1}^{m} l_i w_i h_i / LWH \right\rceil. \tag{5.93}$$

The simplest heuristics for 3-BP problems insert items sequentially into layers parallel to some bin surfaces (e.g., to WH surfaces). In the following, a 3-BP-L procedure, based on this principle, is illustrated.

1: **procedure** 3-BP-L ($\mathbf{l}, \mathbf{w}, \mathbf{h}, L, W, H, T$)

2:     # T is the list of bins used;

3:     Solve the 2-BP problem associated with m items characterized by w_i, h_i, $i = 1, \ldots, m$, and bins characterized by W and H;

4:     # k is the number of bidimensional bins used (referred to as *sections* in the following);

5:     # The length of each section is equal to the length of the longest item loaded into it;

6:     Solve a 1-BP problem with k items, one for each section, with weights equal to their lengths, and bins with capacity equal to L;

7:     **return** T, and the list of items loaded for each bin $j \in T$;

8: **end procedure**

If the items are all available when bin loading starts, it can be useful to sort list S by non-increasing values of the volume. However, unlike one-dimensional problems, more complex procedures are usually needed to improve solution quality.

[McMillan.xlsx] McMillan Company is a motor carrier headquartered in Bristol, United Kingdom. The firm has recently made the procedure for allocating outgoing parcels to semi-automatic vehicles, using a decision support system. This software tool uses the 3-BP-L procedure as a basic heuristic, and then applies a local search procedure. On 26 January, the parcels to be loaded were those reported in Table 5.25. The characteristics of the vehicles are indicated in Table 5.26. The parcels are loaded onto pallets and cannot be rotated. First, the parcels are sorted by non-increasing volumes. Then, the 3-BP-L procedure (in which 2-BP problems are solved through the BL procedure) is used. The solution (see Tables 5.27 and 5.28) is made up of six (2.4×1.8) m^2 sections, loaded as reported in Figure 5.77. Finally, a 1-BP problem is solved by means of the BFD procedure, by taking into account only the length of the sections and checking at each iteration that the weight constraint of the vehicles is not violated. In the solution (see Tables 5.27 and 5.28) three vehicles are used, the most loaded of which carries a weight of 1220 kg, less than the weight capacity. It is worth noting that the lower bound provided by formula (5.93) is $\lceil 35.963/28.08 \rceil = 2$.

Table 5.25 Parcels loaded at McMillan company on 26 January.

Type	Quantity	Length [m]	Width [m]	Height [m]	Volume [m³]	Weight [kg]
1	2	2.50	0.75	1.30	2.4375	155
2	4	2.10	1.00	0.95	1.9950	140

(Continued)

Table 5.25 (Continued)

Type	Quantity	Length [m]	Width [m]	Height [m]	Volume [m³]	Weight [kg]
3	7	2.00	0.65	1.40	1.8200	130
4	4	2.70	0.70	0.80	1.5120	115
5	3	1.20	1.50	0.80	1.4400	110

Table 5.26 Characteristics of the vehicles in the McMillan problem.

Length [m]	Width [m]	Height [m]	Capacity [m³]	Capacity [kg]
6.50	2.40	1.80	28.08	1 230

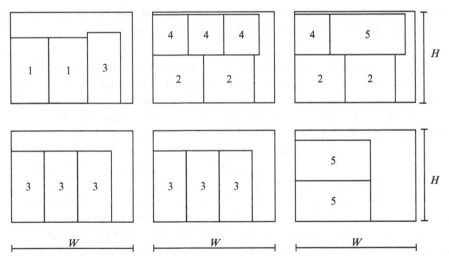

Figure 5.77 Sections generated at the end of the first step (code line 3) of the 3-BP-L procedure in the McMillan problem.

Table 5.27 Length and weight of the sections generated at the end of the first step (code line 3) of the 3-BP-L procedure in the McMillan problem.

Section	Length [m]	Weight [kg]
1	2.50	440
2	2.70	625

(Continued)

Table 5.27 (Continued)

Section	Length [m]	Weight [kg]
3	2.70	505
4	2.00	390
5	2.00	390
6	1.20	220

Table 5.28 Section allocation to vehicles at the end of the second step (code line 6) of the 3-BP-L procedure in the McMillan problem.

Vehicle	Sections
1	2, 3
2	1, 4, 5
3	6

The 3-BP problem can be also tackled using a more sophisticated approach based on the computation of the so-called *extreme points*. The basic idea is to exploit to the maximum the empty space left by the items already loaded into the bins. When a new item $i = 1, \ldots, m$, sized (l_i, w_i, h_i) is loaded into a bin, in such a way that its lower-left vertex (called the *south–west vertex* in the following), has coordinates (x_i, y_i, z_i), it will identify a number of extreme points upon which it may be possible to insert the south–west vertex of a subsequent item (see Figure 5.78). The extreme points are obtained by projecting the vertices $(x_i + l_i, y_i, z_i)$, $(x_i, y_i + w_i, z_i)$, and $(x_i, y_i, z_i + h_i)$ along the orthogonal axes of the bin. These projections encounter the first item already present in the bin or, in absence of it, the plane corresponding to the wall of the bin. The projected points become extreme points. Figure 5.79 depicts as open circles the vertices to be projected and in black the corresponding extreme points.

The following EXTREME_POINTS_UPDATING procedure is used to update the set of available extreme points.

Generally, the heuristics based on extreme points differ from the criterion used to determine at each iteration the extreme point where to place a new item. Below, for the sake of brevity, only the *first-fit decreasing* procedure, called 3-BP-FFD, is illustrated. Note that, in this case, items are sorted considering a specific ordering rule (e.g., by non-increasing volume) before being processed.

1: **procedure** EXTREME_POINTS_UPDATING $(x_i, y_i, z_i, l_i, w_i, h_i, E_j)$

2:     # x_i, y_i, z_i are the coordinates of the extreme point in which the south-west vertex of the item i has been inserted;

3:     # l_i, w_i, h_i are the dimensions of the item i inserted in the bin j;

4:     # E_j is the set of the extreme points associated with bin j;

5:     Remove (x_i, y_i, z_i) from E_j;

6:     **if** $(x_i, y_i, z_i) = (0, 0, 0)$ **then**

7:         # i is the first item inserted in bin j;

8:         Generate three extreme points with coordinates $(l_i, 0, 0), (0, w_i, 0)$ and $(0, 0, h_i)$;

9:         Add the three extreme points to E_j;

10:     **else**

11:         Project the vertices

            • $(x_i + l_i, y_i, z_i)$ towards plane xy and towards plane xz;
            • $(x_i, y_i + w_i, z_i)$ towards plane xy and towards plane yz;
            • $(x_i, y_i, z_i + h_i)$ towards plane xz and towards plane yz;

12:         Add to E_j (if they are not already present) the extreme points given by the intersection of the projections with the first item found in the bin or, in the absence of it, with the corresponding plane;

13:     **end if**

14:     **return** E_j;

15: **end procedure**

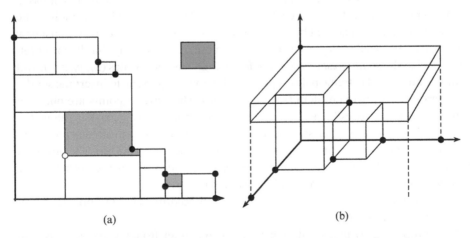

Figure 5.78 Examples of extreme points in: (a) two dimensions; (b) three dimensions. Observe that in (a) the extreme point (open circle) offers the possibility of inserting the new item (top right) in the empty space (in grey) left by the items already loaded, by overlapping the south–west vertex of the item onto the extreme point.

1: **procedure** 3-BP-FFD $(S, \mathbf{l}, \mathbf{w}, \mathbf{h}, L, W, H, T)$
2:     # S is the list of items to be loaded and sorted by their non-increasing volume values;
3:     # V is a list of at least m available identical bins, each of dimensions L, W, and H;
4:     # T is the list of bins used;
5:     # E_j, $j \in T$, is the set of extreme points associated with the bin j used;
6:     Extract the first bin from V;
7:     $T = \{1\}$;
8:     # Initialize the set E_1 with the extreme point corresponding to the south–west corner of the bin;
9:     $E_1 = \{(0, 0, 0)\}$;
10:     **while** $S \neq \varnothing$ **do**
11:         Extract an item i from the top of list S;
12:         Choose the first compatible bin $j \in T$ that can contain item i;
13:         # A bin j is compatible when there is an extreme point $\pi \in E_j$ that allows item i to be placed making its south–west vertex coincident with π without any overlapping with items already present in the bin and without overcoming the bin dimensions (*feasible extreme point*);
14:         Choose the feasible extreme point $\pi \in E_j$ with the lowest coordinates following the z, y, x order;
15:         **if** there is no compatible used bin **then**
16:             Extract a new bin j from V and place it at the bottom of T;
17:             $E_j = \{(0, 0, 0)\}$;
18:             # The unique feasible extreme point corresponds to the south–west corner of the bin;
19:             $\pi_x = 0$;
20:             $\pi_y = 0$;
21:             $\pi_z = 0$;
22:         **end if**
23:         Insert item i in bin $j \in T$ at the extreme point π;
24:         # The south–west vertex of item i is associated with π;
25:         $x_i = \pi_x$;
26:         $y_i = \pi_y$;
27:         $z_i = \pi_z$;
28:         Update E_j by calling EXTREME_POINTS_UPDATING $(x_i, y_i, z_i, l_i, w_i, h_i, E_j)$;
29:     **end while**
30:     **return** T, and the list of items loaded for each bin $j \in T$;
31: **end procedure**

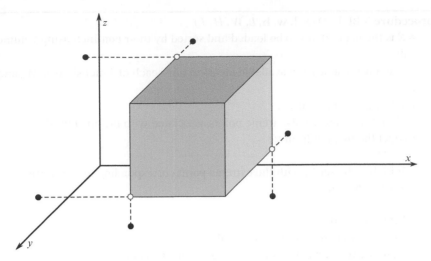

Figure 5.79 Vertices (open circles) whose projections along the orthogonal axes of the bin identify the extreme points (in black).

[LWK.xlsx] LWK ships goods by sea from its warehouse, situated close to the port of Hamburg, to its customers in North America. Products, packed in parallelepipedic shaped parcels, are loaded into 20 ft ISO standard containers (see Section 5.4.5) whose dimensions are reported in Table 5.2.

Until a few months ago, the logistics manager used their expertise and intuition to devise loading plans. Unfortunately, it was not rare that a container, once filled, had to be emptied and then filled in the attempt to insert some additional parcels. The need to speed up and optimize container loading stimulated the company to adopt a decision support system. The basic loading algorithm used by the software tool is the 3-BP-FFD procedure which sorts preliminary the list S of parcels to be loaded (see Table 5.29) in non-increasing values of the volume:

$$S = \{9, 1, 4, 2, 3, 8, 5, 6, 10, 7\}.$$

The first bin, which corresponds initially with the set $E_1^{(0)} = \{(0,0,0)\}$ of extreme points, is selected. Successively, at the first iteration, the first item (9) is placed into the first bin, with its south–west vertex in the lowest-left corner of the bin. Then the list of extreme points is updated (see code lines 6–9 of the EXTREME_POINTS_UPDATING procedure), as follows:

$$E_1^{(1)} = \{(1.63, 0, 0), (0, 2.05, 0), (0, 0, 2.22)\}.$$

Figure 5.80 represents the position of item 9 and the set of the extreme points (in black). At the second iteration, item 1 (with $l_i = 1.20$ m, $w_i = 2.20$ m, $h_i = 2.22$ m) is placed in the first bin as well, with its south–west vertex coincident with the feasible extreme point $\pi = (1.63, 0, 0)$, as shown in Figure 5.81 (see code lines 11–14 of the 3-BP-FFD procedure). As a consequence, in order to update E_1, the following steps are performed (code lines 11–12 of the EXTREME_POINTS_UPDATING procedure):

Figure 5.80 Loading item 9 into the first LWK bin (extreme points are depicted in black).

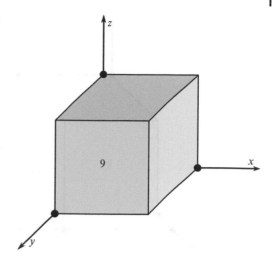

Table 5.29 Items to be loaded in the LWK problem.

Item	Length [m]	Width [m]	Height [m]	Volume [m³]
1	1.20	2.20	2.22	5.860 800
2	1.00	1.40	1.50	2.100 000
3	1.42	1.08	0.84	1.288 224
4	1.20	2.20	2.21	5.834 400
5	0.97	1.03	1.22	1.218 902
6	0.56	1.04	1.90	1.106 560
7	0.83	1.09	1.15	1.040 405
8	1.16	1.37	0.78	1.239 576
9	1.63	2.05	2.22	7.418 130
10	0.83	1.10	1.20	1.095 600

- vertex $(1.63 + 1.20, 0, 0)$ is projected on the planes xy and xz, encountering the bin wall in both cases, so the projected points coincide with the vertex itself, added as extreme point to E_1;
- vertex $(1.63, 0 + 2.20, 0)$ is projected on the plane xy encountering the bin wall (consequently, the vertex itself becomes a new extreme point) and on the plane yz, obtaining a projected point of coordinates $(0, 2.20, 0)$ that is a new extreme point added to E_1;
- vertex $(1.63, 0, 0 + 2.22)$ is projected on the plane xz encountering the bin wall and on the plane yz encountering item 9. In both cases, the projected points coincide with the vertex so $(1.63, 0, 0 + 2.22)$ becomes a new extreme point added to E_1.

Summarizing, the set of extreme points of the first bin (in black in Figure 5.81) at the second iteration becomes

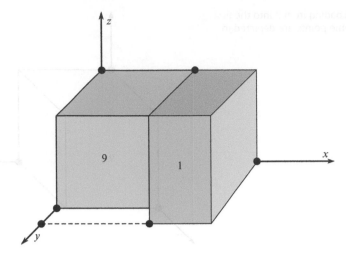

Figure 5.81 Loading item 1 into the first LWK bin (extreme points are depicted in black).

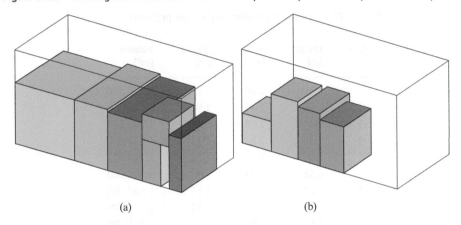

(a)   (b)

Figure 5.82 Full loading plan of (a) first bin; (b) second bin in the LWK problem.

$$E_1^{(2)} = \{(0, 2.05, 0), (0, 0, 2.22), (2.83, 0, 0), (0, 2.20, 0),$$
$$(1.63, 2.20, 0), (1.63, 0, 2.22)\}.$$

The third, fourth, and fifth iterations lead to items 4, 2, and 3 being inserted into the first bin, whereas the successive item (8) is put into a new bin. At this stage, the sets of extreme points associated with the two used bins are:

$$E_1^{(5)} = \{(0, 2.05, 0), (0, 0, 2.22), (0, 2.20, 0), (1.63, 0, 2.22), (1.63, 2.20, 0),$$
$$(2.83, 2.20, 0), (2.83, 0, 2.21), (5.03, 0, 0), (4.03, 1.40, 0), (5.45, 0, 0),$$
$$(5.45, 0, 1.50), (4.03, 1.08, 1.50), (4.03, 0, 2.34), (0, 0, 2.34)\};$$

$$E_2^{(5)} = \{(1.16, 0, 0), (0, 1.37, 0), (0, 0, 0.78)\}.$$

At the end of the procedure, a full loading plan (in which only two bins are used) is obtained (see Figure 5.82). Table 5.30 reports the coordinates of the south–west vertex of each item loaded into the two bins.

Table 5.30 Coordinates of the south–west vertex of the items loaded into the two LWK bins.

		Coordinates		
Item	Bin	(x)	(y)	(z)
1	1	1.63	0.00	0.00
2	1	4.03	0.00	0.00
3	1	4.03	0.00	1.50
4	1	2.83	0.00	0.00
5	2	1.16	0.00	0.00
6	1	5.45	0.00	0.00
7	2	2.96	0.00	0.00
8	2	0.00	0.00	0.00
9	1	0.00	0.00	0.00
10	2	2.13	0.00	0.00

The 3-BP-FFD procedure can be easily adapted to the case in which item rotation is allowed.

5.16   Case Study: Inventory Management at Wolferine

Wolferine is a division of the UOP Limited Industries Group manufacturing copper and brass tubes. The company production takes place in London (Ontario, Canada) in a highly automated factory operating with a very low work-in-process. The raw materials originate from nearby mines: consequently, the firm does not need to stock a large amount of raw materials. As far as finished products' inventories are concerned, Wolferine makes use of EOQ models (see Section 5.12.1). In the autumn of 1980, the firm operated with a production level close to the plant capacity. At that time the interest rate was around 10%. Using this value in the EOQ model, the company set the finished goods inventory level equal to 833 tons. During the subsequent two years, an economic recession hit the industrialized countries. The interest rate underwent continuous and quick variations (up to 20% in August 1981) while the demand of finished products went down by 20% and the price level increased sharply. According to the EOQ model, the finished good inventory level should have been lowered under those conditions. In order to illustrate this conclusion, let n be the number of products; k a reorder fixed cost (assumed to be independent of the product); d_i, $i = 1, \ldots, n$, the annual demand of product i; p (see equation (5.34)) can be expressed as the sum of a bank interest rate p_1 and a rate p_2 associated with warehousing costs ($p = p_1 + p_2$); c_i, $i = 1, \ldots, n$, the price of product i, and $\bar{I}(t)$ the average stock level at time t.

At time t_0 (January 1981), the average inventory level is half of the sum of the EOQ reorder quantities of the single products (see formula (5.45)):

$$\bar{I}(t_0) = \frac{1}{2} \sum_{i=1}^{n} \sqrt{\frac{2kd_i}{pc_i}}.$$

Let δ_1, δ_2 and δ_3 be the percentage variations (supposed equal for all products) of prices, demand and rate p at time period t, respectively. The average stock level at time t is equal to

$$
\begin{aligned}
\bar{I}(t) &= \frac{1}{2} \sum_{i=1}^{n} \sqrt{\frac{2[k(1 + \delta_1)][d_i(1 + \delta_2)]}{[p_1(1 + \delta_3) + p_2][c_i(1 + \delta_1)]}} \\
&= \bar{I}(t_0) \sqrt{\frac{(1 + \delta_2)p}{p_1(1 + \delta_3) + p_2}}.
\end{aligned}
\tag{5.94}
$$

According to formula (5.94), when demand decreased ($\delta_2 < 0$) and rate p increased ($\delta_3 > 0$), the stock level should have been decreased. However, the managers of Wolferine continued to operate as in 1981. As a result, the inventory turnover index (see Section 5.8) decreased dramatically. Moreover, to protect the manufacturing process against strikes at the mines, the firm decided to hold a raw-material inventory as well. As a result, when the recession ended in late 1983, the firm had an exceedingly large stock of both raw materials and finished products.

5.17 Case Study: Airplane Loading at FedEx

FedEx is one of the leading express carriers in the world (see Section 1.6.6). As already illustrated, parcels whose origin and destination exceed a given distance are consolidated in ULDs (see Section 5.4.2) and sent by air. A cargo airplane may fly between a pair of destinations or may follow a multi-stop route where ULDs are loaded or unloaded at intermediate stops.

In order to use cargo airplane capacity efficiently, a key issue is to devise good loading plans, taking into account a number of aspects: the load must be balanced around the centre of gravity of the aircraft, the total weight in the various areas of the aircraft must not exceed given thresholds in order to limit the cutting forces on the plane, etc. These aspects are especially critical for some planes, such as the Airbus A300, characterized by a low fuel consumption, but with more problems of load distribution along the fuselage. In addition, when loading an airplane assigned to a multi-stop route, ULDs to be unloaded at intermediate stops must be positioned close to the exit.

In order to allow airplanes to take off on time, the allocation of ULDs on board must be performed in real time, i.e., ULDs must be loaded on the aircraft as soon as they arrive at the airport. As a matter of fact, 30% to 50% of the ULDs are already on board when the ground staff has a complete knowledge of the features of the ULDs to be loaded. Of course, once some ULDs have been loaded, it may become impossible to load the subsequent ULDs. As a result, when new ULDs arrive at the airport, it is sometimes necessary to define a new loading plan in which some ULDs that were previously loaded are unloaded. This situation arises frequently when the total load is close to the

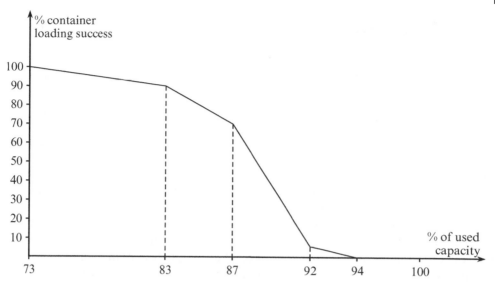

Figure 5.83 Percentage of success in loading an aircraft as a function of the percentage of load capacity in the FedEx case study.

capacity of the aircraft, as shown in Figure 5.83, where the percentage of success in loading an airplane is reported as a function of the percentage of the load capacity used.

The objective pursued by FedEx consists of loading the largest number of ULDs as possible. If no ULD is yet loaded, the following procedure can be followed.

Let m be the number of ULDs to be loaded; n the number of positions in the loading zone; q the number of areas into which the plane is divided; p_i, $i = 1, \ldots, m$, the weight of ULD i; P_j, $j = 1, \ldots, n$, the maximum weight that can be loaded in position j; d_j, $j = 1, \ldots, n$, the distance from position j to the centre of gravity O; M^{min} and M^{max} the minimum and maximum moments of the loads with respect to O; L_k, $k = 1, \ldots, q$, the total maximum weight that can be placed in area k, and f_{jk}, $j = 1, \ldots, n$, $k = 1, \ldots, q$, the fraction of position j contained in area k. Also let x_{ij}, $i = 1, \ldots, m$, $j = 1, \ldots, n$, be a binary decision variable equal to 1 if ULD i is placed in position j, and 0 otherwise, and u_j, $j = 1, \ldots, n$, a binary decision variable equal to 1 if position j is used, and 0 otherwise. A feasible solution is defined by the following set of constraints:

$$\sum_{j=1}^{n} x_{ij} = 1, \ i = 1, \ldots, m; \tag{5.95}$$

$$\sum_{i=1}^{m} x_{ij} \leq mu_j, \ j = 1, \ldots, n; \tag{5.96}$$

$$\sum_{i=1}^{m} p_i x_{ij} \leq P_j u_j, \ j = 1, \ldots, n; \tag{5.97}$$

$$\sum_{i=1}^{m} \sum_{j=1}^{n} d_j p_i x_{ij} \leq M^{max}; \tag{5.98}$$

$$\sum_{i=1}^{m}\sum_{j=1}^{n} d_j p_i x_{ij} \geq M^{\min}; \tag{5.99}$$

$$\sum_{i=1}^{m}\sum_{j=1}^{n} p_i f_{jk} x_{ij} \leq L_k, \ k=1,\dots,q; \tag{5.100}$$

$$x_{ij} \in \{0,1\}, \ i=1,\dots,m, \ j=1,\dots,n; \tag{5.101}$$

$$u_j \in \{0,1\}, \ j=1,\dots,n. \tag{5.102}$$

Constraints (5.95) guarantee that each ULD is allocated to a position. Constraints (5.96) state that if a position $j=1,\dots,n$, accommodates an ULD, then its corresponding u_j variable must be equal to 1. Constraints (5.97) ensure that the total weight loaded in any position does not exceed a pre-established upper bound. Constraints (5.98) and (5.99) impose that the total moment, with respect to point O, is within the pre-established interval. Constraints (5.100) ensure the respect of the weight bounds in each section.

If it is not possible to satisfy all of the constraints (5.95)–(5.102), an ULD $\bar{i}$ is eliminated from the loading list and the set of constraints is updated (in particular, $x_{\bar{i}j} = 0$, $j=1,\dots,n$). The whole procedure is repeated until a feasible solution is found.

If some ULDs have already been loaded, let S be the set of ULDs already loaded and let $j(i), i \in S$, be the position assigned to ULD $i \in S$. Then, additional constraints

$$x_{ij(i)} = 1, i \in S, \tag{5.103}$$

are added to (5.95)–(5.102). If a feasible solution to constraints (5.95)–(5.103) exists, the procedure stops since the loading plan partially executed can be completed. Otherwise, the constraint (5.103) associated with ULD $i' \in S$ allocated to the position closest to the exit of the hold is removed from S (which corresponds to unloading the ULD from the aircraft). This step is repeated until a feasible solution to constraints (5.95)–(5.103) is obtained.

5.18 Questions and problems

5.1 Perform a Web search to determine the cost structure of the main storage and internal transportation systems in your country.

5.2 Consider a warehouse equipped with an AS/RS. The horizontal and vertical speeds of the stacker crane are 225 cm/s and 60 cm/s, while the aisle length and height are 20 m and 12 m, respectively. Assuming that the I/O station of the stacker crane is the leftmost position at one m from the floor, compute the worst case completion time for storing or retrieving a unit load.

5.3 In many industrialized countries, the average inventory turnover index is around 20 for dairy products and around five for household electrical appliances. Discuss these figures.

5.4 [Decoor.xlsx] Decoor is an international company selling home decoration products. The company's warehouse has a storage zone equipped with racks disposed as represented in Figure 5.84. Each storage location measures 1.25×0.91 m² and can host one EPAL-palletized unit load, handled sidewise. Each rack has six

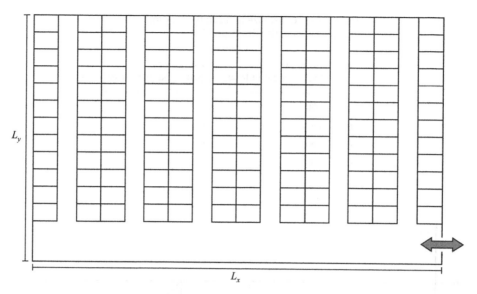

Figure 5.84   Decoor storage zone layout.

Table 5.31   Technical features of the forklift trucks used in the Decoor storage zone.

Parameter	Description	Value
$v^{(u)}$	Travel speed of the unloaded truck	3.75 m/s
$v^{(l)}$	Travel speed of the loaded truck	3.15 m/s
$v^{(au)}$	Unloaded fork ascending speed	0.50 m/s
$v^{(al)}$	Loaded fork ascending speed	0.25 m/s
$v^{(du)}$	Unloaded fork descending speed	0.50 m/s
$v^{(dl)}$	Loaded fork descending speed	0.45 m/s
$t^{(f)}$	Pallet forking and deforking time	70 s

tiers with a gross height of 2.25 m each. All the aisles are 3 m wide. Material handling is performed by three forklifts trucks, whose technical features are shown in Table 5.31. A working day is eight hours. Calculate capacity and throughput (on a daily basis) of the storage zone if the efficiency η of the forklift trucks is equal to one. If $\eta = 0.7$, how many forklift trucks are necessary to guarantee the same throughput?

5.5 Which warehouse layout is the most suitable for a highly seasonable demand pattern?

5.6 Show that a warehouse can be modelled as a queueing system.

5.7 A warehouse stores nearly 20 000 palletized unit loads. Palletized unit loads turn around about five times a year. How much is the required labour force? Assume two eight-hour shifts per day and about 250 workdays per year. (Hint: apply Little's law stating that for a queueing system in steady state, the average length L_Q

Table 5.32 Order size and safety stock for the four SKUs in the Krons problem.

SKU	Order size [palletized unit loads]	Safety stock [palletized unit loads]
B_1	400	100
B_2	700	150
B_3	600	120
B_4	300	80

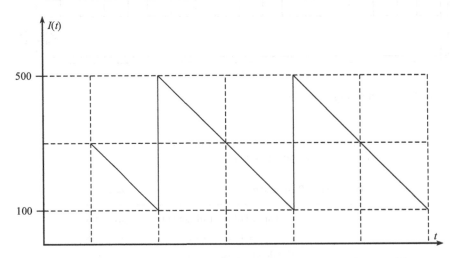

Figure 5.85 Inventory level (in palletized unit loads) for SKU B_1 in the Krons problem.

of the queue equals the average arrival rate λ times the average waiting time T_W, $L_Q = \lambda T_W$.)

5.8 The number of receiving doors in a crossdock can be estimated using Little's law. Taking inspiration from this principle, adapt formula (5.12) to compute the number of necessary receiving doors in a crossdock.

5.9 [Krons.xlsx] Krons produces four types of sodas (B_1, B_2, B_3, and B_4). Its warehouse located in Frankfurt, Germany, is managed according to a reorder point policy. Table 5.32 reports the values of the order sizes (in palletized unit loads) and safety stocks for each SKU. Figures 5.85–5.88 show how the inventory levels (in palletized unit loads) vary over time for the four SKUs. In order to rationalize the warehouse, the logistics manager has to choose a two-class storage policy, where each class is made up of two SKUs. Determine which are the most convenient combination of two classes, with the aim of minimizing the number of storage locations.

5.10 In the Wagner Bros problem, assume that the internal transportation system performs dual cycles. First, derive a closed-form approximation of the dual cycle time as a function of n_x and n_y. Then, determine the optimal storage zone sizing.

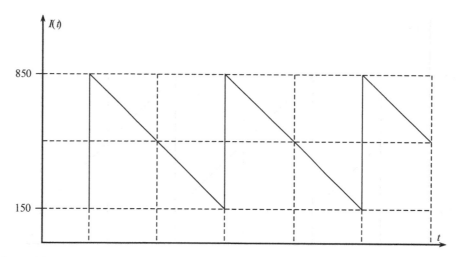

Figure 5.86   Inventory level (in palletized unit loads) for SKU B_2 in the Krons problem.

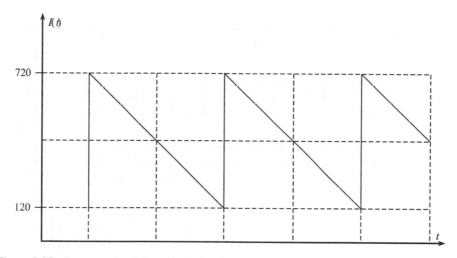

Figure 5.87   Inventory level (in palletized unit loads) for SKU B_3 in the Krons problem.

Finally, compute the increase in throughtput as well as the surface utilization rate.

5.11 Foral Ltd, which distributes electrical products in the UK, plans to build a new warehouse to accommodate 60 000 palletized unit loads. Goods will be stocked on selective racks and handled by means of forklift trucks. Each rack will have five tiers, each of which can store up to five palletized unit loads. Racks will be arranged as in Figure 5.89, where vertical aisles are 1.5 m wide, while the horizontal aisles are 2.5 m wide. Each palletized unit load will require an area equal to 120.0×80.3 cm^2 (taking into account the width clearance). Size the storage zone, under the hypothesis that the probability that the picker enters the storage zone from the first I/O point is p ($0 \leq p \leq 1$), whereas the probability is $(1 - p)$ if they enter from the second I/O point.

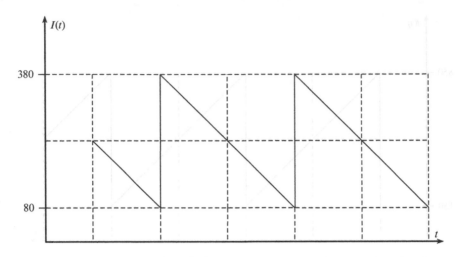

Figure 5.88 Inventory level (in palletized unit loads) for SKU B_4 in the Krons problem.

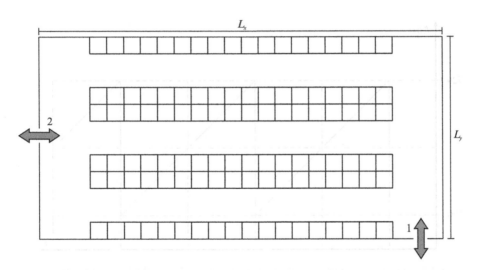

Figure 5.89 Storage zone layout in the Foral problem.

5.12 Let P be the throughput of a warehouse (on a daily basis), l_C and q the length and capacity of a rail car, respectively, and n_C the number of car changes per day. Estimate the length of rail dock l_D needed by the warehouse.

5.13 Explain how to determine the expected dual command travel time of a stacker crane given by formula (5.23).

5.14 [PalletRack.xlsx] Consider the pallet rack depicted in Figure 5.90 in which there are 384 storage locations (arranged on 12 tiers) for palletized unit loads. The upright frames are 17.50 cm wide each, and each beam is 182.50 cm long and 10 cm width, capable to support two palletized unit loads. The clearance between two palletized unit loads and between the upright frame and the palletized unit load is 7.5 cm. The height between two tiers, without considering the beam width is 100 cm. The average speed of the stacker crane serving the rack is 3.00 m/s along

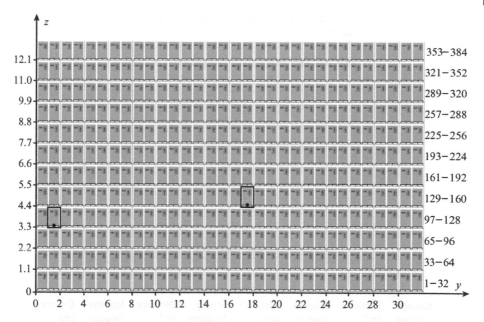

Figure 5.90 Pallet rack of Problem 5.14 served by a stacker crane. For example, the storage locations 98 and 146 reported in the boxes are at coordinates (1.35, 3.3) and (17.35, 4.4), respectively. The relevant stacker crane travel times are $t_{98,146} = \max\{(17.35 - 1.35)/3.00;$ $(4.4 - 3.3)/0.75\} = 5.33$ s, $t_{0,98} = \max\{1.35/3.00; 3.3/0.75\} = 4.40$ s and $t_{0,146} = \max\{17.35/3.00; 4.4/0.75\} = 5.87$ s.

the y direction, whereas the average speed along the z direction is 0.75 m/s. With reference to a Cartesian plane yz whose origin at coordinates $(0,0)$ is coincident with the I/O station of the stacker crane (see Figure 5.90), compute the expected single command travel time T_{SC} by using both (5.20) and (5.22). Then determine the expected dual command travel time T_{DC} by using formulae (5.21) and (5.23). What differences can be appreciated in the two modes of computation?

5.15 [Wert.xlsx] In the Wert problem compute the throughput considering that the stacker cranes perform only dual cycles. How many stacker cranes are required to get a warehouse throughput of 500 palletized unit loads per hour?

5.16 [Motif.xlsx] Motif is a Turkish electrotechnical company which operates a warehouse in Ankara. The storage zone of the warehouse has a single I/O point and 32 storage locations. Tables 5.33 and 5.34 report the features of the six SKUs and the distances between the I/O point and the storage locations, respectively. Assuming that the travel time between the I/O point and the storage location k $(k = 1, \ldots, 32)$ is directly proportional to the corresponding distance, determine the optimal allocation of the SKUs in the storage zone.

5.17 [Beer_demand.py] Implement the beer game (described in Section 1.3.3) in Python. Then, play the game for 24 weeks with three teams of four players each under the hypotheses that, at each facility, the reorder cost is € 100.00, the shortage cost is € 1.00 for one palletized unit load per week, the holding cost is € 1.00 for one palletized unit load per week, the initial inventory is 540 palletized unit loads and the lead time is two weeks. Use the Python function get_demand(t) (in the file Beer_demand.py) to randomly generate the beer weekly demand (in

Table 5.33 Features of the SKUs in the Motif problem.

SKU	Number of storage locations required	Daily number of storages and retrievals
1	6	133
2	5	128
3	8	198
4	5	126
5	4	128
6	3	98

Table 5.34 Distance between the storage locations and the I/O point in the Motif problem.

Storage location	Distance [m]	Storage location	Distance [m]	Storage location	Distance [m]	Storage location	Distance [m]
1	12	9	15	17	17	25	20
2	10	10	11	18	13	26	14
3	13	11	12	19	13	27	16
4	13	12	14	20	17	28	19
5	17	13	14	21	19	29	19
6	14	14	17	22	17	30	20
7	16	15	19	23	14	31	17
8	18	16	16	24	20	32	18

palletized unit loads) by the consumers. Does the simulation show a significant inventory oscillation (i.e., a bullwhip effect) when the game is played without any coordination among the players? Does the bullwhip effect vanish if the players are allowed to share information and coordinate?

5.18 The Scottish manufacturer Lasim needs to determine the order size of a particular clay tile, classified as A-124. It is known that the demand is 2500 units per month. For the production, Lasim spends € 1.13 per unit of product for raw materials, € 0.17 per unit of product for the energy, € 0.72 per unit of product for setting up the equipment; while the labour cost is € 45 000 per year for each of the two workers, that work eight hours per day for 25 workdays in a month. Moreover, about three minutes are needed to produce a single unit of product. To compute the annual holding cost for each product, the reader is referred to equation (5.34), where $p = 6.8\%$. The replenishment rate is 130 units of product per day. The cost for issuing a production order is € 38. Shortages are not allowed.

5.19 In the Lasim problem, determine the reorder point assuming that lead time is 10 days.

5.20 CC Motors needs to order a component named XLL5. The supplier communicates to the company that the price of this item is going to be raised by x. Assuming an instantaneous inventory replenishment, how much of the product CC Motors should buy before the price raises? (Hint: calculate the holding cost incurred by ordering q items now, then the purchase cost savings obtained by ordering q items now; finally, the value of q that maximizes the purchase cost savings minus the extra holding costs).

5.21 Modify the EOQ formula in case the warehouse has a finite capacity Q.

5.22 Kartogroup sells office supplies, including reams of A3 paper. The average demand for reams of paper is 200 units per month. The purchase price of each ream is € 8.70, while the fixed cost of issuing an order is € 1.5. The monthly unit holding cost is estimated as 1.5% of the purchase price. Assuming a lead time of 0.5 months from when the order is issued to when it is received, compute the reorder point. Also assume that the company has a constraint on the stock inventory of 103 pieces. How does the reorder quantity vary?

Starting from the next month, the supplier will be able to guarantee a quantity discount: if Kartogroup buys not less than 90 units, the unit price is € 7.50. Compute the EOQ and the number of orders per month.

5.23 In the Kartogroup problem, consider the possibility to also buy coloured sheets in A4 reams, whose demand is 300 units per month. The purchase price is € 6.50, the fixed cost of issuing an order is € 1.00, and the monthly unit holding cost is € 0.15. The vendor offers the option to place a joint order for both products at a cost of € 1.10. Select the best purchase strategy.

5.24 Each year a drug store sells 25 000 bottles of cough syrup. The cost of each bottle depends on the order size q. The supplier offers a bottle at $ 10 if the amount bought does not exceed 1140 bottles; while the price is reduced to $ 9.50 if more than 1140 are ordered. The unit annual holding cost is estimated as 20% of the selling price, and placing an order costs $ 50.

- Explain why q_2^* is larger than q_1^* by appropriately representing their quantities on a graph, where q_2^* and q_1^* are the EOQ associated with the selling price $ 9.50 and $ 10, respectively.
- Find the EOQ q^*.
- Determine how often the store should place an order.
- Find a value of q_2' such that that the optimal order quantity must be q_1^*.
- Assuming the supplier decides to apply an incremental quantity discount, calculate q^* and evaluate the convenience of the discount strategies.

5.25 [SaoVincente.xlsx] Draw the auxiliary graph used for solving the Sao Vincente Chemical problem as a shortest path problem and solve it by using a shortest path algorithm known to you.

5.26 Modify the Wagner–Whitin model in case the warehouse is capacitated. Does the ZIO property still hold?

5.27 What is the optimal order quantity in the newsboy model if the warehouse is capacitated?

5.28 Pansko, a Bulgarian chemical firm located in Plovdiv, supplies chemical agents to state clinical laboratories. The company knows that the demand of its product Merofosphine has a constant trend and shows no seasonal and cyclical effects. Based on the available data, the forecast amounts to 400 packages per week with a

MSE equal to 2500 (see Section 2.9). Hence, the product demand can be modelled as a normal random variable with an expected value equal to 400 and a standard deviation equal to $\sqrt{2500} = 50$. The production cost is BGN 100 per unit while the profit is BGN 20 per unit (BGN is the code of the Bulgarian currency, i.e., lev). Every time the manufacturing process is set up, a fixed cost of BGN 900 is incurred. The annual unit holding cost can be expressed as a fraction $p = 0.2$ of the unit production cost, see equation (5.34). If the commodity is not available in stock, a sale is lost. In this case, a cost equal to the profit of the lost sale is incurred. The lead time can be assumed to be constant and equal to one week. The inventory is managed by means of an (s, S) policy with a period T of one week. On the basis of the total average costs per week (fixed and variable production costs, holding and shortage costs), select the best values of $s \in \{800, 900, 1000, 1100\}$ and $S \in \{1500, 2000, 2500\}$ by evaluating each such (s, S) policy with a Monte Carlo simulation (see Section 1.10.3).

5.29 In Problem 5.28, assume the two-bin policy is used and determine the best bin capacity among the alternatives 800, 900, 1000, and 1100.

5.30 Extend to three products the joint-order replenishment policy described in Section 5.12.4.

5.31 Modify the formulation of the crossdock door assignment problem (5.75)–(5.78), considering the case of distinct doors for incoming and outgoing vehicles. Assuming that the first four doors of the Ankommet crossdock are of a strip type, and the remaining four are of a stack type, determine the optimal solution of the crossdock door assignment problem thus modified.

5.32 [Ankommet.xlsx] Modify the ALTERNATING_ASSIGNMENT procedure by eliminating the distinction between incoming and outgoing vehicles in the corresponding sorting phases (code lines 4–5). Apply the procedure thus modified to determine a feasible solution of the Ankommet crossdock door assignment problem. Discuss the quality of the obtained solution.

5.33 [Ekom.xlsx] The warehouse in La Valletta, Malta, of Ekom Ltd makes use of order pickers which can hold three roll containers, with a capacity of 15 unit loads each. On 9 January at 10:15, a picking list of eight orders to be retrieved was created (see Table 5.35). The Euclidean distances between the centres of gravity of such orders are reported in Table 5.36. For orders 7 and 8, the Euclidean distance between the centres of gravity can be obtained from Tables 5.37 and 5.38. Determine a lower bound on the number of grouped orders. Identify how to group orders, based on a seed order, choosing orders on the basis of their number of unit loads.

5.34 Determine the picker route in the storage zone of the Guillen warehouse in Figure 5.91, using the S-shape heuristic, the largest-gap heuristic, the combined heuristic, and the aisle-by-aisle heuristic.

5.35 Show that an optimal picker route cannot traverse an aisle (or a portion of an aisle) more than twice. Illustrate how this property can be used to devise a dynamic programming algorithm.

5.36 Demonstrate that the FF and BF procedures for the 1-BP problem take $O(m \log m)$ steps.

5.37 Devise a branch-and-bound algorithm for the 1-BP problem based on formula (5.88).

Table 5.35 Picking list in the Ekom problem.

Order	Number of unit loads
1	17
2	14
3	18
4	9
5	10
6	13
7	7
8	6

Table 5.36 Euclidean distances (in m) between the centres of gravity of orders in the Ekom problem.

Order	1	2	3	4	5	6	7	8
1	0.00	31.56	19.62	31.48	28.60	23.17	25.90	28.77
2		0.00	28.30	20.50	29.49	22.10	25.39	25.34
3			0.00	29.02	21.40	19.57	31.68	21.31
4				0.00	25.06	31.65	28.50	26.33
5					0.00	31.29	22.51	23.98
6						0.00	25.42	19.85
7							0.00	-
8								0.00

Table 5.37 Unit loads and the corresponding Cartesian coordinates for order 7 in the Ekom problem.

Unit load (i)	Abscissa $(x_i^{(7)})$	Ordinate $(y_i^{(7)})$
1	24.06	48.71
2	46.38	46.94
3	53.73	48.37
4	6.45	19.29
5	46.10	48.22
6	6.57	22.58
7	35.98	38.51

Table 5.38 Unit loads and the corresponding Cartesian coordinates for order 8 in the Ekom problem.

Unit load (i)	Abscissa ($x_i^{(8)}$)	Ordinate ($y_i^{(8)}$)
1	1.61	9.47
2	41.71	46.90
3	4.25	34.96
4	32.02	17.49
5	19.24	23.35
6	17.37	26.01

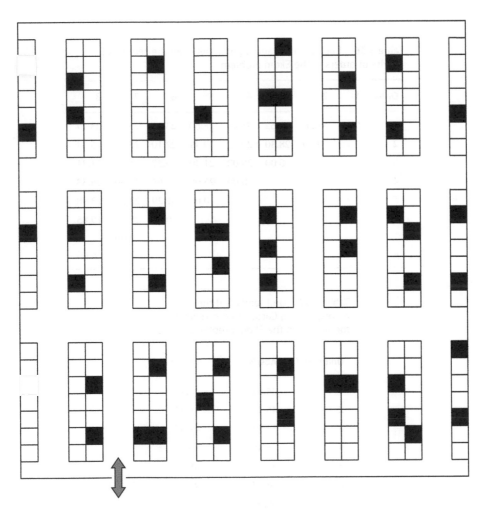

Figure 5.91 Storage zone layout in the Guillen problem.

Table 5.39 List of the parcels to be loaded and corresponding weight in the Brocard problem.

Parcel	Weight [kg]	Parcel	Weight [kg]
1	228	18	170
2	228	19	170
3	228	20	170
4	217	21	170
5	217	22	95
6	217	23	95
7	217	24	95
8	210	25	95
9	210	26	75
10	210	27	75
11	210	28	75
12	195	29	75
13	195	30	75
14	195	31	75
15	170	32	55
16	170	33	55
17	170	34	55

5.38 Devise an improved 1-BP lower bound.

5.39 [Brocard.xlsx] Modify the heuristics for the 1-BP problem in the case where each bin $j = 1, \dots, n$, has a capacity q_j and a cost f_j. Apply the modified version of the BFD procedure to the following problem. Brocard is a carrier mainly operating in France and the Benelux. The vehicle fleet is composed of 14 vans with a capacity of 800 kg and 22 vans with a capacity of 500 kg. The company has to deliver 34 parcels of different sizes from Paris to Frankfurt (the distance between these cities is 592 km). The characteristics of the parcels are reported in Table 5.39. As only five company-owned vans (all having a capacity of 800 kg) will be available on the day of the delivery, Brocard has decided to hire additional vehicles from a third-party company. The following additional vehicles will be available:
- two trucks with a capacity of 3 tonnes each, whose hiring total cost (inclusive of drivers) is 1.4 €/km;
- one truck with trailer, with a capacity of 3.5 tonnes, whose hiring total cost (inclusive of drivers) is 1.6 €/km.

Which trucks should Brocard hire?

5.40 [Brocard.xlsx] Determine a lower bound on the optimal solution cost in the Brocard problem by suitably modifying relation (5.88). Also determine the optimal shipment decision by solving a suitable modification of the 1-BP problem.

Table 5.40 Dimensions of the items to be packed in the Pertax problem.

Item	Length [cm]	Width [cm]	Height [cm]
1	60	70	80
2	40	90	100
3	50	60	120
4	30	20	90
5	100	60	80
6	100	70	120
7	40	80	110
8	100	100	80
9	140	50	100
10	90	70	130
11	80	30	130
12	60	50	120
13	130	90	60
14	100	50	60
15	50	130	70
16	20	60	70
17	90	70	130
18	130	50	120
19	50	30	130
20	60	30	100
21	80	40	110

5.41 Modify the BL procedure to solve a 2-BP problem in which item rotation is allowed.

5.42 Consider a 2-BP problem I with $W = 11, L = 16$ and $n = 6$, where the dimensions (w_i, l_i) of the objects $i = 1, \dots, 6$, are

$$\{(6,8), (7,8), (4,9), (5,7), (4,4), (5,6)\}.$$

Use the BL procedure to determine a feasible solution. Then determine a lower bound $\underline{z}(I)$ on the number of bins. Finally, use $\underline{z}(I)$ to establish whether the BL procedure solution can be deemed to be optimal.

5.43 [Pertax.xlsx] Pertax has to consolidate some items whose characteristics are reported in Table 5.40. The bins to be used have the following dimensions: $L = 200$ cm, $W = 150$ cm and $H = 200$ cm. Use both the 3-BP-L and the 3-BP-FFD procedure to solve this problem. Which procedure provides the best solution? Is it possible to say that this solution is optimal?

6

Managing Freight Transportation

6.1 Introduction

Freight transportation plays a crucial role in bringing together supply and demand among the facilities of a logistics system. It determines the most important part (often between one-third and two-thirds) of the total logistics cost and significantly affects the service level provided to customers.

Efficient and inexpensive freight transportation allows supply chains to operate with a reduced number of facilities in countries where automated manufacturing processes, a low-cost skilled workforce, and low energy prices are available. It also makes it feasible to supply remote markets with perishable products. Finally, it increases competition among companies on a global scale, with obvious advantages for the consumers.

Freight transportation involves not only shippers, carriers, and LSPs (see Section 1.4) but also governments, which construct and operate the transportation infrastructures (such as road and rail networks, port and airport terminals) and define the transportation policies at a regional, national, or international level (an analysis of these policies is beyond the scope of this book).

6.2 Transportation Modes

Transportation modes are different ways by which goods are transported from one place to another through land, air or water. The five basic transportation modes make use of roadways, waterways, railways, airways, and pipelines.

Some transportation modes (e.g., air transportation) do not allow a door-to-door connection between any origin and destination and, therefore, should be used jointly with other modes (*intermodal transportation*). In intermodal transportation, goods may be stored temporarily and then consolidated into different unit loads (e.g., intermodal containers) or vehicles (e.g., railway cars).

Trade over long distances is mainly supported by water, air, and rail transportation.

In 2019, water transportation accounted for 46.0% of the value of goods exported from the EU, and 56.2% of goods imported into the EU. On the other hand, air transportation accounted for 28.6% of the EU exports and 19.4% of its imports.

Introduction to Logistics Systems Management: With Microsoft® Excel® and Python® Examples, Third Edition. Gianpaolo Ghiani, Gilbert Laporte, and Roberto Musmanno.
© 2022 John Wiley & Sons Ltd. Published 2022 by John Wiley & Sons Ltd.

In terms of weight of freight moved, water transportation accounted for 79.1% of the EU exports and 74.6% of the EU imports in 2019, while air transportation accounted for just 1.7% of the EU exports and 0.2% of its imports. These figures indicate that air transportation is the mode of choice for high-value low-weight goods and perishable products.

Trade over medium and short distances is mainly sustained by road, water, and rail transportation depending on geography, and the value and the weight of the cargo.

In 2019, road accounted for nearly half of all tonne-kilometres moved in the EU (53.4%). Water transportation came next (33.8%), followed by rail (12.3%). In terms of tonne-kilometres, air transportation plays only a marginal role in intra EU freight transportation, with a share of 0.4%.

6.2.1 Road Transportation

Road transportation is suitable for almost every type of cargo over short to medium distances. In particular, it is used for parcel collection by couriers and last-mile freight distribution (see Section 1.6.5). Moreover, it plays a pivotal role in intermodal systems where freight has to be transported between (air, water, rail) terminals and industrial facilities (warehouses, manufacturing plants, etc.).

Road transportation services can be categorized into TL and LTL services:

- a TL service moves a single load (usually saturating the capacity of the vehicle) directly from its origin to its destination in a single trip (see Figure 6.1(a));
- LTL services arise when shipments add up to much less than the vehicle capacity (which has become common with the rise of e-commerce), in which case a single vehicle makes more deliveries at different destinations in a single multi-stop trip (see Figure 6.1(b)).

Most LTL services are designed according to a *hub-and-spoke* model (see Figure 6.2): individual small loads are picked up at a shipper's site (usually by a small van) and taken to a terminal (*spoke*) where they are loaded onto a truck and transported to a major terminal (*hub*). From there, loads are transshipped onto other trucks directed to peripheral spokes, close to their final destinations. Finally, goods are delivered to their final destinations (usually by a van along a multistop route). As a rule, LTL services are slower than TL services because they involve more merchandise handling.

Road vehicles can be classified as

- *light-duty vehicles*, including vans, motorbikes (and even bikes are used in last-mile distribution in urban areas, especially in e-commerce, see Problem 1.14);
- *heavy-duty vehicles*, comprising trucks and *long combination vehicles* (LCVs), the latter being tractor–trailer combinations with two or more trailers with a *gross vehicle weight* (GVW) up to over 60 tonnes.

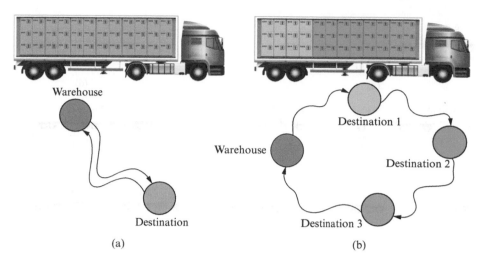

Figure 6.1 (a) TL and (b) LTL transportation services. In both cases it is assumed, for the sake of simplicity, that all the shipments originate from a single warehouse.

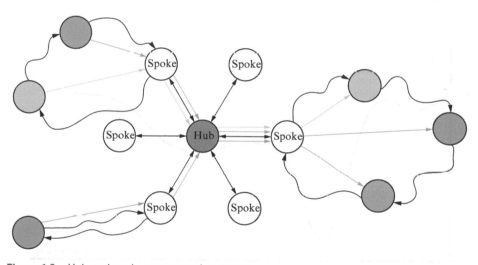

Figure 6.2 Hub-and-spoke transportation system.

Figure 6.3 shows the dimensions and the maximum GVW of the most common LCVs in use in North America and Europe.

The speed of road transportation is quite limited. Speed limits vary depending on the country and municipal regulations. For instance, in Italy the speed limit is 50 km/h in urban areas, 90 km/h on minor out-of-town roads, 110 km/h on major out-of-town roads and 130 km/h on motorways. The actual speed can be much lower based on traffic congestion which depends on the time of the day, the day of the week, and the period of the year. In Figure 6.4 a typical intra-day travel speed pattern is reported for a ring road of a large Italian metropolitan area.

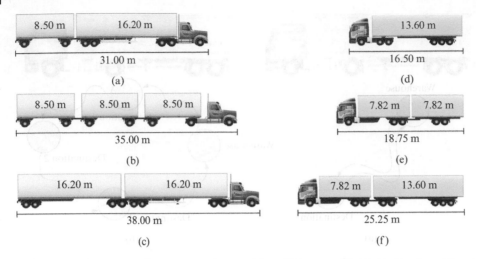

Figure 6.3 Main long combination vehicles: (a), (b), and (c) are used in North America while (d), (e), and (f) are used in Europe; (a) Rocky Mountain double, maximum GVW of 62.5 tonnes; (b) triple trailer, maximum GVW of 53.5 tonnes; (c) turnpike double, maximum GVW of 63.5 tonnes; (d) six axle semi-trailer, maximum GVW of 40.0 tonnes; (e) twin trailer, maximum GVW of 40.0 tonnes; (f) gigaliner, maximum GVW of 60.0 tonnes.

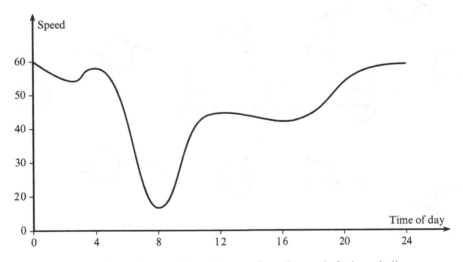

Figure 6.4 A typical intra-day travel speed pattern for a ring road of a large Italian metropolitan area. The speed is in km/h.

6.2.2 Water Transportation

Water transportation is operated by sea or on lakes, canals, or rivers. Traditionally, water transportation was linked to heavy industries (e.g., the steel and petrochemical industries) whose facilities were (and still are) adjacent to ports. Starting in the mid-twentieth century, the introduction of containers (see Section 5.4.5) and intermodality have revolutionized water transportation and have become one of the cornerstones of globalization.

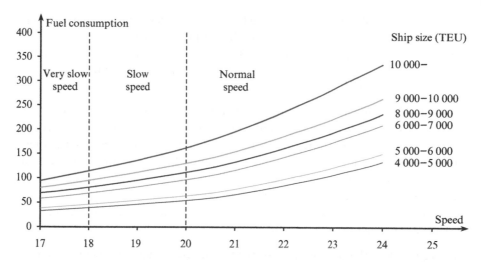

Figure 6.5 Fuel consumption (in tonnes per day) of a merchant vessel as a function of speed (in knots).

Trade over long distances passes mainly through core routes constituting a circum-equatorial band linking major ports in North America, Europe, and Pacific Asia through the Panama Canal, the Suez Canal, the Strait of Hormuz, and the Strait of Malacca. Secondary routes link major ports to smaller markets. In addition, numerous other routes exist for coastal shipping.

Inland waterway systems include the Volga/Don and Danube systems in Europe as well as the St. Lawrence/Great Lakes system and the Mississippi river in North America.

The average speed of a ship is about 20–25 knots (equivalent to 37–46 km/h) which allows to travel about 888–1104 km in 24 hours. As shown in Figure 6.5, fuel consumption is very sensitive to speed, growing according to a roughly cubic law.

Merchant vessels are characterized by their *deadweight* (DWT), defined as the total weight (expressed in tonnes) that they can safely carry. The DWT of most cargo ships ranges from 3000 for small coastal carriers to more than 500 000 for ultra-large crude carriers (see below).

Merchant vessels can be classified primarily according to the type of cargo they are designed to transport:

- *Bulk carriers*, that transport cargo in loose form in the holds (they do not carry cargo on deck). They include
  - *liquid bulk carriers* (or *tankers*), used for transporting oil, liquefied natural gas, and chemicals; their total carrying capacity amounts to 40% of the world's fleet tonnage;
  - *dry bulk carriers* including grains, minerals, cement, etc.; their total carrying capacity amounts to 35% of the world's fleet tonnage;
- *General cargo ships*, having the ability to carry goods in different forms such as boxed, palletized, and containerized. Cargo can be loaded under deck, on deck or between decks. General cargo ships include
  - *container ships*, that carry goods in large containers over the oceans; their total carrying capacity amounts to 13% of the world's fleet tonnage;

– *break-bulk ships*, that transport cargo in bags, boxes, crates, drums, or barrels; their total carrying capacity amounts to 7% of the world's fleet tonnage;

– *roll-on/roll-off* (Ro–Ro) *ships*, that are designed to allow cars, trucks and trains to be loaded directly on board; their total carrying capacity amounts to 5% of the world's fleet tonnage.

Merchant ships can also be classified according to their size and, in particular, whether they can traverse the lock chambers of the Suez and Panama canals. As a rule, large ships can serve a limited number of ports around the world with deepwater terminals and automated loading and unloading facilities. Moreover, small vessels often have cranes or pumps on board for loading and unloading their cargo in small ports without pierside material handling equipment (*geared vessels*).

Bulk carriers (including dry bulk carriers and tankers) are classified into a number of size categories. The most relevant are

- *Handysize* vessels, that are small ships with a capacity of up to 60 000 DWT and a length up to 200 m. They usually operate along regional trade routes and can enter small ports to pick up and drop-off cargoes.
- *Panamax* (see Figure 6.6(a)), *post-Panamax* and *new Panamax* vessels, that are defined in this way considering the maximum vessel sizes able to transit the first, second, and third sets of the Panama Canal's lock chambers, respectively. In particular, a Panamax ship has a length of up to 275 m and a capacity of up to 80 000 DWT, a post-Panamax ship has a length of up to 294 m and presents a capacity of 110 000 DWT, while a new Panamax ship has a maximum length of 366 m, a maximum width of 49 m and a capacity of up to 120 000 DWT.
- *Capesize* ships, that are the largest dry cargo vessels. Since they are too large to transit the Suez Canal and the Panama Canal, they have to pass the Cape of Good Hope or Cape Horn to transit between oceans. Capesize vessels have a maximum length of 270 m and a capacity of up to 400 000 DWT. The term "Capesize" is not applied to tankers.

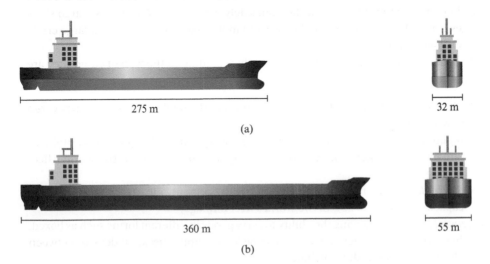

Figure 6.6 (a) Panamax vessel, with a draft of 12 m and (b) Chinamax vessel, with a draft of 24 m.

- *Very large crude carriers* (VLCCs) and *ultra-large crude carriers* (ULCCs), also known as *supertankers*, that are the largest cargo vessels. VLCCs are up to 250 000 DWT and have a length of up to 330 m. ULCCs can be up to 500 000 DWT and have a length of up to 415 m. Both VLCCs and ULCCs can traverse the Suez Canal and are mainly used to transport crude oil from the Persian Gulf to Europe, North America, and Asia.
- *Chinamax* vessels (also known as *Valemax* vessels from the Brazilian multinational mining corporation Vale, see Figure 6.6(b)), that are bulk carriers initially built to transport ores between Brazil and China. They have a capacity of up to 400 000 DWT, a length of about 360 m, a maximum width of 55 m, and require deepwater ports with a draft of 25 m.

Container ships are classified into a number of size categories, according to their dimensions and capacity, the latter being measured in TEU (see Section 5.4.5):

- *Feeder ships* (or, simply, feeders), that are small ships typically taking their cargo from small to large ports where it is transshipped onto larger ships and, vice versa, distributing containers from large ports to smaller peripheral ports. Their capacity is usually under 3000 TEU.
- Panamax, post-Panamax, and new Panamax (meanings already explained above). A Panamax container ship has a maximum capacity of 5000 TEU, while a new Panamax container ship has a capacity of up to 14 000 TEU.
- *Ultra-large container vessels* (ULCVs), that have a capacity greater than 20 000 TEU, a length around 400 m and are able to transit through the Suez Canal.

6.2.3 Rail Transportation

Traditionally, rail transportation is used for moving raw materials (coal, chemicals, etc.) and low-value finished products (steel, paper, sugar, tinned food, etc.). In the last decades, it has also become an integral part of intermodal systems based on the transportation of shipping containers. Figure 6.7 shows some of the most common types of rail cars.

Rail transportation is by far the least expensive land transportation mode for long-distance movements, especially when loads are multiples of a wagon capacity (*carload transfers*, or CL).

Although the maximum possible running speed of a freight train can be as high as 200 km/h (or even more, see Problem 1.11), rail transportation is relatively slow. This is mainly due to the following:

- direct train connections are quite rare; hence, rail cars often have to be transshipped from one convoy to another (possibly of a different railroad company) at special terminals;
- a convoy must include tens of cars in order to be worth operating;
- in several countries (e.g., in the EU), freight trains have low priority compared with passenger trains; a remarkable exception is North America, where the rail network is owned mostly by private companies and governments subsidize passenger transportation.

See the CPR case study in Section 1.6.8 for more details on railway management.

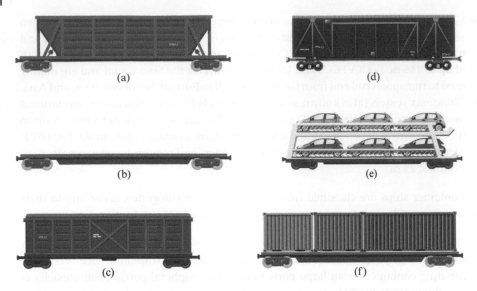

Figure 6.7 Some of the most common types of rail cars: (a) hopper ore car; (b) flat car; (c) hard top gondola; (d) wood chip car; (e) automobile car; (f) container car.

6.2.4 Air Transportation

Air packages are generally transported into ULDs (see Section 5.4.2), either in the belly-hold of scheduled passenger flights or on dedicated freight planes (known as *freighters*, see Figure 6.8). Moreover, in defence and humanitarian logistics, heavy loads, such as tanks and emergency vehicles, are transported in specialized cargo planes.

Air transportation is often used along with road transportation in order to provide door-to-door services. While air transportation is in principle very fast (the cruise speed of commercial flights is approximately 900–1000 km/h), it is slowed down by freight handling at air terminals. Consequently, air transportation is not competitive for short- and medium-haul shipments. In contrast, it is quite popular for the transportation of high-value products (as already observed above) over long distances. The capacity (in terms of both weight and volume) of an aircraft is fairly limited, compared with that of trains and ships.

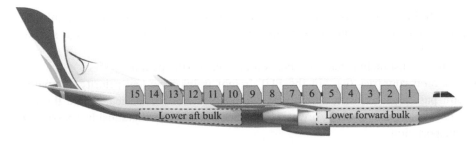

Figure 6.8 A cargo plane with 15 ULDs loaded in the cabin.

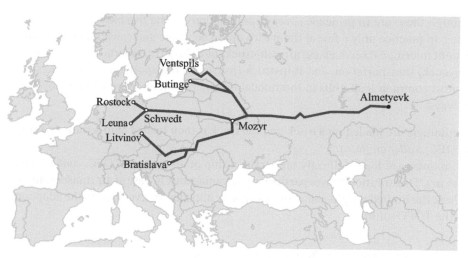

Figure 6.9   The Druzhba pipeline.

6.2.5   Pipeline Transportation

Pipelines are cylindrical pipes made of reinforced concrete, cast iron, steel, etc., used to transport water, oil, and gas over long distances by gravity or pressure. The slowness of the transportation (5–6 km/h) is compensated for, at least in part, by the possibility of continuous supply (24 hours a day) and by the reliability of the pipelines and pumps.

> The West–East gas pipeline in China, with an extension of 8707 km, is the world's longest pipeline. It comprises a main trunkline and eight branches, passing through more than 66 cities in 10 provinces. The gas is used for electricity production in the Yangtze river delta.
>
> The Druzhba pipeline is the world's longest oil pipeline, covering about 4000 km (see Figure 6.9). Druzhba runs from Almetyevsk in central Russia to Mozyr in southern Belarus, where it divides into a northern branch, reaching Schwedt in Germany, and into a southern one to Bratislava. On its path the Druzhba pipeline incorporates 20 pumping stations along the way and supplies different refineries (e.g., Rostock). It has a maximum capacity of around 1.3 million barrels of oil per day. The primacy of the world's longest oil pipeline is challenged by the Eastern Siberia–Pacific Ocean oil pipeline. The construction of the pipeline, which extends for 4857 km, began in 2006 near the town of Taishet in central Siberia, and will connect Taishet to Kozmino on the eastern Siberian coast.

In addition, there exist automated waste collection systems using a pneumatic underground pipe network to collect household waste (see Problem 1.8).

6.2.6   Intermodal Transportation

The possibility of moving freight with more than one transportation mode allows hybrid services to be realized with a reasonable trade-off between cost and transit time.

Although there are in principle several combinations of the five basic transportation modes, in practice only a few of them turn out to be viable and convenient. The most frequent intermodal services are: aircraft–truck (*birdyback*) transportation, train–truck (*piggyback*) transportation, and ship–truck (*fishyback*) transportation. Containers are the most common load units in intermodal transportation and they can be moved in two ways:

- Containers are loaded on a truck and the truck is then loaded onto a train or a ship (TOFC, *trailer on flatcar*).
- Containers are loaded directly on a train, a ship or an aircraft (COFC, *container on flatcar*). This solution proves particularly advantageous in case of air transportation, where the limited weight and volume capacity impedes in several cases the adoption of the TOFC solution.

6.2.7 Comparison Among Transportation Modes

Transportation modes differ with respect to three fundamental parameters: *capacity*, *cost*, and *transit time* (the latter parameter being related to reliability). Competition between transportation modes depends mainly on the distance between origin and destination, the quantities shipped, their value, as well as on the service level that customers expect. In a modern global economy, transportation modes complement one another to provide intermodal transportation services with various mixes in terms of cost, speed, and reliability.

Capacity

The capacity of a vehicle to carry some freight depends to a large extent on the transportation mode. Table 6.1 illustrates such a feature for some of the vehicles previously described.

Table 6.1 Capacity of sample vehicles used in different transportation modes.

Vehicle	Capacity	Truck equivalency
Semi-trailer truck	60 tonnes from 4.0 to 5.3 TEU 30 000 litres	1
100 car train unit	9000 tonnes from 400 to 530 TEU 11 000 000 litres	385
Panamax container ship	5000 TEU	2116
VLCC	250 000 tonnes 2 000 000 barrels of oil	9330
Boeing 747-400F	124 tonnes	5

Cost

Freight transportation cost depends on whether transportation is realized by the shipper, or entrusted to a carrier.

In the former case, the cost is given by the sum of costs associated with the depreciation, maintenance, and insurance of the vehicles (owned or hired), crew wages, fuel consumption, loading, unloading and transshipment operations, administration, and use of vehicle depots. Moreover, there can be additional costs which depend on the specific transportation mode: for example, water shipping leads to port mooring charges, port costs (agencies and piloting), canal transit cost, and so on. Some of the costs depend only on transportation time (e.g., the cost of insurance of owned vehicles), and others depend only on the distance covered (e.g., fuel consumption); still others are dependent both on time and on distance (e.g., vehicle depreciation costs), whereas other costs (such as administration) are customarily allocated as a fixed annual charge. In general, the main determinant of the transportation cost is the distance between the origin and the destination of the shipment, as illustrated schematically in Figure 6.10.

When a shipper uses a carrier, the freight transportation cost can be calculated on the basis of the rates published by the carrier (or, sometimes, as the outcome of a negotiation). For customized transportation, the cost of a TL transportation depends on both the origin and destination of the movement, as well as on the size and equipment of the vehicle required. For an LTL transportation, each shipment is given a rating (called a *class*), which depends on the physical characteristics (weight, density, etc.) of the goods. For example, in North America the railway classification includes 31 classes, while the US National Motor Freight Classification (NMFC) comprises 23 classes. Rates are usually reported in tables, or can be calculated through *rating engines* available on the Internet. In Table 6.2, the LTL rates for freight transportation by road applied in the United States by the National Classification Committee (NCC) are reported; they are related to two classes. Rates are such that costs can be discontinuous, as illustrated in Figure 6.11 (cost may decrease by adding extra weight).

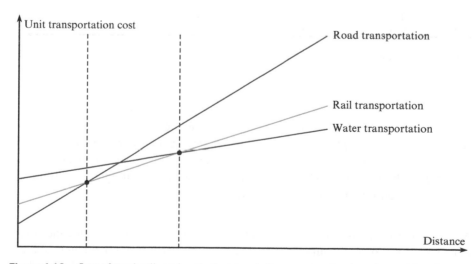

Figure 6.10 Cost of road, rail, and water transportations versus distance covered (qualitative representation).

Table 6.2 LTL rates ($ per 100 pounds) from New York to Los Angeles published by the NCC; classes 55 and 70 correspond to products having densities higher than 15 and 35 pounds per cubic foot, respectively.

Weight (W)	Class 55	Class 70
$0 \leq W < 500$	129.57	153.82
$500 \leq W < 1000$	104.90	124.60
$1000 \leq W < 2000$	89.43	106.10
$2000 \leq W < 5000$	75.17	89.24
$5000 \leq W < 10\,000$	64.82	76.95
$10\,000 \leq W < 20\,000$	53.13	63.05
$20\,000 \leq W < 30\,000$	46.65	55.37
$30\,000 \leq W < 40\,000$	40.15	47.67
$W \geq 40\,000$	37.58	44.64

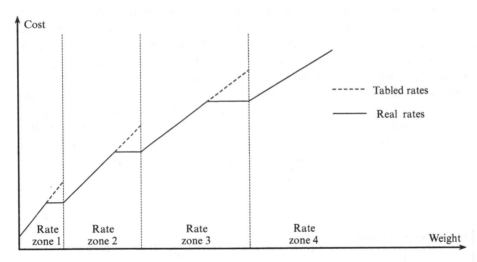

Figure 6.11 Transportation rates for LTL trucking.

Media Action, a US company, wants to transfer 9400 pounds of class 55 freight from New York to Los Angeles. Using a carrier who applies the NCC rates shown in Table 6.2, it should pay $94 \times 64.82 = \$\,6093.08$. If the shipping load were 10 000 pounds, the cost would be $100 \times 53.13 = \$\,5313$. For this reason, the company declares to the carrier (which accepts) that it wishes to ship freight for 10 000 pounds, although the effective weight is less.

Air is the most expensive transportation mode, followed by road, rail, pipeline, and water. According to recent surveys, transportation by truck is approximately seven times more expensive than by train, which is usually four times more costly than by ship and by pipeline.

Transit time

Transit time is the time a shipment takes to move from its origin to its destination. It also includes the time spent for any loading, unloading, and transshipment operations. It is a random variable influenced by weather and traffic conditions, freight loading and unloading procedures, and so on. The coefficient of variation (standard deviation over average transit time, see Section 1.8.3) of the transit time is a common measure of the reliability of a transportation service. On the basis of several statistical investigations carried out on different transportation services, the most reliable transportation mode is, in general, the pipeline, followed by air, train, road, and water transportation.

6.3 Freight Transportation Terminals

In most logistics systems, freight does not travel directly from origin to destination for the sake of economy. Instead, it passes through a number of freight transportation terminals (ports, airports, rail, and road terminals) which are *intermediate* nodes of the logistics system where shipments are consolidated, deconsolidated, or transshipped.

Terminals acting as transshipment points are called *interchange points* if they involve a single transportation mode, or *transfer points* if multiple transportation modes are used. A *hub*, or *gateway*, is a large transshipment terminal, often part of a hub-and-spoke network (see Section 1.6.5).

The main advantage in using freight transportation terminals is that they enable consolidation of cargo thereby allowing the use of vehicles with a high load factor. This generally leads to a cost reduction in the logistics systems.

Freight transportation terminals require specific infrastructures (berths, docks, platforms, loading and unloading bays, yards, sidings, etc.) for managing incoming and outgoing vehicles, and are equipped with a large variety of storage and material handling systems, depending on the type of freight handled. A freight terminal can also act as a warehouse by storing, e.g., in-transit refrigerated goods. Other specific activities depend on the type of terminal. They include, for example, administrative services, such as custom operations and maintenance services (e.g., container washing and repair).

> LNG is a natural gas cooled to −160 °C which shrinks its volume 600 times, making it liquid. LNG is easier and safer to store and transport over long distances, usually by sea in cryogenic tanks.
>
> The Peru LNG port terminal, located in San Vicente de Cañete, contains the first natural gas liquefaction plant in South America. It is used to accommodate large LNG carrier ships. The terminal is characterized by the presence of two storage tanks with a capacity of 130 000 m³ each and a supply pipeline 408 km long.

Apart from the transportation modes involved, the main differences in freight terminals depend on the form of the goods to be handled. Liquid bulk freight generally requires little material handling equipment (such as pumps), but significantly large storage zones; in contrast, dry bulk freight is typically moved with specialized material handling equipment (more specifically, conveyors, see Section 5.6.4); goods of irregular shape (e.g., machinery) require labour intensive freight terminals, because of the great difficulty in handling them automatically; finally, containerized goods can be dealt with by automated material handling equipment (e.g., container cranes and straddle carriers) and are usually stored as stacks.

Special attention is paid in freight terminals (as well as in the vehicles in which they are transported) to the handling of dangerous goods, i.e., goods that pose a risk to health, safety, or the environment. Dangerous goods are classified into nine classes, on the basis of the specific chemical characteristics producing the risk (explosives, gases, flammable liquids, flammable solids, oxidizing agents and organic peroxides, toxic and infectious substances, radioactive substances, corrosive substances, and miscellaneous). Dangerous goods require specific protective packaging during transportation. Moreover, some combinations of different classes of dangerous goods in the same vehicle or in the same storage zone are forbidden. Freight terminals require specific material handling equipment for dealing with dangerous goods (i.e., dedicated storage zones and explosives trace detectors).

> There exist several regulatory schemes for handling dangerous goods in freight terminals and during transportation. The International Civil Aviation Organization has developed dangerous goods regulations for air transportation. Similarly, the International Maritime Organization has developed the International Maritime Dangerous Goods Code for transportation of dangerous goods by sea. Likewise, the Intergovernmental Organization for International Carriage by Rail has developed regulations concerning the international carriage of dangerous goods by rail.

Despite the advantages described above, the use of freight terminals generates costs which represent a significant part of the total transportation costs. They can be broadly classified in infrastructure costs (related to the terminal facilities and material handling equipment), transshipment costs (for loading and unloading freight), and management costs (incurred by the public authority or the private operator which is in charge of the terminal).

6.3.1 Port Terminals

The structure of port terminals has evolved remarkably over the past two centuries. Until late 1950s, port terminals were designed to handle the direct transfer of bulk cargo from ships berthed at docks to trucks or trains (and vice versa). Transshipment operations were very labour-intensive, generally using on-board cranes and took several days. Since the mid-twentieth century, the advent of the intermodal container as a unit of cargo has produced a gradual conversion of most existing port terminals into *container terminals* with a radical change in their layout. It has also influenced the site selection criteria for the construction of new ports. In particular, berths have been redesigned to allow faster loading and unloading operations using *quay cranes* and other container handling and storage equipment. As a result, waiting times to complete loading and unloading operations for a container ship have become several times shorter than for an equivalent break-bulk cargo ship.

All the largest and most modern container terminals provide intermodal transportation, particularly by rail.

The port of Shangai, China, is the world's largest container port, with a volume of 43.3 million TEU handled in 2019. The top-ten list includes six other Chinese container terminal ports (Shenzhen, Ningbo-Zhoushan, Guangzhou, Hong Kong, Qingdao, and Tianjin). The port of Singapore occupies the second position, whereas the first non-Asian container port is Jebel Ali, in Dubai, United Arab Emirates (10th position).

In a container terminal, three distinct areas can be generally distinguished: the quayside, the yard, and the gate (see Figure 6.12).

The *quayside* is made up of berthing positions along the quay and of quay cranes (see Figure 6.13(a)) that load and unload containers from vessels mooring at the berth. Containers are then transferred to the yard by AGVs, *straddle carriers* (see Figure 6.13(b)) or internal trucks. The internal transportation equipment is also used to move containers from the yard to the gate and, when needed, to relocate containers within the storage zone.

The *yard* is used as a storage zone for loading, unloading, and transshipping containers and is typically divided into blocks: each container block is served by one or more yard cranes, such as *rubber-tired gantry cranes* (RTGs, see Figure 6.13(c)) or *rail-mounted* (RMGs). An alternative yard layout is based on lanes served by straddle carriers. The material handling equipment used within the yard allows to distinguish between intensive-use and extensive-use yards. An intensive-use yard ensures a higher storage capacity and is mainly operated by RMGs or RTGs, capable of storing about 100 TEU per 1000 m^2. Container stacks are, on average, five levels high. An extensive-use yard terminal requires a lower storage capacity and is commonly operated by straddle carriers, that can store about 60 TEU per 1000 m^2. Here, the container stacks are generally, on average, three levels high.

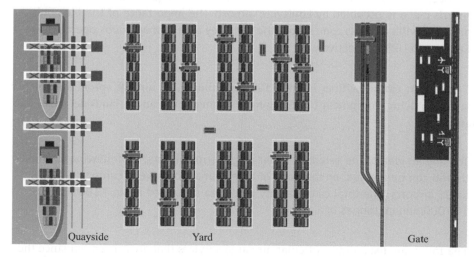

Figure 6.12 Layout of a container terminal.

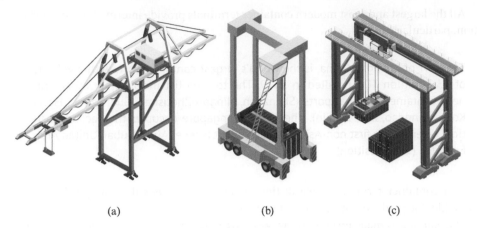

(a)   (b)   (c)

Figure 6.13   (a) Quay crane; (b) straddle carrier; (c) rubber-tired gantry crane.

The *gate* is the area supplied by the yard and used for container pick-up and drop-off operations from or onto trains, trucks, or ships. In the latter case, the container terminal acts as an interchange point, from which smaller (feeder) vessels are used to service peripheral ports. Congestion may arise when the gate shares part of the quay used by incoming vessels, and when incoming and outgoing vessels are simultaneously performing loading and unloading operations in the quay.

6.3.2   Air Cargo Terminals

Air cargo terminals are located within airports, the majority of which are used for transporting both passengers and goods.

> Hong Kong airport is the largest air cargo terminal in the world, handling 4.8 million tonnes of cargo in 2019. Memphis airport is in second place (4.3 million tonnes) and Shanghai Pudong airport is in third position (3.6 million tonnes). The fourth place is occupied by Louisville airport. The importance of the Memphis and Louisville airports comes from the fact they are the major hubs operated by FedEx and UPS, respectively.

The airport cargo facilities include a cargo building, an aircraft apron, an airside ground servicing equipment (GSE) storage/staging zone, and a landside area (see Figure 6.14).

> To have an idea of the extent of an air cargo terminal, the following rule of thumb can be applied, on the basis of a benchmark on several cargo operations at US airports. The total cargo facility land area should be about 72 000 m^2 per 100 000 annual tonnes of air cargo.

The following lists some particular air cargo services that are carried out since the facilities are generally configured for particular types of cargo carrier.

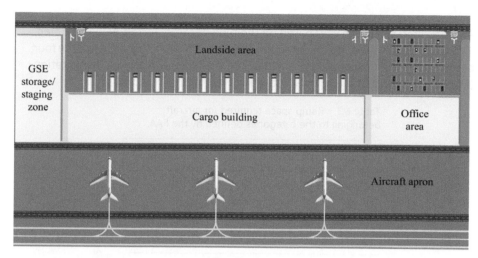

Figure 6.14 Layout of an air cargo terminal.

Integrated air cargo carriers provide full-service logistics operations, handling cargo from shipper to delivery points. This leads to higher efficiency in space utilization of the air cargo facilities, due to a centralized control of all the logistics operations; therefore, integrated cargo facilities are generally capable of processing larger volumes of cargo in less space.

Freight cargo carriers are similar to integrated cargo carriers since cargo is carried on all-freighter aircraft; however, freight cargo carriers do not provide full-service logistics operations and rely on ground handlers to process the cargo and on third-party companies for ground transportation. As a result, the space of the air cargo facilities is not efficiently used when compared to integrated cargo carrier facilities.

In the case of *belly* cargo (i.e., the air cargo carried on passenger aircraft) the cargo facilities also include a truck apron and docks for the loading and unloading of cargo, a warehouse and an office area for the consolidation and deconsolidation, inspection, and storage of cargo, and a container staging area. Belly cargo is usually handled from the passenger terminal area where the aircraft is parked. These facilities, therefore, require access to the airside and efficient connectivity to the terminal.

As far as the *cargo buildings* are concerned, their utilization rates depend on several factors (amount and type of cargo facilities needed at an airport, size of the airport, type of cargo to be moved, etc.) and are more rarely located at smaller airports than at larger airports. In modern air cargo terminals, the cargo buildings are multi-level, which leads to a more efficient utilization rate of the site since more cargo can be processed within a smaller building footprint. In addition, when a third-party cargo handler is under contract with more carriers, it is common to have open spaces which enable the same building to be used to accommodate various operators, and consequently allow better management of the cargo peak variations during the day.

The dimension of the aircraft parking area depends on several factors, including the cargo volume of the cargo air terminal, the fleet composition of cargo planes, the possibility of using the hardstand positions by non-cargo aircraft.

Aircraft ramp space can vary based on the type of cargo aircraft being operated. Table 6.3 reports the ramp space required for aircraft belonging to each of the four categories in which they are classified by the US Federal Aviation Administration (FAA) on the basis of the cargo volume and weight.

Table 6.3 Ramp space required for aircraft belonging to the categories defined by the FAA.

Code category	Reference aircraft	Ramp space [m^2]
C	Boeing 737	1925
D	Airbus A300	3260
E	Boeing 747	5435
F	Boeing 747-8F	7230

The landside area is used for managing the air–truck intermodal freight transportation. The area is characterized by truck manoeuvring and staging areas, employee parking, and customer parking.

Finally, it is worth observing the growing trend in collocating air cargo terminals near fulfilment centres, possibly within the same airport area. In the air cargo terminal the goods generally remain for a limited amount of time, whereas the fulfilment centre works as a warehouse where goods are stored until they are distributed. A fulfilment centre is typically larger (perhaps three or four times) than the corresponding cargo building in the cargo air terminal. The growth of on-airport fulfilment centres is driven by the e-commerce industry (see the Amazon case described in Section 1.6.5).

6.3.3 Rail Freight Terminals

Rail freight terminals are, in most cases, rectangular in shape (see Figure 6.15) and typically occupy a smaller area than port and air cargo terminals. In contrast to air cargo terminals, rail freight terminals are generally distinct from passenger terminals. Passenger rail stations are generally located in the centre of cities, while rail freight terminals are more often located on the outskirts, often near industrial sites, where yard facilities can be more easily located. The capacity of a rail freight terminal strictly depends on the number of tracks available, which is a feature that is difficult to change once the terminal has been built. A peculiarity of the rail freight terminals is the *shunting* (or *switching*, synonym mainly used in the USA) operation, which consists of assembling, sorting, and breaking down the freight trains. This operation is executed in terminals called *classification yards*.

Bailey yard in North Platte, Nebraska, USA, is the largest classification yard in the world, handling about 10 000 railcars per day. It is operated by Union Pacific,

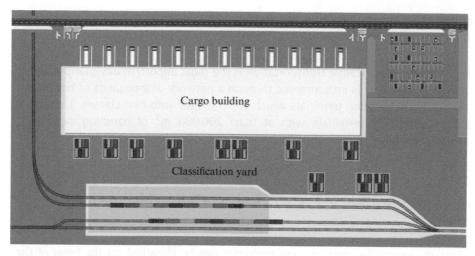

Figure 6.15 Layout of a rail freight terminal.

which is actually the largest rail company in the world, whose value is about $ 75.4 billion.

Shunting can be required more times on the same freight train as the number of railcars and the origin–destination distance increase. In particular, the freight train needs to be assembled at its origin, switched at any intermediary locations and broken down at its destination.

The standard facilities and equipment required by a rail freight terminal depends on the cargo transported (bulk freight, containerized unit loads, see Section 5.4.5) and also by the way the cargo is loaded and unloaded. In roll-on/roll-off rail freight terminals, ramps are required to roll the vehicles (cars, trucks, etc.) into a railcar. Such terminals commonly require a large amount of parking space to store the vehicles to be loaded or unloaded.

Like port and air cargo terminals, rail freight terminals are also widely used as in intermodal freight transportation.

The Interporto Quadrante Europa is located in Italy at the crossroads of the Brenner and the Serenissima motorways and some major railway lines. It extends over an area of 2.5 km^2. About 9 million tonnes of goods transit every year through the terminal by rail and 16 million tonnes by road. Interporto Quadrante Europa is also directly linked to Verona-Villafranca airport. It has been recognized as the best-organized intermodal freight terminal in Europe.

6.3.4 Road Freight Terminals

The most common road freight terminals are warehouses, different types of which have been illustrated in Section 5.2. They also include crossdocks (which have no storage equipment), where the loads on inbound vehicles are moved directly onto outbound vehicles (possibly after a stop of a few hours).

In China, highway freight transportation is the most important basic land transportation mode. It is implemented through a network of thousands of highway freight transportation terminals which are classified into four classes. The first class consists of terminals with at least 200 000 m^2 of handling capacity, registered capital of more than 20 million yuan, or an area of at least 2 km^2.

6.4 Classification of Freight Transportation Management Problems

Freight transportation management problems can be classified on the basis of the distance between the origins and the destinations of the shipments. A distinct role is played by management problems arising in transportation terminals where the transition between different long- or short-haul services takes place.

6.4.1 Long-haul Freight Transportation Management

In *long-haul* freight transportation, goods are moved over relatively long distances (which vary from few hundred to a few thousand kilometres) between terminals or other facilities (plants, warehouses etc.). Long-haul freight transportation services can provide direct or indirect shipments. As mentioned in Section 6.2, with direct shipments, the freight is transferred from the origin to the destination without intermediate transshipments, whereas in the case of indirect shipments the movement is done by means of a sequence of trips which can be carried out with different vehicles if necessary. Direct long-haul freight transportation services can be performed by means of shipper-owned vehicles or by a carrier, whereas indirect services are generally assigned to a carrier. Carriers' decision-making problems are generally more complex than those of companies executing their own transportation. In fact, in the case of direct shipments the number of origins and destinations is usually fairly limited (*few-to-few* or *few-to-many* problems), whereas in indirect services shippers customarily service a large number of customers with many origins and many destinations (*many-to-many* problems).

Carriers can operate according to a schedule (*line services*) or on the basis of customers' requests. Under the former hypothesis, at a tactical level, the carrier should assign the freight traffic to the network of existing transportation lines (*traffic assignment problem*, or TAP, see Section 6.6) on the basis of the transportation demand (which is known). Conversely, when the transportation network has not already been designed or when the existing network proves to be inadequate, the carrier should solve, always at a tactical level, a *service network design problem* (SNDP, see Section 6.7), on the basis of the forecast demand or on the basis of commercial agreements stipulated with the customers. Under the latter hypothesis, at the operational level, the carrier needs to assign its owned fleet dynamically (and the vehicles hired eventually by other shippers) to customer requests, so that these are satisfied, and the cost of using the entire fleet is minimized. At the tactical level, another decision-making problem is the periodic repositioning of empty vehicles, so that the average response time to subsequent requests is kept as low as possible (*vehicle allocation problem*, or VAP, see Section 6.8). Other decision-making problems for carriers are, at the strategic level, the composition of the vehicle fleet to purchase, and, at the tactical level, the optimal crew scheduling (a problem in this category will be provided in Section 6.9). Finally, at the operational level, decision-making problems deal with the reassignment of vehicles and crews to take into account unexpected events, such as changes of the freight orders, vehicle breakdown, strikes, or unfavourable weather conditions.

Regarding companies that act as private shippers, at the strategic level decision-making problems concern the purchase of their own fleet of vehicles, at the tactical level they concern the choice of vehicles to hire to integrate their own fleet (see Section 6.10) and at the operational level they address the optimal consolidation and dispatching of the shipping orders (see Section 6.11).

6.4.2 Freight Transportation Terminal Management

Freight transportation terminals, as already discussed, are complex facilities, the management of which involves the solution of a series of problems at strategic, tactical, and operational levels, with the objective of guaranteeing a certain level of efficiency and effectiveness. An exhaustive description of such problems is beyond the scope of this book; hence, in the following only a list of the most relevant problems is introduced.

Under a strategical perspective, the main decisions involve resource acquisition and, in particular, the construction or modification of the existing infrastructure, and include

- the definition of terminal location and layout (see Chapter 3);
- the sizing of the different functional zones within the freight terminals (see Section 5.10);
- the selection of the necessary storage systems, package identification systems and internal transportation equipment (see Sections 5.4–5.7).

In contrast, the tactical and operational perspectives are more influenced by the peculiarities of the different freight terminals, leading to distinct decisional problems.

The port terminal activities are centred around the management of the flow of inbound containers and their movements within the yard. From a tactical point of view, fundamental decisions are related to

- *berth allocation*, i.e., assigning arriving ships to berths over a given time horizon (considering handling times, ship length, priorities, arrival and departing time windows);
- *quay crane allocation*, i.e., reaching an efficient use of the quay cranes in vessel loading and unloading operations.

From an operational point of view, different activities need to be optimized:

- *yard allocation*, i.e., container storage and stacking policies as well as housekeeping strategies;

- *quay crane scheduling*, by considering constraints about completion times of tasks, overlapping and interferences between quay cranes;
- *ship stowage*, by minimizing the completion loading time and managing the ship stability;
- *vehicle routing* (see Section 6.12) for quay-to-yard transfer operations (or vice versa) and yard-to-gate transportation (or vice versa), by taking into account constraints about the distance travelled to complete the tasks, fleet operating costs, possible delays and, in case of AGV routing, preventing deadlocks and conflicts;
- *workforce management*, i.e., assigning human resources to tasks and pieces of equipment.

Similarly, air cargo terminal tactical activities concern the solution of the

- *gate assignment and scheduling problem*, for planning air cargo arrival position, timing and parking;
- *GSE allocation problem*, for efficiently assigning the equipment to the different gates.

From an operational perspective, optimization leads to

- *scheduling the terminal activities*, by considering constraints related to fuelling, loading and unloading times, aircraft ground movement, and landing;
- *aircraft routing and scheduling*, by taking into account priorities, available fleet, airport capacity and demand constraints;
- *vehicle routing* for landside activities;
- *storing* within cargo buildings;
- *crew assignment* to aircraft and land activities.

In rail freight terminals, tactical decisions concern

- *railcar and locomotive assignment*, to allocate the desired pulling power to the scheduled trains.

From an operational perspective, decisions include

- *yard management*, including the allocation of wagons coming from arriving trains to form the new trains;
- *crew assignment*, to allocate workforce to the various handling operations;
- *track assignment* in order to guarantee that each train can enter and leave the station without delays due to interferences with other trains.

Finally, decision problems related to road freight terminals have been dealt with in Chapter 5 and, in particular, in Section 5.13, regarding the crossdock door assignment problem.

6.4.3 Short-haul Freight Transportation Management

Short-haul freight transportation includes movements having their origin and destination in a relatively small geographic area (e.g., a city or a county). Problems of this type involve companies supplying their customers from RDCs with a fleet of company-owned vehicles (see Figure 6.16); local fast couriers, which transport loads between origin–destination pairs situated in the same area; national or international carriers,

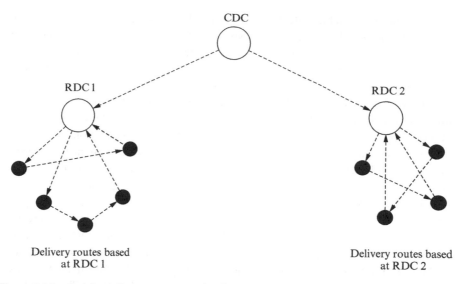

Figure 6.16 Freight delivery routes starting from two RDCs.

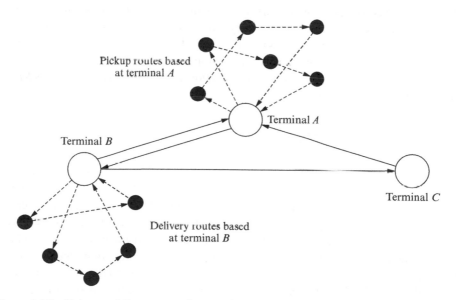

Figure 6.17 Pickup or delivery routes for parcels transported by a national or international carrier.

which need to collect locally outbound parcels before sending them to a remote terminal as a consolidated load, and to locally distribute loads coming from remote terminals (see Figure 6.17).

Short-haul freight transportation problems also arise in garbage collection, mail delivery, appliance repair services, and so on. The main decision-making problems concern, at the strategic level, the location of depots at which the vehicle routes originate (see Chapter 3 for this argument, although in some cases it is also necessary to take vehicle routing explicitly into account, in which case see Section 6.14); at the tactical level, the sizing of the vehicle fleet (see Section 6.10), and, at the operational level, the determination of vehicle routes to satisfy customer requests. In the simplest case of a single service request, the problem consists of determining a least-cost path from an origin to a destination; in the case of several simultaneous requests, a much more complicated *vehicle routing problem* (VRP) has to be solved, as shown in Section 6.12. Sometimes the vehicle routes can be planned in advance. This is the case, for example, of companies that plan the distribution of products on the basis of orders received from customers in previous days. Another case is solid urban waste collection, where the service routes are planned once or twice a year, assuming that the quantity of waste to be collected daily remains almost constant for a certain number of months. In contrast to the static case, there are situations where vehicle routing has to be updated dynamically, as soon as there is a new service request. This is the case, for example, of same-day urban courier companies, which have to pick up and deliver parcels, within a few hours, without previous notice. The availability of low-cost ICT tools (like GIS, GPS, cellular phones, traffic sensors, etc.) allows data acquisition in real time. These data are then used to update vehicle routing dynamically. The main features of the *vehicle routing and dispatching problems* (VRDPs) will be briefly illustrated in Section 6.13.

A different class of operational short-haul freight transportation problem concerns companies that directly control the inventory level of products at the successor nodes of the logistics system (e.g., the retailers). In this case, the company could determine the supply policy (what, how much, and when to supply) and the delivery route, with the objective, among others, of reducing transportation costs. *Inventory routing problems* (IRPs) will be discussed in Section 6.15.

6.5 Transportation Management Systems

A TMS is a software application concerning freight transportation operations. From the point of view of the software architecture (see Figure 1.28), the TMS is placed at the same level as the WMS (see Section 5.9), with which it is directly connected to create continuity and ensure greater speed in the management operations of the supply chains. In addition, TMS provides built-in EDI interfaces to ensure its integration with the ERP system (see Section 1.12). A TMS is adopted by companies that regularly deal with freight transportation management such as manufactures, retailers, e-commerce organizations, and LSPs. It oversees all core phases of freight transportation, i.e., *planning*, *execution*, and *monitoring*.

Transportation planning
A TMS helps select the most appropriate freight shipment, according to several criteria, including cost and time minimization. It is also a tool for carrier selection when transportation activities are outsourced (see Section 1.4). In the latter case, a TMS can also be used to store information on tariffs in such a way that selecting the best rates and service levels is faster (there is no need to search the website of each carrier). A TMS also allows the evaluation of different order management proposals, including shipment batching.

Transportation execution

This phase is devoted to the physical or administrative operations regarding the execution of the transportation plan. These include cargo handling, real-time transportation tracking, routing, real-time communication with carriers (for weather conditions, traffic congestions, route modifications, roadside assistance, etc.), preparation of administrative-accounting documents (i.e., freight billing and custom clearance, etc.).

Transportation monitoring

A TMS offers the possibility of creating reports and dashboards, to evaluate the efficiency and the effectiveness of all the freight transportation operations. To this end, some cost- and service level-related KPIs and performance measures (see Section 1.8) can be used (i.e., cost per unit weight or distance, percentage of on-time shipping operations).

SAP transportation management is a module embedded in S/4HANA (see Section 1.12) for end-to-end freight transportation process support, including freight management (costing, settlement, analytics, and reporting), order management, transportation planning, booking, tendering, and subcontracting. The main key functionalities of the module are: determination of optimal vehicle routes, best carrier, and optimal transportation mode(s); transportation cost calculation; track and trace; load optimization.

6.6 Freight Traffic Assignment Problems

Freight TAPs amount to determining a least-cost routing of goods over an existing network of transportation services from their origins (e.g., manufacturing plants) to their destinations (e.g., retailers). In a sense, the demand allocation problems illustrated in Chapter 3 are particular freight TAPs. From a mathematical point of view, TAPs can be cast as *network flow* problems. Network flow problems include, as special cases, several remarkable network optimization problems, such as the *shortest path problem* and the *transportation problem*.

TAPs can be classified as *static* or *dynamic*. Static models are suitable when the decisions to be made are not affected explicitly by time. They are formulated on a directed graph (or multi-graph) $G = (V, A)$, where the vertex set V often corresponds to a set of terminals or other facilities such as plants or warehouses, and the arcs in the set A represent possible transportation services linking the facilities. Some vertices represent *origins* of transportation demand for one or several products, while others are *destinations*, or act as *transshipment points*. Let K be the set of traffic classes (or, in this case, simply, commodities). Each arc is associated with a cost (possibly dependent on the amount of freight flow on the arc) and a capacity. Cost functions may represent both monetary costs and congestion effects arising at terminals.

In dynamic models, a time dimension is explicitly taken into account by modelling the transportation services over a given planning horizon through a *time-expanded* directed graph. In a time-expanded directed graph, the planning horizon is divided

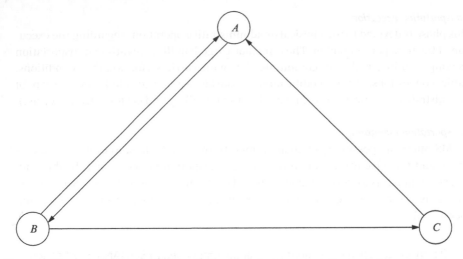

Figure 6.18 A static representation of a three terminal transportation system.

into a number of time periods $t_1, t_2, \ldots$ and the physical network (containing terminals and other material resources) is replicated in each time period. Then, temporal links are added. A temporal link connecting two representations of the same terminal at two different time periods may represent freight waiting to be loaded onto an incoming vehicle, or the time required for freight classification at the terminal. On the other hand, a temporal link connecting two representations of different terminals may describe a transportation service. Further vertices and arcs may be added to model the arrival of commodities at destinations and impose penalties in case of delays. With each link a capacity and a cost may be associated, similar to those used in static formulations. An example of a static transportation service network is shown in Figure 6.18, while an associated time-expanded directed graph is reported in Figure 6.19. In the static network there are three terminals (A, B, C) and four transportation services operating from A to B, from B to A, from B to C, and from C to A. The travel durations are equal to two, two, one, and one days, respectively. If the planning horizon includes four days, a dynamic representation has four vertices for each terminal (A_i, $i = 1, \ldots, 4$, describes terminal A on the ith day). Some arcs (such as (A_1, B_3)) represent transportation services, while others (such as (B_2, B_3)) describe commodities standing idle at terminals. In addition, there may be supersinks (such as terminal C in Figure 6.19), for which the costs on the arcs entering the supersinks represent economic penalties in case of transportation service failure.

6.6.1 Minimum-cost Flow Formulation

Let $O(k)$, $k \in K$, be the set of origins of commodity k; $D(k)$, $k \in K$, the set of destinations of commodity k; $T(k)$, $k \in K$, the set of transshipment points with respect to commodity k; o_i^k, $i \in O(k)$, $k \in K$, the supply of commodity k of vertex i; d_i^k, $i \in D(k)$, $k \in K$, the demand of commodity k of vertex i; u_{ij}, $(i, j) \in A$, the capacity of arc (i, j) (i.e. the maximum flow that arc (i, j) can carry), and u_{ij}^k, $(i, j) \in A$, $k \in K$, the maximum flow of commodity k on arc (i, j). Decision variables x_{ij}^k, $(i, j) \in A$, $k \in K$,

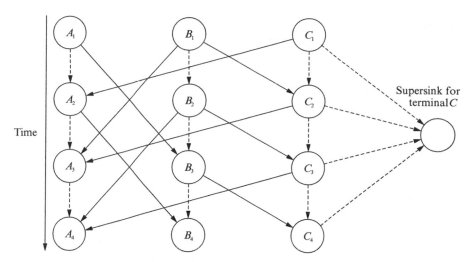

Figure 6.19 Dynamic network representation of the transportation system illustrated in Figure 6.18.

represent the flow of commodity k on arc (i, j). Moreover, let $C_{ij}^k(x_{ij}^k)$, $(i, j) \in A$, $k \in K$, be the cost for transporting x_{ij}^k flow units of commodity k on arc (i, j).

In the following, $G = (V, A)$ is assumed to be a strongly connected directed graph. The extension to the case where G is a multi-graph, or a collection of strongly connected directed subgraphs, is straightforward. A quite general *multi-commodity minimum-cost flow problem* (MMCFP) formulation is as follows:

$$\text{Minimize} \sum_{k \in K} \sum_{(i,j) \in A} C_{ij}^k(x_{ij}^k) \tag{6.1}$$

subject to

$$\sum_{j \in V:(i,j) \in A} x_{ij}^k - \sum_{j \in V:(j,i) \in A} x_{ji}^k = \begin{cases} o_i^k, & \text{if } i \in O(k) \\ -d_i^k, & \text{if } i \in D(k) \\ 0, & \text{if } i \in T(k) \end{cases} \quad \begin{matrix} i \in V, \\ k \in K \end{matrix} \tag{6.2}$$

$$x_{ij}^k \leq u_{ij}^k, \ (i, j) \in A, \ k \in K \tag{6.3}$$

$$\sum_{k \in K} x_{ij}^k \leq u_{ij}, (i, j) \in A \tag{6.4}$$

$$x_{ij}^k \geq 0, \ (i, j) \in A, \ k \in K.$$

The objective function (6.1) is the total cost, and constraints (6.2) correspond to the flow conservation constraints holding at each vertex $i \in V$ for each commodity $k \in K$. Constraints (6.3) impose that the flow of each commodity $k \in K$ does not exceed capacity u_{ij}^k on each arc $(i, j) \in A$. Constraints (6.4) (*bundle constraints*) require that, for each $(i, j) \in A$, the total flow on arc (i, j) is not greater than the capacity u_{ij}.

It is worth noting that o_i^k, $k \in K$, $i \in O(k)$, and d_i^k, $k \in K$, $i \in D(k)$, must satisfy the following conditions:

$$\sum_{i \in O(k)} o_i^k = \sum_{i \in D(k)} d_i^k, \ k \in K,$$

otherwise the problem is infeasible.

In the remainder of this section, some of the most relevant solution methods for some special cases of the MMCFP are illustrated.

6.6.2 Linear Single-commodity Minimum-cost Flow Problems

The *linear single-commodity minimum-cost flow problem* (LMCFP) can be formulated as follows:

$$\text{Minimize} \sum_{(i,j)\in A} c_{ij}x_{ij} \tag{6.5}$$

subject to

$$\sum_{j\in V:(i,j)\in A} x_{ij} - \sum_{j\in V:(j,i)\in A} x_{ji} = \begin{cases} o_i, & \text{if } i \in O \\ -d_i, & \text{if } i \in D, \quad i \in V \\ 0, & \text{if } i \in T \end{cases} \tag{6.6}$$

$$x_{ij} \leq u_{ij}, \ (i,j) \in A \tag{6.7}$$

$$x_{ij} \geq 0, \ (i,j) \in A. \tag{6.8}$$

The LMCFP is a structured LP problem and, as such, can be solved through the simplex algorithm or any other LP procedure. Instead of using a general-purpose algorithm, it is common to employ a tailored procedure, the (primal) *network simplex algorithm*, a specialized version of the classical simplex algorithm which takes advantage of the particular structure of the coefficient matrix associated with constraints (6.6) (corresponding to the vertex–arc incidence matrix of the directed graph G).

The case where there are no capacity constraints (6.7) is first examined. In such a case, it is useful to exploit the following characterization of the basic solutions of the system of equations (6.6), which is stated without proof.

Property. The basic solutions of the system of equations (6.6) have $|V| - 1$ basic variables. Moreover, each basic solution corresponds to a tree spanning G and vice versa.

In order to find a basic solution of problem (6.5), (6.6), (6.8), it is therefore sufficient to select a tree spanning G, set to zero the variables associated with the arcs that are not part of the tree, and then solve the system of linear equations (6.6). The latter step can be easily accomplished through a substitution method. Of course, the basic solution associated with a spanning tree is not always feasible, since the non-negativity constraints (6.8) may be violated (see Figure 6.20).

The network simplex algorithm has the same structure as the standard simplex algorithm. However, the optimality test and the pivot operations are performed in a simplified way. The sketch of the NETWORK_SIMPLEX procedure is reported below.

The particular structure of problem (6.5), (6.6), (6.8) and of its dual,

$$\text{Maximize} \sum_{i\in O} o_i\pi_i - \sum_{i\in D} d_i\pi_i$$

subject to

$$\pi_i - \pi_j \leq c_{ij}, \ (i,j) \in A,$$

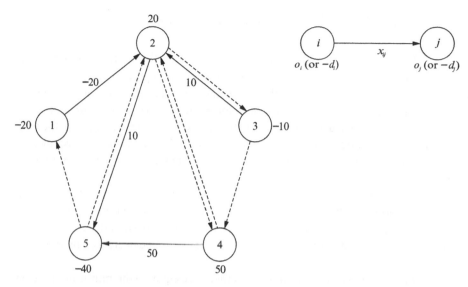

Figure 6.20 A spanning tree (full line arcs) of a directed graph and the associated (infeasible) basic solution (only the basic variables are reported).

1: **procedure** NETWORK_SIMPLEX $(A, \mathbf{c}, \mathbf{o}, \mathbf{d}, \mathbf{x}^*, z^*)$
2:      # $\mathbf{x}^*$ is the optimal solution returned by the procedure;
3:      # z^* is the cost corresponding to the optimal solution;
4:      Find an initial basic feasible solution $\mathbf{x}^{(0)}$;
5:      $h = 0$;
6:      Determine the reduced cost vector $\mathbf{c}'^{(h)}$ associated with $\mathbf{x}^{(h)}$;
7:      **while** $c_{ij}'^{(h)} < 0, (i, j) \in A$ **do**
8:          Choose a variable x_{vw} such that $c_{vw}'^{(h)} < 0$;
9:          Select a variable x_{pq} coming out of the basis;
10:        Make a pivot in order to substitute x_{pq} for x_{vw} in the basis and update the basic feasible solution;
11:        Update the reduced costs associated with the new basic feasible solution;
12:        $h = h + 1$;
13:      **end while**
14:      $\mathbf{x}^* = \mathbf{x}^{(h)}$;
15:      $z^* = \sum_{(i,j)\in A} c_{ij} x_{ij}^*$;
16:      **return** $\mathbf{x}^*, z^*$;
17: **end procedure**

---

enables the execution of the NETWORK_SIMPLEX procedure as follows. At code line 6, the reduced costs can be computed through the formula

$$c_{ij}'^{(h)} = c_{ij} - \pi_i^{(h)} + \pi_j^{(h)}, \ (i, j) \in A, \tag{6.9}$$

where $\pi^{(h)} \in \Re^{|V|}$ can be determined by requiring that the reduced costs of the basic variables be zero:

$$c_{ij}'^{(h)} = c_{ij} - \pi_i^{(h)} + \pi_j^{(h)} = 0, \ (i,j) \in A : x_{ij}^{(h)} \text{ is a basic variable.}$$

If $c_{ij}'^{(h)} \geq 0, (i,j) \in A$, then $\pi_i^{(h)} - \pi_j^{(h)} \leq c_{ij}, (i,j) \in A$, that is, the solution $\pi^{(h)} \in \mathfrak{R}^{|V|}$ is feasible for the dual problem. Then, $\mathbf{x}^{(h)}$ and $\pi^{(h)}$ are optimal for the primal and the dual problems, respectively. On the other hand, if there is a variable x_{vw} whose reduced cost is negative (code line 8) at iteration h, then arc (v, w) does not belong to the spanning tree associated with iteration h. It follows that, by adding (v, w) to the tree, a single cycle Ψ is created. In order to decrease the objective function value as much as possible, the flow on arc (v, w) has to be increased as much as possible while satisfying constraints (6.6) and (6.8).

Let Ψ^+ be the set of arcs in Ψ oriented as (v, w), and let Ψ^- be the set of the arcs in Ψ oriented in the opposite direction (obviously, $\Psi = \Psi^+ \cup \Psi^-$). If the flow on arc (v, w) is increased by t units, then constraints (6.6) require that the flow on all arcs $(i, j) \in \Psi^+$ be increased by t units, and the flow on all arcs $(i, j) \in \Psi^-$ be decreased by the same amount.

The maximum increase of flow on (v, w) is therefore equal to the minimum flow on the arcs oriented in the opposite direction as (v, w), that is,

$$t = \min_{(i,j) \in \Psi^-} \{x_{ij}^{(h)}\}.$$

The arc $(p, q) \in \Psi^-$ for which such a condition holds determines which variable x_{pq} will come out from the basis (code line 9).

The previous description shows that an iteration of the NETWORK_SIMPLEX procedure requires only a few additions and subtractions. As a result, this procedure is much faster than the standard simplex algorithm and, in addition, does not make rounding errors.

In order to find a feasible solution (if any exists), the *big M method* can be used. A new vertex $i_0 \in T$ and $|V|$ dummy arcs between vertex i_0 and all the other vertices $i \in V$ are introduced. If $i \in O$, then a dummy arc (i, i_0) is inserted, otherwise an arc (i_0, i) is added. Let $A^{(a)}$ be the set of dummy arcs. With each dummy arc an arbitrarily large cost M is associated.

The dummy problem is as follows:

$$\text{Maximize} \sum_{(i,j) \in A} c_{ij} x_{ij} + M \sum_{(i,i_0) \in A^{(a)}} x_{i i_0} + M \sum_{(i_0,i) \in A^{(a)}} x_{i_0 i} \qquad (6.10)$$

subject to

$$\sum_{j \in V:(i,j) \in A \cup A^{(a)}} x_{ij} - \sum_{j \in V:(j,i) \in A \cup A^{(a)}} x_{ji} = \begin{cases} o_i, & \text{if } i \in O \\ -d_i, & \text{if } i \in D, i \in V \cup \{i_0\} \\ 0, & \text{if } i \in T \end{cases} \qquad (6.11)$$

$$x_{ij} \geq 0, \ (i,j) \in A \cup A^{(a)}. \qquad (6.12)$$

Of course, the $|V|$ dummy arcs make up a spanning tree of the modified directed graph, corresponding to the following basic feasible solution of problem (6.10)–(6.12) (see Figure 6.21):

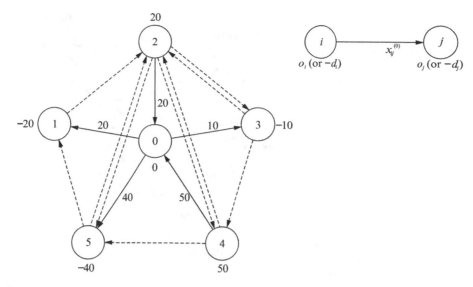

Figure 6.21 Dummy directed graph of the original directed graph in Figure 6.20 (0 is the dummy vertex. Full line arcs belong to the spanning tree. Only the flow associated with a feasible basic variable is reported).

$$x_{ll_0}^{(0)} = o_i, \ i \in O;$$

$$x_{i_0 i}^{(0)} = d_i, \ i \in D;$$

$$x_{i_0 i}^{(0)} = 0, \ i \in T;$$

$$x_{ij}^{(0)} = 0, \ (i, j) \in A.$$

By solving the dummy problem (6.10)–(6.12), a basic feasible solution to the original problem (6.5), (6.6), (6.8) is then obtained.

[NTN.xlsx, NTN.py] NTN is a Swiss freight intermodal carrier located in Lausanne. When a customer needs to transport goods from an origin to a destination, NTN supplies it with one or more empty containers in which the goods can be loaded. Once the goods have arrived at the destination, they are unloaded and the empty containers have to be transported to the pick-up points of new customers. As a result, NTN management needs to reallocate the empty containers periodically (in practice, on a weekly basis). Empty container transportation is very expensive (its cost is nearly 35% of the total operating cost). On 13 May several empty 20 ft ISO containers had to be reallocated among the terminals in Amsterdam, Berlin, Munich, Paris, Milan, Barcelona, and Madrid. The number of empty containers available or demanded at the various terminals is reported, along with transportation costs (in € per container), in Figure 6.22.

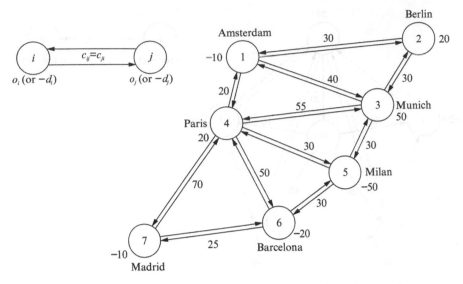

Figure 6.22 Graph representation of the NTN empty container allocation problem.

The problem can be formulated as follows:

Minimize $30x_{12} + 30x_{21} + 40x_{13} + 40x_{31} + 20x_{14} + 20x_{41} + 30x_{23} + 30x_{32}$
$$+55x_{34} + 55x_{43} + 30x_{35} + 30x_{53} + 30x_{45} + 30x_{54} + 50x_{46} + 50x_{64}$$
$$+70x_{47} + 70x_{74} + 30x_{56} + 30x_{65} + 25x_{67} + 25x_{76}$$

subject to

$$x_{12} + x_{13} + x_{14} - x_{21} - x_{31} - x_{41} = -10$$

$$x_{21} + x_{23} - x_{12} - x_{32} = 20$$

$$x_{31} + x_{32} + x_{34} + x_{35} - x_{13} - x_{23} - x_{43} - x_{53} = 50$$

$$x_{41} + x_{43} + x_{45} + x_{46} + x_{47} - x_{14} - x_{34} - x_{54} - x_{64} - x_{74} = 20$$

$$x_{53} + x_{54} + x_{56} - x_{35} - x_{45} - x_{65} = -50$$

$$x_{64} + x_{65} + x_{67} - x_{46} - x_{56} - x_{76} = -20$$

$$x_{74} + x_{76} - x_{47} - x_{67} = -10$$

$$x_{12}, x_{21}, x_{13}, x_{31}, x_{14}, x_{41}, x_{23}, x_{32}, x_{34}, x_{43}, x_{35},$$

$$x_{53}, x_{45}, x_{54}, x_{46}, x_{64}, x_{47}, x_{74}, x_{56}, x_{65}, x_{67}, x_{76} \geq 0.$$

Using the NETWORK_SIMPLEX procedure, the optimal solution illustrated in Figure 6.23 is obtained, the cost of which is equal to € 3900. It can be shown that the optimal solution can easily be found by using the Excel Solver tool under the Data tab menu.

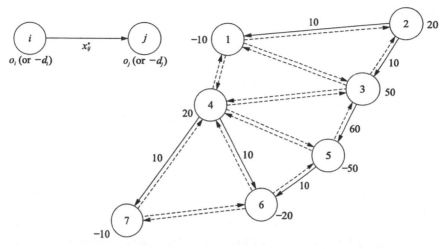

Figure 6.23 Optimal solution of the NTN empty container allocation problem (full line arcs belong to the spanning tree; for each basic variable, the associated flow is reported).

The algorithm, illustrated above, can be easily adapted to the case of capacitated arcs. To this end, constraints (6.7) are rewritten by introducing auxiliary variables $\gamma_{ij} \geq 0$:

$$x_{ij} + \gamma_{ij} = u_{ij}, \ (i,j) \in A.$$

If variable x_{ij} is equal to u_{ij}, then the associated auxiliary variable γ_{ij} takes the value zero and is therefore out of the basis (if the solution is not degenerate). Based on this observation, the following optimality conditions can be derived (the proof is omitted for the sake of brevity).

Theorem. A basic feasible solution $\mathbf{x}^{(h)}$ is optimal for LMCFP if, for each non-basic variable $x_{ij}^{(h)}$, $(i,j) \in A$, the following conditions hold:

$$x_{ij}^{(h)} = 0, \ \text{if } c_{ij}^{'(h)} \geq 0,$$

$$x_{ij}^{(h)} = u_{ij}, \ \text{if } c_{ij}^{'(h)} \leq 0,$$

where $c_{ij}^{'(h)}$ are the reduced costs defined by formula (6.9).

Let $\mathbf{x}^{(h)}$ be the basic feasible solution at iteration h of the NETWORK_SIMPLEX procedure (for simplicity, $\mathbf{x}^{(h)}$ is assumed to be non-degenerate). If the value of a non-basic variable $x_{ij}^{(h)}$, $(i,j) \in A$, is increased, the objective function value decreases if the reduced cost $c_{ij}^{'(h)}$ is negative. On the other hand, if $x_{ij}^{(h)} = u_{ij}$, then a decrease in the objective function value is obtained if the reduced cost $c_{ij}^{'(h)}$ is positive and the value of $x_{ij}^{(h)}$ is decreased.

Let x_{vw} be the variable entering the basis at iteration h. If $x_{vw}^{(h)} = 0$, then $c_{vw}^{'(h)} < 0$ and the arc $(v, w) \in A$ is not part of the spanning tree associated with $\mathbf{x}^{(h)}$. By adding arc (v, w) to the tree, a single cycle Ψ is formed. In the new basic feasible solution, variable

x_{vw} will take a value t equal to

$$t = \min\left\{ \min_{(i,j)\in\Psi^+}\left\{u_{ij} - x_{ij}^{(h)}\right\}, \min_{(i,j)\in\Psi^-}\left\{x_{ij}^{(h)}\right\}\right\}. \tag{6.13}$$

Let (p,q) be the arc outgoing the basis according to (6.13). Then, $x_{pq}^{(h+1)} = u_{pq}$ if $(p,q) \in \Psi^+$, or $x_{pq}^{(h+1)} = 0$ if $(p,q) \in \Psi^-$. Observe that the outgoing arc (p,q) may be the same as the incoming arc (v,w) if $t = u_{vw}$. Similar considerations can be drawn when the variable x_{vw} entering the basis at hth iteration as $x_{vw}^{(h)} = u_{vw}$ with $c_{vw}^{'(h)} > 0$.

[Boscheim.xlsx, Boscheim.py] Boscheim is a German company manufacturing electronics convenience goods. Its KLR-12 HD projector is specifically designed for the British market. KLR-12 is assembled in a plant near Rotterdam, then stocked in two warehouses located in Bristol and Middlesbrough and finally transported to the retailer outlets. The British market is divided into four sales districts whose centres of gravity are in London, Birmingham, Leeds, and Edinburgh.

Yearly demands amount to 90 000, 80 000, 50 000, and 70 000 items, respectively. The transportation costs per item from the assembly plant of Rotterdam to the warehouses of Bristol and Middlesbrough are € 2.45 and € 2.60, respectively, whereas the transportation costs per item from the warehouses to the sales districts are reported in Table 6.4.

Table 6.4 Unit transportation costs (in €) per item from the warehouses to the sales districts in the Boscheim problem.

Warehouse	Sales districts			
	London	Birmingham	Leeds	Edinburgh
Bristol	0.96	0.70	1.52	2.85
Middlesbrough	1.95	1.33	0.60	1.13

Both warehouses have an estimated capacity of 15 000 items and are supplied 10 times a year. Consequently their maximum yearly throughput is 150 000 items.

The annual minimum-cost distribution plan can be obtained by solving the following (see Figure 6.24):

Minimize $2.45x_{12} + 2.60x_{13} + 0.96x_{24} + 0.70x_{25} + 1.52x_{26} + 2.85x_{27}$

$\quad + 1.95x_{34} + 1.33x_{35} + 0.50x_{36} + 1.13x_{37}$

subject to

$$x_{12} + x_{13} = 290\,000$$

$$x_{24} + x_{25} + x_{26} + x_{27} - x_{12} = 0$$

$$x_{34} + x_{35} + x_{36} + x_{37} - x_{13} = 0$$

$$-x_{24} - x_{34} = -90\,000$$

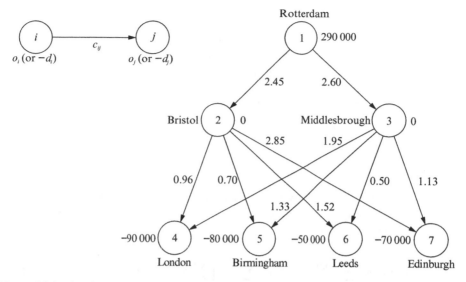

Figure 6.24   Graph representation of the Boscheim problem.

$$- x_{25} - x_{35} = -80\,000$$

$$- x_{26} - x_{36} = -50\,000$$

$$- x_{27} - x_{37} = -70\,000$$

$$x_{12} \leq 150\,000$$

$$x_{13} \leq 150\,000$$

$$x_{12}, x_{13}, x_{24}, x_{25}, x_{26}, x_{27}, x_{34}, x_{35}, x_{36}, x_{37} \geq 0.$$

By using the NETWORK_SIMPLEX procedure or the Excel Solver tool, the optimal solution is determined: $x_{12}^* = 150\,000$, $x_{13}^* = 140\,000$, $x_{24}^* = 90\,000$, $x_{25}^* = 60\,000$, $x_{35}^* = 20\,000$, $x_{36}^* = 50\,000$ and $x_{37}^* = 70\,000$ (as usual, only non-zero variables are reported).

It is worth noting that the district of London will be entirely served by the warehouse of Bristol, while the sales districts of Leeds and Edinburgh will be served by the Middlesbrough warehouse. The sales district of Birmingham is instead supplied by the warehouse of Bristol (75%), and by the warehouse of Middlesbrough (25%). The total transportation cost is € 990 600 per year.

6.6.3   Linear Multi-commodity Minimum-cost Flow Problems

The *linear multi-commodity minimum-cost flow problem* (LMMCFP) can be formulated as the following LP model:

$$\text{Minimize} \sum_{k \in K} \sum_{(i,j) \in A} c_{ij}^k x_{ij}^k$$

subject to

$$\sum_{j \in V:(i,j) \in A} x_{ij}^k - \sum_{j \in V:(j,i) \in A} x_{ji}^k = \begin{cases} o_i^k, & \text{if } i \in O(k) \\ -d_i^k, & \text{if } i \in D(k), \quad i \in V, k \in K \\ 0, & \text{if } i \in T(k) \end{cases}$$

$$x_{ij}^k \le u_{ij}^k, \, (i, j) \in A, \, k \in K$$

$$\sum_{k \in K} x_{ij}^k \le u_{ij}, \, (i, j) \in A \qquad (6.14)$$

$$x_{ij}^k \ge 0, \, (i, j) \in A, \, k \in K.$$

To solve the LMMCFP a tailored Lagrangian procedure can be used. Let $\lambda_{ij} \ge 0$, $(i, j) \in A$, be the Lagrangian multipliers attached to constraints (6.14). The Lagrangian relaxation of the LMMCFP is

$$\text{Minimize} \sum_{k \in K} \sum_{(i,j) \in A} c_{ij}^k x_{ij}^k + \sum_{(i,j) \in A} \lambda_{ij} \left(\sum_{k \in K} x_{ij}^k - u_{ij} \right) \qquad (6.15)$$

subject to

$$\sum_{j \in V:(i,j) \in A} x_{ij}^k - \sum_{j \in V:(j,i) \in A} x_{ji}^k = \begin{cases} o_i^k, & \text{if } i \in O(k) \\ -d_i^k, & \text{if } i \in D(k), \quad \begin{array}{l} i \in V, \\ k \in K \end{array} \\ 0, & \text{if } i \in T(k) \end{cases} \qquad (6.16)$$

$$x_{ij}^k \le u_{ij}^k, \, (i, j) \in A, \, k \in K \qquad (6.17)$$

$$x_{ij}^k \ge 0, \, (i, j) \in A, \, k \in K. \qquad (6.18)$$

Relaxation (6.15)–(6.18), referred to in the following as R-LMMCFP, is made up of $|K|$ independent single-commodity minimum-cost flow problems, since $\sum_{(i,j) \in A} \lambda_{ij} u_{ij}$ in the objective function (6.15) is constant for a given set of Lagrangian multipliers λ_{ij}, $(i, j) \in A$. Therefore, the kth LMCF subproblem, $k \in K$,

$$\text{Minimize} \sum_{(i,j) \in A} \left(c_{ij}^k + \lambda_{ij} \right) x_{ij}^k \qquad (6.19)$$

subject to

$$\sum_{j \in V:(i,j) \in A} x_{ij}^k - \sum_{j \in V:(j,i) \in A} x_{ji}^k = \begin{cases} o_i^k, & \text{if } i \in O(k) \\ -d_i^k, & \text{if } i \in D(k), \quad i \in V \\ 0, & \text{if } i \in T(k) \end{cases} \qquad (6.20)$$

$$x_{ij}^k \le u_{ij}^k, \, (i, j) \in A \qquad (6.21)$$

$$x_{ij}^k \ge 0, \, (i, j) \in A, \qquad (6.22)$$

can be solved through the NETWORK_SIMPLEX procedure.

Let $\text{LB}_{\text{LMCFP}}^k(\lambda)$, $k \in K$, be the optimal objective function value of the kth subproblem (6.19)–(6.22) and let $\text{LB}_{\text{R-LMMCFP}}(\lambda)$ be the lower bound provided by solving the R-LMMCFP.

For a given vector λ of Lagrangian multipliers, $\mathrm{LB_{R\text{-}LMMCFP}}(\lambda)$ is given by

$$\mathrm{LB_{R\text{-}LMMCFP}}(\lambda) = \sum_{k \in K} \mathrm{LB}^k_{\mathrm{LMCFP}}(\lambda) - \sum_{(i,j) \in A} \lambda_{ij} u_{ij}.$$

Of course, $\mathrm{LB_{R\text{-}LMMCFP}}(\lambda)$ varies as Lagrangian multiplier λ changes. As reported in Section 3.9, the Lagrangian relaxation attaining the maximum lower bound value $\mathrm{LB_{R\text{-}LMMCFP}}(\lambda)$ as λ varies is called dual Lagrangian problem. The following property follows from the LP theory.

Property. The lower bound provided by the optimal solution of the Lagrangian problem is equal to the optimal objective function value of the LMMCFP model, that is,

$$\max_{\lambda \geq 0} \{\mathrm{LB_{R\text{-}LMMCFP}}(\lambda)\} = z^*_{\mathrm{LMMCFP}}.$$

Moreover, the dual Lagrangian multipliers λ^*_{ij}, $(i, j) \in A$, are equal to the optimal dual variables π^*_{ij}, $(i, j) \in A$, associated with the relaxed constraints (6.14).

In order to compute the dual Lagrangian multipliers, or at least a set of Lagrangian multipliers associated with a good lower bound, the following LMMCFP_SUBGRADIENT procedure can be used, which is very similar to the procedure already illustrated in Section 3.9.

It is worth observing that the parameters $\beta^{(h)} = \alpha/h$, $h = 1, \dots, \infty$, satisfy the conditions (3.41)–(3.42).

At the end of the LMMCFP_SUBGRADIENT procedure the solution found $\overline{\mathbf{x}}^k$, $k \in K$, for the R-LMMCFP can be feasible for the LMMCFP or not. In the first case, it is not necessarily optimal for the LMMCFP. A *sufficient* (but *not necessary*) condition for it to be optimal is given by the *complementary slackness* conditions

$$\overline{\lambda}_{ij}\left(\sum_{k \in K} \overline{x}^k_{ij} - u_{ij}\right) = 0, \ (i, j) \in A, \tag{6.23}$$

where $\overline{\lambda}_{ij}$, $(i, j) \in A$, are the Lagrangian multipliers associated with the solution $\overline{\mathbf{x}}^k$, $k \in K$. If the complementary slackness conditions (6.23) are not satisfied, then the feasible solution $\overline{\mathbf{x}}^k$, $k \in K$, is simply a *candidate optimal* solution for LMMCFP. In the second case, the solution attaining LB could be infeasible for the LMMCFP.

In both cases, if subproblems (6.19)–(6.22) are solved by means of the NETWORK_SIMPLEX procedure, a basic (feasible or infeasible) solution for the LMMCFP is available. In fact, the basic variables of the $|K|$ subproblems (6.19)–(6.22) make up a basis of the LMMCFP. The basic solution obtained in this way can be used as the starting solution for the primal or dual simplex method depending on whether the solution is feasible for the LMMCFP or not. If initialized in this way, the simplex method is particularly efficient since the initial basic solution provided by the LMMCFP_SUBGRADIENT procedure is a good approximation to the optimal solution.

[Exofruit.xlsx, Exofruit.py] Exofruit imports to EU countries several varieties of tropical fruits, coming mainly from Northern Africa, Mozambique, and Central America. The company purchases the products directly from farmers and transports them by sea to its warehouses in Marseille, France. The goods are then stored in refrigerated cells or at room temperature. Because purchase and selling

```
1:  procedure LMMCFP_SUBGRADIENT (α, β_min, x̄^k, k ∈ K, z̄)
2:      # α, β_min are non-negative user-defined parameters;
3:      # x̄^k, k ∈ K, is the solution of the R-LMMCFP returned by the procedure;
4:      # z̄ is the cost corresponding to the solution of the R-LMMCFP;
5:      LB = −∞;
6:      UB = ∞;
7:      λ_{ij}^{(1)} = 0, (i, j) ∈ A;
8:      β^{(1)} = α;
9:      h = 1;
10:     while (β^{(h)} > β_min) and (LB < UB) do
11:         # Computation of a new LB;
12:         Solve the R-LMMCFP using λ_{ij}^{(h)} = 0, (i, j) ∈ A, as a vector of Lagrangian
            multipliers;
13:         # x_{ij}^{k,(h)}, (i, j) ∈ A, k ∈ K, is the optimal solution of R-LMMCFP;
14:         if LB_{R-LMMCFP}(λ^{(h)}) > LB then
15:             LB = LB_{R-LMMCFP}(λ^{(h)});
16:         end if
17:         if the optimal solution of R-LMMCFP satisfies constraints (6.14) then
18:             # the solution is also feasible for LMMCFP;
19:             if UB > LB_{R-LMMCFP}(λ^{(h)}) then
20:                 # Update UB;
21:                 UB = LB_{R-LMMCFP}(λ^{(h)});
22:                 for (i, j) ∈ A do
23:                     for k ∈ K do
24:                         x̄_{ij}^k = x_{ij}^{k,(h)};
25:                     end for
26:                 end for
27:                 z̄ = UB;
28:             end if
29:         end if
30:         if LB < UB then
31:             # Determine the subgradient of the relaxed constraints;
32:
```

$$s_{ij}^{(h)} = \sum_{k \in K} x_{ij}^{k,(h)} - u_{ij}, \ (i, j) \in A;$$

```
33:         # Update the Lagrangian multipliers;
34:
```

$$\lambda_{ij}^{(h+1)} = \max\left(0, \lambda_{ij}^{(h)} + \beta^{(h)} s_{ij}^{(h)}\right), \ (i, j) \in A;$$

```
35:         # Compute β^{(h+1)};
36:
```

$$\beta^{(h+1)} = \alpha/(h + 1);$$

```
37:             h = h + 1;
38:         end if
39:     end while
40:     if UB = ∞ then
41:         # No feasible solution of LMMCFP has been found;
42:         UB = LB_{R-LMMCFP}(λ^{(h)});
43:         # Return the best solution of the R-LMMCFP;
44:         for (i, j) ∈ A do
45:             for k ∈ K do
46:                 x̄_{ij}^k = x_{ij}^{k,(h)};
47:             end for
48:         end for
49:         z̄ = UB;
50:     end if
51:     return x̄^k, k ∈ K, z̄;
52: end procedure
```

Table 6.5   Demand (in tonnes), maximum amounts available (in tonnes) and purchase prices (in €/tonne), for each macro-product $k = 1, 2$ and for each six-month period $t = 1, 2$, in the Exofruit problem.

	d_{kt} [tonnes]		o_{kt} [tonnes]		p_{kt} [€/tonne]	
	$t = 1$	$t = 2$	$t = 1$	$t = 2$	$t = 1$	$t = 2$
$k = 1$	18 000	18 000	26 000	20 000	500	700
$k = 2$	12 000	14 000	14 000	13 000	600	400

prices vary during the year, Exofruit has to decide when and how much to buy in order to satisfy demand over the year. The problem can be modelled as an LMMCFP. In what follows, a simplified version of the problem is examined. It is assumed that the products are grouped into two homogeneous groups (*macro-products*) and the planning horizon is divided into two six-month periods. Let d_{kt}, o_{kt}, and p_{kt}, $k = 1, 2$, $t = 1, 2$, be the demand, the maximum amount available (in tonnes) and the purchase prices (in €/tonne) of macro-product k in period t, respectively (see Table 6.5). The transportation cost v of one tonne of a macro-product is equal to € 100, while the stocking cost w of one tonne of a macro-product is € 100 per six-month period. Finally, the maximum total quantity q of all goods that can be stored in a six-month period is 8000 tonnes.

The problem can be formulated as an LMMCFP with two commodities (one for each macro-product) on the directed graph $G = (V, A)$ shown in Figure 6.25. In such a representation,

1. vertices 1 and 5 represent the source and the destination of macro-product 1, respectively;
2. vertices 2 and 6 represent the source and the destination of macro-product 2, respectively;
3. vertices 3 and 4 represent the warehouse in the first and in the second six-month period, respectively;
4. arc $(1, 3)$ has a cost unit of flow equal to $c_{13}^1 = p_{11} + v = $ € 600/tonne and a capacity equal to $u_{13} = o_{11} = 26\,000$ tonnes;
5. arc $(1, 4)$ has a cost per unit of flow equal to $c_{14}^1 = p_{12} + v = $ € 800/tonne and a capacity equal to $u_{14} = o_{12} = 20\,000$ tonnes;
6. arc $(2, 3)$ has a cost per unit of flow equal to $c_{23}^2 = p_{21} + v = $ € 700/tonne and a capacity equal to $u_{23} = o_{21} = 14\,000$ tonnes;
7. arc $(2, 4)$ has a cost per unit of flow equal to $c_{24}^2 = p_{22} + v = $ € 500/tonne and a capacity equal to $u_{24} = o_{22} = 13\,000$ tonnes;
8. arc $(3, 4)$ represents the storage of goods for a six-month period and, therefore, has a cost per unit flow equal to $c_{34}^1 = c_{34}^2 = w = $€ 100/tonne and a capacity of $u_{34} = q = 8000$ tonnes;
9. arc $(3, 5)$ has a zero cost per unit of flow and a capacity equal to $u_{35} = d_{11} = 18\,000$ tonnes;

10. arc $(4, 5)$ has a zero cost per unit of flow and a capacity equal to $u_{45} = d_{12} = 18\,000$ tonnes;
11. arc $(3, 6)$ has a zero cost per unit of flow and a capacity equal to $u_{36} = d_{21} = 12\,000$ tonnes;
12. arc $(4, 6)$ has a zero cost per unit of flow and a capacity equal to $u_{46} = d_{22} = 14\,000$ tonnes.

The decision variable x_{ij}^k, $(i,j) \in A$, $k = 1, 2$, represents the flow (in tonnes) of macro-product k from vertex i to vertex j. This means that, e.g., x_{13}^1 is the quantity of macro-product 1 that during the first six-month period is moved from its source to the warehouse. Similarly, x_{35}^1 is the quantity of macro-product 1 that during the first six-month period is moved from the warehouse to its destination (in this case, no stocking cost is paid). The meaning of the remaining variables is straightforward and left to the reader.

The LMMCFP is formulated as follows (for the sake of simplicity, the decision variables that are zero and the flow conservation constraints involving only null decision variables are omitted):

$$\text{Minimize } 600x_{13}^1 + 800x_{14}^1 + 100x_{34}^1 + 700x_{23}^2 + 500x_{24}^2 + 100x_{34}^2$$

subject to

$$x_{35}^1 + x_{45}^1 = 36\,000$$
$$x_{13}^1 - x_{34}^1 - x_{35}^1 = 0$$
$$x_{14}^1 + x_{34}^1 - x_{45}^1 = 0$$
$$x_{36}^2 + x_{46}^2 = 26\,000$$
$$x_{23}^2 - x_{34}^2 - x_{36}^2 = 0$$
$$x_{24}^2 + x_{34}^2 - x_{46}^2 = 0$$
$$x_{13}^1 \leq 26\,000$$
$$x_{14}^1 \leq 20\,000$$
$$x_{35}^1 \leq 18\,000$$
$$x_{45}^1 \leq 18\,000$$
$$x_{23}^2 \leq 14\,000$$
$$x_{24}^2 \leq 13\,000$$
$$x_{36}^2 \leq 12\,000$$
$$x_{46}^2 \leq 14\,000$$
$$x_{34}^1 + x_{34}^2 \leq 8000$$
$$x_{13}^1, \, x_{14}^1, \, x_{34}^1, \, x_{35}^1, \, x_{45}^1, x_{23}^2, \, x_{24}^2, \, x_{34}^2, \, x_{36}^2, \, x_{46}^2 \geq 0.$$

Even though the problem can be easily solved (e.g., by using the Excel Solver), the Lagrangian technique is adopted in the following. By relaxing in a

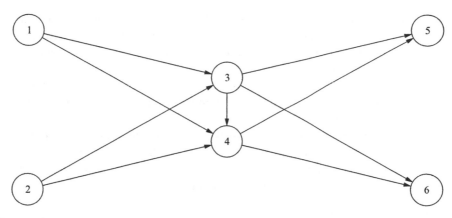

Figure 6.25   Graph representation of the Exofruit problem.

Lagrangian fashion constraint $x_{34}^1 + x_{34}^2 \leq 8000$ with a Lagrangian multiplier λ_{34}, the problem decomposes into two single-commodity linear minimum-cost flow problems. By initializing the LMMCFP_SUBGRADIENT procedure with $\alpha = 0.5$, $\beta_{min} = 0.001$ and $\lambda_{34}^{(0)} = 0$, the procedure, after six iterations, converges to $\lambda_{34} = 100$ and provides a lower bound LB = € 40 200 000.

   Observe that the Lagrangian relaxed constraint $x_{34}^1 + x_{34}^2 \leq 8000$ is satisfied as inequality (i.e., $\bar{x}_{34}^1 + \bar{x}_{34}^2 = 1000 < 8000$) in correspondence of the final solution $\bar{x}^1, \bar{x}^2$ and the solution $\bar{x}^1, \bar{x}^2$ of the R-LMMCFP is feasible for the LMMCFP with a cost corresponding to LB (the exit condition of the LMMCFP_SUBGRADIENT procedure is LB = UB). Consequently, the obtained solution

$$\bar{x}_{13}^1 = 18\,000; \bar{x}_{14}^1 = 18\,000; \bar{x}_{34}^1 = 0; \bar{x}_{35}^1 = 18\,000; \bar{x}_{45}^1 = 18\,000;$$

$$\bar{x}_{23}^2 = 13\,000; \bar{x}_{24}^2 = 13\,000; \bar{x}_{34}^2 = 1000; \bar{x}_{36}^2 = 12\,000; \bar{x}_{46}^2 = 14\,000,$$

is optimal.

   This means that 18 000 tonnes of the first macro-product are sent from its origin to its destination in each six-month period (in this way, no storage costs are provided). In the first six-month period, 13 000 tonnes of the second macro-product are sent from its origin; 12 000 reaches in the same six-month period its destination, whereas 1000 tonnes are stored in the warehouse and then sent to the destination in the second six-month period, together with the 13 000 tonnes sent in the second six-month period from its origin.

6.7   Service Network Design Problems

The design of a network of transportation services is a tactical or operational decision particularly relevant to consolidation-based carriers. Given a set of terminals, the SNDP amounts to determining the features (frequency, number of intermediate stops, etc.) of the routes to be operated, the traffic assignment along these routes, the operating rules at each terminal and possibly the relocation of empty vehicles and containers. The objective is the minimization of a generalized cost taking into account a combination of the carrier's operating costs and customers' expectations. Figure 6.26 shows two alternative service networks for a three-terminal transportation system in which it is assumed that each arc is associated with a line operated once a day. In the former network (see Figure 6.26(a)), each terminal is connected directly to every other terminal (so that each shipment takes one day) but this comes at the expense of a higher operating cost. In the latter network (see Figure 6.26(b)), operating costs are lower but the transportation between certain origin–destination pairs may require two days (unless all lines are synchronized).

In the remainder of this section, the focus is on the basic network design problem, namely, the *fixed charge network design problem* (FCNDP), which can be viewed as a generalization of network flow problems in which a fixed cost f_{ij} has to be paid for using each arc $(i, j) \in A$. Therefore, FCNDPs amount to determining

(a) which arcs have to be employed;
(b) how to transport the commodities on the selected arcs.

Let x_{ij}^k, $(i, j) \in A$, $k \in K$, be a decision variable corresponding to the flow of commodity k on arc (i, j), and let y_{ij}, $(i, j) \in A$, be a binary decision variable, equal to 1 if arc (i, j) is used, 0 otherwise. A quite general formulation of the FCNDP is as follows:

$$\text{Minimize} \sum_{k \in K} \sum_{(i,j) \in A} C_{ij}^k(x_{ij}^k) + \sum_{(i,j) \in A} f_{ij} y_{ij} \tag{6.24}$$

subject to

$$\sum_{j \in V:(i,j) \in A} x_{ij}^k - \sum_{j \in V:(j,i) \in A} x_{ji}^k = \begin{cases} o_i^k, & \text{if } i \in O(k) \\ -d_i^k, & \text{if } i \in D(k), \\ 0, & \text{if } i \in T(k) \end{cases} \quad \begin{array}{c} i \in V, \\ k \in K \end{array} \tag{6.25}$$

$$x_{ij}^k \leq u_{ij}^k, \ (i, j) \in A, \ k \in K \tag{6.26}$$

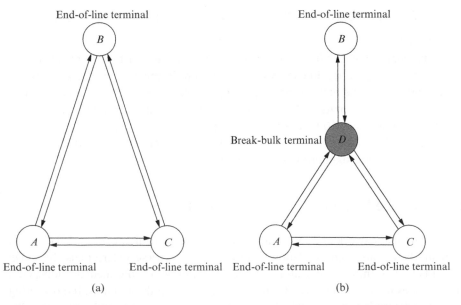

Figure 6.26 Two alternative service networks for a three end-of-line terminal transportation system: (a) each terminal is connected directly to every other terminal; (b) presence of an intermediate break-bulk terminal.

$$\sum_{k \in K} x_{ij}^k \le u_{ij} y_{ij}, (i, j) \in A \qquad (6.27)$$

$$x_{ij}^k \ge 0, \ (i, j) \in A, \ k \in K$$

$$y_{ij} \in \{0, 1\}, \ (i, j) \in A.$$

The objective function (6.24) is the total transportation cost. Constraints (6.25) correspond to the flow conservation constraints holding at each vertex $i \in V$ and for each commodity $k \in K$; constraints (6.26) impose that the flow of each commodity $k \in K$ does not exceed capacity u_{ij}^k on each arc $(i, j) \in A$; constraints (6.27) (*bundle constraints*) require that, for each $(i, j) \in A$, the total flow on arc (i, j) is zero if the arc is not used, or not greater than capacity u_{ij}, otherwise.

In practice, some side constraints may be needed to represent economic and topological restrictions. For example, when several links share a common resource, the following *budget* constraint has to be added to the FCNDP:

$$\sum_{(i,j) \in A} h_{ij} y_{ij} \le b,$$

where $h_{ij}, (i, j) \in A$ is the consumption of resource made by arc $(i, j) \in A$, and b is the total amount of resource available.

The linear fixed-charge network design problem

The *linear fixed-charge network design problem* (LFCNDP) is a particular FCNDP in which the transportation costs per flow unit c_{ij}^k for each arc $(i, j) \subset A$ and for each commodity $k \in K$ are constant.

More formally, the LFCNDP can be formulated as

$$\text{Minimize} \sum_{k \in K} \sum_{(i,j) \in A} c_{ij}^k x_{ij}^k + \sum_{(i,j) \in A} f_{ij} y_{ij}$$

subject to

$$\sum_{j \in V:(i,j) \in A} x_{ij}^k - \sum_{j \in V:(j,i) \in A} x_{ji}^k = \begin{cases} o_i^k, & \text{if } i \in O(k) \\ -d_i^k, & \text{if } i \in D(k), \\ 0, & \text{if } i \in T(k) \end{cases} \quad \begin{array}{l} i \in V, \\ k \in K \end{array}$$

$$x_{ij}^k \le u_{ij}^k, \ (i, j) \in A, \ k \in K \qquad (6.28)$$

$$\sum_{k \in K} x_{ij}^k \le u_{ij} y_{ij}, \ (i, j) \in A$$

$$x_{ij}^k \ge 0, \ (i, j) \in A, \ k \in K$$

$$y_{ij} \in \{0, 1\}, \ (i, j) \in A.$$

[FHL.xlsx, FHL.py] FHL is an Austrian fast carrier located in Lienz, whose core business is the transportation of palletized unit loads. The transportation service is carried out through five crossdocking facilities by using two types of truck. A customer needs to move the palletized unit loads from two RDCs to three retailers. The palletized unit loads are classified into three categories, corresponding to liquid food, dry food, and other goods. The average retailer daily demand in the next trimester is reported in Table 6.6. The palletized unit loads are available each day at the RDCs in the quantities indicated in Table 6.6. The transportation is generally realized by FHL between each pair of facilities using a single truck with a maximum load of 200 palletized unit loads. Only when the terminal is the retailer does FHL use larger trucks of 250-palletized unit load capacity. The potential transportation network is represented by the directed graph $G = (V, A)$ in Figure 6.27. Each arc $(i, j) \in A$ corresponds to a potential transportation service from facility i to facility j. If the service is activated, a fixed transportation cost (in €) f_{ij} and variable costs that are proportional to the number of palletized unit loads of the three categories transported are incurred; c_{ij}^{k}, $k = 1, 2, 3$, indicates the unit variable transportation cost (in €) for each palletized unit load of category k. Variable costs and fixed costs are reported in Table 6.7.

The model can be formulated as an LFCNDP, whose optimal solution, determined by using the Excel Solver tool, leads to a daily cost of € 34 532 and the activation of the transportation services corresponding to the arcs $(1, 3)$, $(1, 4)$, $(2, 4)$, $(2, 5)$, $(3, 6)$, $(4, 6)$, $(4, 9)$, $(5, 7)$, $(6, 8)$, and $(7, 10)$.

Table 6.6 Average daily demand of palletized unit loads of the three categories of products from the three retailers and average daily availability (in palletized unit loads) at the two RDCs in the FHL problem.

	Retailers			RDC	
Category	1	2	3	1	2
Liquid food	80	55	40	80	95
Dry food	74	52	64	90	100
Other	47	42	76	75	90

Table 6.7 Variable costs and fixed costs (in €) in the FHL problem.

	Variable cost			
(i, j)	c_{ij}^{1}	c_{ij}^{2}	c_{ij}^{3}	f_{ij}
$(1, 3)$	22	19	17	450
$(1, 4)$	22	19	17	490
$(2, 4)$	20	21	18	390

(Continued)

Table 6.7   (Continued)

(i,j)	Variable cost			
	c_{ij}^1	c_{ij}^2	c_{ij}^3	f_{ij}
(2,5)	20	21	18	435
(3,6)	22	19	18	296
(4,6)	23	22	21	340
(4,7)	26	25	27	685
(4,9)	23	22	21	625
(5,7)	21	19	20	365
(5,9)	25	23	23	655
(6,8)	24	25	25	325
(6,9)	24	25	25	490
(6,10)	24	25	25	725
(7,8)	26	24	27	710
(7,9)	24	24	25	525
(7,10)	23	23	24	410

Branch-and-bound algorithms can typically only solve instances of the LFCNDP with a few hundreds of arcs and tens of commodities. Since instances arising in applications are much larger, heuristics are often used. To evaluate the quality of the solutions provided by heuristics, it is useful, as already observed in Section 3.9, to compute lower bounds on the optimal objective function value z_{LFCNDP}^*. In the following, two distinct formulations of continuous relaxations and a simple heuristic are illustrated.

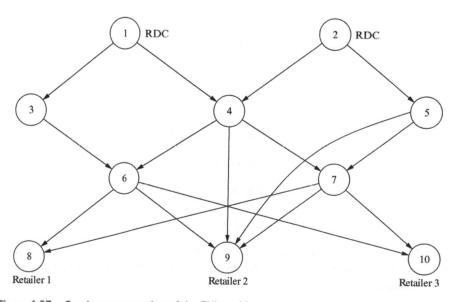

Figure 6.27   Graph representation of the FHL problem.

The Weak Continuous Relaxation

The weak continuous relaxation is obtained by relaxing the integrity requirement on the decision variables y_{ij}, $(i, j) \in A$:

$$\text{Minimize} \sum_{k \in K} \sum_{(i,j) \in A} c_{ij}^k x_{ij}^k + \sum_{(i,j) \in A} f_{ij} y_{ij} \tag{6.29}$$

subject to

$$\sum_{j \in V:(i,j) \in A} x_{ij}^k - \sum_{j \in V:(j,i) \in A} x_{ji}^k = \begin{cases} o_i^k, & \text{if } i \in O(k) \\ -d_i^k, & \text{if } i \in D(k), \\ 0, & \text{if } i \in T(k) \end{cases} \quad \begin{array}{l} i \in V, \\ k \in K \end{array} \tag{6.30}$$

$$x_{ij}^k \leq u_{ij}^k, \ (i, j) \in A, \ k \in K \tag{6.31}$$

$$\sum_{k \in K} x_{ij}^k \leq u_{ij} y_{ij}, \ (i, j) \in A \tag{6.32}$$

$$x_{ij}^k \geq 0, \ (i, j) \in A, \ k \in K \tag{6.33}$$

$$0 \leq y_{ij} \leq 1, \ (i, j) \in A. \tag{6.34}$$

It is easy to verify that every optimal solution of such a relaxation satisfies each constraint (6.32) as an equality since the fixed costs f_{ij}, $(i, j) \in A$, are non-negative. Therefore, decision variables y_{ij}, $(i, j) \in A$, can be expressed as a function of flow variables x_{ij}^k, $(i, j) \in A, k \in K$

$$y_{ij} = \frac{\sum_{k \in K} x_{ij}^k}{u_{ij}}, \ (i, j) \in A.$$

Hence, constraints (6.34) can be replaced by the following conditions:

$$\sum_{k \in K} x_{ij}^k \leq u_{ij}, \ (i, j) \in A.$$

The relaxed problem (6.29)–(6.34) can be therefore equivalently formulated as

$$\text{Minimize} \sum_{k \in K} \sum_{(i,j) \in A} \left(c_{ij}^k + \frac{f_{ij}}{u_{ij}} \right) x_{ij}^k \tag{6.35}$$

subject to

$$\sum_{j \in V:(i,j) \in A} x_{ij}^k - \sum_{j \in V:(j,i) \in A} x_{ji}^k = \begin{cases} o_i^k, & \text{if } i \in O(k) \\ -d_i^k, & \text{if } i \in D(k), \\ 0, & \text{if } i \in T(k) \end{cases} \quad \begin{array}{l} i \in V, \\ k \in K \end{array} \tag{6.36}$$

$$x_{ij}^k \leq u_{ij}^k, \ (i, j) \in A, \ k \in K \tag{6.37}$$

$$\sum_{k \in K} x_{ij}^k \leq u_{ij}, (i, j) \in A \tag{6.38}$$

$$x_{ij}^k \geq 0, \ (i, j) \in A, \ k \in K. \tag{6.39}$$

Model (6.35)–(6.39) is a minimum-cost flow problem with $|K|$ commodities. Let LB_w^* be the lower bound on z_{LFCNDP}^* given by the optimal objective function value of the above relaxation.

The strong continuous relaxation

The strong continuous relaxation is obtained by adding the following *valid* inequalities

$$x_{ij}^k \leq u_{ij}^k y_{ij}, \ (i,j) \in A, \ k \in K, \tag{6.40}$$

to the LFCNDP and removing the integrity constraints on the decision variables y_{ij}, $(i,j) \in A$. Taking into account the fact that constraints (6.28) are dominated by constraints (6.40), and can, therefore, be eliminated, the relaxed problem is

$$\text{Minimize} \sum_{k \in K} \sum_{(i,j) \in A} c_{ij}^k x_{ij}^k + \sum_{(i,j) \in A} f_{ij} y_{ij} \tag{6.41}$$

subject to

$$\sum_{j \in V:(i,j) \in A} x_{ij}^k - \sum_{j \in V:(j,i) \in A} x_{ji}^k = \begin{cases} o_i^k, & \text{if } i \in O(k) \\ -d_i^k, & \text{if } i \in D(k), \\ 0, & \text{if } i \in T(k) \end{cases} \quad \begin{array}{l} i \in V, \\ k \in K \end{array} \tag{6.42}$$

$$x_{ij}^k \leq u_{ij}^k y_{ij}, \ (i,j) \in A, \ k \in K \tag{6.43}$$

$$\sum_{k \in K} x_{ij}^k \leq u_{ij} y_{ij}, \ (i,j) \in A \tag{6.44}$$

$$x_{ij}^k \geq 0, \ (i,j) \in A, \ k \in K \tag{6.45}$$

$$0 \leq y_{ij} \leq 1, \ (i,j) \in A. \tag{6.46}$$

Let LB_s^* be the lower bound on z_{LFCNDP}^* given by the optimal objective function value of the relaxation (6.41)–(6.46). This problem has no special structure and, therefore, is solved by using any general-purpose LP algorithm. By comparing the two continuous relaxations, it is clear that LB_s^* is always at least equal to LB_w^*, that is,

$$\text{LB}_s^* \geq \text{LB}_w^*.$$

This observation leads us to label the former relaxation as *weak*, and the latter as *strong*. Computational experiments have shown that LB_w^* can be as much as 40% lower than LB_s^*.

Add-drop heuristics

Add-drop heuristics are simple constructive procedures in which at each step one decides whether a new arc has to be used (ADD procedure) or an arc previously used has to be left out (DROP procedure). Several criteria can be employed to choose which arc has to be added or dropped. In the following, a very simple DROP procedure is illustrated. In order to describe such a heuristic, it is worth noting that a candidate optimal solution for the LFCNDP is characterized by the set $A' \subseteq A$ of selected arcs. A solution is feasible if the LMMCFP on the directed graph $G = (V, A')$ induced by A' is feasible. If so, the solution cost is made up of the sum of the fixed costs $f_{ij}, (i,j) \in A'$, plus the optimal solution cost of the LMMCFP. Moreover, it is worth noting that the LFCNDP solution associated with $A' = A$, if feasible, is characterized by a large fixed cost and by a low transportation cost. On the other hand, a feasible solution associated with a set A' with a few arcs is expected to be characterized by a low fixed cost and by a high variable cost. Consequently, an improved LFCNDP solution can be obtained by iteratively removing arcs from the set $A' = A$, while the current solution is still feasible and the total cost decreases. The DROP procedure is as follows.

[FHL.xlsx, FHL.py] In order to solve the FHL problem, the DROP procedure is applied. At the first iteration ($h = 0$), the set A is defined as follows:

$$S = A^{(0)} = \{(1,3),(1,4),(2,4),(2,5),(3,6),(4,6),(4,7),(4,9),(5,7),$$
$$(5,9),(6,8),(6,9),(6,10),(7,8),(7,9),(7,10)\}.$$

Initially the LMMCFP on the directed graph $G = (V, A^{(0)})$ is solved (code line 9 of the DROP procedure), obtaining an optimal solution whose value is € 30 341; note that this cost corresponds to the variable transportation cost in the LFNCDP feasible solution. Moreover, arcs $(1,3),(1,4),(2,4),(2,5),(3,6),(4,9),(5,7),(6,8),(6,10),(7,8)$, and $(7,10)$ are activated, determining a total fixed cost of € 5221. As a consequence, $\bar{z} = z^{(0)}_{\text{LFCNDP}} = $ € 30 341 + € 5221 = € 35 562.

Successively, $|S| = 16$ LMMCFPs are solved, each of which defined on a new directed graph G obtained by removing a single arc $(i,j) \in S$ from $A^{(0)}$. In detail, by dropping arcs $(1,3),(1,4),(2,4),(2,5),(3,6)$, the corresponding LMMCFPs are infeasible, while the least-cost LFCNDP feasible solution is determined by dropping arc $(v,w) = (7,8)$ from $A^{(0)}$, obtaining a total cost of € 34 532. At the beginning of the second iteration ($h = 1$), the incumbent is $\bar{z} = 34\,532$, while the sets S and A are updated as follows:

$$S = \{(4,6),(4,7),(4,9),(5,7),(5,9),(6,8),(6,9),(6,10),(7,9),(7,10)\};$$

$$A^{(1)} = \{(1,3),(1,4),(2,4),(2,5),(3,6),(4,6),(4,7),(4,9),(5,7),$$
$$(5,9),(6,8),(6,9),(6,10),(7,9),(7,10)\}$$

At the second iteration a number of $|S| = 10$ LMMCFPs are solved. In detail, by dropping arcs $(4,6)$ and $(6,8)$ the corresponding LMMCFPs are infeasible, while in the other cases the cost $\bar{z}$ of the best current LFCNDP feasible solution is never improved. As a consequence, the DROP procedure ends. The final feasible LFCNDP solution leads to a cost of € 34 532, which corresponds to the optimal cost.

6.8 Vehicle Allocation Problems

VAPs are faced by carriers when transporting full loads over long distances, as in TL trucking and container shipping. Once a vehicle delivers a load, it becomes empty and has to be moved to the pick-up point for another load, or has to be repositioned in anticipation of future demands.

The VAP illustrated in this section is a particular version which amounts to deciding the loads to be accepted and those to be rejected, as well as repositioning empty vehicles. More specifically, in the following the case is considered where all demands of transportation services are known in advance and a single vehicle type exists.

```
 1: procedure DROP (V, A, K, f, x̄, ȳ, z̄)
 2:     # x̄ and ȳ are the feasible solution (if any) of the LFCNDP returned by the procedure;
 3:     # z̄ is the cost corresponding to x̄ and ȳ;
 4:     # S is the list of candidate arcs to be dropped;
 5:     S = A;
 6:     h = 0;
 7:     A^(h) = A;
 8:     exit = FALSE;
 9:     Solve the LMMCFP on the directed graph G = (V, A^(h));
10:     if LMMCFP is infeasible then
11:         # The corresponding LFCNDP is also infeasible;
12:         z̄ = ∞;
13:         exit = TRUE;
14:     else
15:         # x^{k,(h)}, k ∈ K, is the optimal solution of the LMMCFP;
16:         # z_LFCNDP^(h) is the sum of the optimal cost of the LMMCFP and the fixed costs f_{ij}
            associated with the arcs (i, j) ∈ A^(h) for which x_{ij}^{k,(h)} > 0, for some k ∈ K;
17:         z̄ = z_LFCNDP^(h);
18:         exit = FALSE;
19:     end if
20:     while (exit = FALSE) or (S ≠ ∅) do
21:         # ẑ is a support variable used to store the best cost of the LFCNDPs generated in
            the subsequent FOR cycle;
22:         ẑ = ∞;
23:         for (i, j) ∈ S do
24:             A'^(h) = A^(h) \ {(l, j)}
25:             Solve the LMMCFP on the directed graph G = (V, A'^(h));
26:             if LMMCFP is infeasible then
27:                 # The corresponding LFCNDP is also infeasible;
28:                 S = S \ {(i, j)};
29:             else if z_LFCNDP^(h) < ẑ then
30:                 # z_LFCNDP^(h) is the cost of the LFCNDP corresponding to LMMCFP;
31:                 ẑ = z_LFCNDP^(h)
32:                 (v, w) = (i, j);
33:             end if
34:         end for
35:         if ẑ < z̄ then
36:             # A better solution of the LFCNDP has been found;
37:             z̄ = ẑ;
38:             # The arc (v, w) is dropped;
39:             A^(h+1) = A^(h) \ {(v, w)};
40:             S = S \ {(v, w)};
41:             h = h + 1;
42:         else
43:             exit = TRUE;
44:         end if
45:     end while
46:     if z̄ < ∞ then
47:         Determine the LFCNDP feasible solution x̄ and ȳ corresponding to z̄;
48:     end if
49:     return x̄, ȳ, z̄;
50: end procedure
```

The planning horizon is supposed to be made up of T time periods. Let N be the set of points (e.g., terminals) where the (full) loads have to be picked up or delivered; d_{ijt}, $i \in N$, $j \in N$, $t = 1, \dots, T$, the number of requested loads to be moved from origin i to destination j at time period t; τ_{ij}, $i \in N$, $j \in N$, the travel time from point i to point j; p_{ij}, $i \in N$, $j \in N$, the profit (revenue minus direct operating costs) derived from moving a load from point i to point j, and c_{ij}, $i \in N$, $j \in N$, the cost of moving an empty vehicle from point i to point j. Moreover, denote by m_{it}, $i \in N$, $t = 1, \dots, T$, the number of vehicles that enter the system in time period t at point i. The following decision variables are used: x_{ijt}, $i \in N$, $j \in N$, $t = 1, \dots, T$, representing the number of vehicles that start moving a load from point i to point j at time period t (one vehicle for each load); y_{ijt}, $i \in N$, $j \in N$, $t = 1, \dots, T$, representing the number of vehicles that start moving empty from point i to point j at time period t. The deterministic single-vehicle VAP can be formulated as follows:

$$\text{Maximize} \sum_{t=1}^{T} \sum_{i \in N} \sum_{j \in N, j \neq i} (p_{ij} x_{ijt} - c_{ij} y_{ijt}) \tag{6.47}$$

subject to

$$\sum_{j \in N} (x_{ij1} + y_{ij1}) = m_{i1}, \ i \in N \tag{6.48}$$

$$\sum_{j \in N} (x_{ijt} + y_{ijt}) - \sum_{k \in N, k \neq i : t > \tau_{ki}} (x_{ki(t-\tau_{ki})} + y_{ki(t-\tau_{ki})}) - y_{iit-1} = m_{it},$$

$$i \in N, t = 2, \dots, T \tag{6.49}$$

$$x_{ijt} \leq d_{ijt}, \ i \in N, j \in N, t = 1, \dots, T \tag{6.50}$$

$$x_{ijt} \geq 0, \ i \in N, j \in N, t = 1, \dots, T$$

$$y_{ijt} \geq 0, \ i \in N, j \in N, t = 1, \dots, T.$$

The objective function (6.47) is the total profit, minus the cost for moving empty vehicles, over the planning horizon. Constraints (6.48) and (6.49) impose flow conservation at the beginning of each time period (in this case the flow is given by the number of vehicles). Due to these constraints, the decision variables x_{ijt} and y_{ijt}, $i \in N$, $j \in N$, $t = 1, \dots, T$, take integer values implicitly. Constraints (6.50) state that the number of loaded movements of vehicles at each time period $t = 1, \dots, T$ between each pair origin i destination j, $i \in N$, $j \in N$, is bounded above by the demand. It is worth noting that the $d_{ijt} - x_{ijt}$ differences, $i \in N$, $j \in N$, $t = 1, \dots, T$, represent the loads that should be rejected, while the y_{iit} decision variables, $i \in N$, $t = 1, \dots, T$, represent vehicles staying idle (the so-called *inventory movements*). It is easy to recognize that VAP can be modelled as a network flow problem on a time-expanded directed graph in which vertices are associated with (i, t) pairs, $i \in N$, $t = 1, \dots T$, and the flows on arcs represent loaded, empty, and inventory vehicle movements. Since there exists more than a pair of arcs between each pair of nodes, such a network is a directed multi-graph (see Problem 6.9).

[Murthy.xlsx, Murthy.py] Murthy is a motor carrier operating in the Indian region of Andhra Pradesh. Four TL transportation requests were made:

from Chittoor to Khammam on 11 July, from Srikakulam to Ichapur on 11 July and from Anantapur to Chittoor on 13 July (two loads). On 11 July, one vehicle was available in Chittoor and one was available in Khammam. A further vehicle was currently transporting a previously scheduled shipment and would be available in Anantapur on 12 July. Transportation times between terminals are shown in Table 6.8. The revenue provided by a truck carrying a full load is 1.8 times the transportation cost of a deadheading truck, estimated equal to 10 000 rupees for each journey day. The problem to solve for Murthy is a VAP in which $T = \{11\text{ July, }12\text{ July, }13\text{ July}\} = \{1, 2, 3\}$ and $N = \{$Anantapur, Chittoor, Ichapur, Khammam, Srikakulam$\} = \{1, 2, 3, 4, 5\}$. The cost c_{ij} of a journey from city $i \in N$ to city $j \in N$ is simply obtained by multiplying the cost of each journey day by the journey days τ_{ij} from city i to city j. Consequently, the profit p_{ij}, $i, j \in N$, is equal to $0.8 \times 10\,000 \times \tau_{ij}$. Furthermore, the number of available trucks m_{it}, $i \in N$, $t = 1, \ldots, T$, are all zero, except the following: $m_{12} = 1$; $m_{21} = 1$; $m_{41} = 1$. The values d_{ijt}, $i \in N$, $j \in N$, $t = 1, \ldots, T$, are also equal to zero, except the following: $d_{241} = 1$; $d_{531} = 1$; $d_{123} = 2$.

The optimal VAP solution, obtained by using the Excel Solver tool, is $x^*_{241} = 1$, $x^*_{123} = 1$, $y^*_{441} = 1$, $y^*_{112} = 1$, $y^*_{442} = 1$, $y^*_{443} = 2$, while the values of the remaining decision variables are zero. The corresponding optimal cost is 24 000 rupees. It is worth noting that the requests from Srikakulam to Ichapur on 11 July are not satisfied, and those from Anantapur to Chittoor on 13 July are partially satisfied.

Table 6.8 Travel times (in number of days) between terminals in the Murthy problem.

	Anantapur	Chittoor	Ichapur	Khammam	Srikakulam
Anantapur	0	1	2	2	2
Chittoor		0	2	2	2
Ichapur			0	2	1
Khammam				0	2
Srikakulam					0

6.9 A Dynamic Driver Assignment Problem

The *dynamic driver assignment problem* (DDAP) examined in this section arises in TL trucking where full-load trips are assigned to drivers in an ongoing fashion. In TL trucking, a trip may take several days (a four-day duration is not unusual both in Europe and in North America) and customer service requests arrive randomly. Consequently, a single trip is assigned to each driver at a time.

The DDAP can be formulated as a particular single-commodity uncapacitated minimum-cost flow problem (see Figure 6.28) and can be solved efficiently through, for example, the network simplex method.

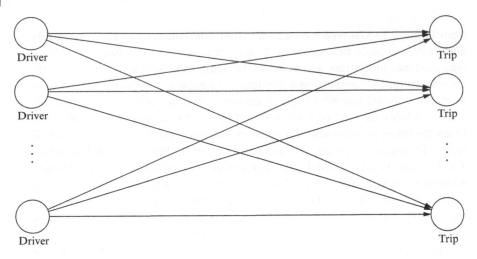

Figure 6.28   An example of a network for a DDAP.

Let D be the set of drivers waiting to be assigned a task and let L be the current set of transportation services to be performed (as each transportation service corresponds to a full-load trip and vice versa). Let $x_{ij}, i \in D, j \in L$, be a binary decision variable equal to 1 if driver i is assigned to transportation service j, 0 otherwise; $c_{ij}, i \in D, j \in L$, is the cost of assigning driver i to transportation service j. The case in which the number of drivers is greater than the number of transportation services required is considered. A possible formulation of the DDAP is the following:

$$\text{Minimize} \sum_{i \in D} \sum_{j \in L} c_{ij} x_{ij}$$

subject to

$$\sum_{j \in L} x_{ij} \leq 1, i \in D$$

$$\sum_{i \in D} x_{ij} = 1, j \in L$$

$$x_{ij} \in \{0, 1\}, \ i \in D, \ j \in L.$$

The DDAP model can be easily modified in the case where the number of transportation services is greater than or equal to the number of drivers. How to deal with this modification is left to the reader as an exercise (see Problem 6.11). In practice, the DDAP model is solved as vehicle locations and customer requests are revealed during the planning horizon (this explains the dynamic component of the model). In addition, penalties or bonuses may have to be added to arc costs $c_{ij}, i \in D, j \in L$, to reflect the cost of taking drivers home after a given number of transportation services have been performed. A dispatcher who wants to take a driver $i \in D$ home at a given point in time can simply reduce the cost of the assignment of that particular driver to trips $j \in L$ whose delivery points are close to the driver home location. This can be accomplished by subtracting a suitable quantity from the c_{ij} costs.

[Planet.xlsx, Planet.py] Planet Transportation is an American motor carrier specialized in TL trucking in Illinois. On 26 January the company had to solve the problem of assigning four full-load transportation services each, three days long. The pick-up points were in Champaign, Danville, Peoria, and Springfield. At that time six drivers were available. The first four were located in Bloomington, Decatur, Mason City and Pekin, and the last two in Springfield. The company formulated and solved a DDAP model with $|D| = 6$ and $|L| = 4$. Costs c_{ij}, $i \in D$, $j \in L$, were assumed to be proportional to distances between driver locations i and full-load trip pick-up points j (see Table 6.9). The optimal driver assignment, computed by using the Excel Solver tool, is reported in Table 6.10. It is worth noting that the driver located in Mason City and one of the two drivers based in Springfield were not assigned.

Table 6.9 Distance (in miles) between driver locations and trip pick-up points in the Planet Transportation problem.

	Location of the drivers			
Pickup point	Champaign	Danville	Peoria	Springfield
Bloomington	51.9	84.1	42.0	67.4
Decatur	46.8	83.3	107.0	40.0
Mason City	97.0	129.1	51.2	47.5
Pekin	95.4	127.6	13.1	69.4
Springfield	85.5	121.9	74.3	0.0

Table 6.10 Optimal driver assignment of the Planet Transportation problem.

Driver	Location of the driver	Trip	Pickup point
1	Bloomington	2	Danville
2	Decatur	1	Champaign
3	Mason City	–	–
4	Pekin	3	Peoria
5	Springfield	4	Springfield
6	Springfield	–	–

6.10 Vehicle Fleet Composition

When demand varies over the year, carriers or shippers usually cover the baseload of demand through an owned fleet, while using hired vehicles to cover peak periods. In what follows, the least-cost mix of owned and hired vehicles is determined under the assumption that all vehicles are identical. Let T be the number of time periods into which the time horizon of a year is decomposed (for example, $T = 52$ if the time period corresponds to a week); v be the decision variable corresponding to the number of owned vehicles, and v_t, $t = 1, \dots, T$, be the required number of vehicles at time period t. Moreover, let c_F and c_V be the fixed and variable cost per time period of an owned vehicle, respectively, and let c_H be the cost per time period of hiring a vehicle. Then, the annual transportation cost as a function of the number of owned vehicles, is

$$C(v) = c_F T v + c_V \left(\sum_{t=1,\dots,T:v_t \leq v} v_t + \sum_{t=1,\dots,T:v_t > v} v \right) + c_H \sum_{t=1,\dots,T:v_t > v} (v_t - v), \quad (6.51)$$

where the right-hand side is the sum of the annual fixed cost, the annual variable cost of the owned vehicles, and the annual cost of hiring vehicles to cover peak demand that may occur during various time periods. It is worth observing that the cost $C(v)$ given by equation (6.51) is defined for discrete values of v included in the interval $[0, \overline{v}]$, where

$$\overline{v} = \max_{t=1,\dots,T} \{v_t\}.$$

To ensure that the range of t in the two summations of equation (6.51) does not depend on the decision variable v, it is possible to construct an equivalent alternative formulation of the problem, left to the reader as an exercise (see Problem 6.13).

Nevertheless, to determine the optimal number v^* of owned vehicles, it is sufficient to calculate the cost $C(v)$ for each $v = 0, \dots, \overline{v}$ and determine v^* by inspection as

$$v^* = \arg \min_{v=0,\dots,\overline{v}} C(v).$$

[FastCourier.xlsx] Fast Courier is a US transportation company located in Wichita, Kansas, that specializes in door-to-door deliveries. The company owns a fleet of 14 vans and hires vans from third parties whenever the service demand exceeds the fleet capacity. Last year, the number of vans weekly used for meeting all the transportation demand is reported in Table 6.11. For next year, the company has decided to redesign its fleet composition with the aim of reducing the annual transportation cost, assuming that the vehicle requests remain identical to those of the year just ended. Assuming that $c_F = \$ 350$, $c_V = \$ 150$ and $c_H = \$ 800$, the transportation costs $C(v)$ calculated by formula (6.51) when v varies in the interval $[0, 32]$ are shown in Table 6.12. As a result, the optimal number of owned vehicles is equal to $v^* = 19$ (or, equivalently, 20), with an annual transportation cost equal to $C(v^*) = \$ 606\,600$. With respect to the currently adopted solution with 14 owned vehicles used, the saving the company will obtain by adopting this new solution is, therefore, equal to $638\,450 - 606\,600 = \$ 31\,850$.

Table 6.11 Weekly number of vans used by Fast Courier during last year.

t	v_t	t	v_t	t	v_t	t	v_t
1	12	14	18	27	23	40	25
2	15	15	17	28	22	41	25

(Continued)

Table 6.11 (Continued)

t	v_t	t	v_t	t	v_t	t	v_t
3	16	16	16	29	24	42	24
4	17	17	14	30	26	43	22
5	17	18	13	31	27	44	22
6	18	19	13	32	28	45	19
7	20	20	14	33	30	46	20
8	20	21	15	34	32	47	18
9	21	22	16	35	32	48	17
10	22	23	17	36	30	49	16
11	24	24	19	37	29	50	16
12	22	25	21	38	28	51	14
13	20	26	22	39	26	52	13

Table 6.12 Annual transportation costs (in \$) at variation of the number of owned vehicles of Fast Courier.

v	C(v)	v	C(v)	v	C(v)
0	853 600	11	682 000	22	613 100
1	838 000	12	666 400	23	620 900
2	822 400	13	651 450	24	629 350
3	806 800	14	638 450	25	639 750
4	791 200	15	627 400	26	651 450
5	775 600	16	617 650	27	664 450
6	760 000	17	611 150	28	678 100
7	744 400	18	607 900	29	693 050
8	728 800	19	606 600	30	708 650
9	713 200	20	606 600	31	725 550
10	697 600	21	609 200	32	742 450

6.11 Shipment Consolidation

This section deals with a consolidation and dispatching problem often faced by companies that carry out their own transportation services. Here, the company has to choose the best way of delivering a set of orders to its customers over a planning horizon made up of T days. The company must decide

- the best transportation mode for each shipment;
- how orders have to be consolidated;
- the features of owned vehicle schedules (start times, intermediate stops, the order in which stops are visited, etc.).

Each order $k \in K$ is characterized by a destination i_k, a weight $w_k \geq 0$, a release day l_k (the day in which order k is ready for shipment; it is assumed, for simplicity, that order k can be shipped at least one day after the release day), and a deadline day d_k (the day within which order k must be delivered to i_k). The company may transport its products by using its owned vehicles, following predetermined routes, or by using a carrier. An owned truck may follow any route r of a pre-established set R. With each route $r \in R$ are associated a set of stops S_r (visited in a given order), a (fixed) cost f_r, and a capacity q_r (the maximum weight that the vehicle operating route r can carry). Moreover, let τ_{kr}, $k \in K$, $r \in R$, be the number of travel days to deliver order k on route r ($\tau_{kr} = 0$ means same-day delivery). Transporting order $k \in K$ to its destination by a carrier costs g_k and takes τ'_k days.

The decision variables are all binary: x_{krt}, $k \in K$, $r \in R$, $t = 1, \ldots, T$, having a value equal to 1 if order k is assigned to route r starting on day t, 0 otherwise; y_{rt}, $r \in R$, $t = 1, \ldots, T$, equal to 1 if route r is operated on day t, 0 otherwise; z_k, equal to 1 if order k is transported by a carrier, 0 otherwise (such a decision variable is defined only if $l_k + \tau'_k \leq d_k$).

The problem is then

$$\text{Minimize} \sum_{r \in R} \sum_{t=1}^{T} f_r y_{rt} + \sum_{k \in K} g_k z_k \tag{6.52}$$

subject to

$$\sum_{k : l_k \leq t \leq d_k - \tau_{kr}, i_k \in S_r} w_k x_{krt} \leq q_r y_{rt}, \ r \in R, t = 1, \ldots, T \tag{6.53}$$

$$\sum_{r : i_k \in S_r} \sum_{t : l_k \leq t \leq d_k - \tau_{kr}} x_{krt} + z_k = 1, \ k \in K \tag{6.54}$$

$$x_{krt} \in \{0, 1\}, \ k \in K, r \in R, t = 1, \ldots, T \tag{6.55}$$

$$y_{rt} \in \{0, 1\}, \ r \in R, t = 1, \ldots, T \tag{6.56}$$

$$z_k \in \{0, 1\}, \ k \in K. \tag{6.57}$$

The objective function (6.52) is the total cost paid to transport orders. Constraints (6.53) state that, for each route $r \in R$ and for each day $t = 1, \ldots, T$, the total weight carried on route r, on day t must not exceed the capacity q_r if y_{rt} is equal to 1, and is equal to zero otherwise. Constraints (6.54) impose that each order is assigned to a route operated by a owned truck or to a carrier. It is easy to show that formulation (6.52)–(6.57) can be transformed into an SNDP on a time-expanded directed graph.

[Oxximet.xlsx, Oxximet.py] Oxximet manufactures semi-finished chemical products in a plant near Milan, Italy. The main customers are located close to Lausanne, Switzerland, and Lyon and Marseille, France. On 11 June 10 customer orders were waiting to be satisfied (see Table 6.13).

Oxximet can use the trucks it owns, according to the set of routes reported in Table 6.14, visiting one or more customers. Each of the truck routes has a duration of one day at most and can be scheduled in every day of the planning horizon.

Truck capacity is 260 quintals. The company can also use carriers, whose cost and delivery time for each order are shown in the last two columns of Table 6.13.

The problem can be formulated by using the model defined by (6.52)–(6.57), with a scheduling time horizon $T = 5$, from 12–16 June inclusive, $|K| = 10$ and $|R| = 6$. The parameters l_k, d_k, w_k, g_k and τ'_k related to the orders $k \in K$ are shown in Table 6.13.

By observing the carrier delivery times, it is found that

$$l_k + \tau'_k \leq d_k, k \in K,$$

that is, the decision variable z_k exists for each $k \in K$. As observed before, $\tau_{kr} = 0$, $k \in K, r \in R$.

The optimal solution, determined by using the Excel Solver tool, provides the delivery scheduling for owned truck routes shown in Table 6.15. In addition,

Table 6.13   Details of the orders received on 11 June by Oxximet.

Order	Customer location	Release day	Deadline day	Weight [quintals]	Carrier shipping cost [€]	Carrier delivery days [days]
1	Marseille	11 June	15 June	207.34	600	1
2	Marseille	11 June	16 June	19.05	300	0
3	Lyon	11 June	16 June	19.59	300	0
4	Marseille	11 June	16 June	35.23	300	1
5	Lausanne	11 June	16 June	61.54	300	1
6	Lausanne	11 June	16 June	38.31	300	1
7	Lyon	13 June	16 June	100.46	600	0
8	Marseille	13 June	16 June	15.44	500	1
9	Lyon	13 June	16 June	56.89	500	0
10	Marseille	13 June	15 June	39.55	500	1

Table 6.14   Routes of the owned vehicles and related fixed costs (in €) of the Oxximet problem.

Id	Route	Cost
1	Milan–Lausanne	800
2	Milan–Lyon	750
3	Milan–Marseille	800
4	Milan–Lausanne–Lyon	830
5	Milan–Lyon–Marseille	870
6	Milan–Lausanne–Marseille	890

Table 6.15 Optimal solution of the owned vehicle routes for the Oxximet problem.

Day	Route	Orders
13 June	Milan–Lyon	3, 7, 9
13 June	Milan–Lausanne–Marseille	2, 4, 5, 6, 8, 10

order 1 should be delivered using carriers ($z_1^* = 1$), with a shipping day that can vary from 12 June to 14 June. The overall optimal cost is € 2240.

6.12 Vehicle Routing Problems

VRPs amount to finding optimal delivery or collection routes from one or several depots to a number of customers. VRPs can be defined on a mixed graph $G = (V, A, E)$, where V is a set of vertices, A is a set of arcs and E is a set of edges. A vertex 0 represents the depot at which m vehicles are based, while a subset $U \subseteq V$ of *required vertices* and a subset $R \subseteq A \cup E$ of *required arcs* and *required edges* represent the customers. VRPs consist of determining a least-cost set of m circuits based at a depot, and including the required vertices, arcs, and edges.

In this graph representation, arcs and edges correspond to road segments, and vertices correspond to road intersections. In addition, isolated customers are represented by required vertices, whereas subsets of customers distributed almost continuously along a set of road segments are modelled as required arcs or edges (this is often the case of mail delivery and solid waste collection in urban areas). See Figures 6.29 and 6.30 for an example. If $R = \emptyset$, the VRP is called a *node routing problem* (NRP), while if $U = \emptyset$ it is called an *arc routing problem* (ARP). NRPs have been studied more extensively than ARPs and are usually referred to simply as VRPs. However, for the sake of clarity, in this book the appellation NRPs is used. If $m = 1$ and there are no side constraints, the

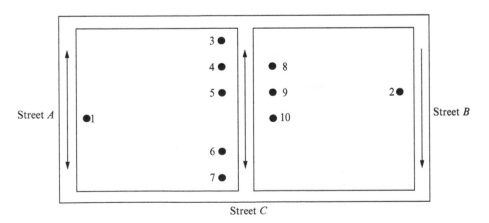

Figure 6.29 A road network where 10 customers (represented by black dots) are to be served. Streets A and C are two-way. Street B is one-way.

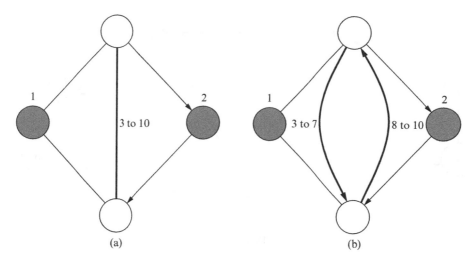

Figure 6.30 (a) A mixed graph representation of a vehicle, traversing street *C* of Figure 6.29, and serving the customers on both sides; and (b) a mixed graph representation of a vehicle, traversing street *C*, and serving the customers on a single side. Required vertices are in grey; required arcs and edges are in bold.

NRP is the classical *travelling salesman problem* which consists of determining a single circuit starting from the depot and spanning the required vertices of *G*, whereas the ARP is the *rural postman problem* (RPP) which amounts to designing a single circuit including the arcs and edges of *R*. The RPP reduces to the *Chinese postman problem* (CPP) if every arc and edge have to be serviced ($R = A \cup E$).

Operational constraints
The most common operational constraints are

- the number of vehicles *m* can be fixed or can be a decision variable, possibly subject to an upper bound constraint;
- the total demand transported by a vehicle at any time may not exceed its capacity;
- the duration of any route may not exceed a work shift duration;
- customers must be served within pre-established time windows;
- some customers must be served by specific vehicles;
- the service of a customer must be performed by a single vehicle or may be shared by several vehicles;
- customers are subject to precedence relations.

When customers impose service time windows or when travel times vary during the day, time issues have to be considered explicitly in the design of vehicle routes, in which case VRPs are often referred to as *vehicle routing and scheduling problems* (VRSPs).

Precedence constraints arise naturally whenever some goods have to be transported between specified pairs of pick-up and delivery points. In such problems, a pick-up and delivery pair is to be serviced by the same vehicle (no transshipment is allowed) and each pick-up point must be visited before the associated delivery point. Another kind of precedence relations has to be imposed whenever vehicles have first to perform a set

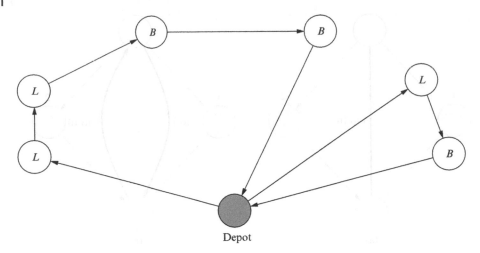

Figure 6.31 Vehicle routing with backhauls (*L*: linehaul customer, *B*: backhaul customers).

of deliveries (*linehaul customers*) and then a set of pick-ups (*backhaul customers*), as is customary in some industries (VRPs with backhauls, see Figure 6.31).

Objective
With each arc and edge $(i, j) \in A \cup E$ are associated a travel time t_{ij} and a travel cost c_{ij}. In addition, each vehicle may be associated with a fixed cost. The most common objective is to minimize the cost of traversing the arcs and edges of the graph plus the sum of the fixed costs associated with using the vehicles.

Travel time estimate
While the computation of the distances between the nodes of a road network is straight-forward, the accurate estimation of the travel times is often difficult, for two reasons. First, the average speed depends on the period of the day as well as on the day of the week and the occurrence of holidays. In particular, in most urban areas a rush hour (or peak hour) happens twice a day, once in the morning and once in the late afternoon, the time intervals when most people commute. Moreover, in some areas the traffic is lighter on weekends than between Monday and Friday. Second, remarkable fluctua-tions around the average speed are caused by weather conditions, accidents, strikes disrupting the public transportation system, and events like sport matches, concerts, political protests, and so on. In the last few decades, there has been a proliferation of online travel time information systems which provide an estimate of the current traversal times through a suitable processing of data coming from a number of sensing technologies (including inductive loops placed in the roadbed, video vehicle detection and, more recently, GPS-based mobile phones). At the moment, such systems cover only a portion of the whole road network, but it is expected that they will be more and more widespread in the coming years. The data collected by the travel time information systems can be used to make forecasts on the future average speeds by using the tech-niques illustrated in Chapter 2. Whenever such data are not available, one can devise a travel time estimate on the basis of the features of the road by using forecasting regres-sion models (see Section 2.7.1). To this end, the factors (regressors) affecting travel time

along a street are identified, and a regression equation is then used to forecast the average travel time as a function of these factors. The most relevant factors are the number of lanes, the street width, whether the street is one-way or two-way, parking regulations, traffic volume, the number of traffic lights, the number of stop signs, and the quality of the road surface.

In a school bus routing and scheduling study, the traversal times of the streets and avenues of Manhattan, New York, were computed by estimating vehicle speed v through the following formula:

$$v = \bar{v} + 2.07x_1 + 7.52x_2 + 1.52x_3 + 1.36x_4 - 3.26x_5 + 4.04x_6,$$

where $\bar{v} = 7.69$ miles per hour is the average bus speed in normal conditions, x_1 is the total number of street lanes, x_2 is the number of street lanes available for buses, x_3 is a binary constant equal to 1 in case of a one-way street and 0 otherwise, x_4 is equal to 1 in case of bad road surface conditions and 2 in case of good road surface conditions, x_5 takes into account the traffic volume (1 = low, 2 = medium, 3 = high) and x_6 is the time fraction of green lights. The coefficients of the regressors $x_1, \dots, x_6$ were estimated through a linear regression model (see Section 2.7.1).

6.12.1 The Travelling Salesman Problem

In the absence of operational constraints, there always exists an optimal NRP solution in which a single vehicle is used (see Problem 6.18). Hence, the NRP reduces to a TSP which consists of finding a least-cost tour including all the required vertices and the depot. In any TSP feasible solution on graph G, each vertex of $U \cup \{0\}$ appears at least once, and two successive vertices of $U \cup \{0\}$ are linked by a least-cost path. As a consequence, the TSP can be reformulated on an auxiliary complete directed graph $G' = (V', A')$, where $V' = U \cup \{0\}$ is the vertex set and A' is the arc set. With each arc $(i, j) \in A'$ is associated a cost c_{ij} equal to that of a least-cost path from i to j in G. These costs satisfy the *triangle inequality*

$$c_{ij} \leq c_{ik} + c_{kj}, \forall (i, j) \in A', \forall k \in V', k \neq i, j.$$

Because of this property, there exists a TSP optimal solution which is a *Hamiltonian tour* in G', defined as a tour in which each vertex in V' appears exactly once. In what follows, the search for an optimal TSP solution is restricted to Hamiltonian tours.

Despite this restriction, the TSP may be a very difficult problem to solve. The application of a *naïve method*, consisting of enumerating all Hamiltonian tours to find the optimal one, requires the generation of all possible permutations of the vertices to be visited and this number is equal to $(|V'| - 1)!$ (or $(|V'| - 1)!/2$ if each tour can be run in either of two directions at the same cost, see in the following). Thus, a TSP with only $|V'| = 20$ vertices has $19! = 121\,645\,100\,408\,832\,000$ possible solutions. This means that the naïve method becomes unusable even for TSPs of not particularly significant size.

If $c_{ij} = c_{ji}$ for each pair of distinct vertices $i, j \in V'$, the TSP is said to be *symmetric* (STSP), otherwise it is called *asymmetric* (ATSP). The STSP is suitable for inter-city transportation, while the ATSP is recommended in urban settings because of one-way streets. Of course, the solution techniques developed for the ATSP can also be applied to the STSP. This method could, however, be very inefficient, as explained later. It is therefore customary to deal with the two cases separately.

The Asymmetric Travelling Salesman Problem

Let x_{ij}, $(i, j) \in A'$, be a binary decision variable equal to 1 if the corresponding arc (i, j) is part of the solution, 0 otherwise. It is worth observing that in the ATSP a solution corresponds to a Hamiltonian circuit (directed tour). The ATSP can then be formulated as follows:

$$\text{Minimize} \sum_{(i,j)\in A'} c_{ij} x_{ij}$$

subject to

$$\sum_{i\in V'\backslash\{j\}} x_{ij} = 1, \ j \in V' \tag{6.58}$$

$$\sum_{j\in V'\backslash\{i\}} x_{ij} = 1, \ i \in V' \tag{6.59}$$

$$\sum_{i\in S}\sum_{j\notin S} x_{ij} \geq 1, \ S \subset V', \ |S| \geq 2 \tag{6.60}$$

$$x_{ij} \in \{0, 1\}, \ (i, j) \in A'.$$

Equations (6.58) and (6.59) are referred to as *degree constraints*. Constraints (6.58) mean that in the solution a unique arc enters each vertex $j \in V'$. Similarly, constraints (6.59) state that a single arc exits each vertex $i \in V'$ in the solution. Constraints (6.60) avoid the generation of solutions consisting of disconnected subcircuits (*connectivity constraints*). In particular, in any feasible solution there should be at least one arc coming out from each proper and a non-empty subset S of vertices in V'. They are redundant for $|S| = 1$ because of constraints (6.59). It is worth noting that the number of constraints (6.60) is $2^{|V'|} - |V'| - 2$. Such constraints can be formulated in an alternative way, algebraically equivalent (see Problem 6.20):

$$\sum_{(i,j)\in A':i,j\in S} x_{ij} \leq |S| - 1, \ S \subset V', \ |S| \geq 2. \tag{6.61}$$

Inequalities (6.61) prevent the formation of subcircuits containing fewer than $|V'|$ vertices (*subtour elimination constraints*), ensuring that, for each possible choice of S, the number of arcs selected in S is smaller than the number of vertices in S.

A lower bound

A good lower bound on the ATSP optimal solution cost z^*_{ATSP} can be obtained by removing constraints (6.60) from ATSP formulation. The optimal solution of the relaxed problem can be obtained by solving the following *assignment problem* (AP):

$$\text{Minimize} \sum_{i\in V'}\sum_{j\in V'} c_{ij} x_{ij} \tag{6.62}$$

subject to

$$\sum_{i \in V'} x_{ij} = 1, \ j \in V' \tag{6.63}$$

$$\sum_{j \in V'} x_{ij} = 1, \ i \in V' \tag{6.64}$$

$$x_{ij} \in \{0, 1\}, \ i, j \in V', \tag{6.65}$$

where, in the objective function (6.62), $c_{ii} = \infty$, $i \in V'$, in order to force $x_{ii}^* = 0$, for all $i \in V'$.

Note that, due to the particular structure of constraints (6.63) and (6.64), the relations (6.65) can be replaced with

$$x_{ij} \geq 0, \ i, j \in V'.$$

The optimal AP solution x_{AP}^* corresponds to a collection of p subcircuits $C_1, \dots, C_p$, spanning all vertices of the directed graph G'. If $p = 1$, the AP solution is feasible (and hence optimal) for the ATSP.

As a rule, z_{AP}^* is a good lower bound on z_{ATSP}^* if the cost matrix is strongly asymmetric (in this case, it has been empirically demonstrated that the deviation $(z_{ATSP}^* - z_{AP}^*)/z_{AP}^*$ from the optimal solution cost is often less than 1%). In contrast, in the case of symmetric costs, the deviation is typically 30% or more. The reason for this behaviour can be explained by the fact that for symmetric costs, if the AP solution contains arc $(i, j) \in A'$, then the AP optimal solution is likely to include arc $(j, i) \in A'$ too. As a result, the optimal AP solution usually shows several small subcircuits of only two vertices and is quite different from the ATSP optimal solution.

[`Bontur.xlsx`, `Bontur.py`] Bontur is a pastry producer, founded in Prague, Czech Republic, in the nineteenth century. The firm currently operates, in addition to four modern plants, a workshop in Gorazdova street where the founder started working. The workshop serves Prague and its surroundings. Every day at 6:30 a fleet of vans carries the pastries from the workshop to several retail outlets (small shops, supermarkets, and hotels). In particular, all outlets of the Vltava River district are usually served by a single vehicle. For the sake of simplicity, arc transportation costs are assumed to be proportional to arc lengths. In Figure 6.32 the road network is modelled as a mixed graph $G = (V, A, E)$, where a length l_{ij} is associated with each arc or edge (i, j). The workshop and the vehicle depot are located at vertex 0. On 23 March seven shops (located at vertices 1, 3, 4, 9, 18, 20, and 22) needed to be supplied. The problem can be formulated as an ATSP on an auxiliary complete directed graph $G' = (V', A')$, where V' is formed by the seven vertices associated with the customers and by vertex 0. With each arc $(i, j) \in A'$ is associated a cost c_{ij} corresponding to the length of the shortest path from i to j on G (see Table 6.16). The optimal AP solution x_{AP}^*, determined by using the Excel Solver tool, is made up of the following three subcircuits (see Figure 6.33):

$$C_1 = ((1, 4), (4, 3), (3, 9), (9, 1)),$$

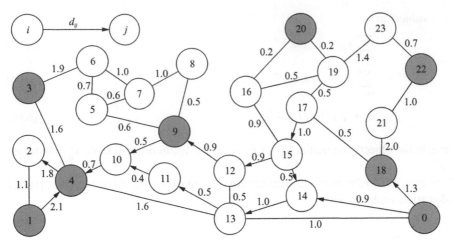

Figure 6.32 A graph representation of a Bontur distribution problem (one-way street segments are represented by arcs, while two-way street segments are modelled through edges). The depot and the service vertices are in grey.

of cost equal to 11.0 km;

$$C_2 = ((0, 18), (18, 0)),$$

of cost equal to 5.2 km;

$$C_3 = ((20, 22), (22, 20)),$$

of cost equal to 4.6 km.

Therefore, the AP lower bound z_{AP}^* on the optimal objective function value of ATSP is equal to

$$z_{AP}^* = 11.0 + 5.2 + 4.6 = 20.8 \text{ km.}$$

Table 6.16 Shortest path length (in km) from i to j, $i, j \in V$, in the Bontur problem.

	0	1	3	4	9	18	20	22
0	0.0	5.5	4.2	2.6	2.4	1.3	2.5	4.3
1	4.7	0.0	3.7	2.1	5.1	6.0	7.2	9.0
3	4.2	4.5	0.0	1.6	3.2	5.5	6.7	8.5
4	2.6	2.9	1.6	0.0	3.0	3.9	5.1	6.9
9	3.8	4.1	2.8	1.2	0.0	5.1	6.3	8.1
18	3.9	7.4	6.1	4.5	3.3	0.0	1.2	3.0
20	3.5	7.0	5.7	4.1	2.9	1.2	0.0	2.3
22	5.8	9.3	8.0	6.4	5.2	3.0	2.3	0.0

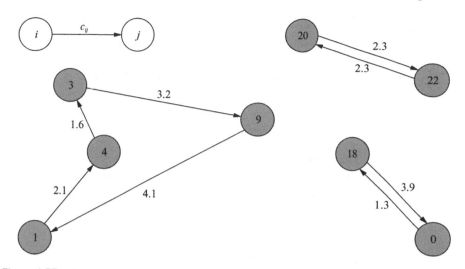

Figure 6.33 A graph representation of the optimal solution of the AP relaxation of the Bontur problem.

Patching heuristic

The patching heuristic works as follows. First, the AP relaxation is solved. If a single circuit is obtained, the procedure stops (the AP solution is the optimal ATSP solution). Otherwise, a feasible ATSP solution is constructed by merging the subcircuits of the AP solution. When merging two subcircuits, one arc is removed from each subcircuit and two new arcs are added in such a way that a new single connected subcircuit is obtained. In the following the PATCHING procedure is described:

[Bontur.xlsx, Bontour.py] In order to find a feasible solution $\bar{x}_{STAP}$ to the Bontur distribution problem, the PATCHING procedure is applied to the AP solution shown in Figure 6.33. At the first iteration, C_1 and C_2 are selected to be merged (alternatively, C_3 could be used instead of C_2). By merging C_1 and C_2 at minimum cost (through the removal of arcs $(3, 9)$ and $(18, 0)$ and the insertion of arcs $(3, 0)$ and $(18, 9)$), the following subcircuit (having length equal to 16.6 km) is obtained (see Figure 6.34):

$$C_4 = ((0, 18), (18, 9), (9, 1), (1, 4), (4, 3), (3, 0)).$$

The partial solution, formed by the two subcircuits C_3 and C_4, is depicted in Figure 6.34. The total length increases by 0.4 km with respect to the initial solution. At the end of the second iteration, the two subcircuits in Figure 6.34 are merged at the minimum cost increase of 0.3 km through the removal of arcs $(18, 9)$ and $(20, 22)$ and the insertion of arcs $(18, 22)$ and $(20, 9)$.

This way, a feasible ATSP solution of cost $\bar{z}_{ATSP} = 21.5$ km is obtained (see Figure 6.35). In order to evaluate the quality of the heuristic solution, the following deviation from the AP lower bound can be computed:

$$\frac{\bar{z}_{ATSP} - z^*_{AP}}{z^*_{AP}} = \frac{21.5 - 20.8}{20.8} = 0.0337,$$

which corresponds to a percentage deviation of 3.37%.

1: **procedure** PATCHING (*C*)
2:     # $C = \{C_1, \ldots, C_p\}$ is the set of the *p* subcircuits in the AP optimal solution;
3:     **if** $p = 1$ **then**
4:         # The AP solution is feasible (and hence optimal) for the ATSP;
5:         exit = TRUE;
6:     **else**
7:         exit = FALSE;
8:     **end if**
9:     **while** exit = FALSE **do**
10:         Identify the two subcircuits $C_h, C_k \in C$ with the largest number of vertices;
11:         Merge C_h and C_k in a unique subcircuit in such a way that the cost increase is kept at minimum;
12:         $p = p - 1$;
13:         **if** $p = 1$ **then**
14:             # an ATSP feasible solution has been determined;
15:             exit = TRUE;
16:         **end if**
17:     **end while**
18:     **return** *C*;
19: **end procedure**

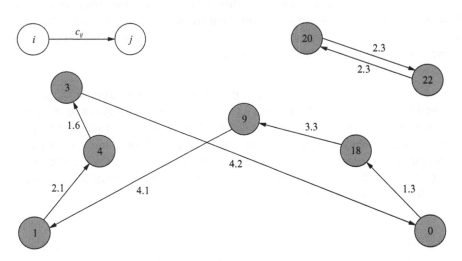

Figure 6.34 A graph representation of the solution at the end of the first iteration of the PATCHING procedure in the Bontur problem.

The Symmetric Travelling Salesman Problem

As explained in the previous subsection, the ATSP lower and upper bounding procedures perform poorly when applied to the symmetric TSP. For this reason, several STSP tailored methods have been developed.

The STSP can be formulated on an auxiliary complete undirected graph $G' = (V', E')$, in which with each edge $(i, j) \in E'$ is associated a transportation cost c_{ij} equal to that of a least-cost path between *i* and *j* in *G*. Hence, the costs c_{ij} satisfy the triangle inequality, and there exists an optimal solution which is a Hamiltonian cycle (undirected tour) in

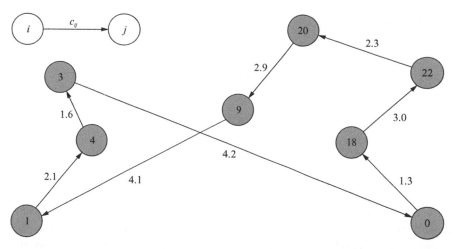

Figure 6.35 Feasible Hamiltonian circuit (obtained at the end of the PATCHING procedure) of the ATSP associated with the Bontur problem.

G'. Let $x_{ij}, (i, j) \in E'$, be a binary decision variable equal to 1 if the corresponding edge $(i, j) \in E'$ is part of the solution, 0 otherwise. The formulation of the STSP is as follows (recall that $i < j$ for each edge $(i, j) \in E'$):

$$\text{Minimize} \sum_{(i,j)\in E'} c_{ij} x_{ij}$$

subject to

$$\sum_{i\in V':(i,j)\in E'} x_{ij} + \sum_{i\in V':(j,i)\in E'} x_{ji} = 2,\ j \in V' \tag{6.66}$$

$$\sum_{(i,j)\in E':i\in S,j\notin S} x_{ij} + \sum_{(j,i)\in E':i\in S,j\notin S} x_{ji} \geq 2,\ S \subset V',\ 2 \leq |S| \leq \lfloor |V'|/2 \rfloor \tag{6.67}$$

$$x_{ij} \in \{0, 1\},\ (i, j) \in E'.$$

Equations (6.66) mean that in the solution exactly two edges must be incident to every vertex $j \in V'$ (*degree constraints*). Inequalities (6.67) state that, for every vertex subset S, there exist at least two edges in the solution with one endpoint in $S \subset V'$ and the other endpoint in $V'\backslash S$ (*connectivity constraints*). Since the connectivity constraints of a subset S and that of its complement $V' \backslash S$ are equivalent, one has to consider only inequalities (6.67) associated with subsets $S \subset V'$ such that $|S| \leq \lfloor |V'|/2 \rfloor$. Constraints (6.67) are redundant if $|S| = 1$ because of (6.66). Alternatively, the connectivity constraints (6.67) can be replaced with the following equivalent *subtour elimination constraints* (see Problem 6.20):

$$\sum_{(i,j)\in E':i,j\in S} x_{ij} + \sum_{(j,i)\in E':j,i\in S} x_{ji} \leq |S| - 1,\ S \subset V',\ 2 \leq |S| \leq \lfloor |V'|/2 \rfloor. \tag{6.68}$$

A lower bound

A lower bound on the optimal solution cost z^*_{STSP} of the STSP can be obtained by solving the following problem (see Problem 6.24):

$$\text{Minimize} \sum_{(i,j)\in E'} c_{ij} x_{ij} \tag{6.69}$$

subject to

$$\sum_{i \in V' : (i,r) \in E'} x_{ir} + \sum_{i \in V' : (r,i) \in E'} x_{ri} = 2 \qquad (6.70)$$

$$\sum_{(i,j) \in E' : i \in S, j \notin S} x_{ij} + \sum_{(j,i) \in E' : i \in S, j \notin S} x_{ji} \geq 1,$$

$$r \notin S, S \subset V', \ 1 \leq |S| \leq \lfloor |V'|/2 \rfloor \qquad (6.71)$$

$$\sum_{(i,j) \in E'} x_{ij} = |V'| \qquad (6.72)$$

$$x_{ij} \in \{0,1\}, \ (i,j) \in E', \qquad (6.73)$$

where $r \in V'$ is arbitrarily chosen (*root vertex*). Model (6.69)–(6.73) corresponds to a minimum spanning r-tree problem (MSrTP), for which the optimal solution is a least-cost connected subgraph spanning G' and such that vertex $r \in V'$ has degree two. The MSrTP can be solved in $O(|V'|^2)$ steps with a procedure that involves, first, the determination of a minimum-cost tree $T^* = (V' \setminus \{r\}, E_T)$ spanning $V' \setminus \{r\}$ and, second, the insertion in T^* of the vertex r, as well as of the two least-cost edges incident to r. Observe that constraint (6.72) is redundant in the STSP formulation, but it is important to ensure that any feasible solution of model (6.69)–(6.73) corresponds to a spanning r-tree.

[SaintMartin.xlsx] Saint-Martin distributes fresh fishing products in Normandy, France. On 7 June the company received seven orders from sales points all located in northern Normandy. It was decided to serve the seven requests by means of a single vehicle sited in Betteville. The problem can be formulated as an STSP on an auxiliary complete undirected graph $G' = (V', E')$, where V' is composed of eight vertices corresponding to the sales points and of vertex 0 associated with the depot. With each edge $(i,j) \in E'$ is associated a cost c_{ij} equal to the shortest distance between vertices i and j (see Table 6.17). The minimum spanning r-tree is depicted in Figure 6.36, to which corresponds a cost $z^*_{MSrTP} = 225.8$ km.

Table 6.17 Distances (in km) between terminals in the Saint-Martin problem.

	Betteville	Bolbec	Dieppe	Fécamp	Le Havre	Luneray	Rouen	Valmont
Betteville	0.0	27.9	54.6	42.0	56.5	37.0	30.9	34.1
Bolbec		0.0	67.2	25.6	28.8	48.4	57.4	21.6
Dieppe			0.0	60.5	95.8	18.8	60.4	52.1
Fécamp				0.0	39.4	43.1	70.2	12.2
Le Havre					0.0	77.2	84.5	44.4
Luneray						0.0	51.6	34.0
Rouen							0.0	59.3
Valmont								0.0

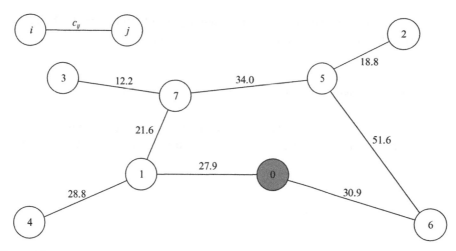

Figure 6.36 Minimum-cost spanning *r*-tree in the Saint-Martin problem.

The MS*r*TP lower bound can be improved in two ways. In the former method, the MS*r*TP relaxation is solved for more choices of the root $r \in V'$ and then the largest MS*r*TP lower bound is selected. In the latter method, $r \in V'$ is fixed but each constraint (6.66) with the only exception of $j = r$ is relaxed in a Lagrangian fashion. Let λ_j, $j \in V' \setminus \{r\}$, be the Lagrangian multiplier attached to vertex $j \in V' \setminus \{r\}$. A Lagrangian relaxation of the STSP is as follows:

$$\text{Minimize} \sum_{(i,j) \in E'} c_{ij} x_{ij} + \sum_{j \in V' \setminus \{r\}} \lambda_j \left(\sum_{i \in V' : (i,j) \in E'} x_{ij} + \sum_{i \in V' : (j,i) \in E'} x_{ji} - 2 \right) \quad (6.74)$$

subject to

$$\sum_{i \in V' : (i,r) \in E'} x_{ir} + \sum_{i \in V' : (r,i) \in E'} x_{ri} = 2 \quad (6.75)$$

$$\sum_{(i,j) \in E' : i \in S, j \notin S} x_{ij} + \sum_{(j,i) \in E' : i \in S, j \notin S} x_{ji} \geq 1, r \notin S, S \subset V', 1 \leq |S| \leq \lfloor |V'|/2 \rfloor \quad (6.76)$$

$$\sum_{(i,j) \in E'} x_{ij} = |V'| \quad (6.77)$$

$$x_{ij} \in \{0, 1\}, (i, j) \in E'. \quad (6.78)$$

Artificially setting $\lambda_r = 0$, the objective function (6.74) can be rewritten as

$$\sum_{(i,j) \in E'} (c_{ij} + \lambda_i + \lambda_j) x_{ij} - 2 \sum_{j \in V'} \lambda_j. \quad (6.79)$$

To determine the optimal Lagrangian multipliers (or at least a set of good multipliers), a suitable variant of the subgradient method illustrated in Section 6.6.3 for the LMMCFP can be used. In particular, at the *h*th iteration the updating formula of the Lagrangian multipliers can be the following:

$$\lambda_j^{(h+1)} = \lambda_j^{(h)} + \beta^{(h)} s_j^{(h)}, j \in V' \setminus \{r\},$$

where:

$$s_j^{(h)} = \sum_{i \in V' : (i,j) \in E'} x_{ij}^{(h)} + \sum_{i \in V' : (j,i) \in E'} x_{ji}^{(h)} - 2, j \in V' \setminus \{r\},$$

$x_{ij}^{(h)}$, $(i, j) \in E'$, is the optimal solution of the Lagrangian relaxation $MSrTP(\lambda)$ (6.79), (6.75)–(6.78) at the hth iteration, and $\beta^{(h)}$ can be set equal to

$$\beta^{(h)} = \alpha/h, h = 1, \dots,$$

where α is a user-defined parameter (see Section 6.6.3 for more details).

[SaintMartin.xlsx, SaintMartin.py] The results of the first three iterations of the subgradient method in the Saint-Martin problem ($r = 0$ and $\alpha = 1$) are

$\lambda_j^{(1)} = 0, j \in V \setminus \{r\};$

$z^*_{MSrTP(\lambda^{(1)})} = 225.8$ km;

$s^{(1)} = [1; -1; -1; -1; 1; 0; 1]^T;$

$\beta^{(1)} = 1;$

$\lambda^{(2)} = [1; -1; -1; -1; 1; 0; 1]^T;$

$z^*_{MSrTP(\lambda^{(2)})} = 231.8$ km;

$s^{(2)} = [1; -1; -1; -1; 1; 0; 1]^T;$

$\beta^{(2)} = 1/2;$

$\lambda^{(3)} = [3/2; -3/2; -3/2; -3/2; 3/2; 0; 3/2]^T;$

$z^*_{MSrTP(\lambda^{(3)})} = 234.8$ km.

After 558 iterations, the lower bounding procedure provides the following spanning r-tree

$$C^* = ((0,1),(1,4),(3,4),(3,7),(5,7),(2,5),(2,6),(0,6)),$$

with a cost of 252.4 km, which, being a Hamiltonian cycle, turns out to be an optimal solution.

Nearest neighbour heuristic

The nearest neighbour heuristic is a simple constructive procedure that builds a Hamiltonian path by iteratively linking the vertex inserted at the previous iteration to its nearest unrouted neighbour. Finally, a Hamiltonian cycle is obtained by connecting the two endpoints of the path. The nearest neighbour heuristic often provides low-quality solutions, since the edges added in the final iterations may be very costly. A more formal description of the NEAREST_NEIGHBOUR procedure is reported below.

[SaintMartin.xlsx, SaintMartin2.py] In order to find a feasible solution to the Saint-Martin problem, the NEAREST_NEIGHBOUR procedure is applied ($r = 0$), and the following Hamiltonian cycle is obtained (see Figure 6.37):

$$C = ((0,1),(1,7),(3,7),(3,4),(4,5),(2,5),(2,6),(0,6)),$$

```
1:  procedure NEAREST_NEIGHBOUR (V', E', C)
2:      Choose (arbitrarily) a vertex r ∈ V';
3:      C = (r);
4:      h = r;
5:      while |C| < |V'| do
6:          Identify the vertex k ∈ V' \ C such that c_hk = min {c_hj, c_jh};
                                                            j∈V'\C
7:          Add k at the end of C;
8:          h = k;
9:      end while
10:     Add r at the end of C;
11:     # C corresponds to a Hamiltonian cycle;
12:     return C;
13: end procedure
```

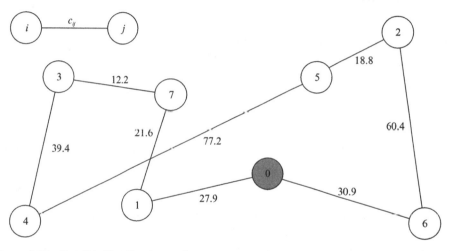

Figure 6.37 Feasible Hamiltonian cycle generated by the NEAREST_NEIGHBOUR procedure in the Saint-Martin problem.

whose cost $\bar{z}_{STSP}$ is 288.4 km. The deviation of this solution cost from the best lower bound available, corresponding to the optimal solution cost equal to 252.4 km, is

$$\frac{\bar{z}_{STSP} - LB}{LB} = \frac{288.4 - 252.4}{252.4} = 0.1442,$$

which corresponds to a percentage deviation of 14.42%.

It is worth noting that the NEAREST_NEIGHBOUR procedure can be easily also applied to determine a Hamiltonian circuit for the ATSP.

Christofides heuristic
The Christofides heuristic is a constructive procedure that works as follows:

1: **procedure** CHRISTOFIDES (V', E', C)

2:     Compute a minimum-cost tree $T = (V', E'_T)$ spanning the vertices of $G' = (V', E')$;

3:     # z_T^* is the cost of T;

4:     Compute a least-cost perfect-matching $M = (V'_M, E'_M)$ among the vertices of odd degree in the tree T;

5:     # $|V'_M|$ is always an even number;

6:     # z_M^* is the optimal matching cost;

7:     # $U = (V', E'_U)$ is the Eulerian subgraph (or multi-graph) of G' induced by the union of the edges of T and M (i.e., $E'_U = E'_T \cup E'_M$);

8:     Determine a Eulerian cycle C_E on $U = (V', E'_U)$;

9:     # $z(C_E) = z_T^* + z_M^*$;

10:    Extract a Hamiltonian cycle C from C_E;

11:    **return** C;

12: **end procedure**

The least-cost perfect-matching problem (code line 4 of the CHRISTOFIDES procedure) corresponds to solving the following problem:

$$\text{Minimize} \sum_{(i,j) \in E'_M} c_{ij} x_{ij} \tag{6.80}$$

subject to

$$\sum_{i \in V'_M : (i,j) \in E'_M} x_{ij} + \sum_{i \in V'_M : (j,i) \in E'_M} x_{ji} = 1, \ j \in V'_M \tag{6.81}$$

$$x_{ij} \in \{0, 1\}, \ (i, j) \in E'_M, \tag{6.82}$$

where the decision variables x_{ij}, $(i, j) \in E'_M$ are binary, each assuming a value equal to 1 if the corresponding edge (i, j) is included in the solution, 0 otherwise. Constraints (6.81) ensure that the matching is such that only one edge is incident in each of the vertices of the set V'_M. The structure of the constraints (6.81) allows the replacement of constraints (6.82) with the following:

$$x_{ij} \geq 0, \ (i, j) \in E'_M. \tag{6.83}$$

The problem (6.80), (6.81), (6.83) can be solved exactly in polynomial time (see Problem 6.25) with an algorithm whose complexity is $O(|V'_M|)^3$.

A Eulerian cycle on a Eulerian undirected (multi-)graph $U = (V', E'_U)$ (code line 8 of the CHRISTOFIDES procedure) can be obtained by using the following END_PAIRING procedure. At code line 2 of the END_PAIRING procedure, to find a spanning of the edges E'_U of U, it is sufficient to visit the (multi-)graph (in width or in depth) until a vertex already visited is reached; in this way, an element of the set C is obtained and, removing such visited edges from E'_U, once again a (multi-)graph is obtained, eventually not connected, in which all the vertices are again of even degree. It is possible to repeat the visit of the (multi-)graph to determine a cycle, until the spanning of the whole (multi-)graph is obtained. At code line 4, the vertex in common between the two cycles acts as a *pivot*. The merge of the two cycles into one (code line 5) can be obtained by visiting the edges of the first cycle up to the pivot, traversing all the edges of the second

---

1: **procedure** END_PAIRING (V', E'_U, C_E)

2:     Determine a spanning of the edges E'_U of U, defined by the set $C = \{C_1, \dots, C_p\}$, $p \geq 1$, of cycles; in each of them every edge is crossed exactly once;

3:     **while** $p > 1$ **do**

4:         Identify two cycles in C which contain at least one common vertex;

5:         Merge the two cycles, so as to obtain a single cycle in which each edge of the two cycles is crossed exactly once;

6:         $p = p - 1$;

7:     **end while**

8:     $C_E = C$;

9:     **return** C_E;

10: **end procedure**

---

cycle, returning to the pivot and continuing the visit with the remaining edges of the first cycle.

At code line 10 of the CHRISTOFIDES procedure, a Hamiltonian cycle C can be easily obtained from the Eulerian cycle C_E by removing any repeated vertices from C_E. In particular, let $C_E = (i_0, i_1, \dots, i_{m-1}, i_m = i_0)$ and suppose that there is at least one repeated vertex, that is, $i_h = i_k$, $h < k$. In this case, the vertex i_k can be removed from C_E, which means, in the sequence of edges that form C_E, the pair of edges $(i_{k-1}, i_k), (i_k, i_{k+1})$ can be replaced by the edge (i_{k-1}, i_{k+1}) (*shortcut*). This technique typically involves a cost reduction (or at least not an increase), since the triangle inequality holds. For this reason, $z(C) \leq z(C_E)$.

Moreover, note that, by simply reversing the order in which the vertices appear in C_E, in some cases, a second Hamiltonian cycle can be obtained whose cost can be compared to that of the first cycle. Finally, it can be shown that the cost of the Christofides solution is at most 50% higher than the optimal solution cost (see Problem 6.26).

[SaintMartin.xlsx, SaintMartin3.py] The Saint-Martin problem is solved by means of the CHRISTOFIDES procedure. The minimum spanning tree is made up of edges $\{(0, 1), (0, 6), (1, 4), (1, 7), (2, 5), (3, 7), (5, 7)\}$, and has a cost of 174.2 km. The optimal matching of the odd-degree vertices (1, 2, 3, 4, 6 and 7) is composed of edges $(1, 4), (2, 6)$ and $(3, 7)$, and has a cost of 101.4 km. The Eulerian undirected multi-graph is shown in Figure 6.38. At code line 10 of the CHRISTOFIDES procedure, the following Eulerian cycle is found:

$$C_E = (0, 1, 4, 1, 7, 3, 7, 5, 2, 6, 0),$$

with cost $z(C_E) = 174.2 + 101.4 = 275.6$ km. Starting from C_E, the following two Hamiltonian cycles are found:

$$C^{(1)} = (0, 1, 4, 7, 3, 5, 2, 6, 0);$$
$$C^{(2)} = (0, 6, 2, 5, 7, 3, 1, 4, 0),$$

with $C^{(2)}$ obtained by removing the vertices repeated from the sequence of vertices $(0, 6, 2, 5, 7, 3, 7, 1, 4, 1, 0)$, corresponding to C_E reversed. The corresponding costs to $C^{(1)}$ and $C^{(2)}$ are:

$$z(C^{(1)}) = 266.5 \text{ km};$$

$z(C^{(2)}) = 267.2 \, \text{km}.$

The chosen Hamiltonian cycle is, therefore, $C^{(1)}$ (see Figure 6.39).

In order to evaluate the quality of the heuristic solution, the following deviation from the best lower bound available, corresponding to the optimal solution cost, can be computed:

$$\frac{\bar{z}_{STSP} - LB}{LB} = \frac{266.5 - 252.4}{252.4} = 0.0559$$

which corresponds to a percentage deviation of 5.59%.

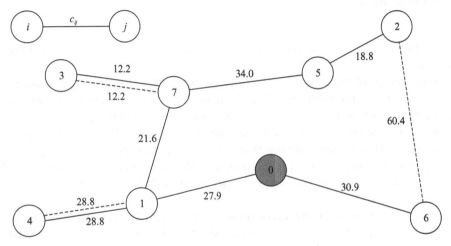

Figure 6.38 Eulerian undirected multi-graph generated at code line 2 of the CHRISTOFIDES procedure in the Saint-Martin problem (edges of the minimum-cost spanning tree are full lines and minimum-cost matching edges are dashed lines).

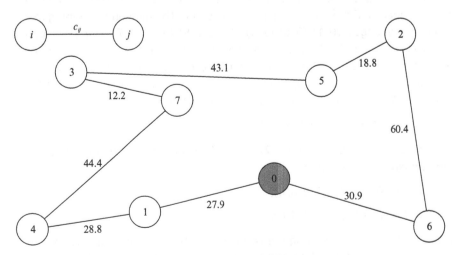

Figure 6.39 Hamiltonian cycle provided by the CHRISTOFIDES procedure in the Saint-Martin problem.

```
 1: procedure LOCAL_SEARCH (x^(0), x̄)
 2:     # x^(0) is the initial feasible solution and c^(0) the corresponding cost;
 3:     # N(x^(0)) is the neighbourhood of x^(0);
 4:     h = 0;
 5:     exit = FALSE;
 6:     while exit = FALSE do
 7:         Enumerate the feasible solutions belonging to N(x^(h));
 8:         Select the best feasible solution x^(h+1) ∈ N(x^(h));
 9:         if c^(h+1) ≥ c^(h) then exit = TRUE;
10:         else
11:             h = h + 1;
12:         end if
13:     end while
14:     # x^(h) is the best solution found;
15:     x̄ = x^(h);
16:     return x̄;
17: end procedure
```

Local search heuristics

Local search heuristics are iterative procedures that try to improve an initial feasible solution $x^{(0)}$. At the hth step, the solutions contained in a neighbourhood of the current solution $x^{(h)}$ are enumerated. If there are feasible solutions less costly than the current solution $x^{(h)}$, the best solution of the neighbourhood is taken as the new current solution $x^{(h+1)}$ and the procedure is iterated. Otherwise, the procedure is stopped (the last current solution is a *local optimum*).

For the STSP, $N(x^{(h)})$ is commonly defined as the set of all Hamiltonian cycles that can be obtained by substituting k edges ($2 \leq k \leq |V'|$) of $x^{(h)}$ for k other edges in E' (*k-exchange*) (see Figure 6.40).

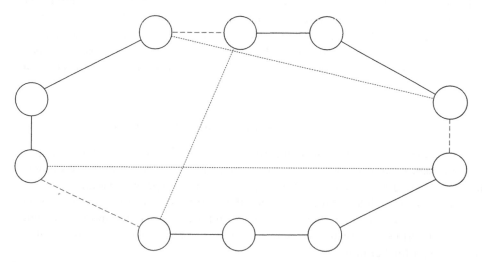

Figure 6.40 A feasible 3-exchange (dashed edges are removed, dotted edges are inserted).

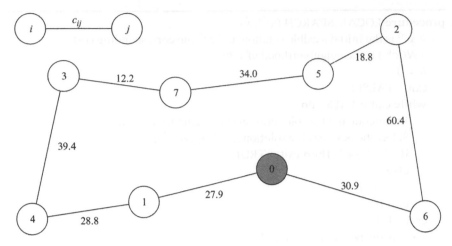

Figure 6.41 Hamiltonian cycle for the Saint-Martin problem with cost (provided by a 2-exchange) better than that reported in Figure 6.39.

In a LOCAL_SEARCH procedure based on k-exchanges, k can be constant or can be dynamically increased in order to intensify the search when improvements are likely to occur. If k is constant, each execution of Step 1 requires $O(|V'|^k)$ operations. In general, k is set equal to 2 or 3 at most, in order to limit the computational effort.

> [SaintMartin.xlsx, SaintMartin4.py] If a 2-EXCHANGE procedure is applied to the solution provided by the CHRISTOFIDES procedure in the Saint-Martin problem, a less costly Hamiltonian cycle (see Figure 6.41) is obtained at the first iteration by replacing edges $(3, 5)$ and $(4, 7)$ with edges $(3, 4)$ and $(5, 7)$. As a consequence, the solution cost decreases by 14.1 km, reaching the value of $(266.5 - 14.1) = 252.4$ km, corresponding to the optimal cost.

Finally, it is worth mentioning that a LOCAL_SEARCH procedure can also be applied to improve the feasible ATSP solutions. The application of the LOCAL_SEARCH procedure based on 2-exchanges to an ATSP solution is left to the reader as an exercise (see Problem 6.22). Note that in this case the application of the LOCAL_SEARCH procedure implies reversing the direction of one of the two paths obtained after removing the arcs.

6.12.2 The Node Routing Problem with Operational Constraints

As previously illustrated, in several settings operational constraints come into play when designing vehicle routes. These restrictions lead to a large number of variants, and the existing algorithms are often dependent on the type of constraints. For this reason, in the remainder of this subsection the most important constrained NRPs are examined and a limited number of techniques, representative of the most used methods, are described. As usual, the interested reader is referred to the relevant scientific literature for further information.

The *node routing problem with capacity constraints* (C-NRP, better known as C-VRP, acronym of *vehicle routing problem with capacity constraints*) consists of designing a set of least-cost vehicle routes starting and ending at the depot, such that:

1. each customer is visited exactly once;
2. a demand d_i is associated with each customer $i \in U$ (demands are either collected or delivered, but not both); then the total demand on each vehicle cannot exceed a given vehicle capacity q (vehicles are assumed to be identical).

This problem can be formulated on an auxiliary complete directed graph $G' = (V', A')$ or on an auxiliary complete undirected graph $G' = (V', E')$ depending on whether the cost matrix is asymmetric or symmetric. In both cases, the vertex set V' is composed of the depot 0 and the customers $i \in U$. In what follows, the focus is on the symmetric version of the problem consisting of delivering demands from the depot to the customers.

Let K be the set of homogeneous vehicles; the number of available vehicles should be sufficient to guarantee the route services (i.e., it cannot be less than $\left\lceil \sum_{i \in U} d_i / q \right\rceil$, see equation (5.88)). With each edge $(i, j) \in E'$ is associated a non-negative cost c_{ij}, corresponding to the cost of a least-expensive path between i and j (so that the triangular inequality holds).

The following decision variables are defined: y_{ijk}, $(i, j) \in E'$, $k \in K$, each of which represents the number of times (at most two, the reason for this will be explained in the following) that vehicle k traverses the edge (i, j), and z_{ik}, $i \in V'$, $k \in K$, each of which of binary type, assuming value 1 if vehicle k visits node i, 0 otherwise.

```
1:  procedure k-EXCHANGE (C^(0), C̄)
2:      # C^(0) is the initial Hamiltonian cycle;
3:      # z_STSP^(0) is the cost of C^(0);
4:      h = 0;
5:      exit = FALSE;
6:      while exit = FALSE do
7:          Identify the best feasible solution C^(h+1) that can be obtained through a k-
            exchange;
8:          Select the best feasible solution x^(h+1) ∈ N(x^(h));
9:          if z_STSP^(h+1) ≥ z_STSP^(h) then exit = TRUE;
10:         else
11:             h = h + 1;
12:         end if
13:     end while
14:     # C^(h) is the best Hamiltonian cycle found for the STSP;
15:     C̄ = C^(h);
16:     return C̄;
17: end procedure
```

The resulting model is as follows:

$$\text{Minimize} \quad \sum_{(i,j)\in E'} \sum_{k\in K} c_{ij} y_{ijk} \tag{6.84}$$

subject to

$$\sum_{i\in U} d_i z_{ik} \le q z_{0k}, \ k \in K \tag{6.85}$$

$$\sum_{k\in K} z_{ik} = 1, \ i \in U \tag{6.86}$$

$$\sum_{i\in V':(i,j)\in E'} y_{ijk} + \sum_{i\in V':(j,i)\in E'} y_{jik} = 2z_{jk}, \ j \in V', \ k \in K \tag{6.87}$$

$$\sum_{(i,j)\in E':i\in S,j\notin S} y_{ijk} + \sum_{(j,i)\in E':i\in S,j\notin S} y_{jik} \ge 2z_{hk},$$

$$S \subseteq U, |S| \ge 2, \ h \in S, \ k \in K \tag{6.88}$$

$$y_{ijk} \in \{0,1\}, \ (i,j) \in E', \ i \ne 0, \ k \in K \tag{6.89}$$

$$y_{0jk} \in \{0,1,2\}, \ (0,j) \in E', \ k \in K \tag{6.90}$$

$$z_{ik} \in \{0,1\}, \ i \in V', \ k \in K. \tag{6.91}$$

The objective function (6.84) represents the total cost generated by the vehicle routes. Constraint (6.85) are the vehicle capacity constraints. In addition, these constraints force variable z_{0k} for vehicle $k \in K$ to be equal to one (i.e., the vehicle k exits from the depot) in case of at least one customer served by the same vehicle (i.e., $z_{ik} = 1, i \in U$). Constraints (6.86) mean that each customer $i \in U$ is served by exactly one vehicle (the case of *split delivery* is examined in Problem 6.28). Constraints (6.87) and (6.88) are routing constraints. They ensure, respectively, that the number of edges incident in a vertex and traversed by vehicle $k \in K$ is even, and that there are no subtours. In this respect, connectivity constraints are defined only for $S \subseteq U$ (i.e., not including vertex 0) such that $|S| \ge 2$, and with the aim of forbidding any subtour within customer subsets. Constraints (6.89)–(6.91) define the integrality constraints of the decision variables. In particular, constraints (6.90) allow the generation of direct routes, serving only one customer. In fact, if $y_{0jk} = 2$, for some $(0, j) \in E'$ and $k \in K$, this means that edge $(0, j)$ is traversed by vehicle k twice, a first time for delivering the quantity d_j from the depot to customer j and a second time for returning the empty vehicle to the depot.

The C-NRP formulation (6.84)–(6.91) can be easily modified by adding a constraint related to the maximum number m of vehicles to be used, that is,

$$\sum_{k\in K} z_{0k} \le m,$$

or, by considering vehicles of different cost and capacity (see Problem 6.28).

Other typical operational constraints arise when a service time s_i is associated with each customer $i \in U$, and the total duration of each route, including service and travel times, must not exceed a given work shift duration. The extension of model (6.84)–(6.91) to the case of duration constraints (namely, CL-NRP) is left to the reader as an exercise, see Problem 6.29.

In the following, an alternative formulation of the CL-NRP can be obtained as follows. Let K be the set of routes in G' satisfying the capacity and length constraints and let c_k, $k \in K$, be the cost of route k. Define a_{ik}, $i \in U$, $k \in K$, as a binary constant equal to 1 if customer i is included into route k, and to 0 otherwise. Let y_k, $k \in K$, be a binary decision variable equal to 1 if route k is activated in an optimal solution, and to 0 otherwise. The CL-NRP can be reformulated as a set partitioning (SP-NRP) problem in the following way:

$$\text{Minimize} \sum_{k \in K} c_k y_k$$

subject to

$$\sum_{k \in K} a_{ik} y_k = 1 , i \in U \tag{6.92}$$

$$y_k \in \{0, 1\}, \ k \in K.$$

Constraints (6.92) establish that each customer $i \in U$ must be served. This formulation is very flexible as it can be easily modified in order to include additional operational constraints. For example, the following constraint

$$\sum_{k \in K} y_k \leq m,$$

limits to m the number of routes that can be activated.

The main weakness of the SP-NRP formulation is the large number of decision variables. However, as already observed (see Section 3.15.2), in some applications the characteristics of the operational constraints can considerably reduce $|K|$.

It is easy to show (see Problem 6.30) that constraints (6.92) can be replaced with the following relations:

$$\sum_{k \in K} a_{ik} y_k \geq 1, \ i \in U,$$

in which case a set covering formulation (SC-NRP) is obtained, which is easier to solve with respect to the corresponding SP-NRP one.

[Bengalur.xlsx, Bengalur.py] Bengalur Oil manufactures and distributes fuel to filling stations in the Karnataka region of India. On 2 July the firm received five orders (see Table 6.18). The distances between the gas stations and the firm's depot are reported in Table 6.19. The vehicles have a capacity of 160 hectolitres.

In order to formulate the problem as an SC-NRP, the feasible routes are enumerated (see Table 6.20; note that, since the problem is symmetric, the routes in which the customers appear in reverse order with respect to those reported in the table are not considered).

The SC-NRP model is

$$\text{Minimize } 180y_1 + 200y_2 + 180y_3 + 160y_4 + 160y_5 + 200y_6 + 200y_7$$
$$+ 180y_8 + 200y_9 + 200y_{10} + 200y_{11} + 220y_{12} + 180y_{13}$$
$$+ 200y_{14} + 180y_{15} + 200y_{16} + 200y_{17} + 200y_{18}$$

subject to

$$y_1 + y_6 + y_7 + y_8 + y_9 + y_{16} + y_{17} + y_{18} \geq 1$$

Table 6.18 Orders (in hectolitres) received by Bengalur Oil on 2 July.

Gas station	Order
1	50
2	75
3	50
4	50
5	75

Table 6.19 Distance (in km) between the gas stations and the firm's depot in the Bengalur Oil problem (depot corresponds to vertex 0).

	Gas station					
	0	1	2	3	4	5
0	0	90	100	90	80	80
1		0	10	20	10	30
2			0	10	20	40
3				0	10	30
4					0	20
5						0

Table 6.20 Feasible routes in the Bengalur Oil problem.

Id	Route	Cost [km]	Id	Route	Cost [km]
1	(0, 1, 0)	180	10	(0, 2, 3, 0)	200
2	(0, 2, 0)	200	11	(0, 2, 4, 0)	200
3	(0, 3, 0)	180	12	(0, 2, 5, 0)	220
4	(0, 4, 0)	160	13	(0, 3, 4, 0)	180
5	(0, 5, 0)	160	14	(0, 3, 5, 0)	200
6	(0, 1, 2, 0)	200	15	(0, 4, 5, 0)	180
7	(0, 1, 3, 0)	200	16	(0, 1, 3, 4, 0)	200
8	(0, 1, 4, 0)	180	17	(0, 1, 4, 3, 0)	200
9	(0, 1, 5, 0)	200	18	(0, 3, 1, 4, 0)	200

$$y_2 + y_6 + y_{10} + y_{11} + y_{12} \geq 1$$

$$y_3 + y_7 + y_{10} + y_{13} + y_{14} + y_{16} + y_{17} + y_{18} \geq 1$$

$$y_4 + y_8 + y_{11} + y_{13} + y_{15} + y_{16} + y_{17} + y_{18} \geq 1$$

$$y_5 + y_9 + y_{12} + y_{14} + y_{15} \geq 1$$

$$y_1, y_2, y_3, y_4, y_5, y_6, y_7, y_8, y_9,$$

$$y_{10}, y_{11}, y_{12}, y_{13}, y_{14}, y_{15}, y_{16}, y_{17}, y_{18} \in \{0, 1\}.$$

A feasible solution to the problem can be determined by applying the CHVATAL procedure illustrated in Section 3.12. Considering that $c_k > 0$, $k = 1, \ldots, 18$, and that the decision variables y_k, $k = 1, \ldots, 18$, appear in all the constraints with unit coefficient, the following ratios are calculated:

$$c_1/n_1 = 180/1 = 180; \qquad c_2/n_2 = 200/1 = 200;$$
$$c_3/n_3 = 180/1 = 180; \qquad c_4/n_4 = 160/1 = 160;$$
$$c_5/n_5 = 160/1 = 160; \qquad c_6/n_6 = 200/2 = 100;$$
$$c_7/n_7 = 200/2 = 100; \qquad c_8/n_8 = 180/2 = 90;$$
$$c_9/n_9 = 200/2 = 100; \qquad c_{10}/n_{10} = 200/2 = 100;$$
$$c_{11}/n_{11} = 200/2 = 100; \qquad c_{12}/n_{12} = 220/2 = 110;$$
$$c_{13}/n_{13} = 180/2 = 90; \qquad c_{14}/n_{14} = 200/2 = 100;$$
$$c_{15}/n_{15} = 180/2 = 90; \qquad c_{16}/n_{16} = 200/3 = 66.66;$$
$$c_{17}/n_{17} = 200/3 = 66.66; \qquad c_{18}/n_{18} = 200/3 = 66.66.$$

Recall that n_k, $k = 1, \ldots, 18$, represents the number of constraints in which the corresponding decision variable y_k appears with a unit coefficient. The ratios with the lowest values are obtained for $k = 16, 17$ and 18; $k = 16$ is chosen arbitrarily, from which $\bar{y}_{16} = 1$ is set, and constraints 1, 3, and 4 are removed from the original model, yielding the following problem:

Minimize $180y_1 + 200y_2 + 180y_3 + 160y_4 + 160y_5 + 200y_6 + 200y_7$

$\qquad +180y_8 + 200y_9 + 200y_{10} + 200y_{11} + 220y_{12} + 180y_{13} + 200y_{14}$

$\qquad +180y_{15} + 200y_{17} + 200y_{18}$

subject to

$$y_2 + y_6 + y_{10} + y_{11} + y_{12} \geq 1$$

$$y_5 + y_9 + y_{12} + y_{14} + y_{15} \geq 1$$

$$y_1, y_2, y_3, y_4, y_5, y_6, y_7, y_8, y_9, y_{10},$$

$$y_{11}, y_{12}, y_{13}, y_{14}, y_{15}, y_{17}, y_{18} \in \{0, 1\}.$$

The coefficients $c_1, c_3, c_4, c_7, c_8, c_{13}, c_{17}$, and c_{18} are greater than zero and the corresponding variables do not appear in any constraint, therefore, $\bar{y}_1 = \bar{y}_3 =$

$\bar{y}_4 = \bar{y}_7 = \bar{y}_8 = \bar{y}_{13} = \bar{y}_{17} = \bar{y}_{18} = 0$. The following ratios are then calculated:

$$c_2/n_2 = 200/1 = 200; \quad c_5/n_5 = 160/1 = 160;$$
$$c_6/n_6 = 200/1 = 200; \quad c_9/n_9 = 200/1 = 200;$$
$$c_{10}/n_{10} = 200/1 = 200; \quad c_{11}/n_{11} = 200/1 = 200;$$
$$c_{12}/n_{12} = 220/2 = 110; \quad c_{14}/n_{14} = 200/1 = 200;$$
$$c_{15}/n_{15} = 180/1 = 180.$$

The ratio with the lowest value is obtained for $k = 12$. For this reason, $\bar{y}_{12} = 1$ is set; hence, constraints 1 and 2 are removed. The problem becomes

$$\text{Minimize } 200y_2 + 160y_5 + 200y_6 + 200y_9 + 200y_{10}$$

$$+ 200y_{11} + 200y_{14} + 180y_{15}$$

subject to

$$y_2, y_5, y_6, y_9, y_{10}, y_{11}, y_{14}, y_{15} \in \{0, 1\}.$$

This means that $\bar{y}_2 = \bar{y}_5 = \bar{y}_6 = \bar{y}_9 = \bar{y}_{10} = \bar{y}_{11} = \bar{y}_{14} = \bar{y}_{15} = 0$ are set. Hence, the feasible solution found provides for the activation of routes 12 and 16, with a total cost of 420 km (see Figure 6.42), which, in this case, by using the `Excel Solver` tool, can be shown to be an optimal solution of the problem.

The set covering formulation can also be used to find a suboptimal solution to the CL-NRP. Let K' be a subset of K of feasible routes, with the condition that each customer $i \in U$ belongs to at least one route $k \in K'$. The following *restricted* formulation of the SC-NRP problem is obtained:

$$\text{Minimize } \sum_{k \in K'} c_k y_k \tag{6.93}$$

subject to

$$\sum_{k \in K'} a_{ik} y_k \geq 1, i \in U \tag{6.94}$$

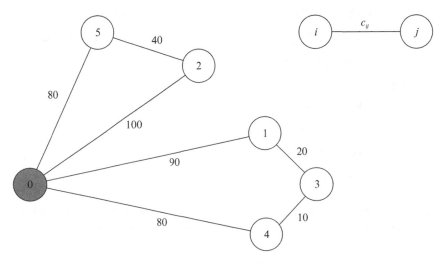

Figure 6.42 Optimal solution of the Bengalur Oil problem.

$$y_k \in \{0, 1\}, k \in K',$$ (6.95)

whose optimal solution y^*, since $K' \subset K$, could provide for some customer $i \in U$ that $\sum_{k \in K'} a_{ik} y_k = p > 1$, in which case it is sufficient to remove customer i from $p - 1$ routes where it is present (this operation does not prejudice the feasibility of the solution). The solution y^* obtained in this way is just a feasible solution of CL-NRP, and its quality clearly depends on the choice of the subset K' of feasible routes.

[Viola.xlsx, Viola.py] On 14 March the Italian oil mill Viola had to plan, starting from the production plant located in Venosa, the delivery of extra virgin oil drums to its customers situated in 10 towns: Ulmeta, Melfi, Rionero in Vulture, Muro Lucano, Avigliano, Pietragalla, Acerenza, Tolve, Genzano di Lucania, and Palazzo S. Gervasio. A vehicle with a capacity of 35 quintals is used for the deliveries. The amounts to be delivered are reported in Table 6.21, whereas the shortest distances between each pair of towns involved are reported in [Viola.xlsx] file.

To find a feasible solution to the routing problem, a subset of 16 feasible routes is generated (reported in Table 6.22). Assigning a binary decision variable y_k, $k = 1, \dots, 16$, to each of the feasible routes reported in Table 6.22, the corresponding restricted formulation (6.93)–(6.95) of the SC-NRP problem is the following:

Minimize $94y_1 + 53y_2 + 147y_3 + 121y_4 + 117y_5 + 108y_6 + 67y_7 + 58y_8$

$$+103y_9 + 167y_{10} + 137y_{11} + 88y_{12} + 118y_{13} + 161y_{14} + 138y_{15}$$

$$+158y_{16}$$

subject to

$$y_1 + y_8 + y_9 + y_{12} + y_{15} \geq 1$$

$$y_1 + y_2 + y_9 + y_{11} + y_{16} \geq 1$$

$$y_2 + y_9 + y_{11} \geq 1$$

$$y_3 + y_{10} + y_{11} \geq 1$$

$$y_3 + y_4 + y_{10} + y_{14} + y_{16} \geq 1$$

$$y_4 + y_{10} + y_{14} \geq 1$$

$$y_5 + y_{13} + y_{14} + y_{15} \geq 1$$

$$y_5 + y_6 + y_{13} + y_{14} + y_{15} + y_{16} \geq 1$$

$$y_6 + y_7 + y_{12} + y_{13} + y_{15} + y_{16} \geq 1$$

$$y_7 + y_8 + y_{12} + y_{13} + y_{15} + y_{16} \geq 1$$

$$y_1, y_2, y_3, y_4, y_5, y_6, y_7, y_8, y_9,$$

$$y_{10}, y_{11}, y_{12}, y_{13}, y_{14}, y_{15}, y_{16} \in \{0, 1\},$$

Table 6.21 Order quantity received by the Viola oil mill from its customers on 14 March.

Vertex	Town	Demand [quintals]
1	Ulmeta	9
2	Melfi	8
3	Rionero in Vulture	9
4	Muro Lucano	10
5	Avigliano	9
6	Pietragalla	12
7	Acerenza	8
8	Tolve	6
9	Genzano di Lucania	7
10	Palazzo San Gervasio	5

Table 6.22 Feasible routes in the Viola problem.

Id	Route	Cost [km]	Id	Route	Cost [km]
1	(0, 1, 2, 0)	94	9	(0, 1, 2, 3, 0)	103
2	(0, 2, 3, 0)	53	10	(0, 4, 5, 6, 0)	167
3	(0, 4, 5, 0)	142	11	(0, 2, 3, 4, 0)	137
4	(0, 5, 6, 0)	121	12	(0, 1, 9, 10, 0)	88
5	(0, 7, 8, 0)	117	13	(0, 7, 8, 9, 10, 0)	118
6	(0, 8, 9, 0)	108	14	(0, 5, 6, 8, 7, 0)	161
7	(0, 9, 10, 0)	67	15	(0, 7, 8, 9, 10, 1, 0)	138
8	(0, 1, 10, 0)	58	16	(0, 2, 5, 8, 9, 10, 0)	158

whose optimal solution (see Figure 6.43), found by using the Excel Solver tool, is the following: $y_2^* = 1$, $y_{10}^* = 1$, $y_{15}^* = 1$ (the value of the remaining decision variables is zero), with $z(y^*) = 358$ km. Note that with this solution (which, as noted above, may not be optimal for the routing problem considered), all the constraints of the model are satisfied for equality (i.e., each customer is served by one activated route).

Constructive Heuristics

In the following, some constructive heuristics procedures for the CL-NRP are illustrated.

CLUSTER_FIRST_ROUTE_SECOND *procedures*

CLUSTER_FIRST_ROUTE_SECOND procedures attempt to determine a good CL-NRP solution in two steps. First, customers are partitioned into subsets $k = 1, ..., m$, each of which is associated with a vehicle k. Second, for each vehicle $k = 1, ..., m$, the STSP on the complete subgraph induced by $U_k \cup \{0\}$ is solved (exactly or heuristically). The

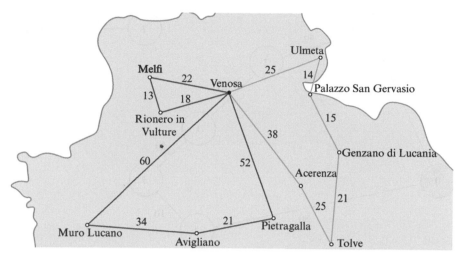

Figure 6.43 Feasible solution of the Viola problem. For each edge the corresponding cost (in km) is reported.

partitioning of the customer set can be made visually or by more formal procedures (as the one proposed by M. L. Fisher and R. Jaikumar). Readers interested in a deeper examination of this subject are referred to the relevant scientific literature.

[Bengalur.xlsx, Bengalur2.py] In the Bengalur Oil problem, customers can be partitioned into two clusters:

$$U_1 = \{2, 5\},$$

$$U_2 = \{1, 3, 4\}.$$

Then, two STSPs are solved on the complete subgraphs induced by $U_1 \cup \{0\}$ and $U_2 \cup \{0\}$, respectively. The result of the CLUSTER_FIRST_ROUTE_SECOND procedure is already shown in Figure 6.42 and corresponds to the optimal solution.

ROUTE_FIRST_CLUSTER_SECOND procedures

ROUTE_FIRST_CLUSTER_SECOND procedures attempt to determine an CL-NRP solution in two stages. A single Hamiltonian cycle (generally infeasible for the CL-NRP) is first generated, eliminating all the operational constraints, through an exact or heuristic STSP algorithm. Then, the cycle is decomposed into m feasible routes, originating and terminating at the depot. The route decomposition can be performed visually or by means of formalized procedures, like the one proposed by J. E. Beasley. Readers interested in this method should consult the relevant scientific literature.

[Bengalur.xlsx, Bengalur3.py] Applying a ROUTE_FIRST_CLUSTER _SECOND procedure to the Bengalur Oil problem, a Hamiltonian cycle, having a cost equal to 240 km, is generated (see Figure 6.44). In the second stage, the cycle

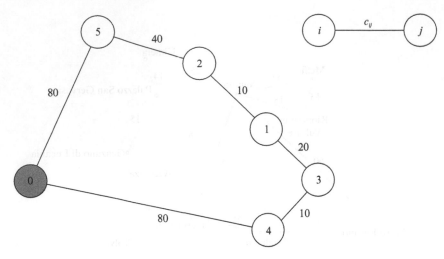

Figure 6.44 Infeasible Hamiltonian cycle generated at the end of the first phase of the ROUTE_FIRST_CLUSTER_SECOND procedure in the Bengalur Oil problem.

is decomposed into two feasible routes, which are the same as those illustrated in Figure 6.42, corresponding in this case to the optimal solution.

Savings heuristic

The savings heuristic is an iterative procedure that initially generates $|U|$ distinct routes each of which serves a single customer. At each subsequent iteration, the heuristic attempts to merge a pair of routes in order to obtain a cost reduction (a *saving*). The cost saving s_{ij} achieved when servicing customers i and j, $i, j \in U$ on one route, as opposed to servicing them individually is (see Figure 6.45):

$$s_{ij} = c_{0i} + c_{0j} - c_{ij}, \ i, j \in U, \ i \neq j.$$

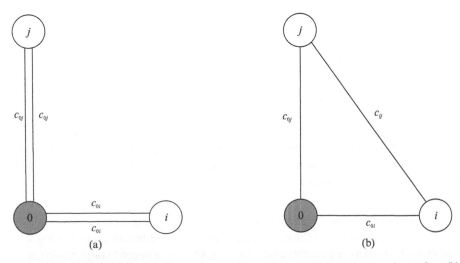

Figure 6.45 Computation of saving s_{ij}: (a) the cost of the two distinct routes is $2c_{0i} + 2c_{0j}$; (b) the cost of the merged route is $c_{0i} + c_{0j} + c_{ij}$.

```
1: procedure SAVINGS (C)
2:      # C is the set of |U| initial routes Ci = (0, i, 0),  i ∈ U;
3:      # L is the list of savings;
4:      L = ∅;
5:      for i ∈ U do
6:          for j ∈ U, i < j do
7:              # Compute the saving sij;
8:              sij = c0i + c0j − cij;
9:              # Update the list L;
10:             L = L ∪ {sij};
11:         end for
12:     end for
13:     Sort L in a non-increasing order of the savings;
14:     while L ≠ ∅ do
15:         Extract a saving sij from the top of L;
16:         if (i and j belong to two separate routes of C in which i and j are
17:            directly linked to the depot, as shown, i.e., in Figure 6.46) and
18:            (the route obtained by replacing edges (0, i) and (0, j) with
19:            edge (i, j) is feasible) then
20:             Merge the two routes;
21:             Update C;
22.         end if
23:     end while
24:     return C;
25: end procedure
```

It is worth noting that s_{ij}, $i, j \in U$, are non-negative since the triangle inequality holds for all costs c_{ij}, $i, j \in E'$. The savings formula still holds if $i \in U$ is the last customer of the first route involved in a merge, and $j \in U$ is the first customer of the second route (see Figure 6.46). The heuristic stops when it is no longer possible to feasibly merge a pair of routes. A more formal description of the SAVINGS procedure is given as follows.

The computational complexity of the SAVINGS procedure is determined by the saving sorting phase (code line 13) and is therefore $O(|V'|^2 \log |V'|^2) = O(|V'|^2 \log |V'|)$. In practice, the heuristic is very fast as it takes less than a second on most computers to solve a problem with hundreds of customers. However, the quality of the solutions can be poor. According to extensive computational experiments, the deviation made by the SAVINGS procedure is typically in the $5 - 20\%$ range. The SAVINGS procedure is very flexible since it can easily be modified to take into account additional operational constraints (such as customer time window restrictions). However, with such variations, solution quality in further reduced. For this reason, tailored procedures are usually employed when dealing with constraints different from capacity and length constraints.

It is worth observing that the SAVINGS procedure can also be applied in the case of directed graphs, assuming that the saving s_{ij} is defined for each $i, j \in U, i \neq j$, and is $s_{ij} = c_{i0} + c_{0j} - c_{ij}$.

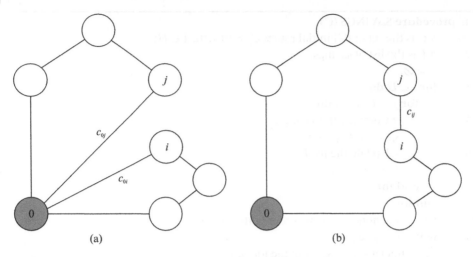

Figure 6.46 Merging (a) two routes in (b) a single route (the saving $s_{ij} = c_{0i} + c_{0j} - c_{ij}$).

[Bengalur.xlsx, Bengalur4.py] By applying the SAVINGS procedure to the Bengalur Oil problem, five individual routes are initially generated:

$C_1 = (0, 1, 0)$;

$C_2 = (0, 2, 0)$;

$C_3 = (0, 3, 0)$;

$C_4 = (0, 4, 0)$;

$C_5 = (0, 5, 0)$.

Then savings s_{ij}, $i, j \in U$, $i \neq j$ are calculated (see Table 6.23) and list L is sorted:

$$L = \{s_{12}, s_{23}, s_{13}, s_{14}, s_{24}, s_{34}, s_{15}, s_{25}, s_{35}, s_{45}\}.$$

Subsequently, routes C_1 and C_2 are merged while savings $s_{23}, s_{13}, s_{14}, s_{24}$ are discarded. Then, saving s_{34} is implemented by merging routes C_3 and C_4. At this stage there are no further feasible route merges. The final solution has a cost of 540 km (see Figure 6.47).

Table 6.23 Cost savings determined for the Bengalur Oil problem.

	1	2	3	4	5
1	–	180	160	160	140
2		–	180	160	140
3			–	160	140
4				–	140

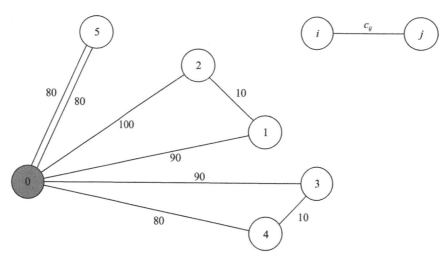

Figure 6.47 Solution provided by the SAVINGS procedure for the Bengalur Oil problem.

6.12.3 The Node Routing and Scheduling Problem with Time Windows

In several settings, customers need to be serviced within specified time windows. This is the case, for example, for retail outlets that cannot be replenished during busy periods. In the simplest version of the *node routing and scheduling problem with time windows* (TW-NRSP), each customer specifies a single time window, while in other variants each customer can set multiple time windows (for example, a time window in the morning and one in the afternoon). Let e_i, $i \in U$, be the *earliest time* at which service can start at customer i, and let l_i, $i \in U$, be the *latest time* (or *deadline*) at which service must start at customer i. Similarly, let e_0 be the earliest time at which vehicles can leave the depot 0, and let l_0 be the deadline within which vehicles must return to the depot.

It is worth noting that even in the case of symmetric travel costs and times, a TW-NRSP solution is made up of a set of directed cycles, because of the time windows that do not allow reversals of route orientations. For this reason, the TW-NRSP is formulated on a complete directed graph $G' = (V', A')$, where V' is composed of the depot 0 and the set U of customers. A travel cost c_{ij} (generally proportional to distance) and a travel time t_{ij} along the quickest path from i to j are associated with each arc $(i, j) \in A'$ (in the symmetric case, $c_{ij} = c_{ji}$ and $t_{ij} = t_{ji}$, for each $i, j \in V', i \neq j$). A service time s_i is associated with each customer $i \in U$. Let K be the set of available vehicles. The decision variables are the following: x_{ijk}, $(i, j) \in A'$, $k \in K$, of binary type, having value equal to 1 if the arc (i, j) is traversed by vehicle k, 0 otherwise; a_{ik}, $i \in U$, $k \in K$, representing the arrival time of vehicle k at customer i; t_k^s, t_k^f, $k \in K$, represent, respectively, the starting and the arrival time of vehicle k at depot 0. The TW-NRSP can be formulated as follows:

$$\text{Minimize} \sum_{(i,j) \in A'} \sum_{k \in K} c_{ij} x_{ijk} \tag{6.96}$$

subject to

$$\sum_{k \in K} \sum_{i \in V' \setminus \{j\}} x_{ijk} = 1, \ j \in U \tag{6.97}$$

$$\sum_{j \in U} x_{0jk} \leq 1, \ k \in K \tag{6.98}$$

$$\sum_{i \in V' \setminus \{h\}} x_{ihk} - \sum_{j \in V' \setminus \{h\}} x_{hjk} = 0, \ h \in U, k \in K \tag{6.99}$$

$$x_{ijk}(a_{ik} + s_i + t_{ij} - a_{jk}) \leq 0,$$
$$(i,j) \in A' : i, j \neq 0, k \in K \tag{6.100}$$

$$x_{i0k}(a_{ik} + s_i + t_{i0} - t_k^f) \leq 0, \ i \in U, k \in K \tag{6.101}$$

$$x_{0jk}(t_k^s + t_{0j} - a_{jk}) \leq 0, \ j \in U, k \in K \tag{6.102}$$

$$a_{ik} \geq e_i, \ i \in U, \ k \in K \tag{6.103}$$

$$a_{ik} \leq l_i, \ i \in U, \ k \in K \tag{6.104}$$

$$t_k^s \geq e_0, k \in K \tag{6.105}$$

$$0 \leq t_k^f \leq l_0, k \in K \tag{6.106}$$

$$x_{ijk} \in \{0,1\}, \ (i,j) \in A', \ k \in K \tag{6.107}$$

The objective function (6.96) represents the cost of the service routes. Constraints (6.97) mean that each customer must be served by exactly one vehicle. Constraints (6.98) state that at most $|K|$ routes originate from the depot. Equations (6.99) are the vehicle flow balance constraints, one for each customer. Each constraint (6.100) states the following condition: if arc (i, j) is traversed by vehicle k, then customers i and j are served by the same vehicle k, hence, $a_{ik} + s_i + t_{ij} \leq a_{jk}$, that is, the arrival time of vehicle k at customer i, plus the service time of customer i and the travel time from i to j cannot be greater than the arrival time of the same vehicle at customer j. Constraints (6.101) and (6.102) are similar to (6.100) and refer to the incoming and outgoing arcs of the depot, respectively. Constraints (6.103)–(6.104) force the beginning of service for each customer to lie within the predefined time windows. Similarly, (6.105) and (6.106) are the time window constraints of the depot. The presence of constraints (6.100) avoids the need to explicitly introduce constraints in the model (6.96)–(6.107) that prohibits the presence of subtours not including the depot. In effect, the case of a feasible solution containing a subtour $\overline{C}$ served by a vehicle $\overline{k}$ and not including the depot would mean, from constraints (6.100), that

$$\sum_{(i,j) \in A'_{\overline{C}}} t_{ij} + \sum_{i \in U_{\overline{C}}} s_i \leq 0,$$

which is an evident contradiction (note that $A'_{\overline{C}}$ and $U_{\overline{C}}$ indicate, respectively, the set of arcs and vertices in $\overline{C}$).

Observe that constraints (6.100) are non-linear, but they can be linearized in the following form:

$$a_{ik} + s_i + t_{ij} - a_{jk} \leq \overline{M}(1 - x_{ijk}), \ (i,j) \in A' : i, j \neq 0, k \in K,$$

where $\overline{M}$ is an arbitrarily large constant. Similarly, constraints (6.101) and (6.102) can be rewritten as

$$a_{ik} + s_i + t_{i0} - t_k^f \leq \overline{M}(1 - x_{i0k}), \ i \in U, k \in K;$$

$$t_k^s + t_{0j} - a_{jk} \le \overline{M}(1 - x_{0jk}), \; j \in U, k \in K.$$

Additional constraints on vehicle capacity can be introduced in the model (6.96)–(6.107), as follows:

$$\sum_{i \in U} d_i \sum_{j \in V' \setminus \{i\}} x_{ijk} \le q_k, \; k \in K,$$

where, as indicated in Section 6.12.2, d_i, $i \in U$, represents the demand of customer i, whereas q_k states the capacity of vehicle k (equal to q in case of identical vehicles).

It is worth observing that model (6.96)–(6.107) imposes that the service beginning at each customer $i \in U$ must lie within the corresponding time window, but this condition is not necessarily guaranteed for the service completion. In this case, constraints (6.104) should be replaced with

$$a_{ik} + s_i \le l_i, \; i \in U, \; k \in K.$$

Finally, the objective function (6.96) can be changed to pursue different goals of practical interest, in particular, the sum (to be minimized) of the arrival times of vehicles at depot 0, i.e., $\sum_{k \in K} t_k^f$, or alternatively, the minimization of the number of vehicles that leave the depot for completing the service, i.e., $\sum_{k \in K} \sum_{j \in U} x_{0jk}$. It is worth observing that, similar to what is illustrated in Section 3.12, these two goals can be combined into a unique objective function, as follows:

$$\sum_{k \in K} \sum_{j \in U} M x_{0jk} + \sum_{k \in K} t_k^f, \tag{6.108}$$

where M is an arbitrarily large positive constant chosen so that the number of used vehicles is as small as possible.

The TW-NRSP is a very hard to solve problem and several heuristic methods for finding good-quality feasible solutions have been proposed over the years, many of them as extensions of the constructive heuristics for the NRP. One of the simplest approaches is, for example, a modified version of the SAVINGS procedure illustrated in Section 6.12.2, in which, at code line 16, the feasibility check of the two merging routes requires that both the vehicle capacity and the time window constraints of the customers and of the depot are verified (see Problem 6.32). The main drawback of this procedure is that the waiting time to start the service to some customers may be very long in the feasible solution obtained. This has a major impact on the minimization of the arrival times at the depot, as well as on the number of vehicles used, since a vehicle could visit other customers instead of waiting for starting a service. In order to mitigate this effect, specific additional feasibility checks for merging two routes can be incorporated in the construction phase, in particular, $w_{jk} \le w_j^{\max}$, where $w_j^{\max}$ is the maximum waiting time allowed for customer $j \in U$ and w_{jk} is the waiting time for starting the service by vehicle $k \in K$ at customer j in case of a route merge.

An insertion heuristic, called the SOLOMON procedure, inspired by a number of heuristics developed by M. Solomon, is introduced in the following. Similar to the NEAREST_NEIGHBOUR procedure illustrated in Section 6.12.1, it builds a set of routes, starting from the depot, by linking at each iteration the vertex inserted at the previous iteration to its "nearest" unrouted neighbour. At each iteration, the evaluation of the possible customer to be inserted in the current route is performed considering the feasibility check in terms of vehicle capacity and time window constraints of the customers and of the depot. A new route is opened when the search for the next customer to be inserted fails, while the procedure ends when there are no more customers to be scheduled. The novelty of this heuristic is the measure of *fitness* γ_{ij} between two vertices i and $j \in V', i \neq j$, defined by an appropriate metric that takes into account the particular structure of the TW-NRSP, balancing different criteria, that is,

$$\gamma_{ij} = \alpha_1 \, t_{ij} + \alpha_2 \, b_{ij} + \alpha_3 \, u_{ij}, i, j \in V', i \neq j, \tag{6.109}$$

where α_1, α_2 and α_3 are parameters chosen in such a way that $\alpha_1 + \alpha_2 + \alpha_3 = 1$. The term b_{ij} is defined as

$$b_{ij} = \max\{e_j, a_{ik} + s_i + t_{ij}\} - (a_{ik} + s_i),$$

where k is the vehicle used to serve vertex i, b_{ij} corresponds to the time difference between the beginning of service at vertex j and the service completion time at vertex i, i.e., the waiting time before the beginning of the service at vertex j. Note that if $i = 0$, the expression becomes $b_{0j} = \max\{e_j, t_k^s + t_{0j}\} - t_k^s$.

The term u_{ij} is computed as

$$u_{ij} = l_j - (a_{ik} + s_i + t_{ij}),$$

and is a measure of the *urgency* of serving customer j; in fact, if $u_{ij} \geq 0$, this means that it is still possible to serve vertex j after vertex i because the service at vertex j can begin before the latest time l_j of the corresponding time window. Note that if $i = 0$, the expression becomes $u_{0j} = l_j - (t_k^s + t_{0j})$.

The result of inserting another vertex after i in the route served by the vehicle k could be that, at the next iteration, vertex j can no longer be inserted in route k and therefore it will be necessary for it to activate a new route. Hence, the sense of "urgency" to serve vertex j mentioned above.

The measure of fitness defined by equation (6.109) aims to balance the three time components through the appropriate choice of the parameters α_1, α_2 and α_3 and at each iteration of the SOLOMON procedure the appropriate minimum value of this measure is selected.

Unlike the NEAREST_NEIGHBOUR, the SOLOMON procedure avoids the insertion of high-cost arcs only at the end of the route, hence mitigating the time windows violation and reducing, at least in principle, the number of vehicles employed to find a feasible solution.

[TW-NRSP.xlsx, TW-NRSP.py] Consider a complete directed graph $G = (V', A')$, where V' is composed of the depot 0 and five customers to be served, i.e., $V' = \{0, 1, 2, 3, 4, 5\}$. The travel times t_{ij} (in minutes) of the quickest path associated with each arc $(i, j) \in A'$ are reported in Table 6.24. The time windows of the depot and of the customers are indicated in Table 6.25. For the sake of simplicity, the time is computed in minutes, starting at the earliest time $e_0 = 0$ (i.e., 00:00) at which vehicles can leave the depot. The service time of every customer is 15 minutes. Consider the case where the first route ($k = 1$) has been partially

Table 6.24 Travel times t_{ij} (in minutes) between the vertices of the graph (depot corresponds to vertex 0).

	0	1	2	3	4	5
0	0	14	56	27	65	51
1		0	42	30	56	37
2			0	56	22	30
3				0	45	42
4					0	35
5						0

Table 6.25 Time windows for the depot and the five customers.

Vertex (i)	Time window	e_i [minutes]	l_i [minutes]
0	00:00 – 06:30	0	390
1	00:00 – 02:45	0	165
2	00:00 – 03:30	0	210
3	00:30 – 02:15	30	135
4	02:30 – 04:30	150	270
5	01:00 – 03:30	60	210

constructed, with vehicle k leaving the depot at $t_k^s = 0$ and visits the vertex 2 ($a_{2k} = 56$) spending 56 minutes for traversing the arc $(0, 2)$. The vehicle arrives at vertex 2 within its time window. After the service completion, vehicle k leaves customer 2 at $a_{2k} + s_2 = 56 + 15 = 71$ minutes from the beginning of the service at the depot. In order to decide the subsequent vertex to serve after customer 2 by the same vehicle, the fitness measure given by (6.109) is used, with different choices of the parameters α_1, α_2 and α_3.

In the first case, assume that $\alpha_1 = 1$, and $\alpha_2 = \alpha_3 = 0$, i.e., $\gamma_{2j} = t_{2j}, j \in V \backslash \{0, 2\}$. The nearest vertex is found considering the minimum value (in minutes) among $\gamma_{21} = 42$, $\gamma_{23} = 56$, $\gamma_{24} = 22$ and $\gamma_{25} = 30$ (note that all the customers can be inserted in the route after customer 2). Vertex 4 is the candidate customer to be inserted in the route k. This means that vehicle k should travel for 22 minutes from customer 2 and arrive at customers 4 after $a_{2k} + s_2 + t_{24} = 56 + 15 + 22 = 93$ minutes from the beginning of the service at the depot. It is worth noting that this choice implies a waiting time of 57 minutes before starting the service, i.e., $a_{4k} = 150$, in order to respect the earliest time e_4 of the corresponding time window.

In the second case, let $\alpha_1 = \alpha_3 = 0$, and $\alpha_2 = 1$, so that $\gamma_{2j} = b_{2j}, j \in V \backslash \{0, 2\}$. The nearest vertex is determined in this case by seeking the minimum value (in minutes) among $\gamma_{21} = \max\{e_1; a_{2k} + s_2 + t_{21}\} - (a_{2k} + s_2) = \max\{0; 56 + 15 + 42\} - (56 + 15) = 42$; $\gamma_{23} = 56$; $\gamma_{24} = 79$; $\gamma_{25} = 30$. As a consequence, vertex 5 is the candidate customer. Vehicle k should move from vertex 2 to vertex 5, travelling for 30 minutes, and arriving at vertex 5 after 101 minutes from the beginning of the service at the depot, and in this case the service could start immediately ($a_{5k} = 101$).

In the third case, $\alpha_1 = \alpha_2 = 0$, $\alpha_3 = 1$ and, consequently, $\gamma_{2j} = u_{2j}, j \in V' \setminus \{0, 2\}$. The nearest vertex corresponds to the customer 3, since $\gamma_{21} = l_1 - (a_{2k} + s_2 + t_{21}) = 165 - (56 + 15 + 42) = 52$; $\gamma_{23} = 8$; $\gamma_{24} = 177$; $\gamma_{25} = 169$. The minimum value is therefore γ_{23}; with this choice, the vehicle k, after serving customer 2, should travel for 56 minutes, arriving at vertex 3 after 127 minutes from the beginning of the service at the depot. The service of customer 3 could start without delay ($a_{3k} = 127$). It is worth noting that this option corresponds to the last possibility to serve customer 3 with vehicle k, otherwise, if served later, a new vehicle should be used.

In the fourth case, a combination of the different time components is obtained considering, for example, the following choice of the parameters: $\alpha_1 = 0.5$, $\alpha_2 = 0.3$ and $\alpha_3 = 0.2$. The fitness values (in minutes) $\gamma_{2j}, j \in V' \setminus \{0, 2\}$ for the unrouted customers are computed as follows: $\gamma_{21} = 0.5 \times 42 + 0.3 \times 42 + 0.2 \times 52 = 44$; $\gamma_{23} = 46.4$; $\gamma_{24} = 70.1$ and $\gamma_{25} = 57.8$. According to this choice, customer 1 is selected to be inserted in route k after customer 2. The vehicle arrives at vertex 1 after 113 minutes from the beginning of the service at the depot. The vehicle can start the service without delay ($a_{1k} = 113$).

It is worth observing that this choice is different from the previous ones and results in a small (though not minimum) travel time distance from vertex 2 to the subsequent vertex, but eliminating the waiting time for the vehicle when serving the new customer.

A more formal description of the SOLOMON procedure is reported in the following.

[McNish.xlsx, McNish.csv, McNish.py] McNish is a chain of supermarkets located in Scotland. On 13 October, the warehouse situated in Aberdeen was required to serve 12 sales points located in Banchory, Clova, Cornhill, Dufftown, Fyvie, Huntly, Newbyth, Newmill, Peterhead, Strichen, Towie, and Turriff. The number of requested palletized unit loads and the time windows within which service was allowed are reported in Table 6.26, whereas distances and travel times on the fastest routes between the cities are reported in the McNish.xlsx file. Each vehicle has a capacity of 30 palletized unit loads. It leaves the warehouse at 9:00 and must return to the warehouse by 14:00. Service time (time needed for unloading a vehicle) can be assumed to be 15 minutes on average for every sales point, regardless of demand. The problem is a TW-NRSP formulated considering the objective function (6.108), constraints (6.97)–(6.106) and the vehicle capacity constraint.

Table 6.26 Number of required palletized unit loads and time windows for the sales point in the McNish problem.

Vertex	Sales point	Demand [palletized unit loads]	Time window
1	Banchory	9	9:00 – 11:30
2	Clova	7	9:00 – 12:30
3	Cornhill	5	9:00 – 12:30
4	Dufftown	4	11:00 – 14:30
5	Fyvie	8	9:00 – 12:30
6	Huntly	8	11:00 – 13:30
7	Newbyth	7	9:00 – 12:30
8	Newmill	6	10:00 – 12:30
9	Peterhead	6	9:00 – 10:30
10	Strichen	6	9:00 – 11:15
11	Towie	4	10:00 – 12:45
12	Turriff	6	10:00 – 12:30

```
 1: procedure SOLOMON (V', A', R)
 2:     # L is the set of the unrouted customers;
 3:     k = 1;
 4:     L = U;
 5:     ncw_route = TRUE;
 6:     while L ≠ ∅ do
 7:         if new_route = TRUE then
 8:             # Initialization of the route R_k;
 9:             R_k = (0);
10:             g = 0;
11:         end if
12:         Identify the feasible vertex h ∈ U \ L such that, according to (6.109),
               γ_gh = min_{j∈U\L} {γ_gj};
13:         if h exists then
14:             Add h at the end of R_k;
15:             g = h;
16:             L = L \ {h}
17:             new_route = FALSE;
18:         else
19:             # Unrouted customers are still present and a new route should be
               initialized;
20:             new_route = TRUE;
21:             k = k + 1;
22:         end if
23:     end while
24:     return R;
25: end procedure
```

The McNish problem is solved by using the SOLOMON procedure (with the choice of parameters $\alpha_1 = 0.5$, $\alpha_2 = 0.3$, $\alpha_3 = 0.2$). Initially, the first route contains only the depot, $R_1 = (0)$.

At the first iteration, for each unrouted customer the values of $\gamma_{0j}^{(1)}$, for each $j \in U$, shown in the second column of Table 6.27 are obtained, from which $R_1 = (0, 1, 0)$. At the second iteration, the values shown in the third column of Table 6.27 impose the selection of vertex 8, and the route becomes $R_1 = (0, 1, 8, 0)$. At the third iteration, the values shown in the fourth column of Table 6.27 are obtained, and the route becomes $R_1 = (0, 1, 8, 5, 0)$. At the next iteration, the values reported in the last column of Table 6.27 imply the selection of vertex 12 and the route $R_1 = (0, 1, 8, 5, 12, 0)$ is completed because of the vehicle capacity constraint. The route R_1 is 155.5 km long and has a completion time equal to $t_1^f = 249$ minutes. This means that the vehicle completes its service arriving at the depot at 13:09.

At the beginning of the fifth iteration, a new route, $R_2 = (0)$, is initialized. After another four iterations (the fitness values are reported in Table 6.28), $R_2 = (0, 9, 10, 7, 3, 0)$ is obtained, with a length of 212.19 km and a completion time equal to $t_2^f = 296$ minutes (i.e., the arrival time of the second vehicle at the depot is at 13:56).

Table 6.27 Fitness values $\gamma_{ij}^{(h)}$ used at iterations $h = 1, 2, 3, 4$ for building the route R_1 considering the unrouted customers in the McNish problem. The infeasible connections are indicated in the table.

j	$\gamma_{0j}^{(1)}$	$\gamma_{1j}^{(2)}$	$\gamma_{8j}^{(3)}$	$\gamma_{5j}^{(4)}$
1	50.4	–	–	–
2	86.4	76.0	57.0	Infeasible
3	95.4	89.8	61.2	37.0
4	152.4	123.1	100.5	74.8
5	72.6	84.7	37.2	–
6	109.5	87.4	58.8	Infeasible
7	84.0	86.2	52.8	28.0
8	71.1	58.6	–	–
9	52.8	Infeasible	Infeasible	Infeasible
10	64.2	Infeasible	Infeasible	Infeasible
11	85.2	95.1	61.2	Infeasible
12	82.2	79.6	46.8	21.4

Table 6.28 Fitness values $\gamma_{ij}^{(h)}$ used at iterations $h = 5, 6, 7, 8$ for building the route R_2 considering the remaining unrouted customers in the McNish problem. The infeasible connections are indicated in the table.

j	$\gamma_{0j}^{(5)}$	$\gamma_{9j}^{(6)}$	$\gamma_{10j}^{(7)}$	$\gamma_{7j}^{(8)}$
2	86.4	97.6	Infeasible	Infeasible
3	95.4	75.4	49.6	34.0
4	152.4	119.2	97.0	79.6
6	109.5	93.4	71.2	53.8
7	84.0	56.8	31.0	–
9	52.8	–	–	–
10	64.2	30.4	–	–
11	85.2	98.2	82.6	Infeasible

Table 6.29 Fitness values $\gamma_{ij}^{(h)}$ used at iterations $h = 9, 10, 11, 12$ for building the route R_3 considering the remaining unrouted customers in the McNish problem. The infeasible connections are indicated in the table.

j	$\gamma_{0j}^{(9)}$	$\gamma_{11j}^{(10)}$	$\gamma_{2j}^{(11)}$	$\gamma_{6j}^{(12)}$
2	86.4	37.0	–	–
4	152.4	77.8	88.2	49.2
6	109.5	59.8	49.4	–
11	85.2	–	–	–

Finally, the last route will be $R_3 = (0, 11, 2, 6, 4, 0)$ (see Table 6.29 for the fitness values) with a cost of 204.8 km and a completion time equal to $t_3^f = 290$ minutes (i.e., the third vehicle arrives at the depot at 13:50). Consequently, the final feasible solution is obtained by using three vehicles, which cover 573.2 km and has a total travel time equal to $249 + 296 + 290 = 835$ minutes, and corresponds to the scheduling shown in Tables 6.30–6.32. Consequently, the objective function of the problem is equal to $3M + 835$.

The McNish problem was also solved to optimality by using McNish.py (see McNish.csv file for more details). The optimal solution leads to a cost of $3M + 802$, that corresponds to a total travel time of 802 minutes obtained using three vehicles. In the optimal solution, the routes for serving the customers are:

$R_1^* = (0, 5, 12, 3, 6, 0)$; $R_2^* = (0, 9, 10, 7, 8, 0)$ and $R_3^* = (0, 1, 11, 2, 4, 0)$, corresponding to a total distance of 558.3 km. For the McNish problem, the SOLOMON procedure proves to be particularly efficient. Indeed the percentage deviation from the minimum total travel time is equal to 4.11% (i.e., $(835 - 802)/802 = 0.0411$) and the number of vehicles used corresponds in this case to the optimal one.

Table 6.30 Schedule of the first route in the McNish problem.

City	Arrival [hh:mm]	Departure [hh:mm]	Cumulated load [Palletized unit loads]
Aberdeen	–	9:00	0
Banchory	9:34	9:49	9
Newmill	10:33	10:48	15
Fyvie	11:16	11:31	23
Turriff	11:47	12:02	29
Aberdeen	13:09	–	

Table 6.31 Schedule of the second route in the McNish problem.

City	Arrival [hh:mm]	Departure [hh:mm]	Cumulated load [Palletized unit loads]
Aberdeen	–	9:00	0
Peterhead	9:58	10:13	6
Strichen	10:43	10:58	12
Newbyth	11:19	11:34	19
Cornhill	12:12	12:27	24
Aberdeen	13:56	–	

Table 6.32 Schedule of the third route in the McNish problem.

City	Arrival [hh:mm]	Departure [hh:mm]	Cumulated load [Palletized unit loads]
Aberdeen	–	9:00	0
Towie	10:07	10:22	4
Clova	10:41	10:56	11
Huntly	11:27	11:42	19
Dufftown	12:08	12:23	23
Aberdeen	13:50	–	

Tabu Search Heuristic

The field of heuristics has been transformed by the development of metaheuristics. One representative method of this class of heuristics is *tabu search* (TS). This is essentially a local search heuristic that generates a sequence of solutions in the hope of generating better local optima. TS differs from classical methods in that the successive solutions it examines do not necessarily improve upon each other. A key concept at the heart of TS is that of neighbourhood. The neighbourhood $N(s)$ of a solution s is the set of all solutions that can be reached from s by performing a simple operation. For example, in the context of the NRP, two common neighbourhood structures are obtained by moving a customer from its current route to another route or by swapping two customers between two different routes. The standard TS mechanism is to move from s to the best neighbour in $N(s)$ (which is not necessarily better than s). This way of proceeding may, however, induce cycling. For example, s' may be the best neighbour of s which, in turn, is the best neighbour of s'. To avoid cycling the search process is prevented from returning to solutions processing some attributes of solutions recently considered. Such solutions are declared "tabu" for a number of iterations. For example, if a customer v is moved from route r to route r' at iteration t, then moving v back to route r will be declared tabu until iteration $t + \theta$, where θ is called the length of the tabu tenure (typically θ is chosen between 5 and 10). When the tabu tenure has expired, v may be moved back to route r at which time the risk of cycling is most likely to have disappeared because of changes that have occurred elsewhere in the solution.

Not only is it possible to accept deteriorating solutions in TS, but also it may be interesting to consider infeasible solutions. For example, in the sequence of solutions s, s', s'', both s and s'' may be feasible while s' is infeasible. If s'' cannot be reached directly from s, but only from s', and if it improves upon s, then it pays to go through the infeasible solution s'. This can occur if, for example, s' contains a route r that violates vehicle capacity due to the inclusion of a new customer in that route. Feasibility may be restored at the next iteration if a customer is removed from route s'. A practical way of handling infeasible solutions in TS is to work with a penalized objective function. If $f(s)$ is the actual cost of solution s, then the penalized objective is defined as

$$f'(s) = f(s) + \alpha Q(s) + \beta D(s) + \gamma W(s),$$

where $Q(s)$, $D(s)$ and $W(s)$ are the total violations of the vehicle capacity constraints, route duration constraints and time window constraints, respectively. Other types of constraints can of course be handled in the same way. The parameters α, β, and γ are positive weights associated with constraint violations. These parameters are initially set equal to 1 and self-adjust during the course of the search to produce a mix of feasible and infeasible solutions. For example, if at a given iteration s is feasible with respect to the vehicle capacity constraint, then dividing α by a factor $(1 + \delta)$ (where $\delta > 0$) will increase the likelihood of generating an infeasible solution at the next iteration. Conversely, if s is infeasible, multiplying α by $(1 + \delta)$ will help the search move to a feasible solution. A good choice of δ is typically 0.5. The same principle applies to β and γ.

TS repeatedly performs these operations starting from an initial solution which may be infeasible. It stops after a preset number of iterations. As is common in TS, this number can be very large (i.e., several thousands).

The application of TS to NRPs with various kinds of constraints has proved the efficiency and robustness of this heuristic over the years.

6.12.4 Arc Routing Problems

As previously defined, an arc routing problem consists of designing a least-cost set of vehicle routes in a mixed graph $G = (V, A, E)$, such that each arc and edge in a subset $R \subseteq A \cup E$ should be visited. A required arc is a one-way street containing customers or a side of a two-way street that cannot be serviced simultaneously with the customers on the other side (a typical example is the solid waste collection along high-speed two-way streets). A required edge is a two-way street for which service can take place driving along the street only once in either direction (a typical example is the solid waste collection along streets with light traffic). Unlike NRPs (that are formulated and solved on an auxiliary complete graph G'), ARPs are generally modelled directly on G.

In this subsection, unconstrained ARPs, namely the CPP and the RPP, are examined. Constrained ARPs can be approached using the algorithmic ideas employed for the constrained NRPs along with the solution procedures for the CPP and the RPP. For example, the ARP with vehicle capacity and length constraints, whose applications arise in garbage collection and mail delivery, can be solved using a "cluster first, route second" heuristic: in a first stage, required arcs and edges are divided into clusters, each of which is assigned to a vehicle, while in the second stage an ARP is solved for each cluster. In this book, constrained ARPs are not tackled for the sake of brevity.

The Chinese Postman Problem

The CPP (formulated first by the Chinese mathematician M. K. Kwan) is to determine a minimum-cost route traversing all arcs and all edges of a graph at least once. Its main applications arise in garbage collection, mail delivery, network maintenance, snow removal, and meter reading in urban areas.

The CPP is related to the problem of determining whether a graph $G = (V, A, E)$ is Eulerian, that is, whether it contains a tour traversing each arc and each edge of the graph exactly once. Obviously, in a Eulerian graph with non-negative arc and edge costs, each Eulerian tour constitutes an optimal CPP solution. In a non-Eulerian graph, an optimal CPP solution must traverse at least one arc or edge twice.

Necessary and sufficient conditions for the existence of a Eulerian tour depend on the type of graph G considered (directed, undirected, or mixed), as stated in the following propositions whose proofs are omitted for brevity.

Property. A directed and strongly connected graph G is Eulerian if and only if it is *symmetric*, that is, for any vertex the number of incoming arcs (*incoming degree*) is equal to the number of outgoing arcs (*outgoing degree*) (*symmetric vertex*).

Property. An undirected and connected graph G is Eulerian if and only if it is *even*, that is, each vertex has an even degree (*even vertex*).

Property. A mixed and strongly connected graph G is Eulerian if and only if:

(a) the total number of arcs and of edges incident to any vertex is even (*even graph*);
(b) for each set S of vertices ($S \subset V$ and $S \neq \emptyset$), the difference between the number of the arcs traversing the cut $(S, V \setminus S)$ in the two directions is less than or equal to the number of edges of the cut (*balanced graph*).

Furthermore, since an even and symmetric graph is balanced, the following proposition holds.

Property. A mixed, strongly connected, even and symmetric graph G is Eulerian.

This condition is sufficient, but not necessary, as illustrated in the example reported in Figure 6.48.

The solution of the CPP can be decomposed in two steps. Defined in the first step is a least-cost set of arcs $A^{(a)}$ and of edges $E^{(a)}$ such that the multi-graph $G^{(a)} = (V, A \cup A^{(a)}, E \cup E^{(a)})$ is Eulerian (if G is itself Eulerian, then $A^{(a)} = \varnothing$ and $E^{(a)} = \varnothing$ and, therefore, $G = G^{(a)}$). In the second step, a Eulerian tour in $G^{(a)}$ is determined.

The first step can be executed in polynomial time if G is a directed or undirected graph, while the complexity grows exponentially if G is mixed. The second step can be performed in $O(|A \cup E|)$ time with the END_PAIRING procedure illustrated in Section 6.12.1.

In the following it is shown how to determine $G^{(a)}$ in case G is directed or undirected.

Directed Chinese postman problem

When G is directed, in an optimal solution the arcs in $A^{(a)}$ form a set of least-cost paths connecting the asymmetric vertices. Therefore $A^{(a)}$ can be obtained by solving a minimum-cost flow problem namely, a transportation problem, on a bipartite directed graph suitably defined. Let V^+ and V^- be the subsets of V whose vertices have a positive and a negative difference between the incoming degree and the outgoing degree, respectively. The bipartite directed graph is $G_T = (V^+ \cup V^-, A_T)$, where $A_T = \{(i, j) : i \in V^+, j \in V^-\}$.

With each arc $(i, j) \in A_T$ is associated a cost w_{ij} equal to that of a least-cost path in G from vertex i to vertex j. Let also $o_i \ (> 0)$, $i \in V^+$, be the supply of the vertex i, equal to the difference between its incoming degree and its outgoing degree. Similarly, let $d_i \ (> 0)$, $i \in V^-$, be the demand of vertex i, equal to the difference between its outgoing degree and its incoming degree. Furthermore, let s_{ij}, $(i, j) \in A_T$, be the decision variable associated with the flow along arc (i, j).

The transportation problem is

$$\text{Minimize} \sum_{(i,j) \in A_T} w_{ij} s_{ij} \tag{6.110}$$

subject to

$$\sum_{j \in V^-} s_{ij} = o_i, \ i \in V^+ \tag{6.111}$$

$$\sum_{i \in V^+} s_{ij} = d_j, \ j \in V^- \tag{6.112}$$

$$s_{ij} \geq 0, \ (i, j) \in A_T. \tag{6.113}$$

Of course, $\sum_{i \in V^+} o_i = \sum_{j \in V^-} d_j$, so that problem (6.110)–(6.113) is feasible. Let s_{ij}^*, $(i, j) \in A_T$, be an optimal (integer) solution of the transportation problem. $A^{(a)}$ is formed by

Figure 6.48 A mixed Eulerian graph which is not symmetric.

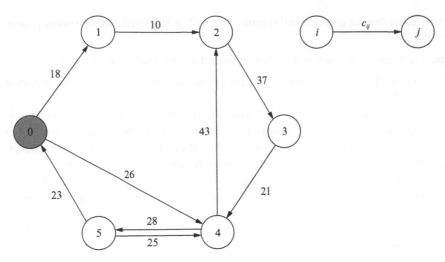

Figure 6.49   A directed graph $G = (V, A)$.

the arcs $(r, s) \in A$ belonging to the least-cost paths associated with the arcs $(i, j) \in A_T$ such that $s_{ij}^* > 0$ (it is worth observing that $(r, s) \in A$ is taken s_{ij}^* times).

[DCPP.xlsx] In the directed graph $G = (V, A)$ shown in Figure 6.49, the differences between the incoming and outgoing degrees of vertices 0, 1, 2, 3, 4 and 5 are −1, 0, 1, 0, 1 and −1, respectively. The least-cost paths from vertex 2 to vertex 0 and from vertex 2 to vertex 5 are given by the sequences of arcs $((2, 3), (3, 4), (4, 5), (5, 0))$ (of cost equal to 109) and $((2, 3), (3, 4), (4, 5))$ (of cost equal to 86), respectively. Similarly, the least-cost path from vertex 4 to vertex 0 is formed by $((4, 5), (5, 0))$, of cost 51, while the least-cost path from vertex 4 to vertex 5 is given by arc $(4, 5)$ whose cost is equal to 28. It is possible, therefore, to formulate the transportation problem on the bipartite directed graph $G_T = (V^+ \cup V^-, A_T)$ represented in Figure 6.50.

The optimal solution, found by using the Excel Solver tool, of the transportation problem is

$$s_{20}^* = 1, s_{25}^* = 0, s_{40}^* = 0, s_{45}^* = 1,$$

of cost equal to 137. Therefore, set $A^{(a)}$ is formed by the arcs of the least-cost paths from vertex 2 to vertex 0 and from vertex 4 to vertex 5. By adding these arcs to the directed graph G, a least-cost Eulerian directed multi-graph is obtained (see Figure 6.51). Hence, an optimal CPP solution of cost 368 is defined by the following arcs:

$$((0, 1), (1, 2), (2, 3), (3, 4), (4, 5), (5, 0), (0, 4),$$

$$(4, 2), (2, 3), (3, 4), (4, 5), (5, 4), (4, 5), (5, 0)).$$

In this solution some arcs are traversed more than once (e.g., arc $(4, 5)$), but on these arcs the service is performed only once.

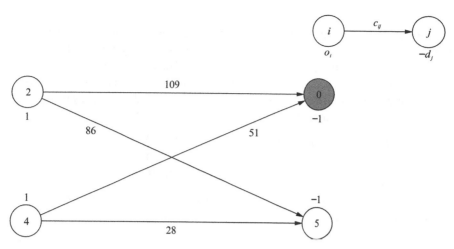

Figure 6.50 Bipartite directed graph $G_T = (V^+ \cup V^-, A_T)$ associated with directed graph G in Figure 6.49.

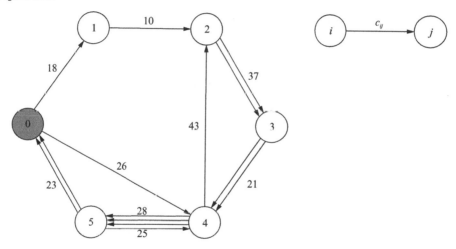

Figure 6.51 Least-cost Eulerian directed multi-graph associated with the directed graph G in Figure 6.49.

Undirected Chinese postman problem

When G is undirected, the set $E^{(a)}$ can be obtained as a solution of a matching problem on an auxiliary undirected graph $G_M = (V_M, E_M)$. V_M is the set of odd vertices in G (V_M is formed by an even number of vertices) and $E_M = \{(i, j) : i \in V_M, j \in V_M, i \neq j\}$. With each edge $(i, j) \in E_M$ is associated a cost w_{ij} equal to that of a least-cost path in G between vertices i and j. The set $E^{(a)}$ is therefore obtained as the union of the edges which are part of the least-cost paths associated with the edges of the optimal matching on G_M.

[Welles.xlsx] Welles is in charge of maintaining the road network of Wales in the United Kingdom. Among other things, the company has to monitor periodically the roads in order to locate cracks and potholes in the asphalt. To this end, the road network has been divided into about 10 subnetworks, each of which has to be visited every 15 days by a dedicated vehicle. The graph representing one such subnetwork is shown in Figure 6.52. In order to determine an optimal CPP solution, a minimum-cost matching problem between the odd-degree vertices (vertices 2, 3, 6, 7, 9, and 11) is solved. The optimal matching, obtained by using the `Excel Solver` tool, is: 2–3, 6–7, and 9–11. The associated set of paths in G is $(2, 3)$, $(6, 7)$, and $(9, 11)$ (total cost is 7.5 km). Adding these edges to G, the Eulerian undirected multi-graph in Figure 6.53 is obtained. Finally, by using the `END_PAIRING` procedure, the optimal CPP solution is obtained:

$$((0, 1), (1, 3), (3, 2), (2, 5), (5, 4), (4, 3), (3, 2), (2, 0),$$
$$(0, 6), (6, 7), (7, 8), (8, 5), (5, 6), (6, 7), (7, 9), (9, 11),$$
$$(11, 10), (10, 8), (8, 9), (9, 11), (11, 12), (12, 9), (9, 0)),$$

the cost of which is 52.8 km, that is, the cost of the edges of G, plus 7.5 km. In this solution, edges $(2, 3)$, $(6, 7)$ and $(9, 11)$ are traversed twice.

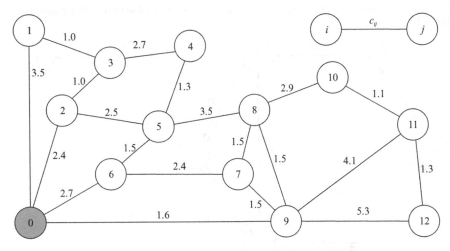

Figure 6.52 Graph representation used in the Welles problem. Costs c_{ij}, $(i, j) \in E$, are expressed in km.

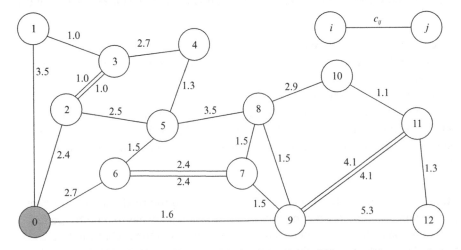

Figure 6.53 Least-cost Eulerian undirected multi-graph in the Welles problem.

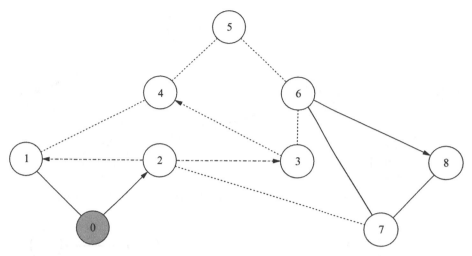

Figure 6.54 A mixed graph $G = (V, A, E)$.

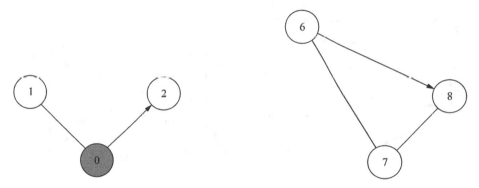

Figure 6.55 Connected components induced by the required arcs and edges of graph G in Figure 6.54.

The Rural Postman Problem

The rural postman problem is to determine in a graph $G = (V, A, E)$ a least-cost route traversing a subset $R \subseteq A \cup E$ of required arcs and edges at least once.

Let $G_1 = (V_1, A_1, E_1), \dots, G_p = (V_p, A_p, E_p)$ be the p connected components of graph $G = (V, R)$ induced by the required arcs and edges (see Figures 6.54 and 6.55).

The RPP solution can be obtained in two steps.

The first step determines a least-cost set of arcs $A^{(a)}$ and edges $E^{(a)}$ such that the multi-graph $G^{(a)} = (V, (R \cap A) \cup A^{(a)}, (R \cap E) \cup E^{(a)})$ is Eulerian (see Figure 6.56).

The computational complexity of this step is exponential even for directed and undirected graphs, if $p > 1$. It is worth observing that, in the case of $p = 1$, the RPP can be reduced to a CPP. In the following, two constructive heuristics for this step are illustrated for directed and undirected graphs, respectively.

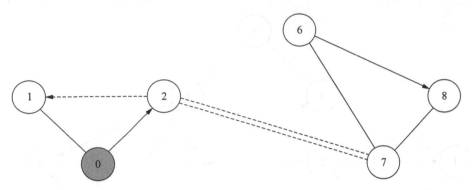

Figure 6.56 Least-cost Eulerian multi-graph associated with graph $G = (V, R)$ of Figure 6.54.

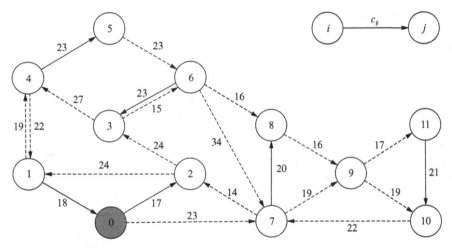

Figure 6.57 Directed graph $G = (V, A)$.

The second step involves determining a Eulerian tour in $G^{(a)}$. This step can be solved in polynomial time with the END_PAIRING procedure.

Directed rural postman problem

The following BALANCE_AND_CONNECT procedure can be applied to construct a directed Eulerian graph.

To construct a directed symmetric graph $G'^{(a)} = (V, R \cup A'^{(a)})$ at code line 2, the procedure employed for the directed CPP can be used. At code line 12, the reduction of the solution cost can be obtained by applying the shortcut method used in the CHRISTOFIDES procedure for the STSP (see Section 6.12.1).

[DRPP.xlsx] The BALANCE_AND_CONNECT procedure is applied to the problem represented in Figure 6.57. The directed graph $G = (V, R)$ contains five

1: **procedure** BALANCE_AND_CONNECT $(V, A, R, G^{(a)})$

2:     Construct a directed symmetric graph $G'^{(a)} = (V, R \cup A'^{(a)})$, by adding to $G = (V, R)$ a suitable set of least-cost paths between non-symmetric vertices;

3:     **if** $G'^{(a)} = (V, R \cup A'^{(a)})$ is connected **then**

4:         # $G'^{(a)}$ is Eulerian;

5:         exit = TRUE;

6:     **end if**

7:     **if** exit = FALSE **then**

8:         # p' $(1 < p' \leq p)$ is the number of connected components of $G'^{(a)} = (V, R \cup A'^{(a)})$;

9:         Construct an auxiliary undirected graph $G^{(c)} = (V^{(c)}, E^{(c)})$, in which there is a vertex $h \in V^{(c)}$ for each connected component of $G'^{(a)}$, and, between each pair of vertices $h, k \in V^{(c)}$, $h \neq k$, there is an edge $(h, k) \in E^{(c)}$; with edge (h, k) is associated a cost g_{hk} equal to:

$$g_{hk} = \min_{i \in V_h, j \in V_k} \{w_{ij} + w_{ji}\},$$

where w_{ij} and w_{ji} are the costs of the least-cost paths from vertex i to vertex j and from vertex j to vertex i in G, respectively;

10:         Compute a minimum-cost tree $T^{(c)} = (V^{(c)}, E_T^{(c)})$ spanning the vertices of graph $G^{(c)}$;

11:         Construct a symmetric, connected and directed graph $G^{(a)} = (V, R \cup A'^{(a)} \cup A''^{(a)})$ by adding to $G'^{(a)} = (V, R \cup A'^{(a)})$ the set of arcs $A''^{(a)}$ belonging to the least-cost paths corresponding to the edges $E_T^{(c)}$ of the tree $T^{(c)}$;

12:         Reduce, if possible, the solution cost;

13:     **end if**

14:     $G^{(a)} = G'^{(a)}$;

15:     **return** $G^{(a)}$;

16: **end procedure**

---

connected components of required arcs. At the end of code line 2 (see Figure 6.58), $A'^{(a)}$ is formed by arcs $(2, 1)$ (the least-cost path from vertex 1 to vertex 2), $(3, 4)$ (the least-cost path from vertex 3 to vertex 4), $(5, 6)$ (the least-cost path from vertex 5 to vertex 6), $(8, 9)$ and $(9, 11)$ (the least-cost path from vertex 8 to vertex 11), and $(10, 7)$ (the least-cost path from vertex 10 to vertex 7).

Subsequently, $V^{(c)} = \{1, 2, 3\}$, and the least-cost paths from vertex 1 to vertex 4 (arc $(1, 4)$) and vice versa (arc $(4, 1)$), and from vertex 6 to vertex 7 (arc $(6, 7)$) and vice versa (arcs $(7, 2)$, $(2, 3)$, and $(3, 6)$), are added to the partial solution (see Figure 6.59). Finally, using the END_PAIRING procedure, the following Eulerian tour of cost 379 is obtained:

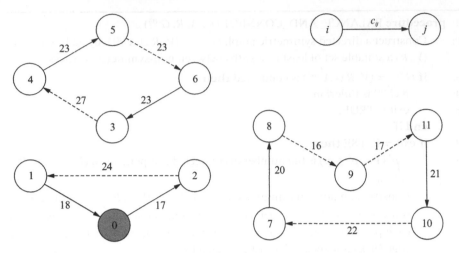

Figure 6.58 Directed graph $G'^{(a)} = (V, R \cup A'^{(a)})$ obtained at the end of code line 2 of the BALANCE_AND_CONNECT procedure.

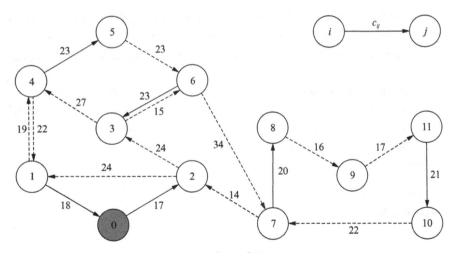

Figure 6.59 Directed graph $G^{(a)} = (V, R \cup A'^{(a)} \cup A''^{(a)})$ obtained at the end of of the BALANCE_AND_CONNECT procedure.

$$((0, 2), (2, 1), (1, 4), (4, 5), (5, 6), (6, 7), (7, 8), (8, 9), (9, 11),$$
$$(11, 10), (10, 7), (7, 2), (2, 3), (3, 6), (6, 3), (3, 4), (4, 1), (1, 0)).$$

It can be shown that in this case, the solution found is optimal.

Undirected rural postman problem

The following FREDERICKSON procedure can be used to construct a Eulerian undirected graph.

1: **procedure** FREDERICKSON $(V, E, R, G^{(a)})$
2:     Construct an even undirected graph $G^{'(a)} = (V, R \cup E^{'(a)})$;
3:     **if** $G^{'(a)} = (V, R \cup E^{'(a)})$ is connected **then**
4:         # $G^{'(a)}$ is Eulerian;
5:         exit = TRUE;
6:     **end if**
7:     **if** exit = FALSE **then**
8:         # p' $(1 < p' \leq p)$ is the number of connected components of $G^{'(a)} = (V, R \cup E^{'(a)})$;
9:         Construct an auxiliary undirected graph $G^{(c)} = (V^{(c)}, E^{(c)})$, in which there is a vertex $h \in V^{(c)}$ for each connected component of $G^{'(a)}$, and between each couple of vertices $h, k \in V^{(c)}$, $h \neq k$, there is an edge $(h, k) \in E^{(c)}$; with each edge (h, k) is associated a cost g_{hk} equal to

$$g_{hk} = \min_{i \in V_h, j \in V_k} \{w_{ij}\},$$

where w_{ij} is the cost of the least-cost path between vertices i and j in G;
10:         Compute a minimum-cost tree $T^{(c)} = (V^{(c)}, E_T^{(c)})$ spanning the vertices of graph $G^{(c)}$;
11:         Construct an even and connected graph $G^{(a)} = (V, R \cup E^{'(a)} \cup E^{''(a)})$ by adding to $R \cup E^{'(a)}$ the set of edges $E^{''(a)}$ (each of which taken twice) belonging to the least-cost paths corresponding to the edges $E_T^{(c)}$ of tree $T^{(c)}$;
12:         Reduce, if possible, the solution cost;
13:     **end if**
14:     $G^{(a)} = G^{'(a)}$;
15:     **return** $G^{(a)}$;
16: **end procedure**

To construct an even undirected graph $G^{'(a)} = (V, R \cup E^{'(a)})$ at code line 2, the matching procedure illustrated for the undirected CPP can be used. At code line 12, the reduction of the solution cost can be eventually obtained by applying the shortcut method used in the CHRISTOFIDES procedure for the STSP (see Section 6.12.1).

[Tracon.xlsx] Tracon distributes newspapers and milk door-to-door all over Wales. In the same road subnetwork considered for the Welles problem, customers are uniformly distributed along some roads (represented by continuous lines in Figure 6.60). The entire demand of the subnetwork can be served by a single vehicle. By applying the FREDERICKSON procedure, the even and connected undirected multi-graph $G^{(a)} = (V, R \cup E^{(a)})$ shown in Figure 6.61 is obtained. Finally, by using the END_PAIRING procedure, the following cycle is generated:

$$((0, 9), (9, 12), (12, 11), (11, 10), (10, 8), (8, 7), (7, 6), (6, 5),$$
$$(5, 4), (4, 3), (3, 1), (1, 3), (3, 2), (2, 5), (5, 6), (6, 0))$$

(total cost is 31.3 km). It is worth noting that edge $(5, 6)$ is traversed twice. It can be shown that the solution found is optimal.

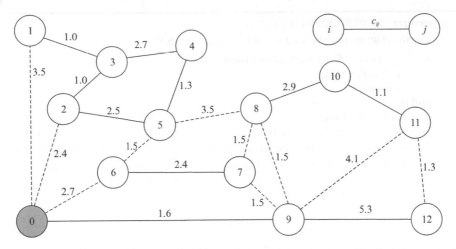

Figure 6.60 Undirected graph $G = (V, E)$ associated with the Tracon problem (costs are in km).

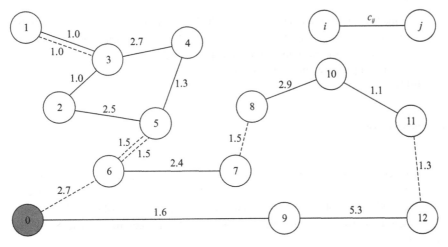

Figure 6.61 Even and connected undirected multi-graph $G^{(a)} = (V, R \cup E^{(a)})$, obtained at the end of the FREDERICKSON procedure applied to the Tracon problem.

6.12.5 Route Sequencing

It is worth observing that in the VRPs illustrated before, each vehicle is used only once during the planning period. This situation is often not representative of real cases, in which, typically, the vehicles could be used for several journeys during the same working shift. This happens, in particular, when each route in the VRP solution has a rather short duration. In these cases it is possible to reduce the number of vehicles used by allocating more routes to the same vehicle, with respect to the total time length constraint. The problem, known as the *vehicle routing problem with multiple trips* (VRPMT), can be treated as a special case of 1-BP problem (see Section 5.15.1), in which the routes to allocate to the vehicles correspond to the objects to be loaded, each of them having a "weight" equal to the time duration, whereas the vehicles correspond to bins of capacity equal to T, where T is the time duration of the working shift.

Table 6.33 Daily truck service route durations of the Han-Cor company.

Route	Duration [hh:mm]	Route	Duration [hh:mm]
1	2:30	8	3:20
2	1:10	9	2:20
3	3:15	10	2:10
4	1:45	11	1:50
5	1:25	12	2:10
6	0:50	13	0:50
7	2:05	14	0:50

[HanCor.xlsx] Han-Cor, a public Korean carrier, has solved a daily routing problem which corresponds to the truck service routes shown in Table 6.33. The daily duration of the transportation service is 6 hours and 20 minutes (or, equivalently, 380 minutes). To allocate the routes to vehicles, the BFD procedure (see Section 5.15.1) can be applied to the 1-BP problem with 14 objects, each of them with "weight" equal to $d_i, i = 1, \dots, 14$ (see the second and fourth columns of Table 6.33). The list S of the routes ordered by non-increasing values of their duration is:

$$S = (8, 3, 1, 9, 10, 12, 7, 11, 4, 5, 2, 6, 13, 14).$$

At the end of the application of the heuristic, it is found that five vehicles $(v_1, \dots, v_5)$ are needed; each of them is allocated one of the following routes:

$v_1 = (8, 1)$, used for 350 minutes overall;

$v_2 = (3, 9)$, used for 335 minutes overall;

$v_3 = (10, 12, 11)$, used for 370 minutes overall;

$v_4 = (7, 4, 5, 6)$, used for 365 minutes overall;

$v_5 = (2, 13, 14)$, used for 170 minutes overall.

6.13 Real-time Vehicle Routing Problems

As pointed out in Section 6.12, several important short-haul freight transportation problems must be solved in real time. In this section, the main features of such problems are illustrated.

In *real-time* VRDPs, uncertain data are gradually revealed during the operational interval, and routes are constructed in an ongoing fashion as new data arrive. The *events* that lead to route modifications can be: (a) the arrival of new user requests, (b) the arrival of a vehicle at a destination, or (c) the update of travel times.

Every event must be processed according to the policies set by the organization operating the fleet of vehicles. As a rule, when a new request is received, one must decide whether it can be serviced on the same day, or whether it must be delayed or rejected. If the request is accepted, it is assigned temporarily to a position in a vehicle route. The request is effectively serviced as planned if no other event occurs in the meantime. Otherwise, it can be assigned to a different position of the same vehicle route, or even dispatched to a different vehicle.

It is worth noting that at any time each driver just needs to know their next stop. Hence, when a vehicle reaches a destination it has to be assigned a new destination. In the past, because of the difficulty of estimating the current position of a moving vehicle, reassignments could not easily be made. However, due to advances in ICT, route diversions and reassignments are now a feasible option and should take place if this results in a cost saving or in an improved service level.

Finally, if an improved estimation of vehicle travel times is available, it may be useful to modify the current routes or even the decision about accepting a request or not. For example, if an unexpected traffic jam occurs, some user services can be deferred. If the demand rate is low, it is somewhat useful to relocate idle vehicles in order to anticipate future demands or to avoid a forecast traffic congestion.

Real-time problems possess a number of particular features, some of which have just been described. In the following, the remaining characteristics are outlined.

Quick response Algorithms for solving real-time VRDPs must provide a quick response so that route modifications can be transmitted in a timely manner to the fleet. To this end, two methods can be used: simple policies (like the "first come, first served"), or more involved algorithms running on parallel hardware. The choice between them depends mainly on the objective, the degree of dynamism and the demand rate.

Denied or deferred service In some applications it is valid to deny service to some users, or to leave them to a competitor, in order to avoid excessive delays or unacceptable costs. For instance, requests that cannot be serviced within a given time window are rejected.

Congestion If the demand rate exceeds a given threshold, the system becomes saturated, that is, the expected waiting time of a request increases to infinity.

The degree of dynamism Designing an algorithm for solving real-time VRDPs depends to a large extent on how much dynamic the problem is. To quantify this concept, the *degree of dynamism* of a problem has been defined. Let $[0, T]$ be the operational interval and let n_s and n_d be the number of static and dynamic requests, respectively ($n_s + n_d = |U|$). Moreover, let $t_i \in [0, T]$ be the *occurrence time* of service request of customer $i \in U$. Static requests are such that $t_i = 0, i \in U$, while dynamic ones have $t_i \in [0, T], i \in U$. The degree of dynamism δ can be simply defined as

$$\delta = \frac{n_d}{n_s + n_d}$$

and may vary between 0 and 1. Its meaning is straightforward. For instance, if δ is equal to 0.3, then three customers out of 10 are dynamic. Such definition can be generalized in order to take into account both dynamic request occurrence times and possible time windows. For a given δ value, a problem is more dynamic if immediate requests occur at the end of the operational interval $[0, T]$. As a result, the measure of dynamism can be generalized as follows:

$$\delta' = \frac{\sum\limits_{i=1}^{n_d} t_i/T}{n_s + n_d}.$$

Again δ' ranges between 0 and 1. It is equal to 0 if all user requests are known in advance, while it is equal to 1 if all user requests occur at time T. Finally, the definition of δ' can be modified to take into account possible time windows on user service time. Let a_i and b_i be the *ready time* and *deadline* of customer $i \in U$, respectively. Then,

$$\delta'' = \frac{\sum\limits_{i=1}^{n_d} [T - (b_i - a_i)]/T}{n_s + n_d}.$$

It can be shown that δ'' also varies between 0 and 1. Moreover, it is worth noting that if no time windows are imposed (i.e., $a_i = t_i$ and $b_i = T$ for each customer $i \in U$), then $\delta'' = \delta'$. As a rule, vendor-based distribution logistics systems (such as those distributing heating oil) are weakly dynamic. Problems faced by long-distance couriers and appliance repair service companies are moderately dynamic. Finally, emergency services exhibit a strong dynamic behaviour.

Objectives In real-time VRDPs, the objective to be optimized is often a combination of different measures. In weakly dynamic systems, the focus is on minimizing the routing cost but, when operating a strongly dynamic system, minimizing the expected *response time* (i.e., the expected time lag between the instant a user service begins and its occurrence time) becomes a key issue. Another meaningful criterion which is often considered (alone or combined with other measures) is throughput optimization, that is, the maximization of the expected number of requests serviced within a given period of time.

6.14 Integrated Location and Routing Problems

Facility location and vehicle routing are two of the most fundamental decisions in logistics. Location decisions that are very costly and difficult to change are said to be strategic, for example, those involving major installations such as factories, airports, fixed transportation links, and so on. Others are said to be tactical because while still being relatively costly, they can still be modified after several years. Warehouse and store locations fall in that category. Finally, operational location decisions involve easily movable facilities such as parking areas, mail boxes, and the like.

Once facilities are located, a routing plan must be put in place to link them together on a regular basis. As already observed in Sections 3.1 and 3.15.2, all too often, facilities are first located without sufficient consideration of transportation costs, which may result in systemic inefficiencies. When planning to locate facilities, it is preferable to integrate in the analysis the routing cost that these will generate. This applies equally well to strategic, tactical, and operational decisions. A strategic location-routing decision is the location of airline hubs the choice of which has a bearing on routing costs. The location of depots and warehouses in a logistics system is a tactical decision influencing delivery costs to customers. A simple example arising at the operational level is mail box location. Locating a large number of mail boxes in a city will improve customer convenience since average walking distance to a mail box will be reduced. At the same time, the cost of emptying a larger number of mail boxes on a regular basis will be higher.

Unfortunately, integrated location-routing optimization models combining these two aspects will often contain too many integer decision variables and constraints to be solvable optimally. Heuristics based on a decomposition principle are often used instead. Facilities are first located, nodes are assigned to facilities and routing is then performed. These three decisions are then iteratively updated until no significant improvement can be reached.

3L Multimedia is a company that distributes newspapers and magazines throughout France. The distribution system, set up in 2021, includes 54 hubs which serve 9542 sales points (supermarkets, newsstands, and booksellers) in France. On the basis of the historical data available about the deliveries carried out using the different fleets of vehicles available, a forecasting study was carried out on the total logistics costs in the next three years, evaluating 10 different potential locations for the hubs.

The data resulting from this study are reported in Table 6.34. Note that the disposal costs refer, for each alternative, to the hubs that should be dismissed. The transportation costs are calculated, simulating the daily delivery plans for the planning horizon (three years), by solving the vehicle routing problem with capacity and length constraints.

The best simulated alternative proves to be the sixth, providing for 40 hubs.

Table 6.34   Logistics costs corresponding to the 10 location alternatives for the 3L Multimedia problem.

Alternative	Number of hubs	Disposal cost [€]	Triennial location cost [€]	Triennial transportation cost [€]	Total cost [€]
1	47	140 000	534 000	856 000	1 530 000
2	55	90 000	610 000	790 000	1 490 000
3	44	122 000	491 000	882 000	1 495 000
4	50	80 000	565 000	844 000	1 489 000
5	38	180 000	443 000	964 000	1 587 000
6	40	75 000	446 000	933 000	1 454 000
7	58	22 000	658 000	781 000	1 461 000
8	60	210 000	663 000	762 000	1 635 000
9	32	180 000	357 000	1 081 000	1 618 000
10	35	55 000	397 000	1 033 000	1 485 000

6.15   Inventory Routing Problems

IRPs are combinatorial problems in which inventory management and transportation decisions have to be made in an integrated fashion and simultaneously.

In many traditional supply chains, a facility receives orders from its own successor nodes (referred to in the following as *customers*, for the sake of brevity) and plans the deliveries in relation to the choices made by the customers themselves. Inventory routing refers to a situation in which the facility (referred to in the following as *supplier*) decides the best replenishment policy of the customers, avoiding stockout and making sure that the quantities delivered are compatible with the storage capacity of the warehouses and with the capacity of the vehicles available for deliveries. The aim is to minimize the total inventory and transportation costs over a given time horizon.

The increasing attention received by the IRPs is strictly related to the development of more and more integrated information systems which allow shared monitoring of inventories. IRPs arose in the 1980s in the petrochemical and gas industries. They have rapidly developed in the automotive industry, and also in the retail industry, where this practice is currently very common, particularly in supply chains characterized by unforeseen variations in demand.

Compared with VRPs, a greater difficulty in dealing with IRPs arises from the fact that such problems integrate a level of tactical decisions, related to the coordinated inventory replenishment of the set of customers typically distributed over a limited geographical area, with operational decisions regarding the best scheduling of the deliveries over the time. Inventory management comes into play whenever each customer consumes products at a fixed or variable rate, and must manage the quantities to be supplied in relation to the inventory levels in its warehouses. This consideration naturally leads to the introduction of the time dimension in the context of IRPs.

In IRPs the planning horizon can be *continuous* or *discrete*. In this section, attention will be focused on the discrete case for which the planning horizon is made up of T time periods (e.g., days). The quantity to be delivered to each customer in each period of the planning horizon is not known (otherwise, the IRP could be decomposed into as many VRPs as the number of time periods of the planning horizon). This implies that in IRPs the decisions to be made are

- in which time periods each customer should be served;
- how much to deliver to a customer when served;
- the delivery routes in each time period of the planning horizon.

Different variants of the IRP can be formulated. In the following, a single-item formulation on a complete undirected graph $G' = (V', E')$ is illustrated, in which $V' = U \cup \{0\}$, where U is the set of customers to be served, 0 represents the supplier and a homogeneous fleet K of vehicles (each of them having capacity q) is available for the deliveries. With each edge $(i, j) \in E'$ is associated a non-negative transportation cost c_{ij}, corresponding to the cost of a least-expensive path between i and j. In this way, the triangular inequality property is satisfied for each cost $c_{ij}, (i, j) \in E$.

The demand of customer $i \in U$ (requested and consumed) in time period $t = 1, \dots, T$ is represented by $d_i^{(t)}$. To the supplier, the product quantity that becomes available at the beginning of time period $t = 1, \dots, T$ is represented by $p^{(t)}$; γ_i and $\mu_i, i \in V'$, indicate

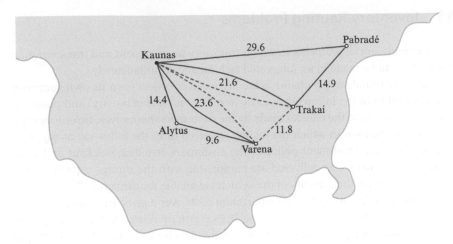

Figure 6.62 Optimal vehicle routes serving the customers of AT24 from the depot in Kaunas. The routes represented by continuous lines refer to the first day of service, while the route represented by dotted lines refers to the second-day service.

the initial inventory level available and the storage capacity at the supplier (when $i = 0$) or at customer i (when $i \in U$), respectively.

Finally, the unit inventory costs at the supplier and at customer $i \in U$ are represented by h_0 and h_i, respectively.

[AT24.xlsx] AT24 is a fuel distribution company with a depot (vertex 0) located in Kaunas, Lithuania. The company must ensure the supply of diesel fuel to some private local companies (i.e., the customers) which, in turn, supply their own fleets of vehicles every day. These customers are situated in the cities of Pabradé, Trakai, Varéna and Alytus (see Figure 6.62), which means that $U = \{1, 2, 3, 4\}$. The transportation costs (in €) between the cities where the depots and the customers are located are shown in Table 6.35. In Table 6.36 the capacity of the customer's tanks and the average daily demand for diesel fuel (both quantities are expressed in quintals) are reported. AT24 has a fleet of several vehicles with a capacity of 7500 quintals each for the supply service of diesel fuel. Storage costs at the customer's tanks are negligible compared with transportation costs. The aim is to find the replenishment policy of the four customers at the minimum transportation cost, to be repeated regularly, by avoiding stockouts and satisfying the daily average demand of the customers, the capacities of the customers' tanks and of the vehicles used for the supply service. It is assumed that the supplier's availability of diesel fuel and the depot capacity are unlimited ($\gamma_0 = \infty$ and $\mu_0 = \infty$) or, alternatively, $\gamma_0 = 0$ and $p^{(t)} = \sum_{i \in U} d_i, \ t = 1, \ldots, T,$

where d_i, $i \in U$, is the average daily demand for diesel fuel required by customer i. Furthermore, the initial quantity of diesel fuel in the tank of each customer is zero (i.e., $\gamma_i = 0$, $i \in U$). Finally, the daily distribution phase of diesel fuel from the supplier to the customers is assumed to be completed before each customer begins to supply its own fleet of vehicles.

Table 6.35 Transportation costs (in €) between the cities where the depot and the customers of AT24 company are located.

City	Kaunas	Pabradė	Trakai	Varėna	Alytus
Kaunas	0.0	29.6	21.6	23.6	14.4
Pabradė		0.0	14.9	26.0	30.2
Trakai			0.0	11.8	15.4
Varėna				0.0	9.6
Alytus					0.0

Table 6.36 Tank capacity and the average daily demand for diesel fuel (both in quintals) of each customer of AT24 company.

Customer (i)	Location	Tank capacity (μ_i)	Average daily demand (d_i)
1	Pabradė	5000	1500
2	Trakai	4500	4500
3	Varėna	3000	3000
4	Alytus	5000	2250

A straightforward daily distribution plan is obtained by taking into account the following considerations. Customers 2 and 3 should be served every day since their maximum inventory levels are equal to their average daily demands and, consequently, it is not possible to stock more than the daily demands of diesel fuel. By taking into account the transportation costs and the geographic distribution of the customers, a feasible solution to the IRP is obtained by serving every day customers 1 and 2 with one vehicle, and customers 3 and 4 with another vehicle. The daily load of the first vehicle is $(1500 + 4500) = 6000$ quintals and the transportation cost is $(29.6 + 14.9 + 21.6) = €\ 66.1$. Similarly, the daily load of the second vehicle is $(3000 + 2250) = 5250$ quintals and the transportation cost is $(23.6 + 9.6 + 14.4) = €\ 47.6$. It is worth observing that the load of both vehicles is less than their capacity. Every day the distribution policy is repeated in the same fashion (i.e., the policy periodicity is one) with a daily transportation cost of $(66.1 + 47.6) = €\ 113.7$.

This distribution planning is not optimal. In fact, it easy to demonstrate that an optimal solution (see Figure 6.62) can be obtained by applying a periodicity of two days (i.e., every two days the distribution policy is repeated). On the first day, customers 1 and 2 are served together with one vehicle, but the quantity transported to customer 1 is twice its average daily demand. This means that the vehicle load is $(2 \times 1500 + 4500) = 7500$ quintals, equal to its capacity. In this way, the quantity transported to customer 1 is sufficient to satisfy the demand of two days (consider that the constraint on the maximum inventory level is still satisfied). Similarly, customers 3 and 4 are served together with another vehicle, but the quantity transported to customer 4 is twice its average daily demand, sufficient to satisfy the demand of two days, without violating inventory constraints. The vehicle load is $(3000 + 2 \times 2250) = 7500$ quintals, again equal to its capacity. On the subsequent day, only customers 2 and 3 should be served, and the cumulative average daily demand of the two customers is compatible with the use of a single vehicle, with a transportation cost of $(21.6 + 11.8 + 23.6) = €57.0$. This means that the transportation cost for the two days is $(113.7 + 57.0) = €170.7$, which is less than the two-day transportation cost of the previous policy, that is, $(2 \times 113.7) = €227.4$.

An IRP formulation can be obtained by using the following decision variables: $I_i^{(t)}$, $i \in V'$, $t = 1, \dots, T + 1$, each of which represents the supplier (if $i = 0$) or the customer (if $i \in U$) inventory level at the beginning of time period t, where the time horizon is extended to time period $T + 1$ in order to evaluate the impact of the decisions made at the end of T; $x_{ik}^{(t)}$, $i \in U$, $k \in K$, $t = 1, \dots, T$, each of which represents the quantity delivered from the supplier at the beginning of time period t to customer i with vehicle k (or, at least, before that the demand $d_i^{(t)}$ begins to be consumed at customer i); $y_{ijk}^{(t)}$, $(i, j) \in E'$, $k \in K$, $t = 1, \dots, T$, each of which represents the number of times (at most two) that vehicle k traverses the edge (i, j) in time period t, and $z_{ik}^{(t)}$, $i \in V'$, $k \in K$, $t = 1, \dots, T$, each of which of binary type, assuming value 1 if vehicle k visits the node (customer or supplier) i in time period t, 0 otherwise. The resulting MIP model is as follows:

$$\text{Minimize} \quad \sum_{i \in V'} \sum_{t=1}^{T+1} h_i I_i^{(t)} + \sum_{(i,j) \in E'} \sum_{k \in K} \sum_{t=1}^{T} c_{ij} y_{ijk}^{(t)} \tag{6.114}$$

subject to

$$I_0^{(1)} = \gamma_0 \tag{6.115}$$

$$I_0^{(t+1)} = I_0^{(t)} + p^{(t)} - \sum_{i \in U} \sum_{k \in K} x_{ik}^{(t)}, \quad t = 1, \dots, T \tag{6.116}$$

$$I_i^{(1)} = \gamma_i, \quad i \in U \tag{6.117}$$

$$I_i^{(t+1)} = I_i^{(t)} + \sum_{k \in K} x_{ik}^{(t)} - d_i^{(t)}, \quad i \in U, t = 1, \dots, T \tag{6.118}$$

$$I_0^{(t)} + p^{(t)} \le \mu_0 \tag{6.119}$$

$$I_i^{(t)} + \sum_{k \in K} x_{ik}^{(t)} \le \mu_i, \quad i \in U, \ t = 1, \dots, T \tag{6.120}$$

$$x_{ik}^{(t)} \le \mu_i z_{ik}^{(t)}, \quad i \in U, \ k \in K, \ t = 1, \dots, T \tag{6.121}$$

$$\sum_{i \in U} x_{ik}^{(t)} \le q z_{0k}^{(t)}, \quad k \in K, \ t = 1, \dots, T \tag{6.122}$$

$$\sum_{i \in V' : (i,j) \in E'} y_{ijk}^{(t)} + \sum_{i \in V' : (j,i) \in E'} y_{jik}^{(t)} = 2 z_{jk}^{(t)},$$
$$j \in V', \ k \in K, \ t = 1, \dots, T \tag{6.123}$$

$$\sum_{(i,j)\in E':i\in S,j\notin S} y_{ijk}^{(t)} + \sum_{(j,i)\in E':i\in S,j\notin S} y_{jik}^{(t)} \geq 2z_{hk}^{(t)},$$

$$S \subseteq U, |S| \geq 2, \ h \in S, \ k \in K, \ t = 1, \dots, T \qquad (6.124)$$

$$I_i^{(t)} \geq 0, \ i \in V', \ t = 1, \dots, T+1 \qquad (6.125)$$

$$x_{ik}^{(t)} \geq 0, \ i \in U, \ k \in K, \ t = 1, \dots, T \qquad (6.126)$$

$$y_{ijk}^{(t)} \in \{0, 1\}, \ (i, j) \in E', \ k \in K, \ t = 1, \dots, T \qquad (6.127)$$

$$y_{0jk}^{(t)} \in \{0, 1, 2\}, \ (0, j) \in E', \ k \in K, \ t = 1, \dots, T \qquad (6.128)$$

$$z_{ik}^{(t)} \in \{0, 1\}, \ i \in V', \ k \in K, \ t = 1, \dots, T. \qquad (6.129)$$

The objective function (6.114) represents the total inventory and the transportation costs over the time horizon. Constraint (6.115) establishes the initial value of the supplier inventory level at the first time period. Constraints (6.116) mean that the inventory level of the supplier at the beginning of time period $t+1$, $t = 1, \dots, T$, is given by the sum of the inventory level at the beginning of the previous time period t, plus the product quantity $p^{(t)}$ available at the beginning of time period t, minus the quantities delivered to customers at the beginning of time period t. Constraints (6.117) set the initial inventory values for each customer. Constraints (6.118) guarantee that the customer inventory level for each customer $i \in U$ at the beginning of each time period $t + 1$, $t = 1, \dots, T$, is equal to the sum of the customer inventory level at the beginning of the previous time period t, plus the quantity delivered at the beginning of time period t, minus the quantity consumed in time period t. Constraints (6.119) and (6.120) ensure the satisfaction of the storage capacity bounds for the supplier and for the customers, respectively, for each time period $t = 1, \dots, T$. Constraints (6.121) express the relationship, for each customer i, $i \in U$, between the decision variables $x_{ik}^{(t)}$ and $z_{ik}^{(t)}$, $k \in K$, $t = 1, \dots, T$. They impose that each customer at time period t should be considered as visited by vehicle k if a quantity greater than zero is delivered from the supplier to the customer itself by using vehicle k. Constraints (6.122) guarantee that the quantities delivered from the supplier to the customers at any time period by using a vehicle are compatible with the vehicle capacity. Moreover, these constraints impose that if one or more customers are visited at time period t by using vehicle k, then also the supplier has to be visited by the same vehicle in time period t (note that each vehicle begins the delivery route starting from the supplier). Constraints (6.123) and (6.124) are the routing constraints. They are similar to the routing constraints (6.87) and (6.88) introduced in the C-NRP formulation (6.84)–(6.91), extended to each period of the planning horizon. Constraints (6.125)–(6.129) define the integrality and non-negativity constraints of the decision variables. In particular, constraints (6.128) allow the generation of direct routes, serving only one customer at time.

Finally, note that, in contrast to the C-NRP formulation, model (6.114)–(6.129) allows that (in a given time period $t = 1, \dots, T$) the quantities to be supplied to the same customer $i \in U$ can be delivered using more than one vehicle (split delivery).

IRPs are very difficult to be solved to optimality, in particular when a large number of customers and a long planning horizon are considered. Therefore, a heuristic procedure is proposed in the following, based on a problem decomposition framework. First, a tactical decision on the delivery schedule for each vehicle and on the quantities

```
 1: procedure IRP_DECOMPOSITION (G', γ, μ, d, h, c, x̄, ȳ, z̄, Ī, w̄)
 2:     # U^(t), t = 1, ..., T, is the set of customers to be replenished in time period t;
 3:     # x̄, ȳ, z̄ and Ī are the feasible solution of the IRP returned by the procedure;
 4:     # w̄ is the cost corresponding to the IRP feasible solution;
 5:     t = 1;
 6:     # Set the initial inventory level of each customer;
 7:     for i ∈ U do
 8:         Ī_i^(1) = γ_i;
 9:     end for
10:     while t ≤ T do
11:         U^(t) = ∅;
12:         for i ∈ U do
13:             Ī_i^(t+1) = Ī_i^(t) + δ_i − d_i^(t);
14:             # δ_i is the quantity to be sent to customer i, according to the feasible
                 replenishment policy adopted, satisfying constraints (6.120) and (6.125);
15:             if δ_i > 0 then
16:                 # Customer i should be replenished at time period t;
17:                 U^(t) = U^(t) ∪ {i};
18:             end if
19:         end for
20:         Solve the C-NRP on the set U^(t) of customers, determining x̄^(t), ȳ^(t) and z̄^(t);
21:         t = t + 1;
22:     end while
23:     w̄ = Σ_{i∈V'} Σ_{t=1}^{T+1} h_i Ī_i^(t) + Σ_{(i,j)∈E'} Σ_{k∈K} Σ_{t=1}^{T} c_{ij} ȳ_{ijk}^(t);
24:     return x̄, ȳ, z̄, Ī, w̄;
25: end procedure
```

to deliver to each customer to avoid stockouts is assumed. Second, the operational problem corresponding to the routing of each vehicle is solved. The described framework is proposed through the IRP_DECOMPOSITION procedure described in the following. Note that the procedure is based on the following hypotheses:

- h_0 is assumed to be negligible and both the initial inventory and the depot capacity at the supplier are unlimited ($\gamma_0 = \infty$ and $\mu_0 = \infty$); in this way, $I_0^{(t)}$ will be always greater than or equal to zero for each time period $t = 1, ..., T + 1$ and the constraints (6.119) will be always satisfied in each time period $t = 1, ..., T$;
- the fleet of homogeneous vehicles is large enough to guarantee the route service. Note that, when split deliveries are allowed, it is demonstrated that the number of vehicles used in each time period $t = 1, ..., T$ cannot be less than $2 \times \left\lceil \sum_{i \in U} d_i^{(t)} / q \right\rceil$.

The feasible replenishment policy (code line 13) could be, for example, the economic order quantity, the reorder point policy or any other (see Section 5.12.2 for more details). The C-NRP (code line 20) is reduced to a TSP in case of $\sum_{i \in U^{(t)}} \delta_i \leq q$.

[Split.xlsx, Split.csv, Split.py] Split is a German company supplying 10 customers each week with a special mixture which is used as a basic
component for preparing a drink sold at retail stores. The product has a demand
recorded every Monday, Wednesday, and Friday. Split intends to adopt an integrated management strategy for the optimal planning of the distribution of this
mixture. The Split distribution problem can be described as an IRP on a complete
undirected graph $G' = (V', E')$, with $V' = \{0, 1, 2, 3, 4, 5, 6, 7, 8, 9, 10\}$ and, con-
sequently, $|E'| = 55$. Vertex 0 represents the Split DC, positioned in Frankfurt,
while the other vertices represent the customers located in other German cities,
enumerated in the following order: Obertshausen, Willmars, Eschborn, Wetten-
berg, Schotten, Rosbach vor der Höhe, Hammelburg, Sprendlingen, Gelnhausen,
and Fulda (see Figure 6.63). The fleet is composed of four vehicles, each having a capacity of 300 quintals, that can operate each day from the Split DC. The
shortest road distances (in km) between the cities are shown in Table 6.37. The
unit transportation cost is 0.45 €/km. The initial inventory level, the maximum
storage capacity, the demand, and the unit storage cost at the customers are
reported in Table 6.38, while the storage capacity and the product availability at the supplier are always sufficient to cover the customer demands (i.e.,
$\gamma_0 = \infty$ and $\mu_0 = \infty$). Note that, the quantities of product consumed per day
by the customers are assumed constant throughout the planning horizon, that is,
$d_i = d_i^{(t)}$, $i = 1, \dots, 10$, $t = 1, 2, 3$ (Monday, Wednesday, and Friday respectively).
The Split distribution problem can be formulated by using model (6.114)–(6.129),
eliminating constraints (6.115), (6.116) and (6.119), as the product availability to
the supplier is unlimited. It can be solved by applying the IRP_DECOMPOSITION
procedure.

Table 6.37 Distances (in km) between the vertices (DC and customers) in the Split
problem.

	0	1	2	3	4	5	6	7	8	9	10
0	0.0	16.5	148.0	12.9	71.6	67.7	27.6	89.0	77.7	45.0	103.0
1		0.0	140.0	28.8	86.5	65.4	43.6	101.0	87.9	33.8	91.8
2			0.0	160.5	145.6	101.0	143.0	71.4	214.0	111.0	49.8
3				0.0	70.5	69.2	26.2	101.2	71.0	57.5	114.4
4					0.0	58.5	49.8	154.0	136.0	87.1	97.9
5						0.0	45.6	99.2	136.4	47.6	51.2
6							0.0	116.0	91.0	53.6	94.5
7								0.0	165.0	68.8	67.8
8									0.0	112.0	174.5
9										0.0	62.5
10											0.0

Table 6.38 Initial inventory level, storage capacity, demand, and inventory cost at the customers of the Split company.

Customer (i)	Initial inventory level [quintals] (γ_i)	Storage capacity [quintals] (μ_i)	Demand [quintals] (d_i)	Inventory cost [€/quintal] (h_i)
1	22	150	40	0.22
2	50	200	50	0.23
3	35	180	34	0.27
4	36	120	30	0.25
5	60	100	25	0.24
6	40	160	20	0.30
7	65	190	60	0.32
8	50	210	50	0.28
9	30	140	56	0.24
10	30	170	30	0.31

In order to reach a good trade-off between inventory and transportation costs, the Split logistics manager establishes that the quantity to deliver to each customer is defined by also taking into account its distance from the depot. To this end, the average distance of the customers from the depot is computed (59.90 km). The replenishment policy (code line 13) is based on the following rules:

- at each time period $t = 1, 2, 3$ in case of risk of stockout (i.e., $\overline{I}_i^{(t)} < d_i$), the customer $i = 1, \ldots, 10$ is replenished by a quantity d_i, if the distance between the customer itself and the Split DC is not greater than 59.90 km; the quantity replenished is doubled, otherwise;
- whenever customer i does not have sufficient space to store the expected quantity in its depot, when it is due to be replenished, the Split DC sends only the necessary quantity to fill the storage capacity μ_i.

The replenished quantities for each customer $i = 1, \ldots, 10$ in each time period $t = 1, 2, 3$ and the corresponding inventory levels $\overline{I}_i^{(t)}$ determined by using the IRP_DECOMPOSITION procedure (code line 14) are reported in Tables 6.39 and 6.40, respectively.

In the first time period, only customers 1 and 9 need to be replenished. They receive a quantity (in quintals) $\delta_1 = 40$ and $\delta_9 = 56$, respectively, in accordance with the defined replenishment policy. It is worth observing that one vehicle is sufficient to serve both customers, since $\delta_1 + \delta_2 = 40 + 56 < q = 300$. For this reason, a TSP is solved on the complete undirected graph induced by the vertex set $U^{(1)} = \{1, 9\} \cup \{0\}$, using the NEAREST_NEIGHBOUR procedure (see Section 6.12.1). Hence, the route obtained is $(0, 1, 9, 0)$ with a total cost of

Table 6.39 Delivered quantities (in quintals) in each time period and for each Split customer.

Customer (i)	Time period		
	$t = 1$	$t = 2$	$t = 3$
1	40	40	40
2	0	100	0
3	0	34	34
4	0	60	0
5	0	0	50
6	0	0	20
7	0	120	0
8	0	100	0
9	56	56	56
10	0	60	0

$0.45 \times (16.5 + 33.8 + 15.0) - €\,42.88$. The inventory level (in quintals) available at the beginning of time period $t = 2$ is computed as $I_1^{(2)} = 22 + 40 - 40$ for customer 1, and $I_9^{(2)} = 30 + 56 - 56$ for customer 9. For all other customers, the inventory level is updated by only subtracting the product demand consumed in the current time interval (see Table 6.40).

In the second time period (see Figure 6.63), the customers to be replenished are $U^{(2)} = \{1, 2, 3, 4, 7, 8, 9, 10\}$, receiving the following quantities (in quintals): $\delta_1 = 40$, $\delta_2 = 100$, $\delta_3 = 34$, $\delta_4 = 60$, $\delta_7 = 120$, $\delta_8 = 100$, $\delta_9 = 56$ and $\delta_{10} = 60$, respectively. It is easy to verify that in this case, at least two vehicles are required to serve all customers. As a consequence, a C-NRP is solved on the complete undirected graph induced by the vertex set $U^{(2)} \cup \{0\}$ using the SAVINGS procedure (see Section 6.12.2). The routes determined are $(0, 1, 9, 4, 3, 8, 0)$ and $(0, 7, 2, 10, 0)$, with total costs of $€\,153.05$ and $€\,140.94$, respectively. Inventory levels are again updated by following the procedure described above. Finally, in the third time period, the set of customers to be replenished is $U^{(3)} = \{1, 3, 5, 6, 9\}$. A TSP is solved using the NEAREST_NEIGHBOUR procedure, obtaining the route $(0, 3, 6, 1, 9, 5, 0)$ with a total cost of $€\,104.31$.

The total storage cost in the planning horizon (also evaluating the effect of the choices made at the end of the three time periods) is equal to $€\,244.88$. Consequently, the total cost of the heuristic solution is $(244.88 + 42.88 + 153.05 + 140.94 + 104.31) = €\,686.06$.

The Split problem was also solved to optimality (see the Split.csv file for the optimal values of the decision variables) and the optimal cost is $€\,595.54$, given by the sum of the inventory cost of $€\,235.18$ and the

Table 6.40 Inventory levels (in quintals) $\bar{I}_i^{(t)}$, $t = 1, 2, 3$, $i = 1, \ldots, 10$, determined applying the IRP_DECOMPOSITION procedure in the Split problem. Note that $\bar{I}_i^{(1)} = \gamma_i$, $i = 1, \ldots, 10$.

	Time period			
Customer (i)	$t = 1$	$t = 2$	$t = 3$	$t = 4$
1	22	22	22	22
2	50	0	50	0
3	35	1	0	0
4	36	6	36	6
5	60	35	10	35
6	40	20	0	0
7	65	5	65	5
8	50	0	50	0
9	30	30	30	30
10	30	0	30	0

transportation cost of € 360.36. In the optimal solution, the sets of customers to be served in each time period are $U^{*(1)} = \{1, 9\}$, $U^{*(2)} = \{1, 2, 3, 4, 5, 6, 7, 8, 9, 10\}$ and $U^{*(3)} = \emptyset$, respectively.

The deviation of the heuristic solution cost from the optimal one, computed as $(686.06 - 595.54/595.54)$, corresponds to a percentage value of 15.20%, which gives an idea of the solution quality obtained by using the IRP_DECOMPOSITION procedure on the Split problem.

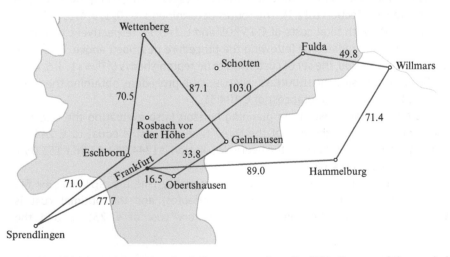

Figure 6.63 Vehicle routes serving the Split customers from the DC in the second time period.

6.16   Case Study: Air Network Design at Intexpress

Intexpress is a firm whose core business is express freight delivery all over North America. In order to guarantee a timely delivery to all destinations, an important part of the transportation activities is performed by plane. The main services offered to the customers are of three types: (a) delivery within 24 hours; (b) delivery within 48 hours, and (c) delivery within three to five days.

Air transportation services are carried out by a fleet of company-owned cargo planes, or by scheduled commercial flights, while road transportation is provided by its own fleet of trucks and vans.

The Intexpress logistics system is made of *shipment centres* (SCs only in this section) and a single hub.

In the SCs and the *hub*, cargo consolidation and deconsolidation operations take place. Some SCs are also air terminals (hereafter referred to as SC/AS, *shipment centre/air stop*) and are used as loading and unloading points for cargo planes, as well as trucks. With each request for transportation service from an origin to a destination, an *origin* SC (SC_o) and a *destination* SC (SC_d) are associated, obviously close to the effective points of origin and destination of the cargo. Freight is first transported to the SC_o, where the consolidation operations take place; then, the consolidated cargo is transferred to the SC_d by a direct transportation performed with commercial flights or truck owned by the company, or using an intermodal transportation with cargo plane and truck, both owned by the company. Finally, freight is moved from the SC_d to the final destination by truck or van. Of course, an SC-to-SC transfer by truck is feasible only if the distance does not exceed a given threshold.

In the case of intermodal transportation, freight from different origins converges at the same SC_o, is consolidated and moved (possibly using a route with multiple intermediate stops at SCs and SC/ASs) in a single load to the hub; similarly, all freight destined to the same SC_d is consolidated and shipped from the hub.

To ensure delivery within 24 hours, the transportation network described works as follows: the outgoing freight is picked up in the late afternoon from the origin and delivered in the morning of the following day. Therefore, a company-owned cargo plane performs three operations every day: it leaves the hub, makes a set of deliveries at SC/ASs and then travels empty from the last delivery point to the first pick-up point, makes a set of pickups at SC/ASs and finally goes back to the hub. All arrivals at the hub must take place before a pre-established arrival time (*cut-off time*, or COT), in such a way that the loads arriving in incoming cargo planes can be unloaded, sorted by destination, and quickly reloaded on the outgoing cargo plane. For the same reasons, each SC/AS is characterized by an "earliest pick-up time" and by a "latest delivery time".

As mentioned above, since it is neither economically feasible nor technically possible for company cargo planes to visit all SC/ASs (road transportation is cheaper than air transportation and the number of company cargo planes is limited), an SC is connected to an SC/AS by truck. Figure 6.64 depicts possible freight routes between an origin–destination pair.

Commercial flights, whose cost depends on the freight quantity and which have a low reliability, are used when at least one of the following conditions occurs:

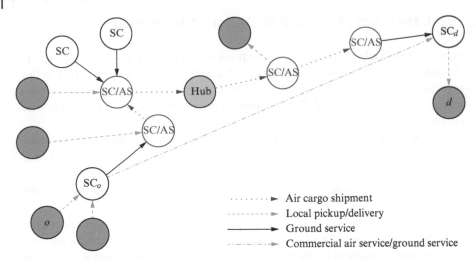

········▶	Air cargo shipment
- - - - -▶	Local pickup/delivery
———————▶	Ground service
·—·—·—·▶	Commercial air service/ground service

Figure 6.64 Possible freight routes between an origin–destination pair for the Intexpress problem. In the intermodal transportation, freight departing from an origin *o* is transported to the nearest SC_o, consolidated, and moved by a company-owned truck to the nearest SC/AS and then, by cargo plane running a route possibly involving multiple intermediate stops, to the hub. Freight destined for *d* is consolidated at the hub and shipped via similar connections.

- the distance between the origin of the route and the closest SC is so large that it is not possible to perform a quick ground connection;
- the company-owned cargo planes have insufficient capacity to satisfy its entire air transportation demand;
- it is not possible to provide an adequate service among the origin–destination pairs by using the company-owned trucks.

Planning the Intexpress service network consists of determining:

- the set of SC/ASs served by each company-owned cargo plane;
- truck routes linking SCs and SC/ASs;
- the transportation tasks performed by commercial flights.

The objective pursued is the minimization of the operational cost subject to "earliest pick-up time" and "latest delivery time" constraints at SCs, to the COT restriction and so on.

The solution methodology used by Intexpress is made up of two stages: in the first stage, the size of the problem is reduced, according to a qualitative analysis; in the second stage, the reduced problem is formulated and solved as an *integer programming* (IP) model. In the first stage,

- origin–destination pairs that can be serviced by truck (in such a way that all operational constraints are satisfied) are allocated to this mode and are not considered afterwards;
- origin–destination pairs that cannot feasibly be served by a dedicated cargo plane or by truck are assigned to the commercial flights;
- low-priority services (deliveries within 48 hours or within three to five days) are made by truck, or by using the residual capacity of the company-owned cargo plane;

- the demands of origin and destination sites are concentrated in the associated SC_o and SC_d.

A route is a partial solution characterized by

- a sequence of SC/ASs of a company-owned cargo plane ending or beginning at the hub (depending on whether it is a collection or delivery route);
- a set of SCs (not ASs) allocated to each SC/AS visited by cargo plane.

Therefore, two routes visiting the same SC/ASs in the same order can differ because of the set of SCs served, or because of the allocation of these SCs to the SC/ASs. If the demand of a route exceeds the capacity of the allocated cargo plane, the exceeding demand is transported by commercial flights. The cost of a route is therefore the sum of costs associated with air transportation, land transportation and possibly a commercial flight if demand exceeds the dedicated cargo plane capacity.

The IP model solved in the second stage is defined as follows. Let K be the set of available cargo plane types; $U^k, k \in K$, the set of the pick-up routes that can be assigned to a cargo plane of type k; $V^k, k \in K$, the set of the delivery routes that can be assigned to a cargo plane of type k; R the set of all routes ($R = \bigcup_{k \in K} (U^k \cup V^k)$); $n^k, k \in K$, the number of company-owned cargo planes of type k; S the set of SCs; $o_i, i \in S$, the demand originating at the ith SC; $d_i, i \in S$, the demand whose destination is the ith SC; $c_r, r \in R$, the cost of route r; $q_i, i \in S$, the cost paid if the whole demand o_i is transported by a commercial flight; $s_i, i \in S$, the cost paid if the whole demand d_i is transported by a commercial flight; $\alpha_i^r, i \in S, r \in R$, a binary constant equal to 1 if route r includes picking up traffic at the ith SC, 0 otherwise; $\delta_i^r, i \in S, r \in R$, a binary constant equal to 1 if route r includes delivering traffic to the ith SC, 0 otherwise, and $\gamma_i^r, i \in S, r \in R$, a binary constant equal to 1 if the first (last) SC/AS of pick-up (delivery) route r is the ith SC, 0 otherwise. The binary decision variables are: $v_i, i \in S$, equal to 1 if demand o_i is transported by commercial flight, 0 otherwise; $w_i, i \in S$, equal to 1 if demand d_i is transported by commercial flight, 0 otherwise, and $x_r, r \in R$, equal to 1 if (pick-up or delivery) route r is selected.

The IP model is

$$\text{Minimize} \sum_{r \in R} c_r x_r + \sum_{i \in S} (q_i v_i + s_i w_i) \tag{6.130}$$

subject to

$$\sum_{k \in K} \sum_{r \in U^k} x_r \alpha_i^r + v_i = 1, \, i \in S \tag{6.131}$$

$$\sum_{k \in K} \sum_{r \in V^k} x_r \delta_i^r + w_i = 1, \, i \in S \tag{6.132}$$

$$\sum_{r \in U^k} x_r \gamma_i^r - \sum_{r \in V^k} x_r \gamma_i^r = 0, \, i \in S, \, k \in K \tag{6.133}$$

$$\sum_{r \in U^k} x_r \leq n^k, \, k \in K \tag{6.134}$$

$$x_r \in F, \, r \in R \tag{6.135}$$

$$x_r \in \{0, 1\}, \, r \in R \tag{6.136}$$

$$v_i \in \{0, 1\}, \ i \in S \tag{6.137}$$

$$w_i \in \{0, 1\}, \ i \in S. \tag{6.138}$$

The objective function (6.130) is the total transportation and handling cost. Constraints (6.131) and (6.132) state that each SC is served by a dedicated route or by a commercial flight; constraints (6.133) guarantee that if a delivery route of type $k \in K$ ends in SC i, $i \in S$, then there is a pick-up route of the same kind beginning in i. Constraints (6.134) set upper bounds on the number of routes that can be selected for each dedicated cargo plane type. Constraints (6.135) express the following further restrictions. The arrivals of the cargo planes at the hub must be staggered in the period before the COT because of the available personnel and of the runway capacity.

Similarly, departures from the hub must be scheduled in order to avoid congestion on the runways. Let n_a be the number of time intervals in which the arrivals should be allocated; n_p the number of time intervals in which the departures should be allocated; $f_r, r \in R$, the demand along route r; a_t the maximum demand that can arrive to the hub in interval $t, \dots, n_a$; TA_t the set of routes with arrival time from t on; TP_t the set of routes with departure time before t and p_t the maximum number of cargo planes able to leave before t. Hence, constraints (6.135) are

$$\sum_{k \in K} \sum_{r \in U^k \cap TA_t} f_r x_r \leq a_t, \ t = 1, \dots, n_a \tag{6.139}$$

$$\sum_{k \in K} \sum_{r \in V^k \cap TP_t} x_r \leq p_t, \ t = 1, \dots, n_p. \tag{6.140}$$

Constraints (6.139) ensure that the total demand arriving cannot exceed the hub capacity in the time intervals from t until n_a, while constraints (6.140) impose that the total number of cargo planes leaving the hub is less than or equal to the maximum number allowed by runaway capacity in each time interval $t = 1, \dots, n_p$.

Other constraints may be imposed. For instance, if the goods are stored in containers, it must be ensured that once a container becomes empty it is brought back to the originating SC. To this end, the cargo plane arriving at an SC/AS and the cargo plane leaving it must be compatible. In the Intexpress problem, there are four types of cargo planes, indexed by 1, 2, 3, and 4. Cargo planes of type 1 are compatible with type 1 or 2, while cargo planes of type 2 are compatible with those of type 1, 2, and 3. Therefore, the following constraints hold:

$$-\sum_{r \in U^1} x_r \alpha_i^r + \sum_{r \in V^1 \cup V^2} x_r \delta_i^r \geq 0, \ i \in S \tag{6.141}$$

$$\sum_{r \in U^1 \cup U^2} x_r \alpha_i^r - \sum_{r \in V^1} x_r \delta_i^r \geq 0, \ i \in S \tag{6.142}$$

$$-\sum_{r \in U^2} x_r \alpha_i^r + \sum_{r \in V^1 \cup V^2 \cup V^3} x_r \delta_i^r \geq 0, \ i \in S \tag{6.143}$$

$$\sum_{r \in U^1 \cup U^2 \cup U^3} x_r \alpha_i^r - \sum_{r \in V^2} x_r \delta_i^r \geq 0, \ i \in S. \tag{6.144}$$

Moreover, some cargo planes cannot land at certain SC/ASs because of noise restrictions or insufficient runway length. In such cases, the previous model can easily be adapted by removing the routes $r \in R$ that include a stop at an incompatible SC/AS.

The variables in the model (6.130)–(6.134), (6.136)–(6.144) are numerous even if the problem is of small size. For example, in the case of four SC/ASs (a, b, c, and d), there are 24 pick-up routes ($abcd$, $acbd$, $adbc$ etc.), each of which has a different cost and arrival time at the hub. If, in addition, two SCs (e and f) are connected by truck to one of the SC/ASs a, b, c, and d, then the number of possible routes becomes $16 \times 24 = 384$ (as a matter of fact, for each SC/AS sequence, each of the two SCs e and f can be connected independently by land to a, b, c, or d). Finally, for each delivery route making its last stop at an SC/AS d, one must consider the route making its last stop at a different SC $g \in S \backslash \{d\}$. Of course, some of the routes can be infeasible and are not considered in the model (in the case under consideration, the number of feasible routes is about 800 000).

The solution methodology is a classical branch-and-bound algorithm in which at each branching node a continuous relaxation of (6.130)–(6.134), (6.136)–(6.144) is solved. The main disadvantage of this algorithm is the large number of decision variables. Since the number of constraints is much less than the number of decision variables, only a few decision variables take a non-zero value in the optimal basic solution of the continuous relaxation. For this reason, the following modification of the method is introduced. At each iteration, instead of the continuous relaxation of (6.130)–(6.134), (6.136)–(6.144), a reduced LP problem is solved (in which there are just 45 000 "good" decision variables, chosen by means of a heuristic criterion); then, using the dual solution of the problem built in this manner, the procedure determines some or all of the decision variables with negative reduced costs, introducing the corresponding columns in the reduced problem. Various additional devices are also used to shorten the execution of the algorithm. For example, in the preliminary stages only routes with an utilization factor between 30% and 185% are considered. This criterion rests on two observations: (a) because of the reduced number of company-owned cargo planes, it is unlikely that an optimal solution will contain a route with a utilized capacity less than 30%, and (b) the cost structure of the air transportation makes it unlikely that along a route more than 85% of the traffic is transported by commercial flights.

The above method was used to first generate the optimal service network using the current dedicated air fleet. The cost reduction obtained was more than 7%, corresponding to a yearly saving of several million dollars. Afterwards, the procedure was used to define the optimal composition of the company's owned fleet. For this purpose, in formulation (6.130)–(6.134), (6.136)–(6.144), it was assumed that n^k was infinite for each $k \in K$. The associated solution shows that five cargo planes of type 1, three of type 2 and five of type 3 should be used. This solution yielded a 35% saving (about $ 10 million) with respect to the current solution.

6.17 Case Study: Dynamic Vehicle-dispatching Problem with Pickups and Deliveries at eCourier

eCourier is a same-day courier company, operating in the Greater London area. It performs pick-ups and deliveries of parcels (documents, packages, palletized unit loads, etc.) by means of a fleet of couriers, who use bicycles, motorcycles, cars, or vans, depending on the unit load to be transported, as well as on the pick-up and delivery locations. Each vehicle type has its own features, such as its speed under diverse traffic conditions, its capacity, or the maximum distance it can cover. Customers ask for service by calling the call centre or by booking via a website. Each customer specifies the pick-up and delivery locations, as well as the time windows during which they want the service to be performed. The traditional model of same-day courier service utilizes human controllers who assign the jobs to the couriers. With this type of organization, however, it can happen that in an environment like the one in which the company operates, this number becomes rather high, with a consequent degradation of the quality of the job assignments as the number of couriers and bookings made every hour increases (in the order of hundreds). In fact, when evaluating a job, a controller must take into account a number of real-time features, among which the parcel type, the vehicle type, the time windows, the travel times, the current locations of the vehicles, possible pending jobs, traffic and weather conditions. Moreover, it can be convenient to reposition idle couriers in high-density zones, to anticipate future demand. Of course, this task is not easy to perform for the controllers. Thus, managing all these aspects creates higher operational costs and lower quality of service (a higher customer inconvenience).

In order to keep the costs as low as possible, the company has decided to develop a decision support system to help the controllers automate the allocation process, thus involving them only as supervisors.

The solution implemented is based on the integration of a number of technologies: GPS and 4G, with each courier being equipped with a GPS device embedded into a 4G palmtop computer; GIS for the navigation system and for tracking couriers' positions; optimization techniques to route the vehicles; and forecasting techniques. The palmtop computers are also used to provide directions to couriers. Travel time forecasting is performed by a neural network that takes into account the vehicle type, as well as weather and real-time traffic data. This information is then used by the allocation system, which assigns each job to the most appropriate courier on the basis of current fleet status, time windows, possible service level agreements, road congestion, as well as weather conditions.

A route is built for each courier according to the jobs allocated, satisfying all the constraints and minimizing the distance covered by the vehicle.

More specifically, at the tactical level, decisions concern shift scheduling, and the number of couriers to be allocated to each shift pattern, subject to constraints on the quality of service to be provided to customers. This problem is usually solved on a weekly or quarterly basis (the demand is usually characterized by significant yearly, weekly, and daily seasonal variations) with the aim of minimizing the staffing cost. On the other hand, at the operational level, a dynamic vehicle dispatching problem with pick-ups and deliveries must be solved in order to allocate each request to a vehicle, to schedule the requests assigned to each vehicle and to reposition idle vehicles.

For the sake of simplicity, the attention will be focused on the operational problem. This problem is defined on a directed graph $G = (V, A)$, where V is a vertex set and A is an arc set. A fleet of m vehicles, located at a depot $i_0 \in V$ at time $t = 0$, has to service a number of pick-up and delivery requests $\{(i_k^+; i_k^-; T_k) : k \in K\}$, where $i_k^+ \in V$, $i_k^- \in V$, $T_k \geq 0$ are the pick-up point, the delivery point and the occurrence time of the kth request, $k \in K$, respectively. Vertices may represent individual customer locations or the zones into which the service territory is divided. Let t_{ij} be the shortest travel time from vertex $i \in V$ to vertex $j \in V$. The aim is to maximize the overall customer service level, rather than to minimize the total covered distance. Let τ_k be the delivery time of the kth request. Each customer is associated with a penalty function $f_k(\tau_k)$.

This definition includes the case for which $f_k(\tau_k)$ represents the customer waiting time (i.e., $f_k(\tau_k) = \tau_k - T_k, \tau_k \geq T_k$) or a more involved penalty function (e.g., $f_k(\tau_k) = 0$, $T_k \leq \tau_k \leq D_k$ and $f_k(\tau_k) = \tau_k - D_k, \tau_k \geq D_k$, where D_k is a *soft deadline* associated with the kth request).

The static version of the problem amounts to determining an ordered sequence of locations on each vehicle route, such that each route starts at the depot; a pick-up and its associated delivery are satisfied by the same vehicle; a pick-up is always made before its associated delivery; and the total penalty incurred by the vehicles $z = \sum_{k \in K} f_k(\tau_k)$ is minimized.

In the dynamic variant, there is also the problem of adequately distributing the waiting time along the routes, since this may affect the overall solution quality, as well as to reposition idle vehicles to anticipate future demand. The objective function to be minimized is the expected customer inconvenience over the planning horizon: $z = \sum_{k \in K} E[f_k(\tau_k)]$, where $E[f_k(\tau_k)]$ is the expected penalty associated with the delivery of the kth request, $k \in K$. Moreover, it is assumed that a vehicle cannot be diverted away from its current destination to service a new request in the vicinity of its current position.

The following anticipatory mechanism embedded in both an insertion and a local search procedure has been devised. Let P_k be the set of pending requests (i.e., the requests that have occurred but have not been serviced) at time T_k, when request (i_k^+, i_k^-, T_k) arrives. A reactive algorithm generates a new solution incorporating i_k^+ and i_k^- with the aim of minimizing the total inconvenience associated with the pending requests, that is $z_k = \sum_{r \in P_k} f_r(\tau_r)$.

The anticipatory algorithms aim at minimizing z_k, plus the expected value (under perfect information) of the total penalty $\xi^{[t_k, t_k + \Delta t_k]}$ associated with the requests arriving in the short-term future $[t_k, t_k + \Delta t_k]$, that is $z'_k = \sum_{r \in P_k} f_r(\tau_r) + E[\xi^{[t_k, t_k + \Delta t_k]}]$, where Δt_k is the short-term duration. Of course, the procedures become reactive if $\Delta t_k = 0$, $k \in K$.

The proposed procedures allowed cCourier to gain significant improvements in terms of lower operational costs. In fact, at the tactical level, the company was able to reduce its costs by approximately 10%. At the operational level, the anticipative procedures allowed the company to improve the quality of service provided to customers by about 60%, and to better distribute requests among the fleet.

6.18 Questions and Problems

6.1 [JKL.xlsx] JKL is a national carrier operating in Australia. The transportation rates for class 25 LTL palletized freight from Melbourne to Darwin are reported in Table 6.41. Determine the break weight formula (the break weight is the weight over which it is convenient to apply the rate of the subsequent range to reduce the overall transportation cost). Compute the break weight for each weight range of Table 6.41.

6.2 Canberra Freight is in charge of transporting auto parts for an US car manufacturer in Australia. Every week a tractor and one or two trailers move from the

Table 6.41   JKL transportation rates
for class 25 LTL palletized freight from
Melbourne to Darwin.

Weight range (*W*) [kg]	Rate [A$/kg]
$0 \leq W < 0.1$	137.13
$0.1 \leq W < 0.2$	112.46
$0.2 \leq W < 0.5$	98.23
$0.5 \leq W < 1$	72.38
$1 \leq W < 10$	60.69
$10 \leq W < 20$	54.21
$20 \leq W < 50$	47.71
$50 \leq W < 100$	46.14
$100 \leq W < 250$	41.23
$W \geq 250$	39.77

port of Melbourne to a warehouse located 430 km away. A tractor costs A$ 140 per day, a trailer A$ 60 and a driver A$ 16.5 per hour, while running costs are A$ 0.75/km. A trailer can contain 36 palletized unit loads. Derive the transportation cost per palletized unit load as a function of shipment size for the case where one or two trailers are used.

6.3 TL trucking rates from Boston, Massachusetts, to Miami, Florida (both in the USA) are usually higher than those from Miami to Boston. Why?

6.4 Freight transportation costs of Class 55 are cheaper than those of Class 70 (see Table 6.2). Why?

6.5 In international freight transportation, a key role is played by the free-trade zones. In such areas, freight may be entered without the intervention of the customs authorities and customs duties are due only when the goods are moved outside. Get more information on the free-trade zones through the Internet.

6.6 [NTN.xlsx] Formulate the minimum-cost flow problem for the Swiss NTN (see Section 6.6.2) intermodal carrier by knowing that the demand of containers in Milan and Madrid is 70 and 25, respectively. What is the optimal solution?

6.7 [Rinaldi.xlsx] Rinaldi is an Italian fast carrier located in Parma, whose core business is the transportation of small-size and high-value refrigerated goods (such as chemical reagents used by hospitals and analysis laboratories). Goods are picked up from manufacturers' warehouses by small vans and carried to the nearest transportation terminal operated by the carrier. These goods are packed onto pallets and transported to destination terminals by means of large trucks. The products are then unloaded and delivered to customers by small vans (usually vans of the same type as those employed for pick-up). In order to make capital investment in equipment as low as possible, Rinaldi has traditionally preferred to concentrate on coordination and management of the terminals, leaving the task of transferring goods to 3PL providers (small hauliers) the task of transferring the goods. Only recently, the company has decided to enter the fast parcel transportation market by opening four terminals in the cities of Bologna, Genoa,

Table 6.42 Forecast transportation demand of refrigerated goods (palletized unit loads per day) in the Rinaldi problem.

	Bologna	Genoa	Milan	Padua
Bologna	–	3	8	2
Genoa	–	–	1	2
Milan	4	2	–	1
Padua	3	1	1	–

Table 6.43 Forecast transportation demand of goods at room temperature (palletized unit loads per day) in the Rinaldi problem.

	Bologna	Genoa	Milan	Padua
Bologna	–	3	4	2
Genoa	1	–	1	–
Milan	6	2	–	2
Padua	1	1	1	–

Padua, and Milan. This choice was made necessary a complete revision of the service network. The decision was complicated by the need to transport the refrigerated goods by special vehicles equipped with refrigerators, while parcels can be transported by any vehicle. The forecast daily average demand of the two kinds of products in the next six-month period is reported in Tables 6.42 and 6.43. Between each pair of terminals, the company can operate one or more lines. Vehicles are of two types:

- trucks with refrigerated compartments, having a capacity of 12 palletized unit loads and a cost (inclusive of all charges) of € 0.4/km;
- trucks with room temperature compartments, having a capacity of 18 palletized unit loads and a cost (inclusive of all charges) of € 0.5/km.

In addition, the company considers the possibility of transporting products at room temperature through another carrier, by paying € 0.1/km for each palletized unit load. A multi-graph representation of the problem is given in Figure 6.65. Distances between terminals are reported in Table 6.44. Formulate the LFCND problem of finding the least-cost service network (hint: $|K| = 22$ commodities, one for each combination of an origin–destination pair with positive demand and a kind of product). Apply the DROP procedure to find a feasible solution of the problem. By using the Excel Solver, determine the optimal solution of the problem and the costs corresponding to the weak and the strong continuous relaxations. What is the quality of the feasible solution provided by the DROP procedure and the two lower bounds?

6.8 Devise a local search heuristic for the SNDP in which an individual move is to remove an existing arc or to add a new arc to the current solution.

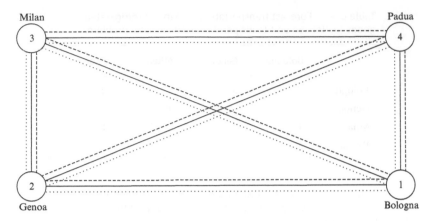

Figure 6.65 Multi-graph representation of the service network design of the Rinaldi problem. In order to simplify the multi-graph, a single edge for each pair of opposite arcs is represented. The dotted edges represent the connections served by truck lines of 12-palletized-unit-load capacity, dashed edges represent the connections served by truck lines of 18-palletized-unit-load capacity and solid lines represent connections served by the external service.

Table 6.44 Distances (in km) between terminals in the Rinaldi problem.

	Bologna	Genoa	Milan	Padua
Bologna	0	225	115	292
Genoa	225	0	226	166
Milan	115	226	0	362
Padua	292	166	362	0

6.9 By using as a reference the problem of Murthy (see Section 6.8), show how the VAP can be modelled as a network flow problem on a time-expanded directed graph. Solve the problem by using the `Excel Solver` tool.

6.10 [`Murthy.xlsx`] Examine how the optimal solution of the Murthy problem (see Section 6.8) changes whenever

- on 11 July there is an empty vehicle in Skrikakulam;
- $d_{533} = 1, d_{531} = 0$;
- $d_{533} = 1, d_{531} = 0$, and $m_{31} = 1$.

6.11 Modify the DDAP model in the case where the number of transportation services is greater than or equal to the number of drivers.

6.12 [`Gare.xlsx`] The travel agency Gare, located in Lausanne, organizes tours by bus in several European capitals. As of April there remain seven different tours to organize whose dates of departure and durations are reported in Table 6.45. The agency has to assign a driver to each tour on the basis of the driver's experience. To this end, a different value is assigned to each driver–tour pair, as reported in Table 6.46: the value 3 means that driver usually makes that trip, the value 1 means that driver has made that trip in the past, and the value 0 means that driver has never made that trip in the past.

Table 6.45 Date of departure and duration (in number of days) of the bus tours organized by the travel agency Gare.

	City						
Tour	Madrid	Paris	Rome	London	Vienna	Budapest	Berlin
Date of departure	2	8	7	9	9	12	13
Trip duration	7	4	4	4	5	4	3

Table 6.46 Ranking value driver–tour of the travel agency Gare problem.

	City						
Driver	Madrid	Paris	Rome	London	Vienna	Budapest	Berlin
1	3	3	0	0	1	3	1
2	1	3	1	0	1	3	1
3	1	3	0	0	1	3	1
4	3	1	1	0	1	3	1
5	1	0	3	0	1	3	1
6	1	0	0	3	1	3	1

6.13 Consider the vehicle fleet composition problem defined by the objective function (6.51) to be minimized and the constraint that requires the integer decision variable v to belong to the interval $[0, \overline{v}]$, where

$$\overline{v} = \max_{t=1,\dots,T} \{v_t\}.$$

Construct an equivalent formulation of the problem to ensure that the range of t in the two summations of equation (6.51) does not depend on the decision variable v (hint: for each time period $t = 1, \dots, T$, consider a binary decision variable x_t, having value 1 if $v_t \le v$, 0 otherwise).

6.14 [FastCourier.xlsx] In the Fast Courier transportation problem (see Section 6.10), assume that the unit cost c_H per time period of hiring a vehicle changes on the basis of the number of hired vehicles (a 10% discount is applied for any additional hired vehicle, up to a total discount of \$ 300). Determine how the original solution changes.

6.15 In which case is the problem (6.52)–(6.57) infeasible?

6.16 [Dungannon.xlsx] Estimate the travel time on an 8.5 km segment of the Dungannon road, between Cookston and Dungannon in Ireland, on the basis of the experimental measures obtained by using a small van and reported in Table 6.47.

6.17 You have an algorithm capable of solving the C-NRP with no fixed vehicle costs, and you would like to solve a problem where a fixed cost is attached to each vehicle. Show how such a problem can be solved using the algorithm at hand.

Table 6.47 Travel time measures (in minutes) on the Dungannon road (traffic volume: 1 = low, 2 = medium, 3 = high; weather conditions: 1 = dry, 2 = light rain, 3 = heavy rain; time slot: 1 = 22:00–6:00, 2 = 6:00–9:00, 3 = 9:00–15:00, 4 = 15:00–22:00).

Traffic volume	Weather conditions	Time slot	Travel time	Traffic volume	Weather conditions	Time slot	Travel time
1	1	1	5.85	2	2	3	6.23
1	1	1	6.23	2	2	3	6.42
1	1	2	6.25	2	2	4	6.61
1	1	2	4.50	2	2	4	6.85
1	1	3	6.18	2	3	1	7.59
1	1	3	5.95	2	3	1	6.74
1	1	4	5.84	2	3	2	6.94
1	1	4	5.84	2	3	2	6.11
1	2	1	5.41	2	3	3	7.80
1	2	1	5.93	2	3	3	7.75
1	2	2	6.06	2	3	4	6.32
1	2	2	4.96	2	3	4	6.85
1	2	3	6.78	2	1	1	7.12
1	2	3	6.80	2	1	1	6.96
1	2	4	6.52	2	1	2	7.55
1	2	4	6.05	2	1	2	7.20
1	3	1	6.04	2	1	3	6.77
1	3	1	7.03	2	1	3	7.88
1	3	2	6.33	2	1	4	6.08
1	3	2	6.82	2	1	4	6.85
1	3	3	6.38	2	2	1	7.91
1	3	3	7.33	2	2	1	7.45
1	3	4	6.70	2	2	2	7.95
1	3	4	7.02	2	2	2	6.56
2	1	1	6.20	2	2	3	7.21
2	1	1	6.07	2	2	3	7.69
2	1	2	5.49	2	2	4	7.37
2	1	2	5.28	2	2	4	7.82
2	1	3	6.96	2	3	1	7.32
2	1	3	5.87	2	3	1	8.67
2	1	4	5.81	2	3	2	8.64
2	1	4	6.55	2	3	2	8.35
2	2	1	5.76	2	3	3	7.80
2	2	1	5.17	2	3	3	8.03
2	2	2	6.58	2	3	4	8.66
2	2	2	5.50	2	3	4	9.07

Table 6.48 Costs associated with the arcs of directed graph G' of Problem 6.22.

	0	1	2	3	4	5	6	7	8	9
0	–	28.55	12.10	43.15	66.10	38.35	79.70	17.00	77.60	55.20
1	15.80	–	23.10	28.85	52.55	53.65	66.90	4.15	71.70	61.90
2	54.95	48.90	–	52.95	88.55	54.75	77.80	42.55	91.45	76.40
3	39.80	27.95	37.85	–	79.00	41.40	79.05	19.15	44.95	35.35
4	49.80	63.95	46.15	43.50	–	71.30	41.60	35.55	67.30	78.60
5	46.50	55.70	28.10	72.30	93.65	–	55.30	54.20	93.70	94.40
6	99.55	84.25	86.15	60.40	63.70	99.85	–	72.30	88.35	81.90
7	31.00	28.80	21.35	34.55	69.55	67.65	67.95	–	77.40	68.30
8	60.35	75.35	46.85	68.15	39.05	75.05	48.90	52.60	–	80.35
9	66.00	61.10	60.75	83.50	87.30	92.50	86.40	62.00	64.00	–

6.18 Show that, if there are no operational constraints, there always exists an optimal NRP solution in which a single vehicle is used (hint: the least-cost path costs satisfy the triangle inequality).

6.19 Show that if the costs associated with the arcs of a complete directed graph G satisfy the triangle inequality property, then there exists in G' an ATSP optimal solution which is a Hamiltonian circuit.

6.20 Show that, in the ATSP and STSP formulations, the connectivity constraints (6.60) and (6.67) are equivalent, respectively, to the subtour elimination constraints (6.61) and (6.68).

6.21 Show that the number of vertices of odd degree in a graph is even.

6.22 [ATSP.xlsx] Let $G = (V, A)$ be a complete directed graph such that $|V| = 10$ and the costs associated with the arcs belong to A, as reported in Table 6.48:

- check whether the triangle inequality property holds for each arc cost;
- formulate the corresponding ATSP;
- solve the relaxed AP;
- determine a feasible ATSP solution and evaluate its quality;
- apply a 2-EXCHANGE procedure in order to find a better feasible solution.

6.23 Formulate the ATSP in an alternative way, by using a polynomial number of constraints (hint: use the continuous decision variables u_i, $i \in U$, each of them representing the position of vertex i in the Hamiltonian cycle. Observe that, if $x_{ij} = 1$, $(i, j) \in A'$, $i, j \neq 0$, then $u_j \geq u_i + 1$).

6.24 Demonstrate that the optimal solution value of MSrTP is a lower bound on the optimal solution value of STSP.

6.25 Show that an optimal solution of the least-cost perfect-matching problem (6.80), (6.81), (6.83) can be obtained by solving the following LP problem:

$$\text{Minimize} \sum_{(i,j) \in E'_M} c_{ij} x_{ij}$$

subject to

$$\sum_{i\in V'_M:(i,j)\in E'_M} x_{ij} + \sum_{i\in V'_M:(j,i)\in E'_M} x_{ji} = 1,\ j \in V'_M$$

$$\sum_{(i,j)\in E'_M,i\in W,j\notin W} x_{ij} + \sum_{(j,i)\in E'_M,i\in W,j\notin W} x_{ji} \geq 1,\ W \subset V'_M,$$

$$|V'_M| > 1, |V'_M|\ \text{odd}$$

$$x_{ij} \geq 0,\ (i,j) \in E'_M.$$

(Hint: observe that, given any subset W of V'_M having odd cardinality, each perfect matching must contain at least one edge incident to the vertices of W.)

6.26 Demonstrate that the cost of the STSP feasible solution produced by the CHRISTOFIDES procedure is within $3/2$ of the optimal cost.

6.27 [GermanExpress.xlsx] The GermanExpress transportation company, based in Cologne, has to schedule a pick-up service to five customers located in the cities of Bonn, Düsseldorf, Essen, Hennef, and Koblenz, respectively. The distance (calculated with respect to the shortest route) between each pair of cities is reported in Table 6.49. The daily average amount of goods to pick up by each customer is provided in Table 6.50. The vans used for transportation have a capacity of 15 quintals each. Determine the number of vehicles to be used and, for each of them, the daily service route starting from the depot located in Cologne.

6.28 Modify the C-NRP formulation (6.84)–(6.91) assuming that (a) each vehicle $k \in K$ has a different fixed cost f_k and a capacity q_k; (b) the demand required by each customer $i \in U$ can be delivered by splitting it among more vehicles.

Table 6.49 Distances (in km) between the depot and the five cities in the GermanExpress transportation problem.

	Cologne	Bonn	Düsseldorf	Essen	Hennef	Koblenz
Cologne	0	40	50	85	52	123
Bonn		0	85	114	30	65
Düsseldorf			0	45	76	134
Essen				0	107	167
Hennef					0	87
Koblenz						0

Table 6.50 Daily average amount of goods (in quintals) to pick-up at the five customers of the GermanExpress.

Bonn	Düsseldorf	Essen	Hennef	Koblenz
4	12	7	5	8

6.29 Extend the C-NRP formulation (6.84)–(6.91) by adding route duration constraints.

6.30 Show that solving the SC-NRP is equivalent to solving the SP-NRP.

6.31 [Bioenergy.xlsx] The Bioenergy is a wood pellet factory located in Austria; it supplies a chain of supermarkets located in Italy and Germany. The orders have to be met by minimizing the transportation cost. The distances among the supermarkets, the distances between the supermarkets and the depot, and the quantity ordered (in palletized unit loads) are reported in Table 6.51. Formulate the problem as a CL-NRP assuming that

- the average unloading time for a palletized unit load is 1.5 minutes;
- the average speed is 90 km/h;
- the work shift is 6 h;
- two type of trucks can be used: those with 22-palletized-unit-load capacity and those with 32-palletized-unit-load capacity.

Determine a feasible solution of the problem.

6.32 [McNish.xlsx] Explain why distances reported in the McNish.xlsx file do not necessarily satisfy the triangle inequality. Solve the McNish problem by applying a modified version of the SAVINGS procedure, considering that the saving $s_{ij}, i, j \in U, i < j$, is defined as $s_{ij} = t_{0i} + t_{0j} - t_{ij}$. The feasibility check (code line 16) of the two merging routes requires that both the vehicle capacity and the time window constraints of the customers and of the depot are verified (hint: because of the presence of the time windows, it is necessary to consider the route orientation in the route merging phase, taking into account that $s_{ij} = s_{ji}, i, j \in U, i \neq j$).

6.33 [MrBread.xlsx] MrBread is a company that produces different types of bread to be distributed to seven supermarkets every morning. The quantity of freight to be delivered and the service time windows for each supermarket are reported in Table 6.52, whereas the travel times on the quickest routes between the supermarkets and the company depot are reported in Table 6.53. Each vehicle has a capacity of 80 quintals. It leaves the warehouse at 9:00 and must return to the warehouse within 13:00. Service time (time needed for unloading a vehicle)

Table 6.51 Distances (in km) between the Bioenergy pellet factory and the supermarkets; in last row, the next-day orders are reported.

	Supermarket			
	1	2	3	4
Depot	100.0	110.0	105.0	92.0
Supermarket 1	0.0	12.0	25.0	24.0
Supermarket 2		0.0	14.5	19.3
Supermarket 3			0.0	17.0
Supermarket 4				0.0
Order [palletized unit loads]	15	32	30	18

Table 6.52 Time windows for the MrBread depot and the seven customers.

Vertex (i)	Time window	Demand [quintals]
0	09:00–13:00	–
1	09:00–12:30	20
2	09:00–11:30	25
3	10:00–12:30	15
4	11:00–14:30	30
5	09:00–12:30	15
6	11:00–13:30	20
7	09:00–12:30	15

can be assumed to be 20 minutes on average for every supermarket, regardless of demand. Formulate and solve to optimality the corresponding TW-NRSP. In order to find a heuristic feasible solution, applying a modified version of the SOLOMON procedure, considering the following rules:

- At each iteration, you should decide which new customer u has to be inserted and between which adjacent customers i and j within the current route. Consider the possibility of inserting a vertex u not only at the end of the current route, but in each feasible position inside it.
- The best feasible insertion place of an unrouted customer is determined by minimizing a measure defined as: $f(i, u, j) = \alpha(t_{iu} + t_{uj} - t_{ij}) + (1 - \alpha)(b_j^u - b_j)$, where b_j^u is the time when the service begins at customer j, provided that customer u is inserted between i and j, while b_j is computed in case u is not inserted.
- Choose $\alpha = 0.7$.

Finally, evaluate the quality of the solution obtained.

Table 6.53 Travel times t_{ij} (in minutes) between the MrBread depot (referred as 0) and the seven supermarkets.

	0	1	2	3	4	5	6	7
0	0	24	56	67	65	51	43	68
1		0	46	50	56	37	40	67
2			0	56	42	70	31	80
3				0	45	42	29	38
4					0	35	22	40
5						0	41	39
6							0	49
7								0

Table 6.54
Distances (in km) between Dortmund and the five cities in the GermanExpress transportation problem.

City	Distance
Bonn	122
Düsseldorf	70
Essen	39
Hennef	119
Koblenz	197

6.34 Devise a local search for the capacitated ARP.

6.35 Illustrate how the CHRISTOFIDES and FREDERICKSON procedures can be adapted to the undirected general routing problem, which consists of determining a least-cost tour including a set of required vertices and edges.

6.36 Analyse the dynamism of an airport bus service.

6.37 [GermanExpress.xlsx] Recall the GermanExpress transportation problem 6.27 and assume that the depot in Cologne has a yearly facility cost of € 155 000. Furthermore, it is assumed that the transportation service is realized 230 times in a year, according to the daily average requests reported in Table 6.50. The van travel cost is about € 1.15 per km. The shipper has the possibility to rent a depot in Dortmund, with a yearly cost of € 130 000, from which the transportation service can alternatively start. Establish which solution is preferable, in terms of yearly cost, by considering the distances between Dortmund and the five customers reported in Table 6.54.

6.38 Modify the mathematical model formulation of IRP (6.114)–(6.127) to take into account the following constraints:

- the supplier's production is unbounded;
- the inventory level of the item at the supplier is $I_0^t = \infty$ for all $t \in T$;
- the initial inventory level of the item at the supplier $I_0^0 = 0$ and the item quantity available in time period $t \in T$ is $p^t = \infty$.

In which case can you suppose that the item consumption rate for each customer is constant during a given planning horizon?

6.39 Modify the IRP_DECOMPOSITION procedure for the Split problem by solving to optimality the TSP and the C-NRP. Does the percentage deviation improve?

6.40 Enumerate some of the most recent research developments in the area of sustainable freight transportation.

Index

Introduction to Logistics Systems Management: With Microsoft® Excel® and Python® Examples, Third Edition. Gianpaolo Ghiani, Gilbert Laporte, and Roberto Musmanno.
© 2022 John Wiley & Sons Ltd. Published 2022 by John Wiley & Sons Ltd.

Printed and bound by CPI Group (UK) Ltd, Croydon, CR0 4YY

27/10/2024

14580308-0001